T0201585

THE LTE/SAE DEPLOYMENT HANDBOOK

THE LTE/SAE DEPLOYMENT HANDBOOK

Edited by

Jyrki T. J. Penttinen
Nokia Siemens Networks Innovation Center (NICE), Spain

A John Wiley & Sons, Ltd., Publication

This edition first published 2012

Registered office
John Wiley & Sons Ltd, The Atrium, Southern Gate, Chichester, West Sussex, PO19 8SQ, United Kingdom

For details of our global editorial offices, for customer services and for information about how to apply for permission to reuse the copyright material in this book please see our website at www.wiley.com.

Library of Congress Cataloging-in-Publication Data

Penttinen, Jyrki T. J.
The LTE/SAE deployment handbook / Jyrki Penttinen. – 1
p. cm.
Includes bibliographical references and index.
ISBN 978-0-470-97726-2 (hardback) – ISBN 978-1-119-95417-0 (ePDF) – ISBN 978-1-119-95418-7 (oBook) – ISBN 978-1-119-96111-6 (ePub) – ISBN 978-1-119-96112-3 (mobi)
1. Long-Term Evolution (Telecommunications) 2. System Architecture Evolution (Telecommunications) I. Title.
TK5103.48325.P46 2012
621.3845′6–dc23

2011033174

A catalogue record for this book is available from the British Library.

Print ISBN: 9780470977262

Set in 10/12pt, Times Roman by Thomson Digital, Noida, India

Contents

List of Contributors

Mohmmad Anas
Adnan Basir
Jonathan Borrill
Francesco D. Calabrese
Luca Fauro
Marcin Grygiel
Jukka Hongisto
Tero Jalkanen
Juha Kallio
Krystian Krysmalski
Sebastian Lasek
Grzegorz Lehmann
Luis Maestro
Krystian Majchrowicz
Guillaume Monghal
Maciej Pakulski
Jyrki T. J. Penttinen
Olli Ramula
Dariusz Tomeczko

Foreword

Manually operated mobile communication networks were a huge success in all the Nordic countries in the 1970s but the popularity of the first-generation automatic networks (NMT) exceeded all expectations in the 1980s. It seemed impossible to estimate realistically the number of base stations needed to respond to the growing demand. Subscribers became accustomed to constantly improving service levels and coverage areas for voice calls. Gradually, during that decade, users adopted wireless voice communication and found that not only did it bring increased efficiency—it was also a highly liberating experience.

Then, along with the second generation in the 1990s (GSM), it became clear that there was a growing demand for more advanced services. International specification work on GSM formed a solid base and a favorable platform for new inventions like Short Message Service (SMS). GSM has been up and running now for more than 20 years. From the number of new innovations in 3GPP standardization it is clear that the evolution of GSM will be secure for a long time.

3G was introduced to the markets in order to provide a base for even more demanding multimedia. It provided additional capacity for voice calls as the 2G systems started to saturate. With its multiple generations and releases, the mobile telecom operators and vendors started to realize the challenges in the field as new services typically require support from both networks and terminals. On the other hand, the terminals' lifecycle is shorter because users consider them to be everyday consumer objects, and more attractive models constantly appear on the market. There is a positive balance between users, operators and equipment vendors as enhanced services typically require updates to terminals and networks.

The deployment of the packet data service as an add-on for GSM, and then its adaptation from the first phase of UMTS, were the important triggers for the use of Internet services via mobile terminals. The rapidly evolving Internet environment itself had a great impact on mobile communications, resulting in the development of multi-usage equipment for services, combining voice connections, messaging, and multimedia.

With the deployment of the third-generation networks, data rates increased in order to provide a smoother user experience. The new business environment started to strengthen. In contrast with the initial model of only few voice service providers in controlled markets, there were now increasing numbers of operators, equipment vendors, service providers, measurement equipment producers, and many other entities contributing to mobile communications. The increasing speed of standardization made development seem unlimited.

Along with the increased data rates associated with the Internet, fixed and mobile communications have also evolved steadily. Open standards, competing operators and multivendor equipment offerings have ensured that the markets developed favorably from the end user's point of view.

Evolution of 2G and 3G is gradually becoming saturated, as happened with the first-generation networks. It is easier to create a new, more efficient platform to provide the required data rate and capacity than to develop existing ones. Statistics from recent years indicate that there has been a huge growth in multimedia data transfer. The exponential growth in the use of data sets higher performance targets for the networks than ever before.

In this context, LTE has been designed as a base for a new 4G era. It paves the way towards 4G by providing a smooth transition from 2G and 3G, including important interworking functionalities as well as higher data rates and capacity than ever before in mobile network environments. In addition to 3GPP networks, LTE/SAE standardization also takes care of the evolution path from CDMA systems.

Evolving technology makes the management of mobile communications businesses more complex. Some operators can build on existing technology; others may have to start from 4G. Fixed networks must also be considered as competition for mobile networks, as their capacity, quality, and flexibility to interwork with wireless technologies increase.

At the same time, the need for relevant information is increasing. Networks are either built from scratch or through designing an evolution path from a previous system. Network planners and other technical people need to know how the systems function, how they can be planned optimally, and how to make sure that user experiences will be positive. Business managers must also understand the basic technology in order to see how they can benefit from it and what they may require from technical staff.

It is a rare to find a person who has a deep understanding of a technology and who can also write about it in an informative, simple, and understandable way. The writer of this book, Jyrki Penttinen, has this skill. This is the right book for those who wish to study LTE and the principles and details of Evolved UTRAN and Evolved Packet Core in a common-sense manner.

Matti Makkonen
CEO, Anvia Plc
Former Vice President, Sonera, Finland

Preface

Long-Term Evolution (LTE) is arguably one of the most important steps in the current phase of the development of modern mobile communications. It provides a suitable base for enhanced services due to increased data throughput and lower latency figures, and also gives extra impetus to the modernization of telecom architectures. The decision to leave the circuit-switched domain out of the scope of LTE/SAE system standardization might sound radical but it indicates that the telecom world is going strongly for the all-IP concept—and the deployment of LTE/SAE is concrete evidence of this global trend.

LTE specifications define evolved radio access for 3GPP's 3G evolution path and so they have an important influence on the core development of the new mobile network system. Along with requirements for high-speed data support for the radio network, the core network specifications have been updated to guarantee end-to-end performance. The specification work under the same 3GPP umbrella ensures that all the relevant aspects are covered in the interworking of the evolved radio and core, as well as between previous generations of 3GPP 2G and 3G networks.

There are many overlapping or similar aspects in LTE and SAE and previous 3GPP systems but the evolved network also brings plenty of novel solutions. Many performance simulations are already available, which indicates the capabilities of LTE/SAE, but the impact of the system on practical network deployment has not been particularly clear until now.

This book aims to address this growing need for information about the practical aspects of the evolved terrestrial radio access network of UMTS (E-UTRAN)—that is, LTE—as well as the evolved packet core network (EPC)—that is, System Architecture Evolution (SAE). The idea of this book is to take a step towards to the preparation of the deployment phase, presenting practical information needed in the designing and building of the LTE/SAE network. The book presents topics and examples that are helpful from the first day of the planning and deployment of LTE/SAE networks, to ensure that the initial phase provides the best possible level of service. It describes the system architecture and functionality, network planning, measurements, security, applications, and other aspects that are important in real telecommunications environments.

The book is written in a modular way. The first module consists of Chapters 1–5, which describe the background and the overall idea of the system. This part includes advice about the practical interpretation of the standards and gives the most important high-level requirements and architectural descriptions of LTE and SAE. This part is thus especially useful for anyone who lacks prior knowledge about the system.

Chapters 6–11 address more specific issues regarding the functionality of LTE/SAE and its services. This part describes the functionality and elements of the system in enough detail to help readers to understand the technical possibilities and challenges of LTE and SAE as a part of the whole mobile communications environment.

The third module consists of Chapters 12–15, which address design-related aspects of the LTE/SAE from a practical perspective. This part contains essential guidelines for the planning, dimensioning, and measurement of LTE/SAE networks. One of the most important parts of this module, and at the same time the core of the whole book, is Chapter 15, which presents valuable recommendations for the transition from other systems to LTE. It gives various technical guidelines and examples as a basis, for example, for refarming strategies.

In general, this book can be used as a central, practical source of information in the deployment phase of LTE/SAE as well as in later phases. The book team would like to remind though that this book gives practical information about the functionality and suggestions for the network deployment, but the correctness of the contents can not be guaranteed by the team. It is encouraged to refer to the specifications and other validated information sources. The team also would like to clarify that the information and opinions presented in this book are solely of the contributors, and our employers may or may not have the same ideas.

If you have any feedback or comments about the content of the book, or suggestions about how it could be enhanced in possible future editions, please do not hesitate to contact the author directly at jyrki.penttinen@nokia.com. Additional information about LTE/SAE, based on developments in the field and feedback, may be found on the author's Internet page at www.tlt.fi.

Jyrki T. J. Penttinen

Acknowledgments

This book is the result of a joint effort by our book team. I would like to thank all the contributors for their challenging work during their spare time. I thank Nokia Siemens Networks, Nokia, TeliaSonera, Anritsu, NetHawk, Rohde & Schwarz, Samsung, and Anite for their collaboration and for providing a practical perspective for this book. Many colleagues assisted us by providing essential comments, documentation and hints. I would like to thank Dr. Harri Holma, Dr. Jorge Hermosillo, Mika Laasonen, Mikko Nurkka, Valtteri Niemi, Timo Saxen, Olli Ramula, and Antti Näykki, and all our other colleagues for their valuable support.

I give my warmest thanks to the Association of Finnish Non-Fiction Writers for their support for this project. I also thank the whole Wiley team for the guidance and professional project management.

Finally, I thank my spouse Elva and our children for their patience and support during the project.

Jyrki T. J. Penttinen

Glossary

128-QAM	128 state Quadrature Amplitude Modulation
16-QAM	16 state Quadrature Amplitude Modulation
1G	First Generation of mobile communication technologies
2G	Second Generation of mobile communication technologies
3G	Third Generation of mobile communication technologies
3GPP	3rd Generation Partnership Project
4G	Fourth Generation of mobile communication technologies
64-QAM	64 state Quadrature Amplitude Modulation
AAA	Authentication, Authorization & Accounting
ABMF	Account Balance Management Function
AC	Admission Control
ACIR	Adjacent Channel Interference Rejection
ACK	Acknowledgment
ACLR	Adjacent Channel Leakage Ratio
ACS	Adjacent Channel Selectivity
ADC	Analogue/Digital Conversion
ADMF	Administration Function
ADSL	Asynchronous Digital Subscriber Line
AF	Africa
AF	Application Function
A-GPS	Assisted Global Positioning System
aGW	Access Gateway
AKA	Authentication and Key Agreement
AMBR	Aggregated Maximum Bit Rate
AMC	Adaptive Modulation and Coding
AMPS	Advanced Mobile Phone System
AMR	Adaptive Multi-Rate
AP	Aggregation Proxy
AP	Asia Pacific
APAC	Asia Pacific
APN	Access Point Name
APN-AMBR	APN aggregate maximum bit rate
AR	Aggregation Router
ARFCN	absolute radio-frequency channel number
ARP	Allocation Retention Priority

ARP	Automatic Radio Phone
ARPU	Average Revenue Per User
ARQ	Automatic Repeat reQuest
AS	Application Server
AS SMC	AS Security Mode Command
ATB	Adaptive Transmission Bandwidth
ATCA	Advanced Telecommunications Computing Architecture
ATM	Asynchronous Transfer Mode
AuID	Application Usage ID
AUTN	Authentication token
AVC	Advanced Video Codec
AWS	Advanced Wireless Services
BCCH	Broadcast Control Channel
BCH	Broadcast Channel
BD	Billing Domain
BE	Best Effort
BER	Bit Error Rate
BICC	Bearer Independent Call Control
BLER	Block Error Rate
BPSK	Binary Phase Shift Keying
BQS	Bad Quality Samples
BS	Base Station
BSC	Base Station Controller
BSR	Buffer Status Report
BSS	Business Support System
BTS	Base Transceiver Station
BW	Bandwidth
C/I	Carrier per Interference
CA	Certification Authority
CAMEL	Customised Applications for Mobile networks Enhanced Logic
CAPEX	Capital Expenditure
CAZAC	Constant Amplitude Zero AutoCorrelation
CC	Content of Communication
CCCH	Common Control Channel
CCN	Cell Change Notification
CCO	Cell Change Order
CDF	Charging Data Function
CDMA	Code Division Multiple Access
CDR	Call Drop Rate
CDR	Charging Data Record
CEO	Chief Executive Officer
CET	Carrier Ethernet Transport
CFB	Call Forwarding Busy
CFNRc	Call Forwarding Not Reachable
CFNRy	Call Forwarding No Reply
CFU	Call Forwarding Unconditional
CGF	Charging Gateway Function
CLIP	Calling Line Presentation

CLIR	Calling Line identity Restriction
CMAS	Commercial Mobile Alert System
CMP	Certificate Management Protocol
CN	Core Network
COLP	Connected Line Presentation
COLR	Connected Line identity Restriction
CoMP	Coordinated multipoint
CP	Cyclic Prefix
CPICH	Common Pilot Channel
CPM	Converged IP Messaging
CQI	Channel Quality Indicator
CR	Carriage Return
CRC	Cyclic Redundancy Check
CS	Circuit Switched
CSFB	Circuit Switched Fall Back
CSI	Channel State Information
CT	Core Network and Terminals (TSG)
CTF	Charging Trigger Function
CTM	Cellular Text Telephony Modem
DAB	Digital Audio Broadcasting
DCCA	Diameter Credit Control Application
DCCH	Dedicated Control Channel
DD	Digital Dividend
DFCA	Dynamic Frequency and Channel Allocation
DFT	Discrete Fourier Transform
DFTS-OFDM	Discrete Fourier Transform Spread-OFDM
DHCP	Dynamic Host Configuration Protocol
DHR	Dual Half Rate (voice codec)
DL	Downlink
DLDC	Downlink Dual Carrier
DL-SCH	Downlink Shared Channel
DMR	Digital Mobile Radio
DoS	Denial of Service
DPI	Deep Packet Inspection
DRB	Data Radio Bearer
DRX	Discontinuous Reception
DSCP	DiffServ Code Point
DSL	Digital Subscriber Line
DSMIPv6	Dual-Stack Mobile IPv6
DTCH	Dedicated Traffic Channel
DTM	Dual Transfer Mode
DTMF	Dual Tone Multi-Frequency
DTX	Discontinuous Transmission
DUT	Device Under Test
DVB-H	Digital Video Broadcasting, Handheld
DVB-T	Digital Video Broadcasting, Terrestrial
ECM	EPS Connection Management
E-CSCF	Emergency Call State Control Function

EDGE	Enhanced Data Rates for Global Evolution
EF	Expedited Forwarding
EFL	Effective Frequency Load
E-GPRS	Enhanced GPRS
EHPLMN	Equivalent HPLMN
eHRPD	Evolved High Rate Packet Data
EMM	EPS Mobility Management
EMR	Enhanced Measurement Reporting
eNB	Evolved Node B
ENUM	E.164 Number Mapping
EPC	Evolved Packet Core
ePDG	Evolved Packet Data Gateway
EPS	Evolved Packet System
ETSI	European Telecommunications Standards Institute
ETWS	Earthquake and Tsunami Warning System
EU	European Union
E-UTRAN	Evolved UMTS Radio Access Network
EV-DO	Evolution-Data Only
EVM	Error Vector Magnitude
FACCH	Fast Associated Control Channel
FCC	US Federal Communications Commission
FCCH	Frequency Correction Channel
FDD	Frequency Division Duplex
FDPS	Frequency-Domain Packet Scheduling
FER	Frame Erasure Rate
FFS	For Further Study
FFT	Fast Fourier Transform
FH	Frequency Hopping
FMC	Fixed Mobile Convergence
FNO	Fixed Network Operator
FPLMTS	Future Public Land Mobile Telecommunications System
FR	Frame Relay
FR	Full Rate (voice codec)
FR-AMR	AMR Full Rate
GAA	Generic Authentication Algorithm
GAN	Generic Access Network
GBR	Guaranteed Bit Rate
GCF	Global Certification Forum
GERAN	GSM EDGE Radio Access Network (TSG)
GGSN	GPRS Gateway Support Node
GMLC	Gateway Mobile Location Centre
GMSK	Gaussian Minimum Shift Keying
GPRS	General Packet Radio Service
GRE	Generic Routing Encapsulation
GRX	GPRS Roaming Exchange
GSM	Global System for Mobile communications
GSMA	GSM Association
GTP	GPRS Tunnelling Protocol

GTT	Global Text Telephony
GTT-CS	Global Text Telephony over video telephony
GTTP	GPRS Transparent Transport Protocol
GTT-Voice	Global Text Telephony over voice
GW	Gateway
HARQ	Hybrid Automatic Retransmission on request/Hybrid Automatic Repeat Request
HD	High Definition
HDSL	High-bit-rate Digital Subscriber Line
HeNB GW	Home eNB Gateway
HeNB	Home eNB
HLR	Home Location Register
HO	Handover
hPCRF	Home Policy and Charging Rules Function
HPLMN	Home PLMN
HR	Half Rate (voice codec)
HR-AMR	AMR Half Rate
HRPD	High Rate Packet Data
HSCSD	High Speed Circuit Switched Data
HSDPA	High Speed Downlink Packet Access
HSPA	High Speed Packet Access
HSS	Home Subscriber Server
HSUPA	High Speed Uplink Packet Access
IBCF	Interconnection Border Control Functions
ICE	Intercepting Control Element
ICI	Inter-Carrier Interference
ICIC	Inter Cell Interference Control
ICS	IMS Centralized Services
I-CSCF	Interrogating Call State Control Function
IDFT	Inverse Discrete Fourier Transform
IEEE	Institute of Electrical and Electronics Engineers
IETF	Internet Engineering Task Force
IFFT	Inverse Fast Fourier Transform
I-HSPA	Internet HSPA
IMEI	International Mobile Equipment Identity
IMS	IP Multimedia Sub-system
IMSI	International Mobile Subscriber Identity
IMS-MGW	IMS-Media Gateway
IMS-NNI	IMS Network-Network Interface
IM-SSF	IP Multimedia – Service Switching Function
IMT-2000	International Mobile Telecommunication requirements (ITU)
IMT-Advanced	Advanced International Mobile Telecommunication requirements (ITU)
IN	Intelligent Network
INAP	Intelligent Network Application Protocol
IOT	Inter-Operability Testing
IP	Internet Protocol
IPsec	IP Security

IP-SM-GW	IP-Short Message-Gateway
IPv4	IP version 4
IPv6	IP version 6
IPX	IP eXchange
IQ	In-phase (I) and out of phase (Q) components of modulation
IRI	Intercept Related Information
ISC	IMS Service Control
ISI	Inter-Symbol Interference
ISIM	IMS Subscriber Identity Module
ISR	Idle Mode Signaling Reduction
ISUP	ISDN User Part
ITU	International Telecommunication Union
ITU-R	ITU's Radiocommunication Sector
ITU-T	ITU's Telecommunication sector
IWF	Interworking Function
JSLEE	JAIN Service Logic Execution Environments
KDF	Key Derivation Function
KPI	Key Performance Indicator
LA	Latin America
LA	Link Adaptation
LA	Location Area
LAU	Location Area Update
LBO	Local Breakout
LCS	Location Service
LEA	Law Enforcement Agencies
LEMF	Law Enforcement Monitoring Facilities
LI	Lawful Interception
LIG	Legal Interception Gateway
LRF	Location Retrieval Function
LSP	Label Switch Path
LTE	Long Term Evolution
LTE-A	LTE-Advanced
LTE-UE	LTE User Equipment
MA	Mobile Allocation
MAC	Medium Access Control
MAIO	Mobile Allocation Index Offset
MAN	Metropolitan Area Network
MBMS	Multimedia Broadcast Multicast Service
MBR	Maximum Bit Rate
MCC	Mobile Country Code
MCCH	Multicast Control Channel
MCH	Multicast Channel
MCS	Modulation and Coding Scheme
MC-TD-SCDMA	Multi-Carrier Time-Division Synchronous-Code-Division Multiple Access
MC-WCDMA	Multi-Carrier Wide-band Code-Division Multiple Access
ME id	Mobile Equipment Identifier
ME	Middle East

MEA	Middle East and Africa
MER	Modulation Error Rate
MGCF	Media Gateway Control Function
MGW	Media Gateway
MIMO	Multiple Input Multiple Output
MME	Mobility Management Entity
MMS	Multimedia Messaging Service
MMTel	Multimedia Telephony
MNC	Mobile Network Code
MO	Mobile Originating
MOBSS	Multi-Operator Base Station Subsystem
MOCN	Multi-Operator Core Network
MORAN	Multi-Operator Radio Access Network
MOS	Mean Opinion Score
MPLS	Multi-Protocol Label Switching
MRF	Media Resource Function
MRFC	Media Resource Function Controller
MRFP	Media Resource Function Processor
MS	Mobile Station
MSC	Mobile services Switching Center
MSC-B	Second (another) MSC
MSISDN	Mobile Station ISDN number
MT	Mobile Terminating
MTCH	Multicast Traffic Channel
MT-LR	Mobile Terminating Location Request
MTM	Machine-to-Machine (communications)
MVNO	Mobile Virtual Network Operator
MWC	Mobile World Conference
MWI	Message Waiting Indication
NA	Network Assisted
NA	North America
NACC	Network Assisted Cell Change
NACK	Negative Acknowledgment
NAS SMC	NAS Security Mode Command
NAS	Non Access Stratum
NB	Node B
NCCR	Network Controlled Cell Reselection
NDS	Network Domain Security
NE Id	Network Element Identifier
NGMN	Next Generation Mobile Networks (Alliance)
NGN	Next Generation Network
NH	Next Hop parameter
NITZ	Network Initiated Time Zone
NMT 450	Nordic Mobile Telephone in 450 MHz frequency band
NMT 900	Nordic Mobile Telephone in 900 MHz frequency band
NMT	Nordic Mobile Telephone
NNI	Network-Network Interface
NOC	Network Operations Centre

NRT	Near Real Time
NVAS	Network Value Added Services
OAM&P	Operations, Administration, Maintenance, and Provisioning
OCF	Online Charging Function
OCS	Online Charging System
OFCS	Offline Charging System
OFDMA	Orthogonal Frequency Division Multiple Access
OLLA	Outer Loop Link Adaptation
OLPC	Open Loop Power Control
OMS	Operations and Management System
OoBTC	Out of Band Transcoder Control
OPEX	Operating Expenditure
OSC	Orthogonal Sub Channel
OSPIH	Internet Hosted Octect Stream Protocol
OSS	Operational Support System
OTA	Over The Air
OTT	Over The Top
PAPR	Peak-to-Average Power Ratio
PBCH	Physical Broadcast Channel
PBR	Prioritised Bit Rate
PC	Personal Computer
PC	Power Control
PCC	Policy and Charging Control
PCCH	Paging Control Channel
PCEF	Policy and Charging Enforcement Function
PCEP	Policy and Charging Enforcement Point
PCH	Paging Channel
PCI	Physical Cell Identifier
PCRF	Policy and Charging Rules Function
P-CSCF	Proxy Call State Control Function
PD	Packet delay
PDCCH	Physical Downlink Control Channel
PDCP	Packet Data Convergence Protocol
PDH	Plesiochronous Digital Hierarchy
PDN	Packet Data Network
PDN-GW	Packet Data Network Gateway
PDP	Packet Data Protocol
PDSCH	Physical Downlink Shared Channel
PDU	Packet Data Unit
PDV	Packet Delay Variation
PGC	Project Co-ordination Group
P-GW	Packet Data Network Gateway
PHB	DiffServ Per Hop Behavior
PHICH	Physical Hybrid ARQ Indicator Channel
PHR	Power Headroom Report
PKI	Public Key Infrastructure
PLMN	Public Land Mobile Network
PLR	Packet Loss Ratio

PMCH	Physical Multicast Channel
PMI	Precoding Matrix Indicator
PMIP	Proxy Mobile IP
PMIPv6	Proxy Mobile IP version 6
PPP	Point to Point Protocol
PRACH	Physical Radio Access Channel
PRB	Physical Resource Block
PS	Packet Switched
PS	Presence Server
PSAP	Public Safety Answering Point
PSD	Packet Switched Data
PSN	Packet Switched Network
PTCRB	PCS Type Certification Review Board
PTP	Point-to-Point
PUSCH	Physical Uplink Shared Channel
PWS	Public Warning System
Q	Quality
QAM	Quadrature Amplitude Modulation
QCI	QoS Class Identifier
QoS	Quality of Service
QPSK	Quadrature Phase Shift Keying
RA	Registration Authority
RA	Routing Area
RACH	Random Access Channel
RAN	Radio Access Network (TSG)
RAND	Random challenge number
RAT	Radio Access Technology
RAU	Routing Area Update
RB	Resource Block
RBG	Radio Bearer Group
RCS	Rich Communication Suite
RES	Response
RF	Radio Frequency
RF	Rating Function
RFSP	RAT/Frequency Selection Priority
RI	Rank Indicator
RLC	Radio Link Control
RLT	Radio Link Timeout
RMS	Root Mean Square
ROHC	Robust Header Compression
RoI	Return of Investment
RRC	Radio Resource Control
RRH	Remote Radio Head
RRM	Radio Resource Management
RRU	Remote Radio Unit
RS	Reference Signal
RSCP	Received Signal Code Power
RSRP	Reference Signal Received Power

RSRQ	Reference Signal Received Quality
RSSI	Received Signal Strength Indicator
RT	Real Time
RTCP	RTP Control Protocol
RTP	Real Time Transport Protocol
RX	Receiver
RX-D	Diversity Receiver
RXLEV	RX Level
RXQUAL	RX Quality
S/P-GW	Serving Gateway and PDN Gateway (combined), see SAE GW
SA	Service and System Aspects (TSG)
SACCH	Slow Associated Control Channel
SAE	System Architecture Evolution
SAE-GW	Combined S-GW and P-GW
SAIC	Single Antenna Interference Cancellation
SAU	Simultaneously Attached Users
SBC	Session Border Controller
SCC AS	Service Centralization and Continuity Application Server
SC-FDMA	Single Carrier Frequency Division Multiple Access
SCH	Shared Channel
SCIM	Service Control Interaction Management
SCP	Service Control Point
S-CSCF	Serving Call State Control Function
SCTP	Stream Control Transfer Protocol
SDCCH	Stand-alone Dedicated Control Channel
SDF	Service Delivery Framework
SDH	Synchronous Digital Hierarchy
SDP	Session Description Protocol
SEG	Security Gateway
SEL	Spectral Efficiency Loss
SEM	Spectral Emission Mask
SFN	Single Frequency Network
SGSN	Serving GPRS Support Node
S-GW	Serving Gateway
SIB	System Information Block
SIM	Subscriber Identity Module
SINR	Signal-to-Interference-and-Noise Ratio
SIP	Session Initiation Protocol
SISO	Single Input Single Output
SLF	Subscriber Locator Function
SM	Short Message
SMG	Special Mobile Group
SMS	Short Message Service
SMSC	Short Message Service Centre
SN ID	Serving Network's Identity
SNR	Signal-to-Noise Ratio
SON	Self Organizing/Optimizing Network
SR	Scheduling Request

SRS	Sounding Reference Signal
SRVCC	Single Radio Voice Call Continuity
SS	Signal Strength
STM	Synchronous Transfer Mode
S-TMSI	Temporary Mobile Subscriber Identity
STN-SR	Transfer Number for Single Radio
SU-MIMO	Single User MIMO
SUPL	Secure User Plane Location
TA	Tracking Area
T-ADS	Terminating Access Domain Selection
TAS	Telephony Application Server
TAU	Tracking Area Update
TBF	Temporary Block Flow
TBS	Transport Block Size
TCH	Traffic Channel
TCP	Transmission Control Protocol
TDD	Time Division Duplex
TDM	Time Division Multiplex
TDMA	Time Division Multiple Access
TD-SCDMA	Time Division Synchronous Code Division Multiple Access
TEID	Tunnel Endpoint Identifier
TFO	Tandem Free Operation
THIG	Topology Hiding
TISPAN	Telecommunications and Internet converged Services and Protocols for Advanced Networking
ToP	Timing over Packet
TR	Technical Recommendation
TrFO	Transcoder Free Operation
TrGW	Transition Gateway
TRX	Transceiver
TS	Technical Specification
TSG	Technical Specification Group
TSL	Timeslot
TTCN3	Testing and Test Control Notation Version 3
TTI	Transmission Time Interval
TU3	Typical Urban 3 km/h
TX	Transmitter
UDP	User Datagram Protocol
UE	User Equipment
UL	Uplink
UL-SCH	Uplink Shared Channel
UMA	Unlicensed Mobile Access
UMTS	Universal Mobile Telecommunications System
UNI	User-Network Interface
UPE	User Plane Entity
URI	Uniform Resource Identity (SIP)
URL	Uniform Resource Locator
USB	Universal Serial Bus

USIM	Universal Subscriber Identity Module
USSD	Unstructured Supplementary Service Data
USSDC	USSD Centre
UTRAN	UMTS Terrestrial Radio Access Network
UWB	Ultra Wide Band
VHF	Very High Frequency
VLAN	Virtual Local Area Network
VoIP	Voice over IP
VoLGA	Voice over LTE via Generic Access
VoLTE	Voice over LTE
vPCRF	Visited PCRF
VPLMN	Visited PLMN
VPLS	Virtual Private LAN Service transport
WB	Wideband
WB-AMR	Wideband Adaptive Multi Rate
WCDMA	Wideband CDMA
WE	West Europe
WI	Work Item
WiMAX 2	IEEE 802.16m-based evolved WiMAX
WiMAX	Worldwide Interoperability for Microwave Access
WLAN	Wireless Local Area Network
WRC	World Radiocommunication Conference
XCAP	XML Configuration Access Protocol
XDM	XML Document Management
XDMS	XML Document Management Server
XML	Extensible Markup Language
XRES	Expected Response

1

General

Jyrki T. J. Penttinen

1.1 Introduction

This chapter introduces the contents of the book. It includes high-level information about the LTE system and design, and instructions how to use the modular structure of the chapters in an efficient way in different phases of LTE/SAE network planning, rollout, operation and optimization. As the book is practical, it is suitable for network operators, equipment manufacturers, service providers, and educational institutes at many levels.

1.2 The LTE Scene

Long-Term Evolution (LTE), as its name indicates, has been planned to meet the ever growing demands of mobile communications network customers in the forthcoming years. This new system provides considerably higher data rates and lower latency in order to deliver multimedia content in an efficient way, which benefits the end-users (who experience improved data transfer and communications) as well as the operators, who can optimize the network infrastructure to provide high-capacity and high-speed data communications.

Telecommunications and the content of the information are based increasingly on the Internet Protocol (IP), via multiple transport solutions. Information delivery was traditionally based on the circuit-switched (CS) domain until the late 1990s for data, and up to now for voice traffic. The LTE concept indicates that the trend for all content is definitely towards IP because the LTE specifications do not even define circuit-switched interfaces any more. The lack of a definition for the circuit-switched domain of the telecommunication networks is one of the strongest proofs of the direction of the current packet-based evolution. The decision to leave this interface out of the LTE specifications might be considered drastic but, on the other hand, it will definitely speed up the process for moving the telecom traffic towards the packet-switched domain, which supports the idea of delivering most communications over IP, including the voice service.

Long-Term Evolution defines the radio interface in this evolved phase of 3G. It provides considerably higher data rates in a more advanced and efficient way than other earlier large-scale

The LTE/SAE Deployment Handbook, First Edition. Edited by Jyrki T. J. Penttinen.

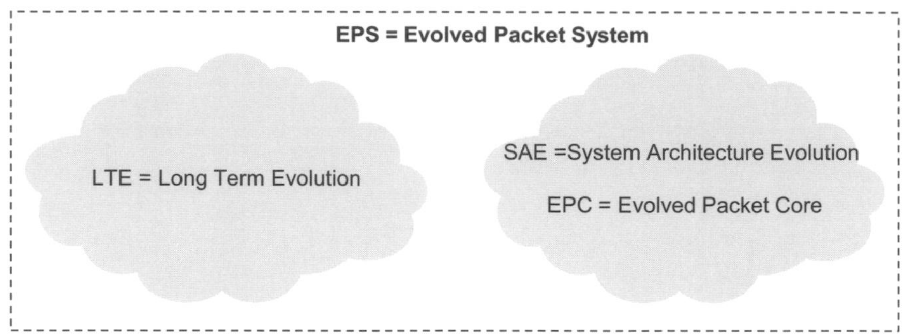

Figure 1.1 The overall division of LTE and SAE.

mobile communications systems. This means challenges for the rest of the network—which could be considered as a positive development. In order to handle all the potential capacity that LTE can deliver, the core network side also has to be modified. This definition is called System Architecture Evolution (SAE) (see Figure 1.1).

Common functionality between LTE and SAE and earlier mobile networks is needed to guarantee smooth continuity of calls in locations where LTE coverage is still missing. Even though the LTE/SAE standards lack CS domain definitions, there are practical means for managing the connection during the call, or during the idle mode. The voice call is obviously the most important mobile—and telecom—service, and this can be handled by the VoIP connections via the packet domain. The challenge arises when the LTE service ends suddenly. One way to handle the call without breaks is to hand the connection over to the 2G or 3G networks, which still include CS interfaces to the fixed telephone network.

The LTE/SAE solution brings faster data services to telecommunications than ever before via this type of large mobile network. Moreover, LTE/SAE will reduce delays in the data communications considerably. One of the most interesting aspects of the functionality of the LTE/SAE is its scalability. This permits LTE/SAE networks to be deployed in many scenarios, from the stand-alone network to the small-scale initial add-on as a part of the frequency refarming, and in a growing network that delivers more capacity as the frequency bands of previous networks are reduced. There are many possible mobile telecommunications deployment strategies for which the LTE/SAE fits as a logical solution.

Like any other mobile communications system, LTE/SAE has its own evolution path. Planning has already taken place for the evolved version, which can be called LTE-Advanced. It will provide considerably higher data rates due to the wider frequency bandwidth and other enhancements. Furthermore, LTE-Advanced is one of the systems that comply with fourth-generation requirements, defined by ITU-R.

1.3 The Role of LTE in Mobile Communications

Traditionally, during the 2G era and in the beginning of the 3G system deployment, data service utilization was at a low level, representing typically a maximum of 2% of the whole traffic. The circuit-switched voice service and short message service have been the dominating teleservices. Even the introduction of the first packet data solutions—GPRS (General Packet Radio Service) and its evolved version, EGPRS (Enhanced GPRS) or EDGE (Enhanced Data Rates for Global Evolution)—did not increase the data service utilization level considerably,

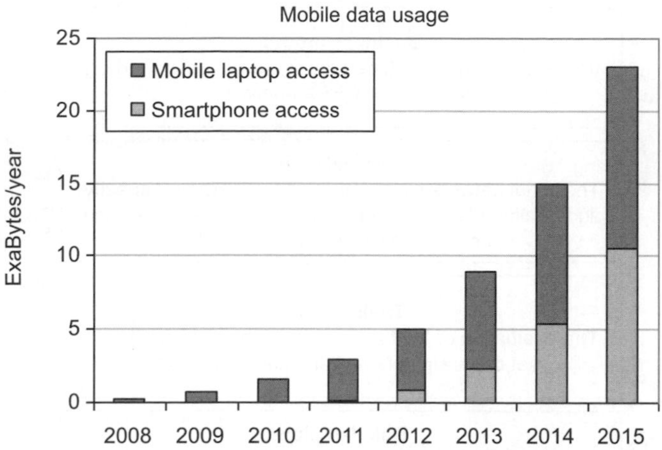

Figure 1.2 The estimate of the near-future data usage [5].

although they were necessary steps to provide a cost-optimized method for handling the bursty traffic of the Internet Protocol. At the moment, circuit-switched data is considered old-fashioned and expensive for both users and operators, and it is thus disappearing from the service sets of the operators.

The utilization level of the packet data has increased recently as a result of considerably higher data rates and lower latency, which makes mobile data communications comparable to, or in some cases even more attractive than, typical Internet subscription. As a result, more applications have been developed both for leisure purposes and for business use. One of the main drivers for the future data utilization is the growth of the smart-phone penetration. For example, Informa has estimated that, during 2010, 65% of global mobile data traffic was generated by 13% of mobile subscribers who use smart phones, with average traffic per user of 85 MB per month. Japan is the most active in the mobile data usage, with 199 MB per month. Figure 1.2 shows the predicted data growth until 2015.

1.4 LTE/SAE Deployment Process

A typical LTE/SAE deployment largely involves the same steps as previous mobile communications network deployments. The new aspects are related to contents of the projects due to the more advanced data rates and use cases.

Figure 1.3 shows an example of the most important tasks prior to and during the LTE/SAE deployment.

In a typical LTE project, the business model is created on both the operator side and the vendor side. This dictates whether the project is feasible and if it can be carried out with the given assumptions and within the project time frame. Chapter 2 presents some high-level aspects that should be taken into account in business modeling.

The nominal plan gives a first-hand estimate about technical issues like the number of sites needed in order to meet the required capacity and quality levels of the network. The nominal plan is thus tightly related to the business plans as the amount of technical material dictates the final CAPEX and OPEX of the network, and thus influences directly the return on investment

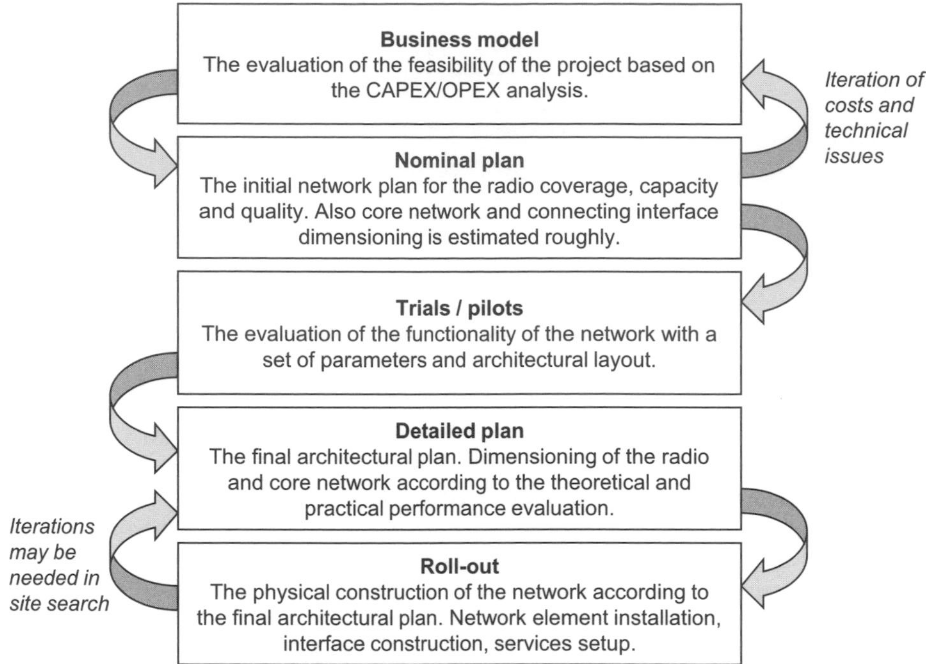

Figure 1.3 An example of possible LTE deployment project phases.

(ROI) estimates. Business and nominal planning is thus an iterative process in the most accurate type of feasibility analysis.

Trials and pilots provide important proof of concept and of the practical performance of the networks, especially in the early stages of the technologies. Even when the technology matures, prior testing of solutions is important as there are always local differences in the radio environment, use cases, and traffic profiles. Field tests typically give more accurate information about the functionality and performance of the network under realistic conditions. Prior to the field tests, a comprehensive set of laboratory tests is typically carried out as part of the system verification process. This is an essential phase of system development to ensure that new functionalities are backwards compatible with earlier solutions. For example, the interworking of LTE/SAE should be validated for functioning with the 2G and 3G networks in all use cases such as CS voice call fallback. An important part of this phase is to ensure interoperability between different vendors, according to standard requirements.

The detailed planning phase includes the final architectural plan of the network. It also involves the coverage plan, which should take into account the special characteristics of the local environment—that is, the type of area (rural, suburban, urban, or dense urban) and distribution of different cluster types. The expected traffic profiles, on the other hand, dictates the detailed capacity plan. All the relevant interfaces are dimensioned.

Rollout is the first phase in the commercial network deployment. The project typically occurs as quickly as possible, which means that a number of parallel work groups install the equipment. If reutilization of physical sites is not possible in some or all instances (as is the case for greenfield operators), the physical preparation of the site is also needed. There might be a need to build towers if renting is not a feasible option. In order to prepare the site, an important

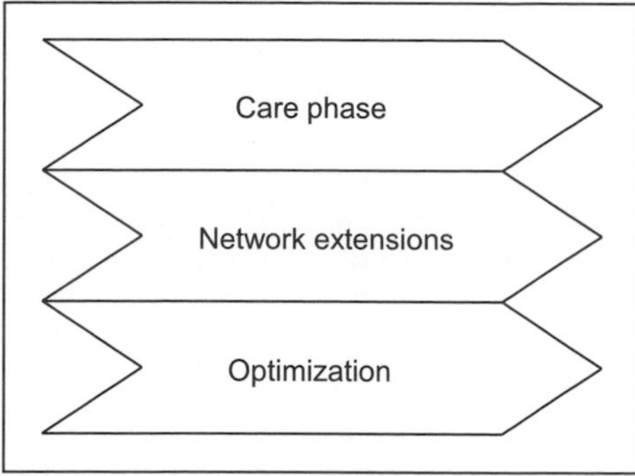

Figure 1.4 Example of the phases in the mature phase of the network.

site-hunting task has to be executed. Typically, a preferred area (ring) is located based on the outcome of the detailed planning. If the preferred site location cannot be obtained due to technical or commercial issues, a modified plan is needed to seek the best available option. The modification may have a wider effect, influencing the location of neighboring sites, which means that confirmation of site locations is advisable in order to avoid modifications in the rather hectic phase of the rollout. The number of sites during this phase normally varies between a few hundred in a limited area or a small country, up to thousands for a very wide area or throughout a large country.

After the rollout, there is a more stable phase in network operation when the fine-tuning of the network begins. The optimization of network quality is one of the important tasks in this care phase of the network. Network coverage and capacity need to be extended as the traffic and the network utilization grow.

Care activities can be done either by the operator or third party, who can also be a network vendor. Care tasks include maintenance work on the network, fault management, performance monitoring, backup and restoring of the network data, inventory management, and other daily routines, in order to make sure that the network functions correctly.

Figure 1.4 summarizes the main tasks in the constructed network and in its mature phase. These tasks are executed in a parallel way during the life cycle of the network.

As a last task of the operator, the network is ramped down at some point when new and more cost-effective technologies are available. This has already been the case for the first generation analog networks practically worldwide. Figure 1.5 presents a high-level summary of the generations of the mobile networks and their use since the existence of mobile communications.

As can be seen from Figure 1.5, the first generation of the mobile networks—the analog representatives of various systems—has already ended; for example, NMT (Nordic Mobile Telephone in Nordic countries, Switzerland, Russia), Netz C (in Germany), AMPS (in England), to mention some, have already practically disappeared due to low spectral efficiency and incompatibility between the systems. GSM, as the widest spread digital 2G mobile communications representative, together with the other 2G systems like IS-95/CDMA (in the Americas), have clearly demonstrated the market need for mobile communications. This phase is still very much ongoing due to better spectral efficiency, international compatibility, roaming, and data communications. One of the most useful novelties of 2G compared to 1G is the short message service (SMS), which has been the basis of all kinds of personal communications and service information flow.

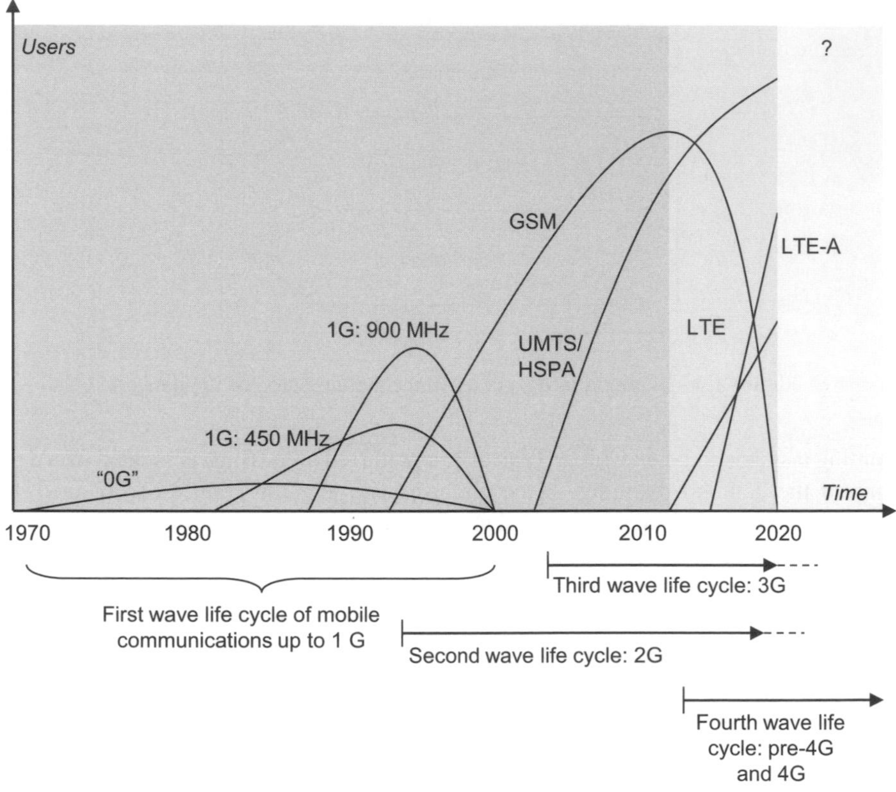

Figure 1.5 The idea of the mobile evolution path, including LTE-Advanced.

The third-generation networks were designed to offer much higher data rates in order to deliver real multimedia content. The startup of, for example, UMTS, was slow due to the slow introduction of the handsets that really would have been considerably more useful in the multimedia or high-speed data transfer. This affected the markets—strong expectations were not met. In any case, the 3G systems with their evolution path, for example, via the HSPA (High Speed Packet Access)-capable mobiles and networks, are finally opening the doors for the multimedia era.

Long-Term Evolution, and the respective core network evolution, SAE, are more efficient in terms of the shorter waiting period at the beginning of the data transfer, and due to the plans to offer higher data rates with low latency [1]. Long-Term Evolution can be considered as part of the evolution path of the 3G systems. According to the ITU's definition, LTE represents the third generation of mobile communications. According to [2], the third generation requirements are listed in IMT-2000 while the fourth generation requirements are included in IMT-Advanced. Furthermore, the IMT-2000 technologies are defined in ITU-R recommendation M.1457, which includes, for example, LTE.

At the European Union level, the licenses in the frequency bands of 800 MHz, 2500 MHz and 3500 MHz are not tied to mobile network types. Their usage has been defined for terrestrial systems that are suitable for electrical communications services.

The utilization of the frequencies can be defined further at a national level, naturally respecting the higher level regulation of the ITU. As an example, in Finland, LTE is subject to state regulations. These allow, for example, a telecommunications company that has a right to operate a GSM network on the 880–915 MHz, 925–960 MHz, 1710–1785 MHz and 1805–1880 MHz frequency bands, to use these frequencies for the operation of the UMTS network. A telecommunications company that has a right to operate third-generation mobile communications on the 1710–1785 MHz and 1805–1880 MHz frequency bands can also use these frequencies for the provision of services via third-generation, long-term technology, that is, LTE [3].

This book contains information for the deployment of LTE/SAE networks, illustrating the most important aspects that should be taken into account in planning and deployment, as well as providing examples from different phases of the deployment process. Each LTE/SAE deployment is naturally an individual task with sometimes hard-to-estimate details, perhaps in some instances arising from techno-economical aspects. In any case, this book is meant to be a useful resource for technical personnel involved in the deployment process, and a central source of basic information as well as a source of useful references for further studies. Furthermore, the approach of this book is as practical as possible. Its main focus is on 3GPP release 8, although aspects of Release 9 and 10 are also revised in order to take the evolution path into account in the early phase of the networks.

1.5 The Contents of the Book

Long-Term Evolution and SAE are described here in such a way that the book can be used as supporting background material for LTE/SAE deployment and prior preparations. The book supports hands-on tasks when deploying and maintaining the LTE/SAE network. The book describes the principles and details of LTE/SAE, with the most relevant aspects of the functioning, planning, construction, measurements and optimization of the radio and core networks of the system. The book focuses on the practical description of LTE/SAE, LTE functionality and planning, and realistic measurements of the system. In general, the book describes points that are useful when planning and constructing the LTE/SAE network and services, and it thus completes the conceptual descriptions of the LTE/SAE found in other titles. It has been designed especially to complement the book *LTE for UMTS—Evolution to LTE Advanced* [4].

The contents of this book include a general view of the evolution path, network architecture, and business models, technical functioning of the system, signaling, coding, different modes for channel delivery and for ensuring security of the core and radio systems, and the in-depth planning of the core and radio networks, the field-test measurement guidelines, hands-on network planning advice, and suggestions for parameter adjustments. The book also gives an overview of the next generation LTE—LTE-Advanced—which represents the fourth generation mobile system. One of the most concrete descriptions of deployment can be found in Chapter 15, which gives guidelines on the recommended evolution paths from the previous mobile network systems towards the LTE era.

The topic is relatively new, and although various LTE books exist, they do have a rather limited focus, concentrating, for example, on the radio interface, and giving rather theoretical information on how to interpret LTE specifications and what performance we can expect from LTE networks and services. The most practical hands-on level description of end-to-end functionality, network planning and physical construction of the LTE networks is thus still lacking. This book aims to correct this need, by describing the complete picture and providing practical details with plenty of examples from the operational networks in order to serve as a handbook and guide in the planning and operational phase of the networks.

The book is modular, giving an overall description for telecom personnel who are not yet familiar with LTE, as well as more detailed and focused practical guidelines for telecom specialists. It contains an introductory module that is suitable for initial study and revision. The latter part of the book is useful for experienced professionals who may benefit from the practical descriptions of the physical core and radio network planning, end-to-end performance measurements, physical network construction and optimization of the system.

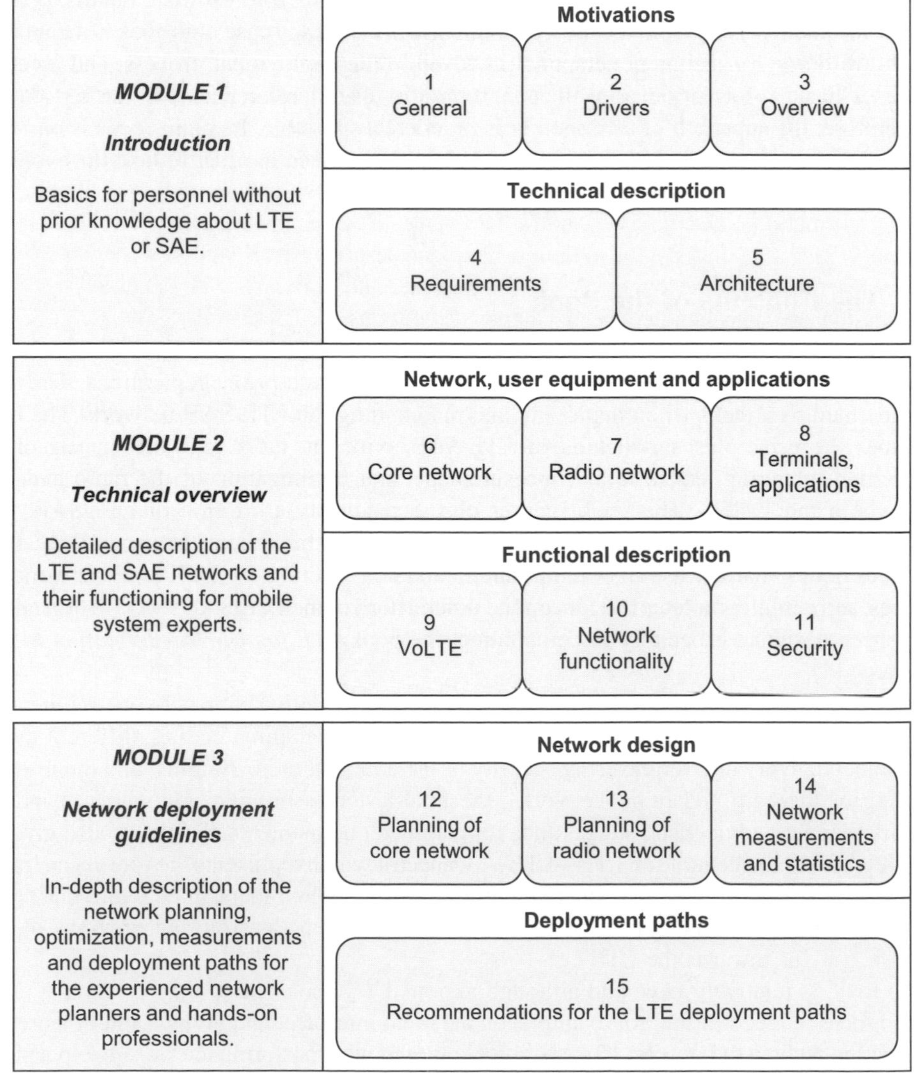

Figure 1.6 The contents of the book.

The contents of the modules is as follows. Part I presents general information about LTE and SAE, whereas modules II and III are useful for professionals who will work in the rollout of LTE/SAE deployment.

- Module I (Chapters 1–5): Principles and architecture of LTE/SAE.
- Module II (Chapters 6–11): LTE/SAE network and functionality.
- Module III (Chapters 12–15): LTE/SAE planning, optimization, measurements and deployment guidelines.

The book is especially suitable for hands-on technical training for technicians and installation engineers. It can also be used in technical institutes and at university level for theoretical and laboratory training.

The contents of the book are presented in Figure 1.6 and are as follows. Chapter 1 gives a general description of LTE and SAE. Chapter 2 identifies the drivers for LTE/SAE by presenting the reasons for the development of advanced data rates. Chapter 3 gives an overview of LTE/SAE with a short guide to the standardization. The further evolution of LTE is also described in this chapter. Chapter 4 presents a practical interpretation of the standards' requirements and their effect on the network planning, deployment and optimization. Chapter 5 describes the LTE/SAE architecture, including the functional blocks, interfaces, and protocol layers. Information is also given about the hardware, software, and high-level circuit diagrams of the most important parts of the modules. Chapter 6 describes the SAE core network, together with a description of the core elements, hardware, and software. Site-specific issues and transmission are also handled here. Chapter 7 contains the LTE radio network with LTE spectrum allocations and a common-sense description of the OFDM and SC-FDMA—that is, the downlink and uplink parts of the LTE radio. Radio resource management and handovers are presented thoroughly.

Chapter 8 presents the terminals, Chapter 9 voice service, and Chapter 10 the functionality of LTE/SAE. Chapter 11 presents security-related aspects of the network and terminal. Chapter 12 presents the core—SAE—planning. Chapter 13 presents the radio—LTE—planning, with LTE link budget dimensioning examples and practical aspects of the installation. Chapter 14 describes issues related to LTE/SAE measurements, with case examples. Chapter 15 wraps up the book by presenting a practical set of recommendations for different deployment scenarios, with practical use cases and examples about network planning issues and rollout strategies.

References

[1] 3GPP TS 36.101 v8.12.0. (2010) *User Equipment (UE) Radio Transmission and Reception*, 3rd Generation Partnership Project, Sophia-Antipolis.

[2] ITU (2010) Press Release, www.itu.int/net/pressoffice/press_releases/2010/40.aspx (accessed 29 August 2011).

[3] Finlex (2009) Frequency Regulation in Finland, www.finlex.fi/fi/laki/alkup/2009/20091169.29 (accessed August 2011).

[4] Holma, Harri and Toskala, Antti (2011) *LTE for UMTS. Evolution to LTE-Advanced*, 2nd edn, John Wiley & Sons, Ltd, Chichester.

[5] Nokia Siemens Networks (2009) Mobile Network Statistics of Nokia Siemens Networks. Nokia Siemens Networks report.

2

Drivers for LTE/SAE

Jyrki T. J. Penttinen

2.1 Introduction

This chapter gives an overview of the reasons for the standardization and deployment of LTE/SAE. First, a short revision of the mobile generations and their characteristics is provided. Then a short summary about the development of the data services follows.

2.2 Mobile System Generations

The use of mobile communications has grown exponentially in recent decades. The early walkie-talkie terminals that were able to provide user-controlled point-to-point radio connections for voice showed the potential of the wireless world.

Larger scale mobile communications development began when the pre-first-generation networks, sometimes referred as generation 0, were deployed. One of the first commercial radio mobile systems, Automatic Radio Phone (ARP), was brought into commercial use in Finland in 1971, after a couple of years of construction. Despite its low capacity by today's standards, it was useful enough to last 30 years in the commercial market. It functioned in the VHF band, which provided very large coverage areas per base station. It was thus especially useful for basic voice calls in all the regions of the country. This analog network was operated by Telecom Finland (nowadays knows as Telia Sonera Finland), and it had tens of thousands of customers in its peak years.

The first fully automatic mobile communications system, and also the most international variant of the first generation mobile communications network deployed at that time in Finland and other Nordic countries, was NMT 450 (Nordic Mobile Telephone, the number indicating the frequency band in MHz). It was commercialized at the beginning of 1980s, and was initially designed for a vehicular environment. Like ARP, it was meant only for an analog voice service. Later in the same decade, an advanced version was commercialized—NMT 900. This system included handheld terminals from the beginning of the launch. Later, NMT 450 was developed to support handheld devices. These systems showed for the first time the usefulness of international functionality, although still on a relatively small scale. The system was adopted

The LTE/SAE Deployment Handbook, First Edition. Edited by Jyrki T. J. Penttinen.

later in Switzerland, Russia, and some other countries. In the 1990s, the system was even used for data connection with a separate data adapter or with a specially designed data terminal called DMR (Digital Mobile Radio). Even moving surveillance pictures could be transmitted on the road via an NMTImage solution [1]. The popularity of the system could be observed from the number of customers, which was more than ten time greater than ARP. There were various mobile networks similar to NMT in Europe, the Americas and Japan.

One of the reasons why the second-generation mobile communication systems, Global System for Mobile Communications (GSM) were the most popular variant, was that they addressed a need for international roaming that had been identified. GSM was designed in the European Telecommunications Standards Institute (ETSI). This generation contains more technical definitions compared to the previous systems, and as the technology in general was developing towards the digital era, the GSM system was designed to be fully digital. This provided clear advantages over analog systems, including constant voice quality and separation of the terminal and subscriber modules (Subscriber Identity Modules or SIM). Circuit-Switched (CS) data in GSM was introduced some years after the first commercial launches. It was included in the phase 2 GSM standards, which functioned at a maximum of 9.6 kbps. At the same time, the Short Message Service (SMS) became available as the networks and terminals started to support the functionality.

SMS opened the way to the ever expanding value-added services. After the initial circuit-switched data service evolution, packet-switched data came into the picture. GPRS (General Packet Radio Service) and its extension, EDGE (Enhanced Data Rates for Global Evolution), were the arrowheads for the global wireless data development. The trend is clear—basically all the telecommunications will be based on IP. Other second-generation systems also appeared in the market, like CDMA-based IS-95.

The success story of GSM continues, and it is still being standardized in the Third Generation Partnership Project (3GPP). Its evolution path includes the Dual Half Rate (DHR) mode, which provides four times more voice capacity than the original full rate codec, and Downlink Dual Carrier (DLDC), which uses two separate frequencies for the downlink data channels, providing around 500–600 kbps data rates with a 5 + 5 time-slot configuration in the uplink/downlink.

The 3GPP standardization name of 2G evolution is called GERAN (GSM/GPRS/EDGE). GERAN is represented in the standardization body as one complete Technical Specification Group (TSG).

During the constantly developing standards and new versions for the network releases, produced in the Special Mobile Groups (SMGs), the limits of the GSM platform were identified. This triggered a parallel process for the development of the third-generation mobile communication system. The idea of this generation was to offer even more capacity by other radio technologies, and to provide new multimedia capabilities for the growing demand of the mobile data usage.

The standardization of the Universal Mobile Telecommunications System (UMTS) was initiated in ETSI, first under the name Future Public Land Mobile Telecommunications System (FPLMTS). The standardization was moved to 3GPP along with the GSM evolution in 1999.

As in GSM, UMTS standards were created under a process of constant development, which triggers new releases on a regular basis. Development can be seen especially in the enhancement of data rates. The first theoretical UMTS data rate was 2 mbps, from which the practical top speed was a somewhat decent 384 kbps compared with today's maximum data rates. Evolving applications and customer habits required more data rates and this has been tackled by the introduction of the HSPA and HSPA+.

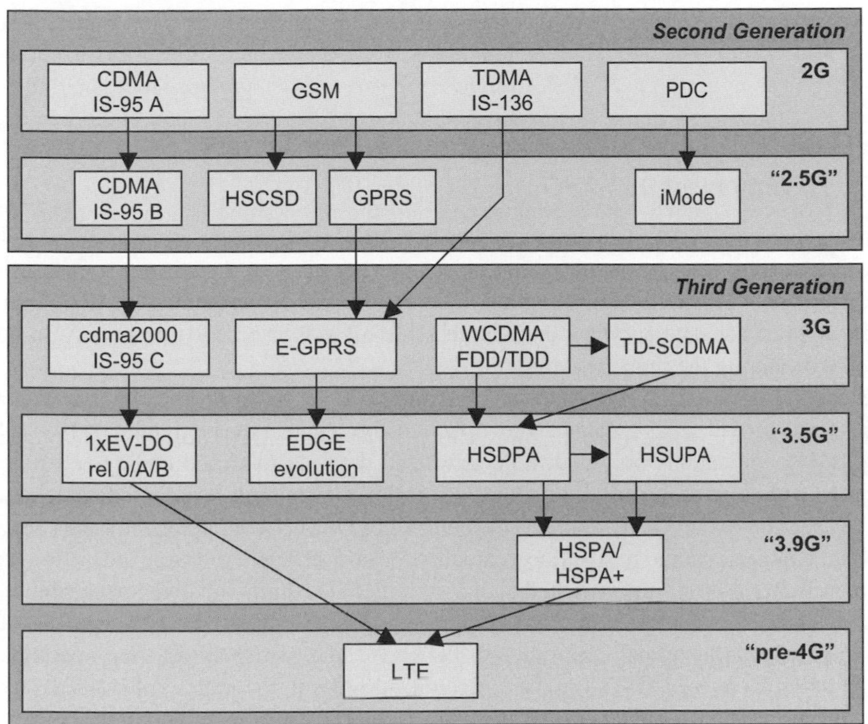

Figure 2.1 The contents of the 2G and 3G generations.

There has been market positioning by manufacturers and operators globally, and the division of the current and near-future radio access technologies into different generations has been somewhat confusing. It seems that the interpretation of the first (analog) and second (TDMA) generations is in accordance with practice and ITU principles, but with the evolution of the third generation it looks less clear how to name the generations.

Figure 2.1 shows the idea of the generations, interpreted from the ITU-R web pages [2]–[4]. Please note that the terms "3.5 G" and "3.9 G" originate from the industry and are not defined as such by ITU.

In general, current work on LTE and LTE-Advanced is considered as constituting the 4G era. This means that the mobile telecommunications industry has mostly taken the view that the performance of LTE is already within 4G. The practical explanation for this could be that LTE is, in fact, much closer to 4G than to 3G, and thus 4G has already been adopted in the marketing of LTE.

ITU's Radiocommunication Sector ITU-R completed the assessment of six candidate submissions for the global 4G mobile wireless broadband technology October 21, 2010. According to ITU-R terminology, the fourth generation refers to IMT-Advanced and contains various technical requirements, for example for the data rates. The proposals resulted in two 4G technologies: "LTE-Advanced" and "WirelessMAN-Advanced." Both of these solutions have thus been recognized officially by ITU-R as true 4G technologies, although the interpretation of the mobile telecommunications industry is somewhat looser in the case of the first, Release 8 LTE. Regardless of whether LTE is classified as the last step in 3G, or as a first or preliminary step in the 4G era, it paves the way towards the ITU-defined 4G, and LTE will be considered as a fully 4G system at the latest with the deployment of its advanced version that is based on Release 10.

2.3 Data Service Evolution

2.3.1 Development up to 3G

The first-generation systems did not have data bearers as such, although it was possible to use data to some extent via a data modem and data adapter. In this way, the first generation provided peak data rates of 9.6 or 14.4 kbps in good radio conditions. Nevertheless, the first generation was never used for data services on a large scale—it was used for this mainly in special telemetry or among the most active users.

The second generation took data services into account from an early phase. The GSM specifications provided basic data bearers up to 9.6 kbps even at the beginning of 1990s. Since then, GSM specifications provided more advanced data via modified coding schemes, first making 14.4 kbps possible, and soon up to near 60 kbps via circuit-switched data. It was only after the creation of the packet data service concept of GSM (General Packet Radio Service—GPRS) that data use started to soar due to much more cost-efficient resource utilization that the permanently reserved circuit-switched data could offer. The multislot concept combined with adaptive channel coding schemes via Enhanced Data Rates for GSM/Global Evolution (EDGE) provides theoretical data rates of 384 kbps (DL), which in practice is comparable with the first data rates of UMTS in the downlink. At present, the most evolved version of the GSM data services is called the Downlink Dual Carrier (DLDC), and combines timeslots from two separate frequencies in the downlink direction. Combining the practical multislot concept with adaptive channel coding scheme, the data rate can be around 500 kbps with the 5 + 5 timeslot configuration, and close to 1 mbps in the theoretical case of 8 + 8 timeslots.

The third-generation mobile communications system was designed to be capable of handling a multimedia environment from day one. The first basic data rate of 384 kbps (DL) has recently increased considerably through the introduction of High Speed Downlink Packet Access (HSDPA), High Speed Uplink Packet Access (HSUPA), and nowadays via High Speed Packet Access (HSPA) and its evolved stage, HSPA+.

The current versions of the third-generation data services are already clearly taking off compared to the first UMTS data service data speeds. Figure 2.2 presents the principle of the evolution of GSM and UMTS data services along with the new releases.

The 3G data evolution is a result of the 3GPP standardization. As a rule of thumb, commercial solutions (vendor releases) appear in the markets within two or three years from the freezing of the respective releases of the standards. Figure 2.3 shows the approximate durations of the standardization of the 3G releases up to freezing, and the development time of the releases until they were commercially available.

2.3.2 Demand for Multimedia

The packet data service of GSM and UMTS has opened the way for the real multimedia era. As mobile data usage grows faster than ever at a global level, spectral efficiency has been noted to be one of the most critical items for the operators. Not only the enhanced spectral efficiency is enough, but also the utilization of larger bandwidth is one of the most important issues in order to guarantee the further evolution of mobile networks as the data rate is directly dependent on that.

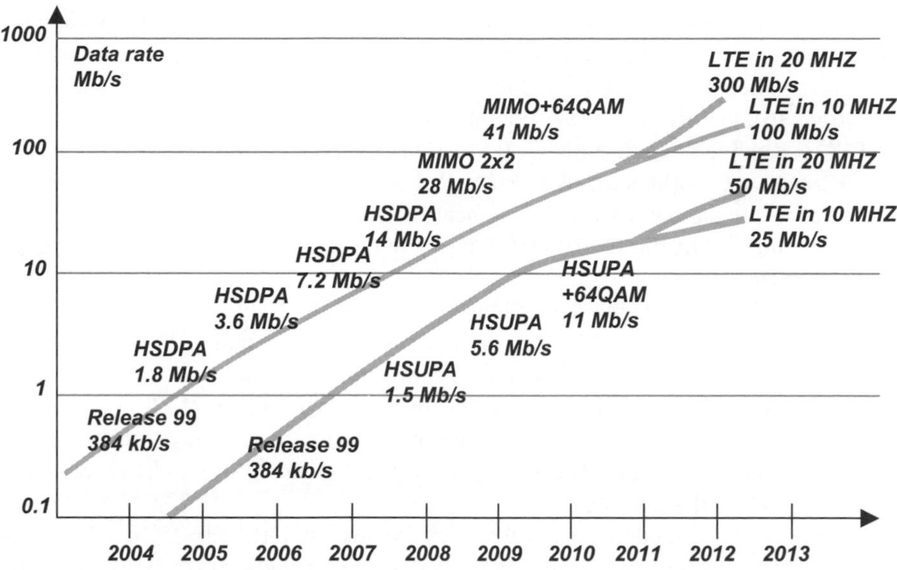

Figure 2.2 The evolution path of 3G data services.

After the increased data rates provided via the HSPA evolution, 3GPP started to evaluate its successor candidates at the end of 2004. The goal for new radio performance was set higher that in any of the previous WCDMA solutions for 3GPP. The peak data-rate requirement was decided to be at least 100 mbps in DL, and over 50 mbps in UL. In addition, latency had to be improved considerably. The working name for this new idea was called as Long Term Evolution (LTE), which, after this initial study phase, also became the public name for the respective radio interface. Nevertheless, the 3GPP standardization calls the radio interface Evolved UMTS Radio Access Network (E-UTRAN).

The evolved radio interface with higher data rates also requires significant enhancements on the 2G/3G packet data network side—that is, the GPRS core—of the mobile networks. This 3GPP study item was called System Architecture Evolution (SAE), and this is currently used

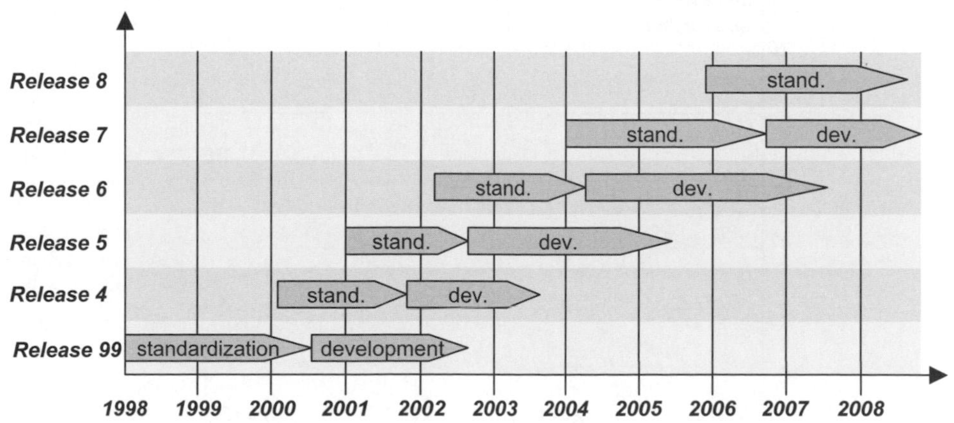

Figure 2.3 The 3GPP releases for the data enablers with the respective releases.

as a practical name for the evolved packet core. In fact, SAE is a term that is used in practice to describe the Evolved Packet Core (EPC). SAE and EPC are thus parallel terms to describe the same item. The EPC is documented in the Release 8 Technical Report [5] and Technical Specifications [6], [7]. Release 8 was completed at the beginning of 2009.

The EPC is capable of connecting GERAN, UTRAN, LTE, Femto Access Point and other non-3GPP access networks such as CDMA, WiMAX and WLAN. The definitions for the handover procedures allow a rapid deployment of LTE in various scenarios. The handover between LTE and CDMA2000 eHRPD (Evolved High Rate Packet Data) is a special case that has been optimized.

The EPC and the access network that it uses are called the Evolved Packet System (EPS). The main difference between EPS and previous solutions is that EPS does not contain definitions for the circuit-switched domain connections any more, clearly indicating the evolution path towards an all-IP environment. In this environment, IMS has an important role. In all-IP architectures, voice and SMS are handled in alternative ways such as via session continuity, which combines 2G/3G and LTE functionalities using system handovers or using the Voice over IP (VoIP) solution.

LTE/SAE provides a modern means for using Point-to-Point (PTP) IP-based multimedia services, which require more bandwidth and lower latency, including mobile TV/audio, online gaming and other applications that works with high data rates, constant connectivity and needs service continuity on the move. This is actually an iterative evolution, as the provision of more capacity, increased speed and reliability of the packet delivery also increases the utilization of the data transmission of new solutions, including machine-to-machine (MTM) communications. Figure 2.4 shows an estimate of data traffic to 2013.

Nowadays, IP packet traffic is the dominant data transfer method. It is the most practical way to send and receive data due to the bursty nature of modern data communications, Internet being the most popular data delivery method.

UMTS releases have brought new architectural solutions to handle round-trip delays more efficiently and thus improve the throughput that the end user experiences.

As the architecture of the Release 7 Internet-HSPA (I-HSPA) indicates, the functions of the Radio Network Controller (RNC) have already been moved to the base station, or nodeB.

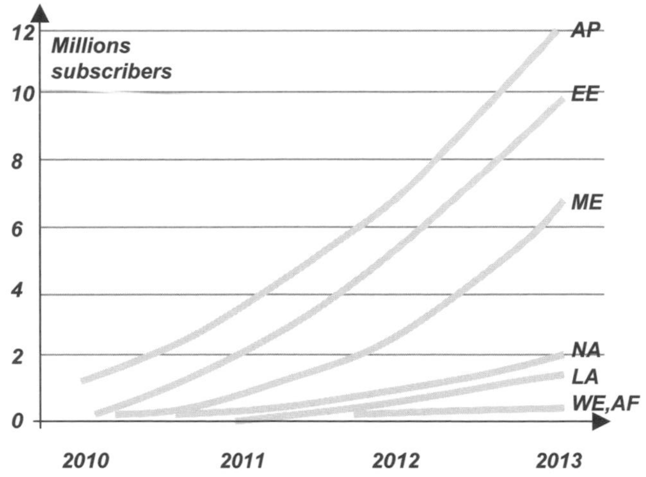

Figure 2.4 The estimated data growth as a function of the world regions, as interpreted in [8].

Table 2.1 The LTE deployment projects by the end of 2010

Operator	Location and other information
NTT DoCoMo	Japan
MetroPCS	USA, 9 cities
Telecom Austria	Austria, Vienna
TeliaSonera	Denmark, Finland, Norway, Sweden
Verizon	USA, 38 cities
Vodafone	Germany, rural areas

The packet connection chain thus contains fewer elements than in Release 6 and previous phases of UMTS and GSM. The benefit of this simplification can be seen in the shorter signaling connections and thus in smaller round trip delays, which benefits the throughput values directly.

2.3.3 Commercial LTE Deployments

According to TeliaSonera, the first LTE deployment was made in Sweden [9]. According to [10], there were nine commercial launch efforts by the end of 2010. Table 2.1 shows the principle of these networks, sorted in alphabetical order.

2.3.4 LTE Refarming Eases Development

One might ask why LTE networks are needed if Release 7 already handles data traffic in quite an efficient way. The answer is that LTE/SAE uses the same architectural principle but further enhances performance.

One of the key drivers for LTE is its highly adaptive usage in different network deployment and evolution scenarios. The trend is that, with higher capacity demand for the mobile networks, earlier releases of UMTS and GSM are no longer so spectral efficient.

This leads to the need to refarm the frequencies. One possible evolution path is to reduce the GSM capacity and add UMTS. This is justified for the more efficient spectrum usage of 3G.

The UMTS uses a 5 MHz frequency band (or, in certain vendor-dependent solutions, lower bands like 4.2 MHz) whereas GSM divides the frequency into 200 kHz blocks. Each GSM frequency contains eight physical time slots that can be shared between eight full-rate voice codec users or 16 half-rate voice codec users, although part of the capacity is reserved for the signaling. GSM evolution has also brought a dual half-rate codec feature, which doubles the previous half-rate capacity in those areas where the signal strength and quality are sufficient for the pairing of HR users into the same HR time slot.

GSM data service usability is still evolving. The latest addition is a DLDC, Downlink Dual Carrier, which uses the time slots of two separate carriers and is able to combine a total of 10 time slots per user. The total maximum data rate for a single user is thus about 600 kbps. The feature will be further enhanced in the second phase of this Advanced EDGE, providing about 1 mbps for a single GSM user.

There are also other features introduced with GSM, which enhance the spectral efficiency of the 2G network. One example is DFDA, which dynamically allocates frequencies depending on the interference levels.

Together with the DHR, DFCA and DLDC features, the current GSM traffic can be handled within a considerably smaller frequency band, still maintaining the same or lower blocking

Table 2.2 The GSM voice codec modes and respective number of channels as a function of available bandwidth

GSM mode	Bandwidth (MHz)					
	1.4	3.6	5	10	15	20
FR	56	144	200	400	600	800
HR	112	288	400	800	1200	1600
DHR	224	576	800	1600	2400	3200

rate. This leads to one of the relevant deployment scenarios, which is a GSM/LTE network without 3G services.

The real benefit of LTE is dynamic frequency allocation. The band can be defined between 1.4 and 20 MHz as shown in Table 2.2. The table also shows how many GSM FR/HR voice channels the LTE band would provide if GSM was utilized in it.

Depending on the current GSM frequency band, the refarming of its frequencies for LTE can be done in the smoothest way by utilizing HR and DHR of the GSM.

As an example, let us assume that there is a 10 MHz band in use for the current GSM 900. The current peak-hour blocking rate is 2%. Let us calculate the benefit of LTE in different GSM development scenarios if LTE is utilized purely for voice traffic.

The GSM roadmap can include DHR functionality. When activated, it brings double capacity compared with the previous HR mode in the functional areas of DHR. The functional area depends on the radio conditions, load of the network, and DHR selection algorithm. Figure 2.5 shows the principle of DHR, which indeed brings an additional coverage area near the base station, where capacity utilization is double compared to the HR mode.

We can assume a simplified version that can activate the DHR call pair when the quality value is <0.2% BER (quality Q0) and the received power level is better than −90 dBm. Then, the unpairing of the DHR calls to two separate HR calls can be triggered when the quality value is below Q5 or the received power level becomes worse than −90 dBm. Based on real field data from laboratory and field tests of OSC [11], [12], the proportion depends on the SAIC handset penetration and the overall quality of the GSM network. As a rough general estimate, the DHR proportion of the voice calls can be, for example, 30% with 60% SAIC handset penetration. We can thus estimate that in a typical GSM network, the voice channels can have a share of 30%

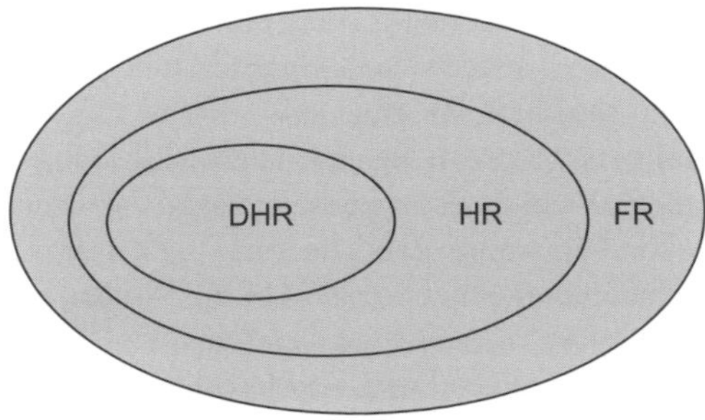

Figure 2.5 DHR increases the GSM capacity in the coverage area where the carrier per interference level is sufficiently high. This capacity gain can be used directly for refarming the frequencies for the LTE system.

DHR (with four times capacity per TSL), HR 20% (with two times capacity per TSL) and FR 10% (with single capacity per TSL). The packet switched data traffic can be assumed to use only the left-over time slots that would not affect the capacity dimensioning of the voice traffic. The TRX can be assumed to deliver on average 90% of the traffic to voice while the rest is reserved for the signaling. In the 10 MHz GSM band, we have 10 MHz/200 kHz channels (TRXs), each containing 8 TSLs, and 90%*8 TSLs for the voice calls (peak hour maximum of physical TSLs).

When LTE 900 is activated, it can use the 1.4 MHz, 3 MHz, 5 MHz, or 10 MHz bandwidths. As the original GSM that utilized 10 MHz band can now be squeezed into a much smaller bandwidth, the LTE can be used in a parallel way. It is very possible that, with the same original GSM blocking probability without DHR, the activation of DHR facilitates the delivery of the same traffic within 5–7 MHz. This means that LTE can use the 1.4–3 MHz, or even 5 MHz block at the same time. The terminals that do not support LTE can use at maximum the GSM band, and the new LTE terminals benefit from the high-speed additional capacity within the same frequency band.

Comprehensive refarming examples can be seen in Chapter 15, which shows the deployment paths from previous technologies towards LTE.

2.4 Reasons for the Deployment of LTE

2.4.1 General

The general drivers for the selection of LTE include cost, as the competition and open standard provides attractive costs for the terminals and networks. The systematic approach of 3GPP also guarantees that the system will be future proofed during the evolution of the system and during its complete life cycle. One of the more important aspects from the beginning has been the possibility for international roaming.

Trends seem to indicate the reduction of the role of, for example, CDMA. There are several examples of operators who have selected WCDMA and more specifically, HSPA/HSPA+ for the evolution. LTE/SAE provides smooth continuity for operators who already have the 3GPP network. Logically, LTE/SAE can be deployed without prior 2G/3G infrastructure.

2.4.2 Relationship with Alternative Models

Alternatives to LTE could be, for example, a USB stick-model of HSPA+ terminal, or WiMAX. There has been a lot of discussion in the operator community of the advantages and disadvantages of LTE compared to the alternative models. These alternative models might sometimes concentrate on 3G networks (in practice HSPA and HSPA+) instead of investing in new LTE/SAE architecture, or they might go in another direction entirely by deploying WiMAX technology.

Some operators concentrate on HSPA because they have failed to obtain a LTE license or LTE licenses have not been allocated yet in their particular country. Others are investing into HSPA as their short-term strategy while waiting for LTE to mature—for example, for network vendors to iron out any bugs/interoperability issues that are there, as in any new product, and also for terminal vendors to produce a large number of LTE-capable handsets and dongles. It is quite clear that using HSPA+, for example offering 42 mbps, does offer an attractive value proposition for a 3G operator running a radio network that can be easily updated (e.g. just installing a new software) to HSPA. However, for many of the operators heavily investing in HSPA in short/mid-term, LTE is still the top long-term target due to the benefits of LTE over HSPA, such as higher bandwidth, lower latency, and the eventual lower production cost of LTE.

The LTE versus WiMAX story has been well documented elsewhere but, based on recent developments, it looks as if WiMAX has failed to make a major dent on the LTE plans of operators globally. Thus it is safe to say that LTE is the winner of these two technologies.

2.4.3 TD-LTE versus FD-LTE

In addition to the "normal" FDD-LTE deployed by the first LTE/SAE operators around the world, there is also another LTE variant called TDD-LTE. Regulators around the world have already granted TDD licenses for WiMAX. These frequency bands are now also considered as a feasible option for TDD-LTE.

Basically this depends on the operator, whether it has been awarded TD- or FD-LTE license, and also the level of device support for these technologies. Due to FD-LTE being more widely deployed, there is more support for it at the moment. This is something that has to be taken into account when actually deploying LTE.

2.5 Next Steps of LTE/SAE

LTE-Advanced will lead to greater convergence in telecommunications, as Figure 2.6 emphasizes. LTE paves the way towards greater usability of multiple technologies, which can provide optimal performance and user experience, depending on the situation.

The 4G telecommunications system is characterized by the following aspects:

- convergence of the networks, technologies, applications and services;
- a personalized and pervasive network, which is transparent for users;
- the All-IP concept and a single core network;
- separation of services, applications, transport, and access.

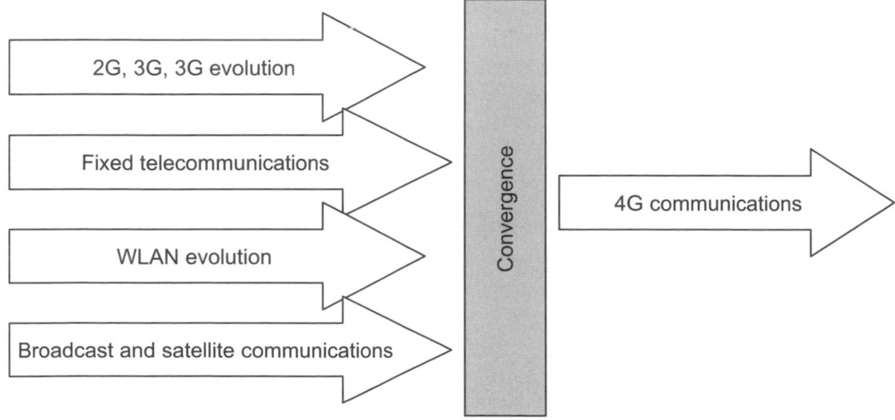

Figure 2.6 The path towards 4G happens via the convergence of telecommunications systems.

Table 2.3 Summary of the benefits of LTE

Functionality	Observations
OFDMA and CP	In LTE DL. Simple equalization. Good scheduling time performance. Inter-Symbol Interference (ISI) suppression
SC-FDMA and CP	In LTE UL. Energy savings. Optimal Peak-to-Average Power Ratio (PAPR)
QPSK, 16-QAM, 64-QAM	Adaptive usage of modulation, which provides optimal bit rates
Scalable bandwidth	In LTE, the band can be set to 1.4, 3.6, 5, 10, 15 and 20 MHz. This optimizes the band usage, and eases the refarming scenarios
TTI	In LTE, TTI is 1 ms. This leads to a better response to the channel variation, and to the higher bit rates
Flat architecture	Lower latency and simpler architecture.
All IP	No need for CS domain which simplifies the architecture
MIMO	Provides optimal performance
Frequency reuse of 1/1	Enhanced spectral efficiency

The result of this evolution provides smooth provision of the traditional and totally new services, and more attractive user experiences for the customers.

2.6 Summary of the Benefits of LTE

The 3GPP Long-Term Evolution represents a major advance in cellular technology, as can be inferred from Table 2.3. It is designed to fulfill the demand for the high-speed data and media transport in the forthcoming years. LTE provides mobile network operators with a base for high-performance services due to high uplink and downlink bit rates as well as low latency.

The LTE infrastructure is simpler than in the previous network solutions. LTE provides flexible frequency allocation, which can be used in the optimization of traffic delivery in different scenarios. It supports both FDD paired and TDD unpaired spectrums. One of the most important aspects is related to compatibility, as LTE–SAE will with GSM, WCDMA/HSPA TD-SCDMA and CDMA.

References

[1] NMTImage (1994) Conference proceedings. Digital Mobile Radio conference (DMR), Stockholm.

[2] ITU (2000) Mobile network generation discussions, www.itu.int/ITU-D/imt-2000/Revised_JV/IntroducingIMT_item3.html (accessed August 29, 2011).

[3] ITU (n.d.) LTE network launch information, www.itu.int/ITU-D/ict/newslog/CategoryView,category,4G.aspx (accessed August 29, 2011).

[4] ITU (2010) ITU's 4G advances, www.itu.int/net/pressoffice/press_releases/2010/40.aspx (accessed 29 August 2011).

[5] 3GPP TS 23.882. (n.d.) *System Architecture Evolution (SAE): Report on Technical Options and Conclusions*, 3rd Generation Partnership Project, Sophia-Antipolis.

[6] 3GPP TS 23.401. (n.d.) *General Packet Radio Service (GPRS) Enhancements for Evolved Universal Terrestrial Radio Access Network (E-UTRAN) Access*, 3rd Generation Partnership Project, Sophia-Antipolis.

[7] 3GPP TS 23.402. (n.d.) *Architecture Enhancements for Non-3GPP Accesses*, 3rd Generation Partnership Project, Sophia-Antipolis.

[8] ABI (2008) Forecast LTE Subscribers, World Market Forecast: 2010–2013, ABI Research, unpublished report.

[9] Nokia Siemens Networks (n.d.) Information about the first LTE initiations, www.nokiasiemensnetworks.com/file/5516/teliasonera-builds-europe%E2%80%99s-first-multi-city-lte4g-network (accessed 29 August 2011).

[10] Nokia Siemens Networks (2009) Long Term Evolution (LTE) will meet the promise of global mobile broadband. Nokia Siemens Networks, White paper.

[11] Penttinen, J.T.J., Calabrese, F.D., Maestro, L. *et al.* (2010) Capacity Gain Estimation for Orthogonal Sub Channel. Proceedings of the Sixth International Conference on Wireless and Mobile Communications (ICWMC), Valencia. CPS, Halifax, pp. 62–67.

[12] Penttinen, J.T.J., Calabrese, F.D., Niemelä, K., *et al.* (2010) Performance Model for Orthogonal Sub Channel in Noise-Limited Environment. Proceedings of the Sixth International Conference on Wireless and Mobile Communications (ICWMC), Valencia, 2010, pp. 56–61. CPS, Halifax.

3

LTE/SAE Overview

Jyrki T. J. Penttinen and Tero Jalkanen

3.1 Introduction

LTE refers to the long-term evolution of the 3GPP radio network—it can be considered as a synonym for the evolved UMTS Terrestrial Radio Access Network (UTRAN), as it defines the new LTE radio interface. It enhances the previous performance of the IP domain of the networks, and increases the flexibility of radio network design. It provides several channel bandwidths, which is useful in general network evolution, including refarming strategies. LTE can use either Frequency Division Duplex (FDD) or Time Division Duplex (TDD) modes. It is based on Orthogonal Frequency Division Multiple Access (OFDMA) technology in the downlink, and SD-TDMA in the uplink, which optimizes the radio interface for the mobile environment, especially in presence of multipath radio components and fast fading of multiple signals. It can also use the latest technologies on top of the basic solution, including 64 QAM (64 state Quadrature Amplitude Modulation) and different configurations of the Multiple Input Multiple Output (MIMO) antennas.

Evolved UTRAN contains only one type of base station, called eNodeB—referring to the evolved NodeB in previous network element solutions of 3GPP networks. In other words, there is no separate controller element in LTE but the respective functionality has been concentrated on the base station side.

The rest of the network, which is not related to the radio, is called SAE, referring to the System Architecture Evolution of the 3GPP core network. In fact, the general term for the evolved core part of the network is nowadays EPC, referring to the Evolved Packet Core. The name indicates that the system is based purely on IP traffic, and LTE thus does not have any connections defined for the Circuit-Switched domain.

The complete solution of the evolution is called EPS (Evolved Packet System), and it contains both LTE and EPC. The term SAE, is not in official use in standardization terminology any more. Nevertheless, it was introduced to the market with such enthusiasm that it has remained in spoken terminology. For this reason, either SAE or EPC can be used to describe the evolved core, although the standard term EPC is recommended for uniformity.

The LTE/SAE Deployment Handbook, First Edition. Edited by Jyrki T. J. Penttinen.

3.2 LTE/SAE Standards

LTE and SAE are standardized in the 3GPP (the 3rd Generation Partnership Project), which works, producing 3GPP specifications, on GSM, W-CDMA and LTE specifications.

GSM is a part of the second-generation mobile network solutions. It includes terms like GPRS (General Packet Radio Service) and EDGE (Enhanced Data rates for GSM/Global Evolution). The general term for the 2G radio network of 3GPP is GERAN (GSM/EDGE Radio Access Network). The third generation includes standardization terms to describe 3G networks like the UTRAN (UMTS Terrestrial Radio Access Network), the European UMTS (Universal Mobile Telecommunications System) and the Japanese FOMA (Freedom of Mobile Multimedia Access).

The standardization work of GSM and UMTS is a continuous process. The work was initiated in ETSI where the GSM and first UMTS standards were created. In 1999, the 3GPP took over both the GSM and UMTS and has been evolving the standards ever since. The standardization work is ongoing and the outcome is a set of new releases every now and then. Revised versions of many of the specifications are produced up to four times a year. The organization is democratic, and the solutions are selected thus via voting in subgroups and in the quarterly TSG plenary meetings of 3G. The TSG GERAN meets five times a year.

Figure 3.1 shows the current setup of the standardization groups of 3GPP. The RAN groups have defined the LTE specifications while SA has made respective definitions for the SAE.

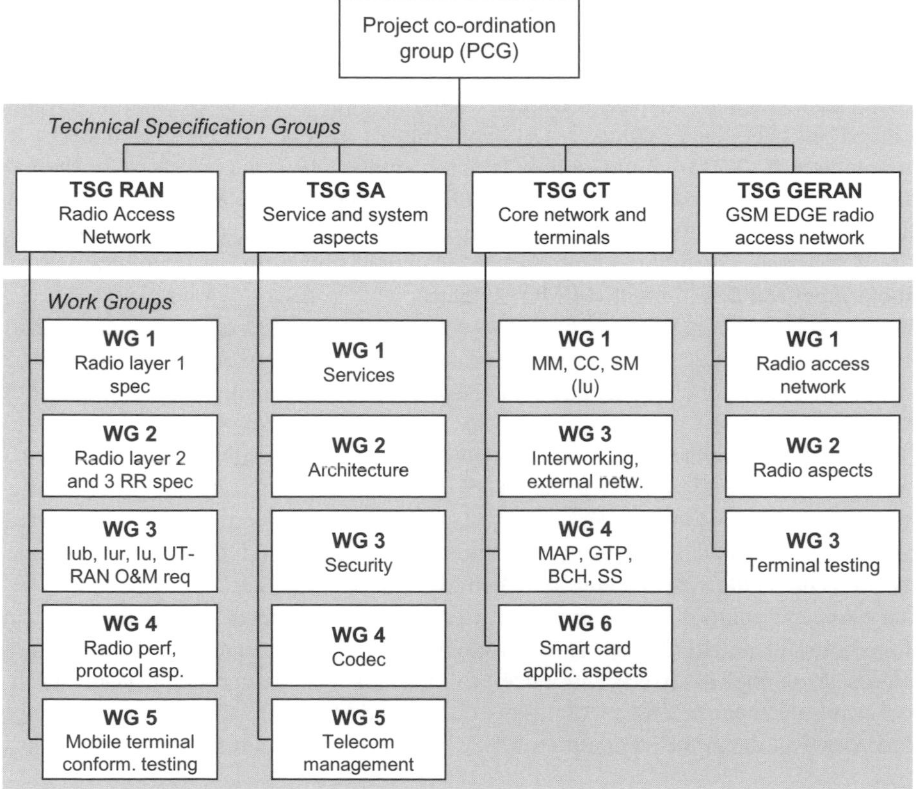

Figure 3.1 3GPP standardization is done in Technical Specification Groups (TSG) that include Radio Access Networks (RAN), System Architecture (SA), Core Networks and Terminals (CT) and GSM EDGE Radio Access Network (GERAN). The final standards are approved or rejected in plenary meetings.

3.3 How to Find Information from Specs?

All the relevant information about LTE/SAE and previous 3GPP definitions can be found on the 3GPP Web page (www.3gpp.org). The 3GPP standards and specification numbering is logical—the numbering indicates the main topic, and the latter part indicates the more specific item for each release.

The LTE (Evolved UTRA) and LTE-Advanced radio technology specifications can be found in the 3GPP 36 series.

A straightforward way to find the main topics is to investigate the overall specification numbering scheme first. The 3GPP specifications are divided into thematic topics that is indicated by the first part of the specification, which can be a technical specification (TS), or technical recommendation (TR), as shown in Table 3.1.

The LTE/SAE is defined in 3GPP specification number 36, although there are many interdependencies between the 36 series and other specification numbers. This is logical as LTE uses a base defined by previous 3G standards, the main differences being in the radio channels, access methods, and other radio-related new or modified definitions.

The specification number further consists of the subclassification (the decimal part) and version with the date when the specification was formalized in the 3GPP acceptance processes.

Table 3.1 The 3GPP specification division

Item	Specification/recommendation numbering	
	GSM specific, original ETSI nr/Release 4 and beyond	3G and GSM, Release 99 and later
Requirements	01/41	21
Service aspects of stage 1	02/42	22
Technical realization of stage 2	03/43	23
Signaling protocols of stage 3, UE—network	04/44	24
Radio aspects	05/45	25
Voice codecs	06/46	26
Data	07	27
Signaling protocols of stage 3, RSS—CN	08/48	28
Signaling protocols of stage 3, intra-fixed-network	09/49	29
Program management	10/50	30
Subscriber Identity Module (SIM), Universal Subscriber Identity Module (USIM), IC cards, test specifications	11/51	31
Operations, Administration, Maintenance, and Provisioning (OAM&P) and charging	12/52	32
Access requirements and test specifications	13	
Security aspects	—	33
UE, SIM and USIM test specifications	11	34
Security algorithms	Not public	35
Evolved UTRA (LTE) and LTE-Advanced radio technology	—	36
Multiple radio access technology aspects	—	37

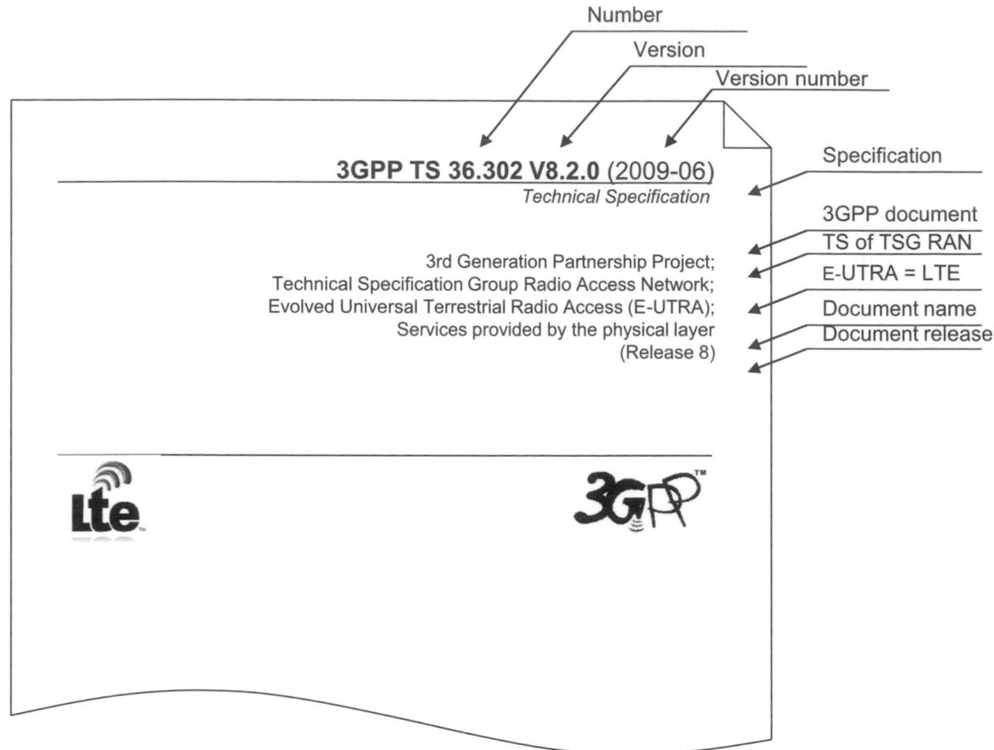

Figure 3.2 An example of the 3GPP standard's cover page. Reproduced by permission from 3GPP. 3GPP and ETSI are trade marks of ETSI.

The front page of 3GPP LTE/SAE specification consists of the information as seen in the example in Figure 3.2.

Release 8 is the first one where definitions for the LTE and SAE are found. The standards are divided into the following categories at the moment:

- Release 8, the base for LTE.
- Release 9, relatively minor additions for LTE.
- Release 10, the base for LTE-Advanced.
- Release 11, enhancements for LTE-Advanced.

The LTE definitions for Release 8 and Release 9 have already been frozen, and LTE-Advanced standardization is ongoing. In fact, LTE and LTE-Advanced standardization will continue in a parallel way.

The completion of 3GPP Release 9 happened in the 3GPP Plenary meeting TSG#48 in Korea. The current focus is now in progress towards always-on, always-broadband aspects.

The publication of Release 8 and Release 9 provides the telecommunications industry with a logical evolution towards the next generation mobile networks, which in terms of 3GPP refer to LTE. Common standardization makes LTE compatible with legacy networks, which is one of the important arguments of the evolution. As 3GPP work is continuous,

Table 3.2 The main division of the LTE specifications

Technical Standard (TS)	Main topics
TS 36.1xx	Equipment requirements for terminals and eNodeB elements
TS 36.2xx	Physical layer (1) specifications
TS 36.3xx	Layer 2 and 3 specifications
TS 36.4xx	Network signaling specifications
TS 36.5xx	User equipment and conformance testing specifications

there will be further milestones, including Release 11. Its focus can be expected to be on the further maturation of LTE and LTE-Advanced, including System Architecture Evolution (SAE). The work also continues due to LTE deployments, which will reveal further development items.

The new LTE-related Release 8 definitions can be found in the 3GPP 36 series. Table 3.2 lists the main topics of the specifications.

3.4 Evolution Path Towards LTE

The need for higher data rates and faster signaling response times than GSM and UMTS/HSPA could offer was the main trigger for LTE/SAE standardization. Behind this need are the forecasts for data evolution, which indicate increasing importance of mobile data services. Cost reduction is also one of the main drivers for network operators to introduce LTE/SAE.

In order to offer considerably enhanced user experiences, the optimization of delay is essential. This means that the minimized latency and round-trip-delay will lead to higher TCP traffic throughput and lower UDP/RTP traffic jitter, and eventually to higher quality real-time services. Fast service availability is essential, and is accomplished via low bearer setup times. These benefits increase the awareness of LTE and increase the utilization of the networks. Forecasts indicate that over 90% of the traffic will be data traffic in 2012 with mobile data traffic four to five times more in 2012 compared to 2010.

As the trends clearly show, the number of mobile subscribers using Internet services—Web browsing, e-mail, social networks with text and multimedia chat–is increasing heavily. At the same time, data transfer that was dominated traditionally by the business user community is nowadays going towards the younger user profiles who have grown accustomed to using the same services in fixed Internet over several years.

In order to provide smooth user experiences it is essential to reduce the latency of the signaling, or control plane messaging. The round-trip delay that previous 2G and 3G networks contained is seen as a limiting factor in the provisioning of smooth services.

The benefit of pure IP-based mobile network architecture is the possibility to simplify functionality. The IP technology makes it possible to concentrate functionalities in just a few elements. This results in the deployment of significantly lower number of elements in LTE/SAE compared to the traditional mobile networks. LTE/SAE thus offers clear benefits for the CAPEX and OPEX optimization of the network.

The pure IP solution means that EPC is optimized for delivering data in the PS domain. In the previous solutions of 2G/3G, data delivery has been something of a compromise due to its support for both circuit and packet-switched domains. Nevertheless, LTE/SAE is designed in such a way that it is backwards compatible with the previous solutions. In practice, there are mechanisms that support service continuity via, for example, roaming between 2G, 3G and LTE networks.

An important aspect of the planning of EPC is backward compatibility. EPC and EPS provide mechanisms in order to support mobility when the terminals change between 2G, 3G, and LTE networks. The basis for mobility is either the GTP or PMIP mobility protocol. Within the EPC, a new GTP version has been introduced, GTPv2 [1], for control-plane signaling. The previous GTPv1-U is still employed without changes for the user-plane transport.

For the non-3GPP access systems, for example, WLAN and WiMAX, LTE/SAE supports mobility based on the generic mobility protocols of IETF, like PMIPv6 and DSMIPv6. For the special case of the CDMA2000 eHRPD system, the handover mechanism to LTE was designed by introducing special control-plane and user-plane interfaces to the EPC. This facilitates the transfer of information from one access network to another one prior to the handover procedure, which optimizes and harmonizes the actual handover procedure.

3.5 Key Parameters of LTE

LTE offers the possibility of introducing services at a variety of frequencies as described in Chapter 7. One of the main benefits of the LTE system over the previous versions of mobile communications systems is the possibility of defining various bandwidths, up to 20 MHz.

The LTE capacity offering is based on the resource blocks (RB). The number of RBs depends on the bandwidth, as in Table 3.3. More detailed principles regarding RBs are given in Chapter 7.

Other essential LTE parameters are the following, valid both for FDD and TDD bands of UMTS:

- Multiple access methods: in the downlink OFDMA (Orthogonal Frequency Division Multiple Access), and SC-FDMA (Single Carrier Frequency Division Multiple Access) in the uplink.
- In the downlink, LTE can use a wide choice of MIMO configurations in order to benefit from the transmission diversity, spatial multiplexing, and cyclic delay diversity. In the uplink, there is the possibility of employing multi-user collaborative MIMO.
- The peak rate of LTE is up to 150 mbps, which is obtained by using the UE category 4 with 2×2 MIMO in the full 20 MHz bandwidth. Alternatively, a speed of 300 mbps can be achieved with the UE category 5 and 4×4 MIMO in the 20 MHz band. In the uplink, the maximum speed of 75 MB7s can be achieved in the 20 MHz band.

Table 3.3 The amount of LTE resource blocks (RB) per bandwidth

	LTE Band width (MHz) 1.4	3.6	5.0	10	15	20
RBs:	6	15	25	50	75	100

3.6 LTE vs WiMAX

It is interesting to investigate if LTE/SAE and WiMAX are competing technologies. In fact, both are designed to transfer packet data instead of the voice service. Furthermore, both systems are based on OFDM technology in the downlink.

WiMAX is defined in IEEE standard 802.16. Similarly, as in the case of the previous generation of the IEEE-defined WiFi, it is an open standard with extensive revisions by the engineer community prior to having been ratified as a standard. This provides a means to introduce WiMAX equipment on such a scale that the cost for the end user is relatively low. Similarly, the 3GPP standardization provides LTE/SAE equipment interoperability, which results in economies of scale.

At this stage, from the end-user point of view, one of the differences between LTE/SAE and WiMAX is that the latter offers higher data rates. The timing is also different between these technologies. LTE deployment takes a certain time period while the current generation of WiMAX is available in the markets. According to the statistics presented by the WiMAX Forum [2], [3], there were 582 WiMAX deployments in 150 countries by February 2011, which indicates that WiMAX is deployed largely at the global level. It can be assumed in general that the mass rollout of LTE/SAE will take place during 2012. It is not clear, though, how popular the basic version of WiMAX will be compared to LTE, as the benefit of the latter is the possibility to integrate it as a logical part of the rest of the 2G/3G infrastructure, which is attractive for incumbent operators as well as for the end users due to the service continuity aspect.

For the mobile operator, the most important difference is that LTE/SAE can be deployed smoothly as an extension in the existing GSM and/or UMTS infrastructure. WiMAX, in turn, requires a completely new network. The main benefit is thus that the 2G/3G infrastructure is very largely used at a global level, and the use of new LTE terminals would be a straightforward step for the end users while the operators benefit from the existing infrastructure.

3.7 Models for Roaming Architecture

3.7.1 Roaming Functionality

Specific services used on top of LTE/SAE have their own demands. For example, voice has a major impact on the roaming functionality. If CS Fallback is used then voice roaming is handled via existing 2G/3G CS voice-roaming arrangements. If VoLTE is used then there are a number of ways to do it—this is also related to the question of IMS roaming arrangements in general because it is expected that all IMS-based services would be using the same roaming functionality, which should simplify the deployment.

In case of the OTT voice solution it is expected that it will be handled as normal LTE/SAE data roaming. Further details of roaming in LTE/SAE are described in Chapter 10.

3.7.2 Operator Challenges

As a part of general LTE/SAE architecture considerations one needs to take into account important interoperator aspects. The main issue here is that 3GPP has defined multiple architectural models of how the traffic between VPLMN (Visited PLMN) and HPLMN (Home PLMN) can be handled. This mostly relates to the area of EPC; for example the PGW node can be located either in VPLMN or in HPLMN.

The architecture selected for LTE/SAE roaming requires careful coordination with the other operators, otherwise the situation can easily lead to an unmanageable mess, with various different noninteroperable solutions deployed by individual operators based on their own commercial and technical analysis. This does not really benefit anybody, apart from roaming brokers and vendors of Interworking Function (IWF) nodes trying to offer the necessary mapping between the different solutions. Major operators typically have a few hundred 2G/3G roaming partners, so logically it can be expected that in LTE/SAE roaming the number of connected partners can be tens or eventually even hundreds.

From the RAN point of view, the main difficulty with the LTE/SAE roaming area relates to the fact there is quite an extensive number of potential frequency bands used for LTE, ranging from 700 MHz to 3500 MHz. In practice this means that for a device to really work globally it would need to support the different bands deployed in various parts of the world. Obviously the device does not have to support all the possible frequency bands but even ten bands would mean the device has to double the capability of today's high-end phones such as iPhone 4 or Nokia N8, which are pentaband devices—capable of supporting five 3G bands. At least in the beginning it is not likely that LTE devices are that much more capable, thus a device designed for the US market might not function when roaming, for example, within Europe or Asia using LTE.

From the end-user perspective this unfortunate situation is somewhat similar to the early days of 2G or 3G roaming when people traveling globally had to take into account that their regular device might not be able to roam in every destination. However, a typical LTE device is capable of using the 2G/3G network, which means that the user is able to use his/her devices through the more widely existing 2G/3G roaming agreement between VPLMN and HPLMN, although at a lower speed than would be possible with LTE roaming. It will take some time for LTE roaming to be really widely deployed commercially, just as it took quite a while for operators to build a wide 3G roaming coverage.

Further details of the various aspects of LTE/SAE roaming are explained in Chapter 10.

3.7.3 CS Fallback

Even though LTE/SAE is a pure IP network, there are connections to the "legacy" networks. One example of this is CS Fallback for voice as defined in [4]. Basically CS Fallback means that the "good old" 2G/3G CS network is used to produce voice calls for a LTE/SAE user—that is, for the duration of the call the device is switched from the LTE network to the 2G/3G network, and once the call is over the device switches back to the LTE access. The reason for this is the lack of a CS voice domain on the LTE/SAE side, which means that voice has to be produced via some other mechanism such as VoLTE or OTT. If these other mechanisms are not deployed, the only solution for producing voice in LTE/SAE is to use the existing 2G/3G network via CS Fallback.

In practice, for mobile-originating calls, the User Equipment (UE) makes a service request to the network, which then moves the UE to the 2G/3G network. For mobile-terminating calls, when receiving the call, the MSC-S pages the UE through the LTE network and the UE is then moved to the 2G/3G network in order to receive the CS voice call from the MSC-S. The 3GPP has defined options on how this switch between the networks actually happens—for example, towards 3G either PS Handover or Redirect mobility mechanism can be used.

As a part of the work on CS Fallback, 3GPP has defined "SMS fallback," which is called SMS over SGs. This allows the LTE/SAE device to send and receive short messages via the

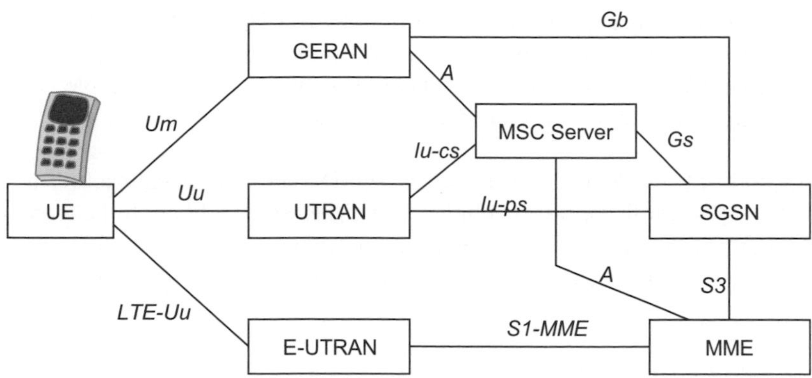

Figure 3.3 EPS architecture for CS fallback and SMS over SGs as interpreted in [4].

MME⟺MSC-S interface. It should be noted that the device is still attached to the LTE/SAE network when using SMS over SGs, which is a major difference compared to CS Fallback voice, which requires the device to attach to the 2G/3G network. Figure 3.3 shows the high-level architecture of CS Fallback.

From the deployment point of view, CS Fallback requires underlying 2G/3G network coverage, support for it in the core network, and also devices have to support CS Fallback explicitly. In addition to having the requirement of supporting the 2G/3G with the LTE/SAE network, there are some other drawbacks to the CS Fallback solution. For example, the data service being used by the end user is also switched from LTE to 2G/3G when an incoming/outgoing voice call takes place. Depending on the service used this could have little impact or a major impact. In the case of LTE ⇒ 2G fallback it is quite clear that the bandwidth will drop dramatically or, if DTM is not supported, then the whole data session will be suspended for the duration of the voice call. From the end-user perspective this will be likely to cause irritation.

Perhaps the biggest question being analyzed by operators and vendors at the moment is the performance of CS Fallback. It is clear that, due to the additional steps involved in delivering the voice call via CS Fallback, namely the actual handover from LTE to 2G/3G, there has to be an increased delay compared to "native" CS voice in 2G/3G network or VoLTE in LTE/SAE. A number of paper calculations or laboratory studies have been performed, little data is publicly available, but generally figures being discussed by experts suggest that there may be an additional delay of something like 2–5 s.

It should be noted that in the worst-case scenario of a CS Fallback UE calling another CS Fallback UE, even the normal end users will probably notice something strange due to both ends requiring that additional delay, which combined can lead to suboptimal performance. How often this scenario happens is really a question for the individual operators to consider based on, for example, their network and device roadmaps. From the deployment point of view it is quite safe to say that it should be avoided if possible.

However, until CS Fallback is fully tested on real live networks, it is very hard to actually predict the increased latency. The latency might vary quite a bit due to different radio conditions, and the load situation—and naturally the parameters used in that particular radio and core network—have a major impact too. Another aspect is the mechanism used for CS Fallback—for example PS Handover has its own advantages and disadvantages over Redirect. 3GPP is still working on enhancing the performance of CS Fallback, so the base delay caused by Rel-8 versus Rel-9 implementation will probably be different.

It is safe to say that SMS support—that is, use of SMS over SGs—will be widely used because performance issues are not a major problem for SMS and due to the fact that, when using SMS over SGs, the device stays attached to the LTE/SAE network. So, for example, an LTE/SAE data dongle can receive and send short messages while it is used for accessing the Internet services.

The operators of GSM Association as well as NGMN consider CS Fallback as the common intermediate step towards the target solution of VoLTE, so it is likely to be a part of many vendors' roadmaps (www.ngmn.org/news/ngmnnews/newssingle2/article/ngmn-alliance-delivers-operatorss-agreement-to-ensure-roaming-for-voice-over-lte-357.html).

3.7.4 Inter-Operator Security Aspects

As a part of the more general security review of the whole LTE/SAE environment, it is important to consider interoperator aspects, namely roaming and interconnection. GSMA document IR.77 describes the general guidelines for the GRX/IPX network, which are also valid when deploying, for example, LTE/SAE roaming regardless of the service/application used.

The main issues relate to the fact that GRX/IPX is a completely separated network—that is, it is not visible or accessible from the Internet. This places some restrictions on deployment. In practice one needs to use core network nodes that can access both GRX/IPX and the Internet and can support multihoming or otherwise be capable of having two completely separate interfaces: one for the Internet and another for GRX/IPX. IP addressing used for these interfaces needs to be also separate—that is, one cannot reuse an IP address in GRX/IPX that is already utilized in Internet.

Additional guidance is required, especially in case some other network, for example the Internet, is used for connecting with the other operators. This is highly relevant for the security level of the original network: GRX/IPX can be considered to be a secure network since it is only accessible for operators that are part of the "club", whereas the Internet by its definition is open for everybody. So one needs more security-related functions, such as SBC to guard incoming access and IPSec tunnel securing the traffic, in Internet compared to GRX/IPX.

In practice these issues entail some additional work for the deployment of LTE/SAE due to the different networks that are present. However, it should be noted that this work, while somewhat onerous, is considered to be absolutely essential in order to ensure that the vital interoperator environment is kept as simple and manageable as possible to allow, for example, fault, misuse and fraud discovery to be more efficient than in the completely open Internet.

3.7.5 Selection of Voice Service Method

One of key questions, if not the biggest question, for LTE/SAE operator is how to solve the issue of providing voice service to the customers as voice is still the application that is most used. Operators and vendors have spent countless hours studying this vital issue and it seems that there is no simple solution that would be commercial and technically optimal, taking into account, for example, the need for wide support in networks and devices in the near future.

An operator must also consider other related aspects such as regulatory demands (emergency calls, lawful interception), support for supplementary services and other required items such as support for numbering plans and especially number portability. These issues might appear to be minor but it is well known that for example fulfilling lawful interception requirements has affected many implementation projects in a drastic way. Thus, from the deployment point they are certainly something to keep in mind.

When deploying a new voice service it probably would be wise to consider enhancing it to give it some advantage over the existing CS voice. For example "HD Voice" – use of more advanced wideband codec such as AMR-WB—has been widely noted as something that many operators are thinking about. Some deployment of AMR-WB has taken place already in the CS domain but this would make particular sense for the new PS domain. From the customer perspective the advantage of "HD Voice" over the traditional narrowband codec is clear: significantly better voice quality.

Basically these are the main options for the LTE/SAE voice solution:

- CS Fallback;
- VoLTE;
- OTT;
- VoLGA.

CS Fallback will be described in more detail in Chapter 9. For a greenfield LTE/SAE operator without an underlying 2G/3G network this does not make much sense but for an existing 2G/3G operator deploying the new LTE/SAE network it offers a means to ensure that the LTE/SAE customer would be able reuse the current CS voice capability of the 2G/3G network. In practice the possibility of reusing the existing implementation reduces the need to deploy some other LTE/SAE-specific solution for the voice service.

VoLTE means Voice over LTE as documented in GSM Association IR.92 "IMS Profile for Voice and SMS." This is an IMS-based solution crafted out of 3GPP-defined MMTel specifications [5], [6].

VoLTE was formerly known as "One Voice." Deployment of a full-scale VoLTE solution requires quite many features from the radio and the core networks: IMS core system, Telephony AS, SRVCC (for handling voice during 2G/3G handovers), QoS support for radio bearers (ensuring quality for voice over data) and dynamic PCC architecture. Once this rather extensive arrangement is in place, VoLTE is expected to give a common global interoperable solution on LTE/SAE voice.

VoLTE is being further developed and enhanced in GSMA. For example, support for the advanced functionality of 3GPP Rel-10 compliant SRVCC, addition of PS video-call service to the existing PS voice call functionality, and extension of IR.92 to function in the HSPA access network in addition to LTE/SAE have been proposed. Thus it is possible that, in addition to the basic voice service, VoLTE would provide the operator with video call functionality. For instance, 3G CS video call has not been used much—but lately the introduction of Apple's FaceTime has demonstrated that there is an interest to offer this kind of service commercially.

One additional point to notice is that the VoLTE approach is aligned with the RCS (Rich Communication Suite) approach, as RCS uses VoLTE as the voice solution in LTE/SAE environment. Both VoLTE and RCS use IMS, so from the core network point of view they are very much complementary services.

From the deployment perspective, the voice solution for LTE/SAE needs to be fully integrated with the existing voice solution of 2G/3G, otherwise LTE/SAE customers will not be able to call "legacy" users of 2G/3G and PSTN networks and vice versa seamlessly. This requires functionality such as PS/CS voice conversion performed by the network, supported for example, by the IMS core system via MGW/MGCF nodes in the case of VoLTE. For CS Fallback this legacy support is in place natively due to reuse of legacy CS voice functionality.

For the OTT solution this functionality needs to be arranged on a case-by-case basis as there is no general mechanism available. Typically VoIP providers have their own particular methods of handling this requirement, such as Skype offering their Skype-In/Out service for interacting with the CS domain.

Over The Top (OTT) is a general term used to describe "Internet players" such as Google and Skype. Since LTE/SAE is a pure IP network, it is entirely possible to use those OTT VoIP solutions to cover the needs of the end users for the voice service. Generally speaking, an OTT solution is very much related to the bigger question of whether it is wise, technically or commercially speaking, for an operator to do everything by itself or instead to try to outsource everything possible. Going for an OTT solution is basically outsourcing the voice solution, depending on the actual details of the arrangement between the operator and the OTT provider. In the most simple scenario the OTT player provides everything related to the service and the operator just offers an interface which its customers can use to reach this OTT player. In a nutshell this is the "bit pipe" scenario, which typically makes an operator CEO see red. However, a number of operators have done that or at least are considering doing that. For example Sprint has teamed with Google to "launch an integrated Google Voice experience" on all phones, including LTE [7].

In addition to the solutions described above there is also VoLGA (Voice over LTE via Generic Access), which can be described as UMA/GAN over LTE. That is, the existing CS voice core network would be integrated with LTE access networks to offer the voice call functionality. This might sound as an interesting solution due to reuse of CS voice but VoLGA has never gained a major support in the operator community, apart from Deutsche Telekom. One of the problems related to VoLGA is the fact that there is no 3GPP specification for it. Some players have also seen CS Fallback as a simpler (less expensive) intermediate step than VoLGA towards to the long-term solution of purely IP-based voice over LTE/SAE. Currently the interest in VoLGA can be described as minimal, thus it is not expected to really compete with the other solutions documented in this chapter for the widely deployed LTE/SAE voice solution.

At the moment it seems that first doing CS Fallback and then going for VoLTE is seen as the common solution for the mobile industry, as documented, for example, by GSMA and NGMN press releases. Individual operators can choose to do something else, of course, such as making an alliance with Skype, Google or some other VoIP provider to offer voice services for their LTE/SAE customers.

3.7.6 Roaming and Interconnection Aspects of LTE/SAE

Interconnection between LTE/SAE operators in the simplest way means just reusing the existing technical and commercial arrangements in place for the services used in the 2G/3G environment. However, as LTE/SAE is an "All IP" network, it would make sense to ensure that the interconnection environment is based on IP, especially if purely IP-based services such as VoLTE or RCS are used on top of LTE/SAE access.

Typically TDM connections are used today for voice interconnection. This is fine for CS voice but in case of, for example, two VoLTE operators it would mean that the originating operator converts IP-based voice to CS-based voice, which is used over the TDM interconnection interface, and then the terminating operator converts the voice back to IP-based voice so that the end user using VoLTE can understand it. Naturally this PS/CS conversion, occurring twice, just adds delay, potentially degrades voice quality and, generally speaking, adds unnecessary costs to the voice interconnection due to both operators having to use their CS/PS conversion nodes (such as MGW/MGCF). In a nutshell it can be said that exchanging traffic between IP-based core systems using a TDM-based interconnection network is not really reasonable either from the technical or commercial point of view.

A technically and commercially preferable approach would probably be to go for the IP-based interconnection–that is, both signaling and media would be run over an IP based network. From the deployment perspective there are a number of options for this IP interconnection solution:

1. Leased line;
2. Connection point;
3. Internet;
4. IPX.

Within a country a dedicated leased line (whether a physical or a logical connection) is quite often used between the operators. The downside of this approach is that it is not a valid option for international connections because arranging bilateral connections to tens or even hundreds of interconnection partners simply does not work.

Connection point is a slightly evolved version of (1). This basically requires a single point where all the interconnected parties bring their own leased lines. This helps to avoid the main problematic issue of solution (1), which is arranging and operating multiple separate connections, as all the interconnected operators would have just a single line to the connection point regardless of the number of interconnection partners connected there. This kind of approach is used today, for example, by some VoIP providers to arrange the peering of traffic within a "club" of providers. However, this still does not provide a feasible option to produce the global connectivity required for an operator.

The Internet, obviously, is a wonderful thing but, when it comes to the carrier-grade connectivity that is typically seen as core issue of an operator-offered service, it lacks certain important features such as predictable latency and availability. From a security point of view, the Internet might not be the first choice of a person responsible for the security aspects of an operator service such as voice. Another aspect is the desire of many operators to use the multilateral model instead of bilaterally arranging connections to each interconnection partner, which requires a lot of manual work. In practice there is a need for some service provider(s) to take care of "brokering" the technical and commercial requirements between the operators so that an operator can have only a single connection to this "broker", which then forwards the traffic to other destinations accordingly.

For the reasons listed above, the GSM Association developed GRX (GPRS Roaming eXchange), which has been commercially used since the year 2000 for arranging the IP connections needed in PS roaming. Basically GRX is a dedicated IP network completely separated from Internet, open only for the members of GSMA. GRX consists of around 20 GRX carriers connected to form an "IP cloud", which is necessary for the global connectivity required by hundreds of 2G/3G mobile operators around the world that are using GRX to provide, for example, routing of traffic between visited and home operator in 3G data roaming. The concept of GRX has been further developed to include full QoS support, hubbing model (the multilateral connectivity option) and inclusion of non-GSMA operators, mainly FNOs (Fixed Network Operators). This new concept is called IPX (IP eXchange) and it is a potential candidate for the interconnection needs of the LTE/SAE operator, in addition to supporting the roaming scenarios. GRX and IPX are described in a number of GSMA documents, the main paper being IR.34 [8]. Further information on GRX/IPX can be found in the roaming-related Section 10.4.

It is entirely possible to use a combination of these alternatives—for example an operator could choose to handle national interconnection via leased lines while deploying IPX for all the international connections. This kind of mix does place an additional burden on the design and operation of an operator's internal and external IP core network; for example, outgoing traffic towards Operator C needs to be routed differently from the traffic going towards Operator D if those are reached via different interoperator IP network. Therefore it would probably make sense to try to minimize the number of different solutions used simultaneously, if possible.

It is important to notice that, if the intention is to deploy new IMS-based services such as RCS (using Presence, Video Share, Image Share, IM, etc.) one needs to deploy an IP-based interconnection in order to support these new services. Running for example, Presence traffic over TDM is not a feasible option.

Note that it is also possible to run CS voice over an IP interconnection, using, for example, a 3GPP Release 4 core architecture of MSC-S/MGW "soft-switch" nodes. Signaling between MSC-S nodes would use SIP-I while the media runs via RTP/RTCP. Both control and user planes are natively IP based. This kind of architecture is widely deployed by the operators around the world—for example Nokia Siemens Networks has provided their MSC-S product to more than 260 customers. This would, for example, allow interconnections related to CS Fallback to run over IP-based NNI, even though UNI would still be CS based. This kind of general TDM replacement by IP-based networks is an ongoing trend in the operator community at the moment, even for pure 2G/3G operators.

Finally it is important to notice that in multi-operator environments, such as interconnections, other operators have an impact to the solution deployed individual operators. The reason is simply that there is no point in selecting the Internet as the interconnection mechanism, for example, if one's the major interconnection partners are going for IPX.

3.8 LTE/SAE Services

LTE/SAE is designed to provide fast data transmission. In addition, the LTE terminal may support the previous 2G and 3G radio access technologies (RATs) of 3GPP. Handover between LTE/SAE and other, non-3GPP networks is also possible, the CDMA2000 being one of the most important ones. Figure 3.4 shows the architectural layout that supports this multi-RAT functionality.

3.8.1 Data

From the user's viewpoint, the data service of LTE does not differ much from the previous principles of packet-switched data of 2G and 3G. The typical situation is the USB dongle as an LTE terminal, which is used as a "virtual" modem in laptops. Figure 3.5 shows the setup.

The terminal is initiated using AT commands as defined in 3GPP standard 27.007. In practice, the user normally does not have to physically type the commands as the respective terminal wizard can be used. Once the settings have been typed in, the process can be initiated by opening the wizard. A separate communications terminal application can also be used to type the AT commands.

The basic principle of the AT commands used in LTE follows the standardized AT commands found in [7], with extended definitions for subparameters. In this sense, the AT

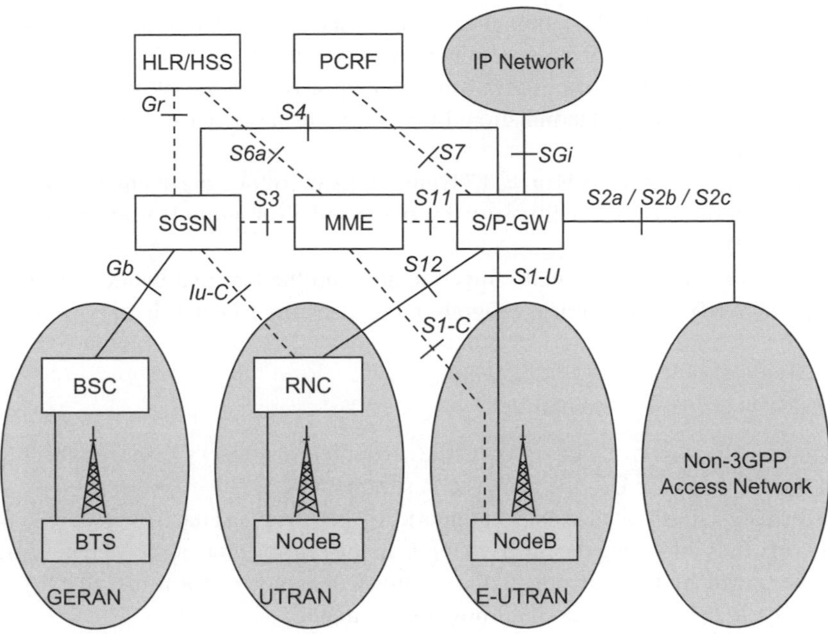

Figure 3.4 The LTE/SAE system supports multi-RAT functionality.

Figure 3.5 A typical use case of LTE and LTE-Advanced terminal assumes a connection to the USB port of a laptop. The ZTE terminal model printed by courtesy of TeliaSonera.

command execution of the LTE terminal is similar to the basic modems utilized back in 1980s and 1990s in the circuit-switched data communications. The AT concept was first used in GPRS in the Release 97 specifications of ETSI in order to provide a means to handle the GPRS terminal for call initiation and termination. Ever since, the new releases have introduced more commands.

The principle of AT commands in the LTE environment follows the definitions of those used for the UMTS terminals. The UE is seen by the terminal program (for example, Windows Hyper Terminal).

When the LTE-UE is connected to the computer and the terminal window is opened, the fastest way to verify if the terminal program recognizes the LTE-UE is to type

$$AT < CR >$$

In this notation, the symbol <CR> refers to the carrier return (enter). On successful connection between the PC and LTE-UE, the response is simply "OK."

When the LTE-UE is switched on, it automatically performs an attach procedure to the LTE/SAE network, thus informing the MME that it is in the routing area of the LTE coverage area, ready to be paged by the network. If the terminal does not perform the attach procedure automatically, it can be done via the following command:

$$+CGATT = n$$

In this format, the value of n = 1 refers to the attachment, and n = 0 leads to the detachment procedure. As is the case with many other AT commands, this command executed with a parameter of "?" shows the state (1 indicating that the terminal is attached to LTE, and 0 that it is detached). The command executed with " = ?" shows the options this command has, that is, "0" and "1."

The next step for creating the LTE packet data call is to perform an activation request for an EPS bearer resource, which is equivalent to the PDP context activation in the GPRS. More specifically, this creates the connection between LTE-UE and P-GW by defining the utilized protocol. The command line would thus be:

$$+CGACT = n$$

The parameter n indicates the state of the EPS bearer; it is activated with "1" or deactivated with "0." The command also includes (optional) parameters indicating the more specific EPS bearer. If it is left out, the default bearer is activated.

In case of the Mobile Originated (MO) data call, the EPS bearer is now selected with the following command:

$$+CGDCONT =$$

This command includes parameters in the following way:

- <cid> specifies a particular EPS bearer definition, beginning with the value of 1.
- <PDP_type> is the protocol type utilized in this specific EPS bearer. The possible options in Release 8 are: X.25, IPv4, IPv6, IPv4v6 (virtual PDP_type in order to handle dual IP stack

UE capability), OSPIH (Internet Hosted Octect Stream Protocol), and PPP (Point to Point Protocol).

- <APN> is a logical name in string format that indicates the P-GW, GGSN or the external packet data network.
- <PDP_addr> identifies the mobile terminal in the address space applicable to the PDP.
- <d_comp> is meant to select the PDP compression.
- <h_comp> is meant for the PDP header compression.

Optionally, the following can be defined:

- <IPv4AddrAlloc> is a parameter that indicates how the MT/TA requests to get the IPv4 address information. The possible values are: 0 (default value to indicate that the IPv4 address is allocated through NAS Signaling), and 1 (indicate that the IPv4 address is allocated through DHCP).

The command can be repeated by entering <CR> and defining a command line for the parallel EPS bearer activation—that is, multiple different bearers can be activated via the same terminal.

The EPS bearer is removed with the following command:

+CGDSCONT

In case of the Mobile Terminated (MT) call, the LTE-UE can be set to answer automatically to the network originated call:

+CGAUTO = n

In this notation, "n = 1" means that the auto answer is switched on, and "n = 0" disables the auto answering. This command thus enables the network initiated EPS Bearer Activation (or Modification) Request messages in EPS.

For a more detailed description of all the defined AT commands in LTE/SAE, please refer to [10].

3.8.2 Voice

For the LTE voice service, basically all the applications that provide the voice delivery via a packet data service can be applied either by combining USB-dongle LTE-UE with laptop or an integrated mobile phone that contains the applications for voice communication. The setup and utilization of voice is thus related to the application level, and the LTE/SAE network is seen as a bit pipe for service delivery. The more detailed description of the VoIP call is presented in Chapter 9, on the functionality of LTE/SAE.

3.8.3 MBMS

The LTE MBMS concept (Multimedia Broadcast Multicast Service) is an additional definition in LTE. It has been left for the Release 9 set of definitions. The solution can be called E-MBMS, and it is an integral part of LTE.

The MBMS transmission of the LTE network has two options. It can be used in a single-cell or multi-cell mode. In multi-cell transmission, the cells and content are synchronized in order to make the LTE-UE soft-combine the energy of multiple radio transmissions. This results in the use of a superimposed signal, which is seen as a multipath-propagated radio path for the terminal. This concept is also known as Single Frequency Network (SFN).

The E-UTRAN can be configured on a cell-by-cell basis in order to form SFN areas in the MBMS service. The MBMS service can be adopted in the same, shared carrier as unicast traffic, or it can be deployed in a totally separate carrier.

An extended cyclic prefix is utilized in the MBMS concept. In the case where subframes are carrying SFN data of MBMS, specific reference signals are used. The MBMS data is delivered on the MBMS traffic channel (MTCH).

3.9 LTE-Advanced—Next Generation LTE

3.9.1 Key Aspects of LTE-Advanced

The basic version of LTE, defined in the Release 8 series of the 3GPP specifications, can be considered as a "beyond 3G, pre-4G" system, sometimes referred to as 3.9 G technology in nonstandard communications. In practice, the operators are already interpreting LTE as 4G. Officially, though, interpreting ITU definitions literally, the initial version of LTE does not meet with the IMT-Advanced (international 4G) requirements. For example, LTE is not able to provide the 1 gbps data rates required in IMT-Advanced.

In order to respond to the ITU request for proposals for 4G candidates, 3GPP defined a compatible radio-interface technology. This work culminated in the Release 9 definitions, with a set of requirements for the LTE-Advanced system. The requirements are found in the 3GPP Technical Report 36.913 [11]. Logically, this document is based on both the ITU requirements for 4G and on the requirements for operators to enhance LTE.

The first full 4G system could be provided via the further development of LTE, which is called LTE-Advanced. It has been defined for the first time in Release 10 of the 3GPP specifications. LTE-Advanced is not the only 4G technology, though. ITU has also approved IEEE 802.16m, which is commonly known as "WiMAX 2," as one of the 4G technologies in the IMT-Advanced family. Figure 3.6 summarizes the actual situation of 4G technologies.

LTE-Advanced requirements are listed in 3GPP specification number 36.913 (*Requirements for Further Advancements for E-UTRA—LTE-Advanced*) and in [12], [13]. The Release 10 series

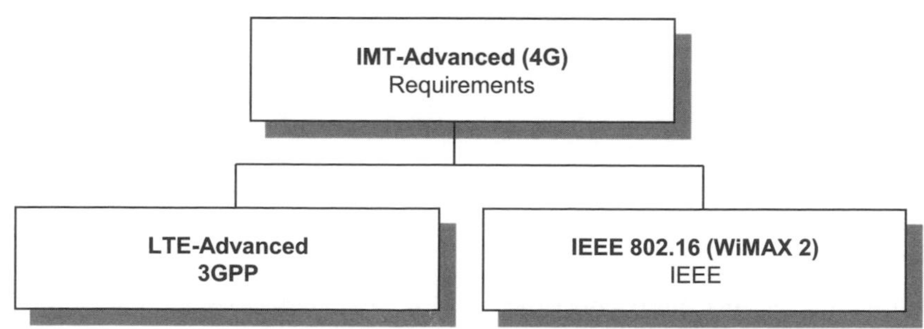

Figure 3.6 The 4G systems approved by ITU-R.

Table 3.4 Comparison of 3G and 4G performance

System	3GPP Release	Maximum data rate (DL)	Maximum data rate (UL)	Round Trip Time (RTT)
WCDMA	99 and 4	384 kbps	128 kbps	150 ms
HSPA	5 and 6	14 mbps	5.7 mbps	100 ms
HSPA+	7	28 mbps	11 mbps	50 ms
LTE	8 and 9	100 mbps	50 mbps	10 ms
LTE-Advanced	10	1 gbps	500 mbps	5 ms

of 3GPP specifications was frozen in March 2011. So far, LTE-Advanced field tests and trials have been carried out using prototype equipment.

The ITU-R has evaluated the technological candidates for the systems that comply with ITU definitions for the fourth-generation mobile telecommunications system. The key requirements are the following:

- 1 gbps peak data rates in downlink;
- 500 mbps peak data rate in uplink;
- three times higher spectrum efficiency than in the LTE system;
- 30 b/s/Hz peak spectrum efficiency in the downlink;
- 15 b/s/Hz peak spectrum efficiency in the uplink;
- support for scalable bandwidth and spectrum aggregation where a noncontiguous spectrum needs to be used;
- the latency requirement for the transition from idle to connected mode should be faster than 50 ms, and after that, less than 5 ms (one way) for an individual packet transmission;
- two times higher user data throughput in the cell edge than in LTE;
- three times higher average user data throughput than in LTE;
- the same mobility performance as LTE;
- LTE-Advanced must be able to interwork with LTE and previous 3GPP systems.

3.9.2 Comparison of 3G and 4G

Table 3.4 summarizes the main performance indicators of 3G and 4G of the 3GPP systems. The maximum data rate increases constantly while the transmission delay reduces.

3.9.3 Enablers for the LTE-Advanced Performance

In order to comply with the ITU-R requirements for IMT-Advanced–that is, for the 4G generation–LTE-Advanced contains a number of technologies. One of the important bases for the high data rates is MIMO but the combination of separate methods makes it possible to achieve adequate performance.

The radio bearer is still formed via the same OFDMA (Orthogonal Frequency Division Multiple Access) in the downlink and SC-FDMA (Single Channel Orthogonal Frequency Division Multiple Access) in the uplink. LTE-Advanced also provides the possibility to use a multicarrier concept for the provision of up to 100 MHz bandwidths, which result in data rates that are compatible with IMT-Advanced requirements.

One of the key functional aspects of LTE-Advanced is its backward compatibility with LTE. The LTE-UE is thus able to work in an LTE-Advanced network and vice versa. This is logically essential in order to guarantee a smooth transition within the 3GPP evolution path.

The World Radiocommunication Conference (WRC-07) has considered suitable frequencies for possible use. Decisions have been made about the reservation of the new IMT-Advanced spectrum and the organization of existing frequency bands, in order to facilitate higher bandwidth utilization—that is, over 20 MHz blocks. It is clear that he LTE-Advanced should be able to share frequencies with the basic version of LTE.

There have been various ideas for improving the performance of the LTE-Advanced system. Some examples about the study items are presented in TR36.912, and include, for instance, the following ideas:

- Self Organizing Networks (SON) items;
- evolved, collaborative smart antennas;
- advanced network architectures;
- flexible spectrum usage;
- carrier aggregation of contiguous and noncontiguous spectrum allocations;
- dual transmitter antennas for SU-MIMO and diversity MIMO;
- interference suppression;
- hybrid OFDMA and SC-FDMA in uplink;
- relay node base stations;
- coordinated multipoint (CoMP) transmission and reception;
- scalable system bandwidth between 20 and 100 MHz;
- local area optimization of air interface;
- nomadic and local area network solutions;
- cognitive radio;
- autonomous network testing;
- enhanced precoding and forward error correction;
- asymmetric bandwidth assignment for FDD;
- inter-eNB coordinated MIMO in downlink and uplink;
- multiple carrier spectrum access.

The LTE-Advanced system has been designed to be flexible. With some additions to basic LTE, the advanced version can use 8×8 MIMO and 128 QAM. Combined, these could in theory, under ideal radio conditions and at 100 MHz aggregated band, provide a maximum peak data rate of over 3 gbps in the downlink direction.

The performance enhancers of LTE-Advanced can be considered as a set of features that can be deployed flexibly on top of the Release 8 network.

References

[1] 3GPP 29.274. *3GPP Evolved Packet System (EPS); Evolved General Packet Radio Service (GPRS) Tunnelling Protocol for Control plane (GTPv2-C); Stage 3*, V. 8.10.0. 2011-06-15, 3rd Generation Partnership Project, Sophia-Antipolis.

[2] WiMAX (n.d.) Deployment pages, www.wimaxforum.org (accessed August 29, 2011).

[3] WiMAX (n.d.) Deployment statistics, www.wimaxforum.org/resources/monthly-industry-report (accessed August 29, 2011).

[4] 3GPP 23.272. (2010) *Circuit Switched (CS) Fallback in Evolved Packet System (EPS); Stage 2*, V. 8.10.0. 2010-12-20, 3rd Generation Partnership Project, Sophia-Antipolis.

[5] 3GPP TS 24.173. (2009) *IMS Multimedia Telephony Communication Service and Supplementary Services; Stage 3*, V. 8.6.0. 2009-12-17, 3rd Generation Partnership Project, Sophia-Antipolis.
[6] 3GPP TS 26.114. (2010) *IP Multimedia Subsystem (IMS); Multimedia Telephony; Media Handling and Interaction*, V. 8.7.0. 2010-12-21, 3rd Generation Partnership Project, Sophia-Antipolis.
[7] Sprint (2011) 4G device news. http://newsroom.sprint.com/article_display.cfm?article_id=1831 (accessed August 29, 2011).
[8] GSM Association (2008) Inter-Service Provider IP Backbone Guidelines 4.4. 19 June, www.gsmworld.com/documents/ir3444.pdf (accessed August 29, 2011).
[9] ITU (n.d.) Draft new Recommendation V.250: "Serial asynchronous automatic dialling and control", www.itu.int/rec/T-REC-V.250/en (accessed August 29, 2011).
[10] 3GPP TS 27.007. (2010) *Technical Specification Group Core Network and Terminals. AT command set for User Equipment, Release 8*, V8.13.0, 2010-09, 3rd Generation Partnership Project, Sophia-Antipolis.
[11] 3GPP TR 36.913. (2009) *Requirements for Further Advancements for Evolved Universal Terrestrial Radio Access (E-UTRA) (LTE-Advanced)*, V. 8.0.1, 2009-03-16, 3rd Generation Partnership Project, Sophia-Antipolis.
[12] 3GPP TS 36.806 (2010) *Evolved Universal Terrestrial Radio Access (E-UTRA); Relay architectures for E-UTRA (LTE-Advanced)*, V. 9.0.0, 2010-04-21, 3rd Generation Partnership Project, Sophia-Antipolis.
[13] 3GPP (n.d.) Feature and study item list, www.3gpp.org/ftp/Specs/html-info/FeatureList--Rel-10.htm (accessed August 29, 2011).

4

Performance Requirements

Jyrki T. J. Penttinen

4.1 Introduction

This chapter presents the performance requirements of LTE. It contains practical interpretations of the standards' requirements and their effect on network planning, deployment and optimization, the effects of LTE and SAE on system interworking, handover procedures, and mixed user profiles, issues related to network synchronization and the Timing over Packet functionality.

4.2 LTE Key Features

4.2.1 Release 8

According to 3GPP, the LTE Release 8 Key Features are the following:

High spectral efficiency. This is accomplished via the OFDM in the downlink, making the system robust against multipath interference and providing high affinity to advanced techniques such as frequency domain channel-dependent scheduling and MIMO. In the uplink, DFTS-OFDM, or Single-Carrier FDMA, is used, which provides a low Peak-to-Average Power Ratio (PAPR) and user orthogonality in the frequency domain. It is also possible to use the multi-antenna application.

Very low latency means that there is a short setup time as well as short transfer delay. The handover latency and interruption time are short, as well as the TTI and RRC procedures. To support fast signaling, the RRC states are simply defined.

Support for variable bandwidth means that it is possible to define some of the following bands: 1.4 MHz, 3 MHz, 5 MHz, 10 MHz, 15 MHz and 20 MHz.

Simple protocol architecture means that the communication is shared, channel based, and that only the packet-switched domain is available, yet it is capable of supporting Voice over IP calls.

Simple architecture has been achieved by introducing eNodeB as the only E-UTRAN node. This leads to a smaller number of RAN interfaces between the eNodeB and MME/SAE-Gateway (S1) and between two eNodeB elements (X2).

The LTE/SAE Deployment Handbook, First Edition. Edited by Jyrki T. J. Penttinen.

Table 4.1 The main LTE parameters

Parameter	Value
Modulation	Selectable form QPSK, 16-QAM, 64-QAM
DL access scheme	OFDMA
UL access scheme	DFTS-OFDM (i.e., SC-FDMA)
Bandwidth	Selectable from 1.4/3.0/5.0/10/15/20 MHz
Minimum TTI	1 ms
Sub-carrier spacing	15 kHz
Cyclic prefix length	short: 4.7 us/long: 16.7 us
Spatial multiplexing	DL: maximum of 4 layers per UE UL: 1 layer per UE MU-MIMO for DL and UL

The essential functionality of LTE is compatibility with earlier 3GPP releases, providing smooth interworking. In addition, interworking is defined with other systems, for example, cdma2000.

The LTE standard supports both FDD and TDD modes within a single radio access technology. Efficient multicast/broadcast functionality is included in the standard, which makes it possible to use a single frequency network SFN concept as an option that is available in OFDM.

LTE also supports the Self-Organizing Network (SON) operation, which can be highly efficient in dynamic and automatic network tuning as a function of selected network performance indicators, including feedback from fault management.

Table 4.1 shows the most important LTE Release 8 parameters and their values.

4.2.2 *Release 9*

LTE—both the evolved radio as well as the evolved core—has been on the market since late 2010. Release 8 was frozen in December 2008, which is the basis for the first generation of the LTE equipment. The Release 8 LTE specifications are noted to be stable. In addition to the basic set of Release 8 definitions, relatively small enhancements were introduced in Release 9, which was functionally frozen in December 2009.

As the Release 9 brings only minor enhancements, the respective LTE requirements have not been changed since the original set.

4.2.3 *Release 10*

The Release 10 for the first time bring LTE to a level that complies with the ITU requirements for the fourth generation of the mobile technologies.

ITUs Radiocommunication Sector (ITU-R) defines the fourth generation requirements under the name IMT-Advanced. More specifically, ITU-R Working Party 5D is charged with defining IMT-Advanced global 4G technologies [1], [2].

ITUs Radiocommunication Sector (ITU-R) completed the assessment of six candidate submissions for the global 4G mobile wireless broadband technology on October 21, 2010. Harmonization among the proposals resulted in two technologies, "LTE-Advanced" (developed by 3GPP as LTE Release 10 and Beyond under the name LTE-Advanced) and

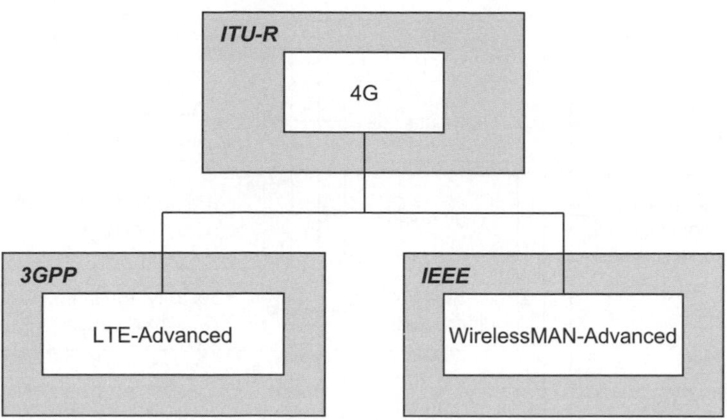

Figure 4.1 The current 4G technology field is divided between the 3GPP and IEEE solutions that comply with the ITU-R requirements for the fourth generation of the mobile communications.

"WirelessMAN-Advanced" (developed by IEEE as the WirelessMAN-Advanced specification incorporated in IEEE Std 802.16 beginning with approval of IEEE Std 802.16m) being accorded the official designation of IMT-Advanced, which qualifies them as official 4G technologies on behalf of ITU. These technologies successfully met all of the criteria established by ITU-R for the first release of IMT-Advanced. Figure 4.1 shows the current situation in the 4G field.

Following the success of the ITU-defined IMT-2000 (3G) systems, ITU-R launched the IMT-Advanced (4G) initiative with its strategic IMT future vision in 2002 [3]. It subsequently established the services, spectrum and performance requirements for IMT-Advanced as well as a detailed evaluation process. In partnership with the industry, the six proposals received by ITU in October 2009 were evaluated against the requirements. Industry consensus and harmonization fostered by ITU-R among these six proposals have resulted in consolidation of the proposals into two agreed IMT-Advanced technologies. These technologies moved into the final stage of the IMT-Advanced process, the development of LTE-Advanced.

IMT-Advanced has been defined in order to provide a global platform on which to build the next-generations of interactive mobile services, which will result in faster data access, enhanced roaming capabilities, unified messaging and broadband multimedia.

The high-level definition of ITU for the Next Generation Network (NGN) from December 2004, which today is called IMT-Advanced, is the following [4]:

> An NGN is a packet-based network able to provide telecommunication services and able to make use of multiple broadband, QoS-enabled transport technologies and in which service-related functions are independent from underlying transport-related technologies. It enables unfettered access for users to networks and to competing service providers and/or services of their choice. It supports generalized mobility which will allow consistent and ubiquitous provision of services to users.

More specific LTE-Advanced requirements include the following, according to [5]: "Systems beyond IMT-2000 will be realized by functional fusion of existing, enhanced and newly developed elements of IMT-2000, nomadic wireless access systems and other wireless

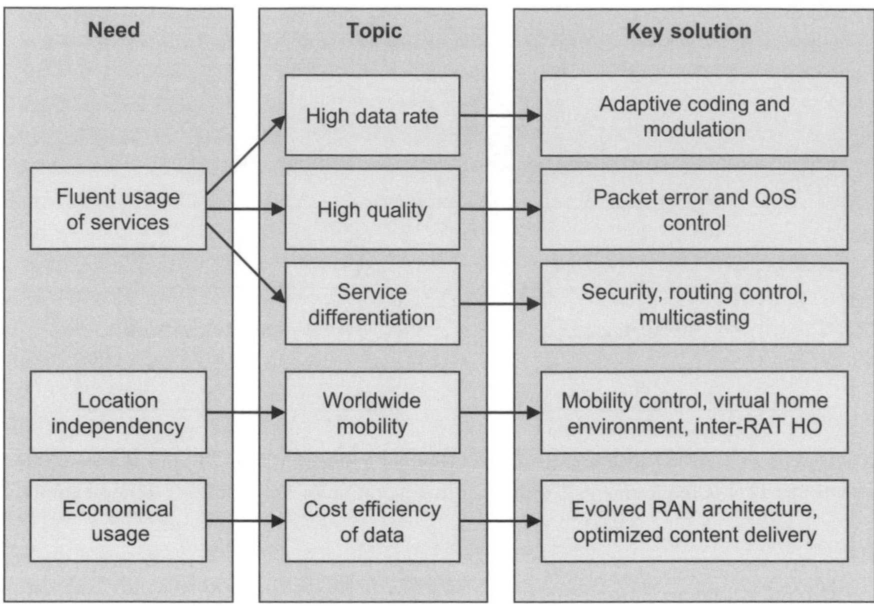

Figure 4.2 High-level demand and solutions of 4G.

systems, with high commonality and seamless interworking." This means that 4G is actually a mix of various interacting systems, not only one single standard, as emphasized in Figure 4.2.

The targeted data rates of 4G, with wide-area coverage and significant mobility, are in the range of 50–100 mbps. The definition of the 4G system states that it is a completely new fully IP-based integrated system of systems and network of networks achieved after convergence of wired and wireless networks, computers, consumer electronics, and communication technology, and several other convergences that will be capable of providing 100 mbps and 1 gbps, respectively in outdoor and indoor environments, with end-to-end QoS and high security, offering any kind of service at any time as per user requirements, anywhere, with seamless interoperability, always on, at affordable cost, with one billing and fully personalized.

The technical requirements for the Beyond IMT-2000 system are:

- high data-rate transmission: for example, DL:100 mbps/UL:20 mbps;
- larger system capacity—for example, while 3G provides with 1.2 mbps in the 1 MHz band and 6-sector-BTS, 4G offers that five to ten times more efficiently;
- lower cost/bit;
- wireless QoS control: NRT Service, RT Service, Multi-Cast Service.

In order to provide the services with the new data rates and lower latencies, the suitable radio technologies for B3G systems are:

- MIMO (Multiple-Input Multiple Output);
- link adaptation techniques;
- multi-carrier based modulation and access (OFDM/OFDMA);
- iterative (multi-user) processing;

Table 4.2 Comparison of the main characteristics of 3G and 4G systems

Topic	3G	4G
Priority	Voice as priority, and data as secondary	Data as priority, VoIP being only as one application for voice
Data rate	384 kbps–2 mbps (UMTS), up to 14 mbps in DL (HSDPA) and 5.8 mbps in UL (HSUPA), up to 84 mbps in DL and 22 mbps in UL (HSPA +)	1 gbps (DL) 500 mbps (UL)
Switching method	Circuit and packet switched	Packet switched
Frequency band	900 MHz–2.5 GHz, possibly for 450 MHz	600 MHz–8 GHz
Bandwidth	5 MHz (UTRAN) 1.4–20 MHz (E-UTRAN)	Up to 100 MHz

- "cross-layer" optimization and design principles;
- possibly Ultra-WideBand (UWB).

In order to function in an optimized way, 4G networks might require, for example, the following, high-level items:

- wider frequency bandwidth than in previous systems;
- truly global mobility and service portability;
- spectrally more efficient solutions like modulation schemes;
- completely digital network to fully utilize IP and converged video and data.

Table 4.2 shows the comparison of the most essential performance related items of 3G and 4G.

4.3 Standards LTE Requirements

LTE is defined in the 36 series in the 3GPP numbering system. The most up-to-date documentation that is referred to in this book is based on the September 2009 version. The LTE Release 8 specifications can be found in the web pages of 3GPP online in [6].

4.3.1 Early Ideas of LTE

The first ideas of LTE occurred in the 3GPP RAN Evolution workshop, on November 2–3, 2004 in Toronto, Canada. This event triggered a further study of the evolved 3G mobile system path. In this event, operators, equipment manufacturers and research institutes presented over 40 contributions that contained proposals for the further evolution of the Universal Terrestrial Radio Access Network (UTRAN).

The outcome of the workshop was a set of overall requirements for the evolved system, including the following:

- reduced cost per bit;
- increased service provisioning, in order to get more services at lower cost with better user experience;

- flexibility for the usage of the frequency bands;
- simplified architecture with clearly open interfaces;
- reasonable terminal power consumption.

It was noted that the Evolved UTRAN should bring significant improvements compared with previous systems in order to justify the standardization work involved. Furthermore, optional solutions should be minimized. It was also noted that not only the radio interface but the whole system has such large impacts during the enhancements that collaboration with 3GPP SA work groups was essential. This would guarantee that the new performance of the access networks as well as core network would be designed in a controlled way.

A feasibility study on UTRA and UTRAN Long-Term Evolution was started in December 2004. The aim of this work was to develop a framework for the evolution of the 3GPP radio access technology towards a high-data rate, low-latency and packet-optimized radio-access technology. It should be noted that the feasibility study focused on the packet-switched domain. The set of study items were the following:

- Radio-interface physical layer, including both downlink and uplink, in order to seek for means to support flexible transmission bandwidth up to 20 MHz, introduction of new transmission schemes, and advanced multi-antenna technologies.
- Radio interface layer 2 and 3, for example, signaling optimization.
- UTRAN architecture in order to identify the optimum UTRAN network architecture and functional split between RAN network nodes.
- RF issues.

As a continuum of the initiation, all radio access network work groups participated in the study. SA WG2 also participated in the selected area of the network architecture. As a result of the feasibility study, 3GPP Technical Report 25.913 [7] contains detailed requirements for the criteria in Table 4.3.

It may be noted that the original performance targets of LTE were well kept until the final freezing of the Release 8 specifications. The only notable difference between the original and further development was the bandwidth values, which are 1.4, 3, 5, 10, 15 and 20 MHz in the final version of the LTE standards.

4.3.2 *Standard Radio Requirements of LTE*

The preparation of LTE standardization was initiated after the creation of the initial list of requirements. First, the suitable technology for the radio interface was studied under the RAN, and can be seen in TR 25.814 [8]. The options were reduced with time, until the decision about the OFDM in DL and SC-FDMA was taken in December 2005. One details is that it was decided that inter-Node-B macrodiversity would not be employed in LTE. For those who are interested in this phase of the selection process, more information can be found in report RAN#30.

LTE can use QPSK, 16-QAM, and 64-QAM modulation schemes dynamically based on radio conditions. The possible uplink data modulation schemes are (π/2-shift) BPSK, QPSK, 8-PSK and 16-QAM. It was also agreed that it would be possible to use Multiple Input Multiple Output (MIMO), with possibly up to four antennas at the UE, and four antennas at the cell site. One of the important decisions was related to the coding scheme, which was selected based on turbo codes.

Table 4.3 The criteria for the 3GPP Long Term Evolution

Peak data rate	Radio Resource Management	Spectrum efficiency
Instantaneous DL peak data rate of 100 mbpsin 20 MHz DL spectrum (5 bps/Hz).	Enhanced support for end to end QoS.	Downlink: In a loaded network, target for spectrum efficiency (bits/sec/Hz/site), 3 to 4 times Release 6 HSDPA).
Instantaneous uplink peak data rate of 50 mbps (2.5 bps/Hz) within a 20 MHz uplink spectrum allocation).	Efficient support for transmission of higher layers.	Uplink: In a loaded network, target for spectrum efficiency (bits/sec/Hz/site), 2 to 3 times Release 6 Enhanced Uplink
	Support of load sharing and policy management across different Radio Access Technologies	
Control-plane latency: Transition time $<$ 100 ms from a camped state to an active state.	*Control-plane capacity:* At least 200 users per cell should be supported in the active state for spectrum allocations up to 5 MHz.	*Coverage:* Throughput, spectrum efficiency and mobility targets should be met for 5 km cells, and with a slight degradation for 30 km cells. Cells range up to 100 km should not be precluded.
Transition time $<$ 50 ms between a dormant state and an active state.		
User-plane latency: Less than 5 ms in unload condition (i.e., single user with single data stream) for small IP packet	*User throughput:* Downlink: average user throughput per MHz, 3 to 4 times Release 6 HSDPA.	*Complexity:* Minimize the number of options.
	Uplink: average user throughput per MHz, 2 to 3 times Release 6 Enhanced Uplink.	No redundant mandatory features.
Mobility: E-UTRAN should be optimized for low mobile speed from 0 to 15 km/h.	*Spectrum flexibility:* E-UTRA shall operate in spectrum allocations of different sizes, including 1.25 MHz, 1.6 MHz, 2.5 MHz, 5 MHz, 10 MHz, 15 MHz and 20 MHz in both the uplink and downlink. Operation in paired and unpaired spectrum shall be supported. NOTE: the requirement was later changed to 1.4, 3, 5, 10, 15 and 20 MHz blocks.	*Spectrum flexibility (cont.):* The system shall be able to support content delivery over an aggregation of resources including Radio Band Resources in the same and different bands, in both uplink and downlink and in both adjacent and non-adjacent channel arrangements. A "Radio Band Resource" is defined as all spectrum available to an operator

(*continued*)

Table 4.3 (*continued*)

Peak data rate	Radio Resource Management	Spectrum efficiency
Higher mobile speed between 15 and 120 km/h should be supported with high performance.		
Mobility across the cellular network shall be maintained at speeds from 120 km/h to 350 km/h (or even up to 500 km/h depending on the frequency band).		
Co-existence and Inter-working with 3GPP Radio Access Technology:	Architecture/migration:	Further Enhanced Multimedia Broadcast Multicast Service (MBMS):
Co-existence in the same geographical area and colocation with GERAN/UTRAN on adjacent channels.	Single E-UTRAN architecture.	While reducing terminal complexity: same modulation, coding, multiple access approaches and UE bandwidth than for unicast operation.
E-UTRAN terminals supporting also UTRAN and/or GERAN operation should be able to support measurement of, and handover from and to, both 3GPP UTRAN and 3GPP GERAN.	The E-UTRAN architecture shall be packet based, although provision should be made to support systems supporting real-time and conversational class traffic.	Provision of simultaneous dedicated voice and MBMS services to the user.
The interruption time during a handover of real-time services between E-UTRAN and UTRAN (or GERAN) should be less than 300 msec.	E-UTRAN architecture shall minimize the presence of "single points of failure."	Available for paired and unpaired spectrum arrangements.
	E-UTRAN architecture will support an end-to-end QoS.	
	Backhaul communication protocols should be optimized.	

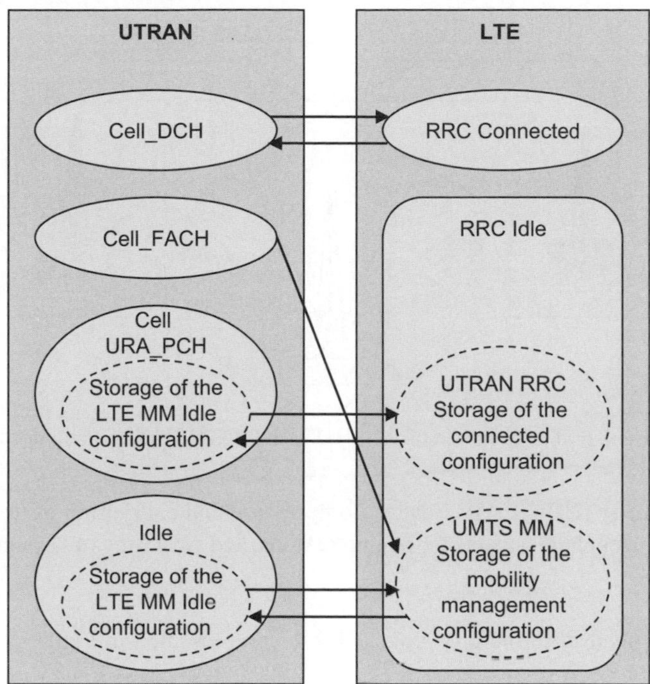

Figure 4.3 The mapping of the Radio Resource Control states of UTRAN and LTE.

RAN also further processed the link layer of Evolved UTRAN, assuming that the protocol architecture should be simplified. It was also hoped to simply the MAC layer, without a dedicated channel. The optimization involved avoiding similar types of functionality in the radio and core sides of the network. More specifically, the Transmission Time Interval (TTI) of 1 ms was fixed, in order to reduce the signaling overhead.

The Radio Resource Control (RRC) stages were simplified, and only RRC-Idle and RRC-Connected states were included in the assumptions. The exception for this is the possible legacy UTRAN RRC state interfacing. Figure 4.3 shows the RRC states as interpreted from Technical Report 25.813 [9].

The architectural definitions by RAN WG3 consisted of a single Evolved UTRAN element—that is, Evolved NodeB—which can be called E-NodeB or eNB. This element provides the user and signaling plane connections between the LTE core and UEs. Furthermore, unlike in previous solutions, the eNBs can be connected together via the X2 interface. This is used, for example, for the handover procedure between the eNBs.

On the other hand, the eNBs are connected to the Evolved Packet Core (EPC) network elements via the S1 interface. The E-UTRAN architecture can be found in [10]. Figure 4.4 shows the basic principle of the architecture. As can be seen, the S1 interface supports hash-topology between the eNbs and core elements.

According to the requirements, the elements have a functional split in the following way:

- **eNodeB:** Radio Resource Management (RRM), which contains Radio Bearer Control, Radio Admission Control, Connection Mobility Control, Dynamic Resource Allocation.
- **MME:** Distribution of paging messages to the eNodeB elements.

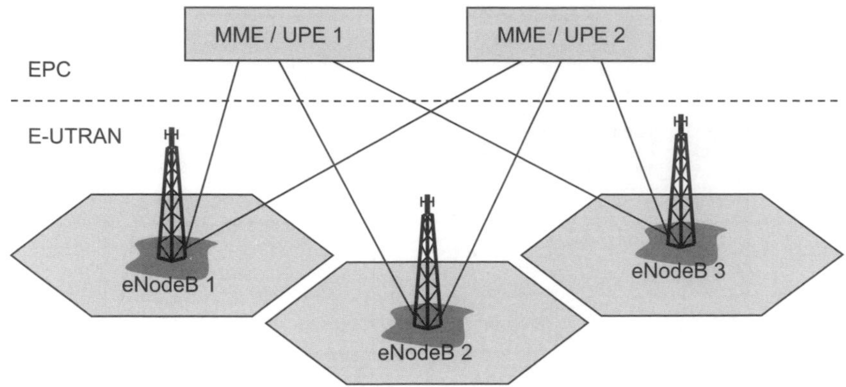

Figure 4.4 The principle of the E-UTRAN/EPC (LTE/SAE) architecture.

- **User Plane Entity (UPE):** IP Header Compression and encryption of user data streams, termination of user plane packets for paging reasons, and switching of U-plane for support of UE mobility.

These requirements lead to the protocol stack scheme presented in Figure 4.5 [9].

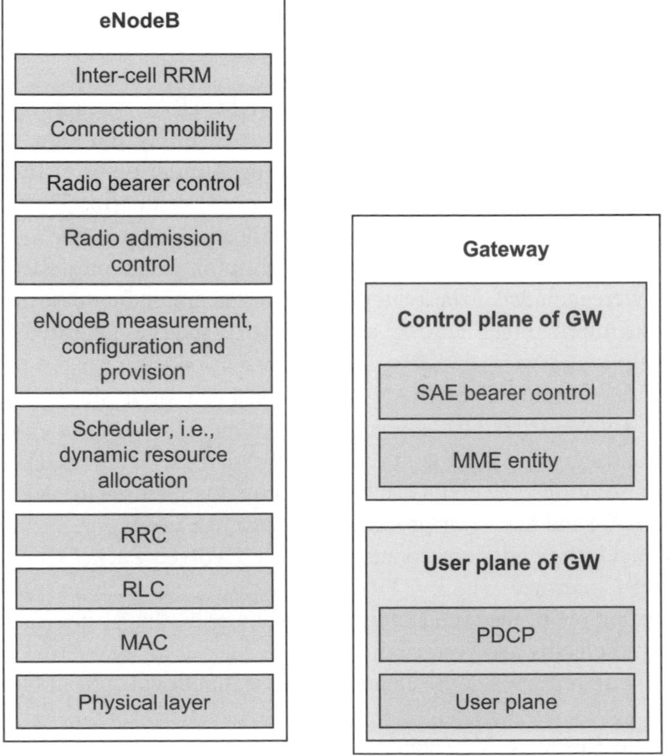

Figure 4.5 The functions of eNodeB and gateway.

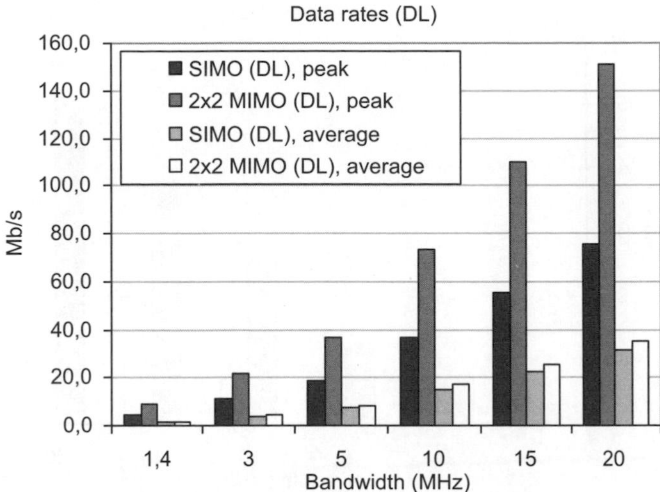

Figure 4.6 The theoretical peak and average data rates of LTE release 8 in the downlink.

The study phase of LTE was finalized by September 2006. One of the most significant outcomes of this work was that E-UTRA would provide clearly higher data rates compared with previous systems. The main items that provide the enhancement are higher frequency bandwidth and the adaptation of MIMO.

As a next step, the actual standardization work began, and the respective Work Item (WI) was created to introduce E-UTRAN into the 3GPP Work Plan officially.

4.3.3 Data Performance

Figures 4.6 and 4.7 show the maximum theoretical peak data rates of cell per TTI that can be achieved in Release 8 LTE network in the ideal conditions. The highest data rates in the 20 MHz bandwidth are a result of the clearly improved spectral efficiency compared to the previous systems. These figures are applicable to the situation with the LTE-UE close to the eNodeB (within 20% of the maximum radius), and the terminal located in a noise-limited environment. The layer 1 net data throughput has been taken into account as defined in TS 36.213, with 64-QAM in the downlink and 16-QAM in the uplink.

In the dimensioning of the network, the available capacity depends on the overhead. With a typical traffic profile the overhead can be about 50% for small packets (60 bytes), and 25% for medium-sized and large packets (600/1500 bytes). The radio interface overhead is due to the PDCP and RLC, and represents about 5%. The transport overhead depends on the usage of the IPSec, being approximately 25% with and about 15% without it.

4.3.4 LTE-UE Requirements

The LTE-UE categories and respective capabilities are shown in Table 4.4. It should be noted that all the categories support the frequency bandwidths up to the maximum possible of 20 MHz. Reception diversity is mandatory for all the classes. For eNodeB, transmission diversity is defined for 1–4 transmitter antennas. The 64-QAM modulation is not mandatory in uplink direction for the UE classes 1–4, but otherwise both downlink and uplink should support

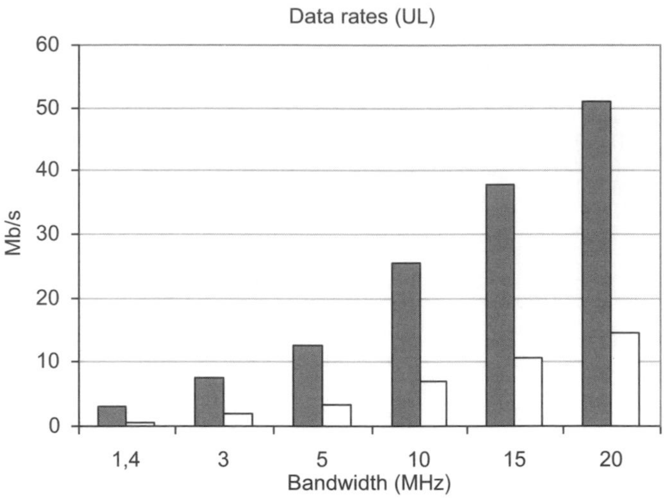

Figure 4.7 The theoretical peak and average data rates of Release 8 LTE in the uplink.

all the modulation schemes from the set of QPSK, 16-QAM and 64-QAM. It should also be noted that downlink 2 × 2 MIMO is mandatory in the classes 2–5.

4.3.5 Delay Requirements for Backhaul

The QoS requirements of the transport network are principally determined by the user service aspect. The delay requirement for the interactivity is defined as a maximum allowed response time. Moreover, if the connection is based on the TCP, throughput performance is taken into account. The delay requirements that are set by the radio network layer protocols are given for the handover and ANR in the S1 and X2 interfaces. As a comparison, the delay requirements for the Iub and Iur interfaces of the WCDMA system As a comparison, the delay requirements

Table 4.4 Key characteristics of LTE-UE classes

Functionality	Class 1	Class 2	Class 3	Class 4	Class 5
Peak data rate in downlink (mbps)	10	50	100	150	300
Peak data rate in uplink (mbps)	5	25	50	50	75
Modulation in downlink	QPSK, 16-QAM, 64-QAM	QPSK, 16-QAM, 64-QAM	QPSK, 16-QAM, 64-QAM	QPSK, 16-QAM, 64-QAM	QPSK, 16-QAM, 64-QAM
Modulation in uplink	QPSK, 16-QAM	QPSK, 16-QAM	QPSK, 16-QAM	QPSK, 16-QAM	QPSK, 16-QAM, 64-QAM
MIMO in downlink	optional	2 × 2	2 × 2	2 × 2	4 × 4

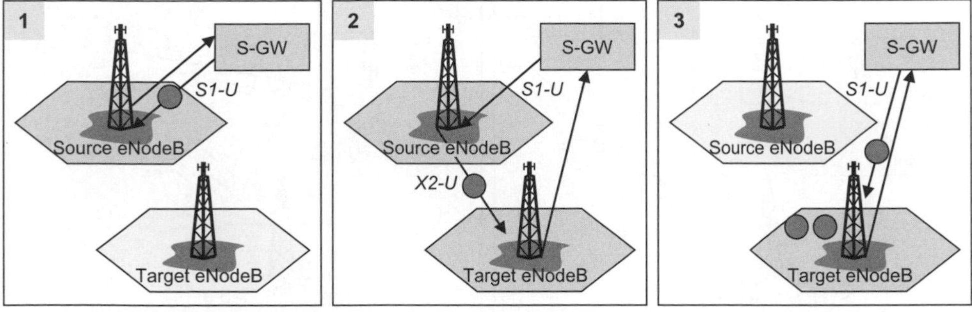

Figure 4.8 The handover between two eNodeB elements.

for the Iub and Iur interfaces of the WCDMA system are related to the macro-diversity combining, outer loop power control, frame synchronization and packet scheduler.

Logically, the end user experiences the delay differently depending on the application. The use of, for example, video streaming, FTP or web browsing would require less rapid response times, delays of more than 1 s being acceptable for users. The quality of video telephony or audio streaming would suffer from delays greater than 200 ms, the optimal value being around 100–200 ms. The most critical applications like real-time gaming, video conferencing, video broadcast and machine-to-machine remote control would work well if the delay is between 20–100 ms.

The internal LTE handover, which happens via signaling of the direct X2 interface between eNodeB elements, causes a 30–50 ms radio link interruption until the handover procedure has been completed. Data transmission is suspended during this time period. If data is arriving from the network to LTE-UE, it is buffered accordingly. The effect of this should be taken into account in network deployment by designing the latency in the X2 interface to be equal or lower than the radio interruption time in order to avoid a bottleneck in this part, in the events presented in Figure 4.8. This is thus one of the optimization tasks of the operator.

The capacity can be increased for inter-eNodeB handovers when signaling via the X2 interface. At the stage of activating the X2 signaling, the source eNodeB fills the uplink path slightly more, if we assume asymmetric user traffic and symmetric backhaul capacity in both directions. On the other hand, the target eNodeB experiences a short increase in downlink traffic at the same time, which is still in practice less than 2% in average. For this reason, the additional capacity is not needed in the typical cases of the deployment of the network. Nevertheless, the extra capacity should be considered in areas where high handover performance is required. Figure 4.9 clarifies the method.

The transport service attributes—the values for service-level parameters—are typically driven by the user service requirements. The values depend on the overall situation of the network but the following recommendations could be considered as general guidelines:

- Packet delay (PD) is recommended to be equal to or less than 10 ms. It should be noted that in the user plane, the packet delay affects latency and throughput of the TCP type of services. For the control plane of the WCDMA, the maximum value of 20 ms is still acceptable.
- Packet delay variation (PDV) is recommended to be equal to or less than ±5 ms. In the user plane, this value is recommended especially for the voice service. In the control plane, this value is recommended for the timing over packet (ToP) concept, based on the IEEE 1588–2008 definitions.

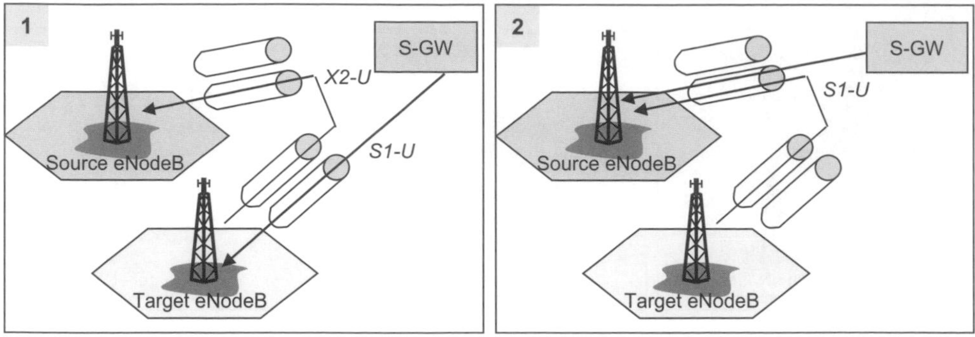

Figure 4.9 The capacity can be increased for the X2 handover purposes in the highest demand areas of LTE.

- Packet loss ratio (PLR) is recommended to be equal to or less than 10^{-4}. It should be noted that, in the user plane, the packet loss performance affects the throughput values of TCP-based services.

4.3.6 System Architecture Evolution

System architecture requirements needed to be created with the planned evolution of LTE. The System Architecture Working Group 2 thus initiated a study called "System Architecture Evolution (SAE)" with the objective of developing a framework for the evolution or migration of the 3GPP system to a higher data rate, lower latency, and packet optimized system that supports multiple Radio Access Technologies. It should be noted that the focus of this work was on the PS domain with the assumption that voice services are supported in this domain. SA2's SAE work was conducted under the Work Item "3GPP system architectural evolution," which was approved in December 2004. It was initiated when it became clear that the future was clearly IP with everything (the "all-IP" network, AIPN–details can be found in TS 22.978 [11]), and that access to the 3GPP network would ultimately not only be via UTRAN or GERAN but by WiFi, WiMAX, or even wired technologies.

The main objectives of SAE investigations were to investigate:

- impact on overall architecture resulting from RANs LTE work;
- impact on overall architecture resulting from SA1's All IP Network concept;
- overall architectural aspects resulting from the need to support mobility between heterogeneous access networks.

Figure 4.10 shows the evolved system architecture, possibly relying on different access technologies, as interpreted from [12].

The new reference points were identified as listed below:

- S1 provides access to Evolved RAN radio resources for the transport of user-plane and control-plane traffic. The S1 reference point will enable separation of MME and UPE and also deployment of a combined MME and UPE solution.
- S2a provides the user plane with related control and mobility support between a trusted non-3GPP IP access and the SAE Anchor.

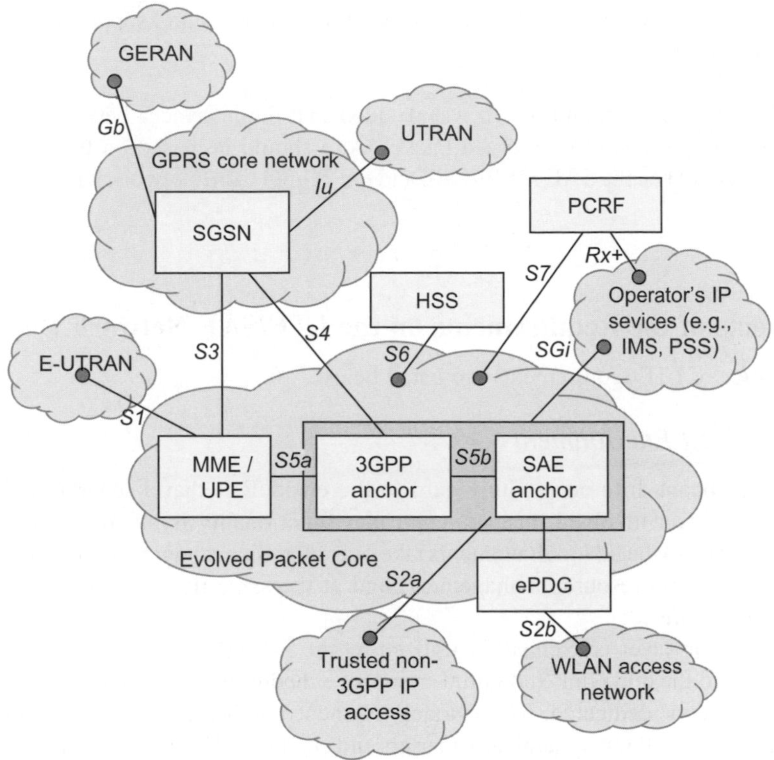

Figure 4.10 The evolved packet system architecture.

- S2b provides the user plane with related control and mobility support between ePDG and the SAE Anchor.
- S3 enables user and bearer information exchange for inter 3GPP access system mobility in idle and/or active state. It is based on Gn reference point as defined between SGSNs. User data forwarding for inter 3GPP access system mobility in the active state.
- S4 provides the user plane with related control and mobility support between GPRS Core and the 3GPP Anchor and is based on a Gn reference point as defined between SGSN and GGSN.
- S5a provides the user plane with related control and mobility support between MME/UPE and 3GPP anchor. It is FFS whether a standardized S5a exists or whether MME/UPE and 3GPP anchor are combined into one entity.
- S5b provides the user plane with related control and mobility support between 3GPP anchor and the SAE anchor. It is FFS whether a standardized S5b exists or whether 3GPP anchor and SAE anchor are combined into one entity.
- S6 enables transfer of subscription and authentication data for authenticating and authorizing user access to the evolved system (AAA interface).
- S7 provides transfer of (QoS) policy and charging rules from PCRF to Policy and Charging Enforcement Point (PCEP). The allocation of the PCEP is FFS.
- SGi is the reference point between the Inter-AS Anchor and the packet data network. The packet data network may be an external operator public or private packet data network or an intra-operator packet data network—for example, for provision of IMS services.

This reference point corresponds to Gi and Wi functionalities and supports any 3GPP and non-3GPP access systems.

In order to reuse earlier definitions, it was decided that the interfaces between the SGSN in 2G/3G Core Network and the Evolved Packet Core should be based on the GTP protocol. The interfaces between the SAE MME/UPE and the 2G/3G Core Network will be based on the GTP protocol.

4.4 Effects of the Requirements on the LTE/SAE Network Deployment

The key issues of LTE requirements are listed below.

4.4.1 Evolved Environment

The most significant difference with regard to the eNodeB is that a separate radio control element is no longer involved. Instead, controller functionality is integrated into the same eNodeB. This means that eNodeB elements take over all radio management functionality. This results in faster radio resource management, and at the same time it simplifies the overall network architecture.

The IP transport layer is exclusively utilized by the E-UTRAN.

With regard to the downlink and uplink resource scheduling, in UMTS physical resources are either shared or dedicated. The eNodeB element handles all physical resource via a scheduler and assigns them dynamically to users and channels. This provides greater flexibility than the older systems could offer.

In the frequency domain, there is scheduling based on the resource carrier bandwidth. The scheduling uses those resource blocks that are not faded as shown in the example of Figure 4.11. It is good to note that this method is not possible in CDMA-based systems.

The other LTE functionality includes the following items:

- Hybrid Automatic Retransmission on request/Hybrid Automatic Repeat Request (HARQ) has already been used for HSDPA and HSUPA. Its functionality increases LTE performance with regard to delay and throughput performance, especially for cell-edge users. It implements a retransmission protocol on layers 1 and 2 that allows retransmitted blocks to be sent with different coding from the first one.
- Awareness of QoS can be accomplished in such a way that the scheduler handles and distinguishes different QoS classes. In fact, this is done in LTE because otherwise real time services would not be possible via E-UTRAN. Differentiated services are possible with the system.
- Self-configuration is a relatively large area, which is currently under further investigation. The basic idea of this item is to allow eNodeB elements to configure themselves automatically with minimum or no human interaction. It eases the physical work and reduces human errors in the tuning of the parameter values, although it cannot replace human work completely.
- Packet Switched Domain only means that no circuit switched domain is provided. If circuit switched applications are required, like a voice service, they should thus be implemented via the IP domain.

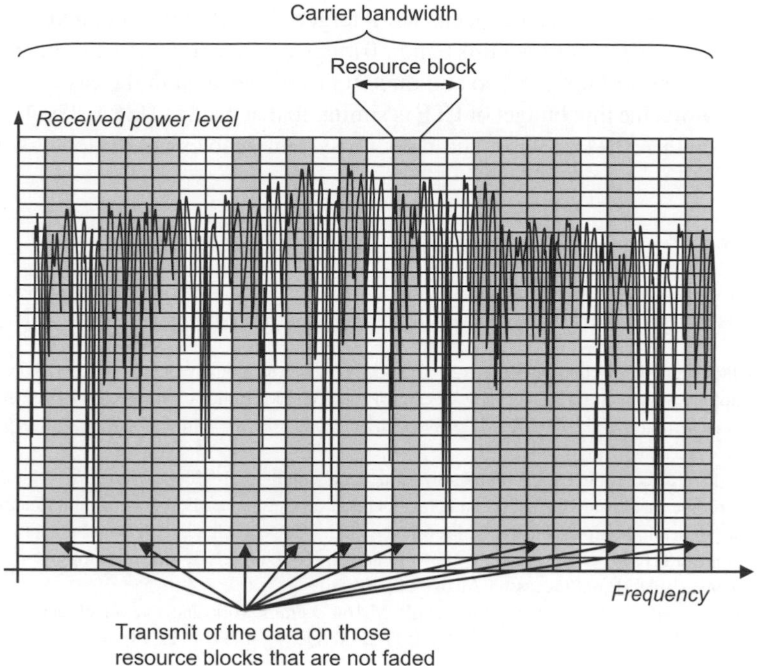

Figure 4.11 The resource blocks that are not faded are selected for the transmission.

- Non-3GPP access means that the EPC will also be prepared to be used by non-3GPP access networks (e.g., LAN, WLAN, and WiMAX). This will provide true convergence of different packet radio access systems.
- Multiple Input Multiple Output (MIMO) is supported by LTE as an option. It allows utilization of multiple transmitter and receiver antennas. According to the 3GPP specifications, up to four antennas can be used by a single LTE cell, which provides a performance gain due to spatial multiplexing. MIMO is one of the most important technologies that increase the spectral efficiency of LTE compared to previous systems.

4.4.2 Spectral Efficiency

Generally, LTE provides higher spectral efficiency than other similar systems. Simulations show that Release 7 HSPA and WiMAX perform close to each other, but LTE clearly outperforms both technologies. For example, the ITU contribution from the WiMAX Forum shows a value of 1.3 and 0.8 b/s/Hz/cell for the downlink and the uplink, respectively, whereas the values are 1.75 and 0.75 b/s/Hz/cell for LTE Release 8. Release 7 HSPA shows values of 1.2 and 0.4 b/s/Hz/cell respectively. Compared to the older solutions, LTE has, for example, three times more efficient spectral utilization than Release 6 HSPA in the downlink, although with a UE equalizer the Release 6 HSPA produces values of 1.05 and 0.35 b/s/Hz/cell. The references used in this analysis are from 3GPP R1-071992 for HSPA Release 6 and LTE Release 8, whereas the results for HSPA Release 6 with an equalizer are from 3GPP R1-063335, and the values for the HSPA Release 7 and WiMAX are based on NSN/Nokia simulations.

For the round-trip delay—that is, for the latency—the values for the small IP packets are 40 ms, 25 ms and 10 ms for HSPA (10 ms variant), HSPA (2 ms variant) and LTE, respectively.

In conclusion, LTE performance meets 3GPP targets, the most important key performance values being over 100 mbps (downlink) and 50 mbps (uplink) for the peak bit rate, and the spectral efficiency being between two and three times higher than in the case of Release 6 of HSPA. Furthermore, the link budget of LTE is similar to that used in HSPA. The VoIP capacity is 40–50/MHz with AMR12.2 kbps, and the LTE system enables latency of less than 10 ms.

References

[1] Tabbane, S. (2007) Mobile next generation network, evolution towards 4G. ITU-D/ITU-T Seminar on Standardization and Development of Next Generation Networks for the Arab Region, April 29–May 2, Manama (Bahrain).

[2] Lazhar Belhouchet, M., Hakim Ebdelli, M. (2010) ITU/BDT Arab Regional Workshop on "4G Wireless Systems" LTE Technology Session 3: LTE Overview—Design Targets and Multiple Access Technologies, January 27–29.

[3] ITU (2010) Description of the ITU-R IMT-Advanced 4G standards, www.itu.int/net/pressoffice/press_releases/2010/40.aspx (accessed August 29, 2011).

[4] ITU-T Y.2011. Telecommunication Standardization Sector of ITU (10/2004). Series Y: Global Information Infrastructure, Internet Protocol Aspects and Next Generation Networks—Frameworks and functional architecture models. General principles and general reference model for Next Generation Networks. http://docbox.etsi.org/STF/Archive/STF311_OCG_ECN&S_NGN_RegulAspects/Public/NGN_Terminology_docs/T-REC-Y.2011-200410-I!!PDF-E.pdf (accessed August 29, 2011).

[5] Rec. ITU-R M.1645 1. Recommendation ITU-R M.1645. Framework and overall objectives of the future development of IMT-2000 and systems beyond IMT-2000 (Question ITU-R 229/8). www.ieee802.org/secmail/pdf00204.pdf (accessed August 29, 2011).

[6] 3GPP (n.d.) Standards, www.3gpp.org (accessed August 29, 2011).

[7] 3GPP TR 25.913. (2009) *Requirements for Evolved UTRA (E-UTRA) and Evolved UTRAN (E-UTRAN)*, V. 8.0.0, 2009-01-02, 3rd Generation Partnership Project, Sophia-Antipolis.

[8] 3GPP TR 25.814. (2006) *Physical Layer Aspect for Evolved Universal Terrestrial Radio Access (UTRA)*, V. 7.1.0, 2006-10-13, 3rd Generation Partnership Project, Sophia-Antipolis.

[9] 3GPP TR 25.813 (2006) *Evolved Universal Terrestrial Radio Access (E-UTRA) and Evolved Universal Terrestrial Radio Access Network (E-UTRAN); Radio interface protocol aspects*, V. 7.1.0, 2006-10-18, 3rd Generation Partnership Project, Sophia-Antipolis.

[10] 3GPP TR 25.912. (2009) *Feasibility Study for Evolved Universal Terrestrial Radio Access (UTRA) and Universal Terrestrial Radio Access Network (UTRAN)*, V. 8.0.0, 2009-01-02, 3rd Generation Partnership Project, Sophia-Antipolis.

[11] 3GPP TS 22.978. (2009) *All-IP Network (AIPN) Feasibility Study*, V. 8.0.0, 2009-01-02, 3rd Generation Partnership Project, Sophia-Antipolis.

[12] 3GPP TR 23.882. (2008) *3GPP System Architecture Evolution (SAE): Report on Technical Options and Conclusions*, V. 8.0.0, 2008-09-24, 3rd Generation Partnership Project, Sophia-Antipolis.

5

LTE and SAE Architecture

Jyrki T. J. Penttinen

5.1 Introduction

The evolved 3G system consists of radio and core networks. The core side of the new network is called System Architecture Evolution (SAE) and the radio part is called LTE (Long Term Evolution). It should be noted that SAE was used at the beginning of the standardization, but nowadays the core part is called EPC (Evolved Packed Core). As the term SAE has been established in common terminology, it can be used in parallel with EPC. This book uses both the official standard term, EPC, and the nonstandard SAE.

This chapter presents the functional blocks and interfaces of LTE/EPC. The new architecture is illustrated using comparisons of the solutions with the earlier mobile communications systems.

A protocol layer structure and the functioning of the LTE/EPC protocols are described, and examples are given in order to clarify the principles of each protocol.

The LTE architecture is defined in 3GPP specification 36.401 (E-UTRAN architecture description).

5.2 Elements

LTE is based on flat architecture, meaning that there is only one element type for the radio network, and one element type for the core network [1]. Figure 5.1 shows the high-level architecture of LTE and compares it with the packet-switched domain of the earlier systems.

The overall LTE/EPC division is the following. LTE refers to the E-UTRAN (Evolved UMTS Radio Access Network), whereas EPC means the evolved core network. Figure 5.2 clarifies the division.

The difference between the 3G basic architectural concept is shown in Figure 5.3.

Figure 5.4 shows the more detailed distribution of the elements with respective interfaces. As the figure indicates, LTE can be connected to the packet core of GERAN and UTRAN as a

The LTE/SAE Deployment Handbook, First Edition. Edited by Jyrki T. J. Penttinen.

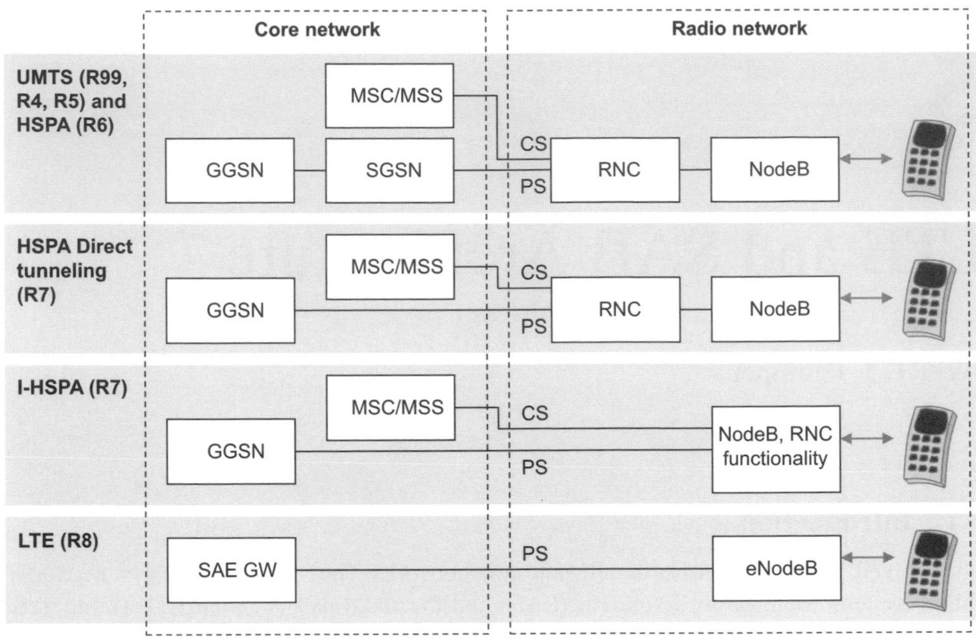

Figure 5.1 The evolution of network architectures. LTE simplifies the general layout of the network via the flat architecture by providing only packet switched connections.

new additional radio technology via the EPC. The SGSN acts as a centralized connection point for all of these technologies.

E-UTRAN contains only one type of node, eNB, which provides the air interface to UE. The eNB elements are connected to MMEs and S-GWs via the S1 interface. Furthermore, eNBs can be connected to each other via the X2 interface. It is worth noting that, unlike in previous solutions, in LTE the single eNB can be connected to multiple MMEs and multiple S-GWs. This ability provides additional reliability flexibly, and is referred to as S1-flex.

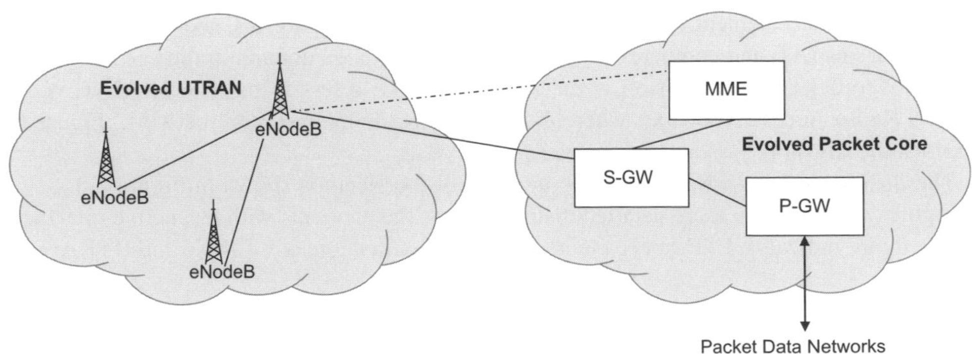

Figure 5.2 The Evolved Packet System (EPS) consists of Evolved UTRAN (LTE) and EPC.

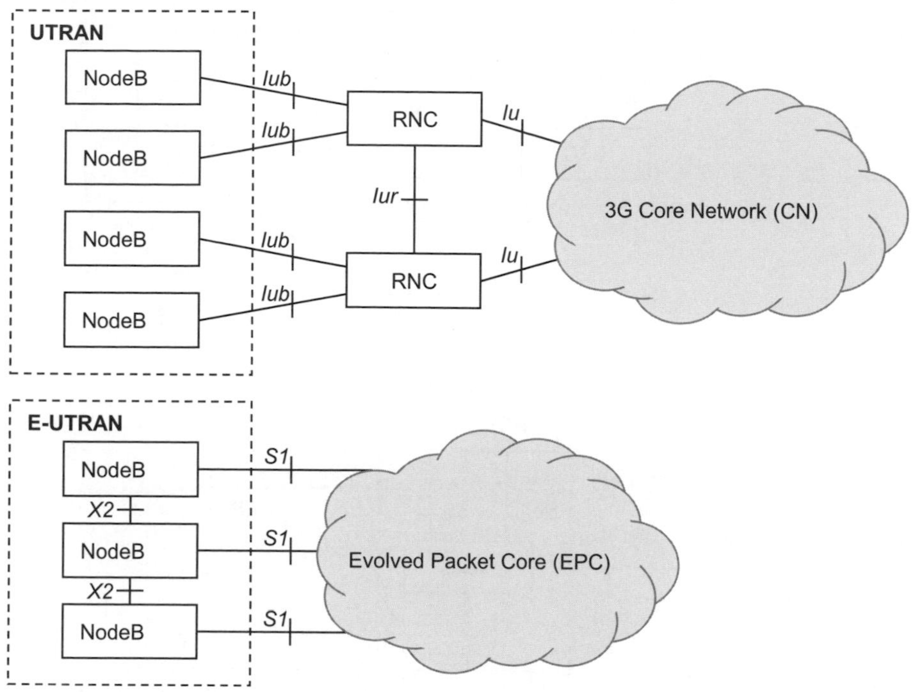

Figure 5.3 The architectural difference between UTRAN and E-UTRAN.

5.2.1 eNodeB

The eNB element of LTE is responsible for radio transmission and reception with UE. eNB provides the required functionality for the radio resource management (RRM), including admission control, radio bearer control, scheduling of user data, and control signaling over the air interface. In addition, eNB takes care of the ciphering and header compression over the air interface.

The clearest difference between UTRAN and E-UTRAN can be seen in the role of the base station. The eNodeB of LTE now includes basically all the functionalities that were previously concentrated on the RNC of the UTRAN system. In addition, the traditional tasks of the NodeB are still included in the new eNodeB (eNB) element. eNB works thus as the counterpart of the UE in the radio interface but includes procedures for decision making related to the connections. This solution results in the term "flat architecture" of LTE, meaning that there are less interfaces and only one element in the hierarchy of the architecture.

As control has been moved closer to the radio interface, the respective signaling time has also been reduced. This is one of the key issues for the reduction of the latency of LTE compared to the previous solutions of the 3G [2]–[4].

More specifically, the eBN element handles the following tasks:

- Radio Resource Management (RRM);
- Radio Bearer Control;
- Radio Admission Control;
- Connection Mobility Control;

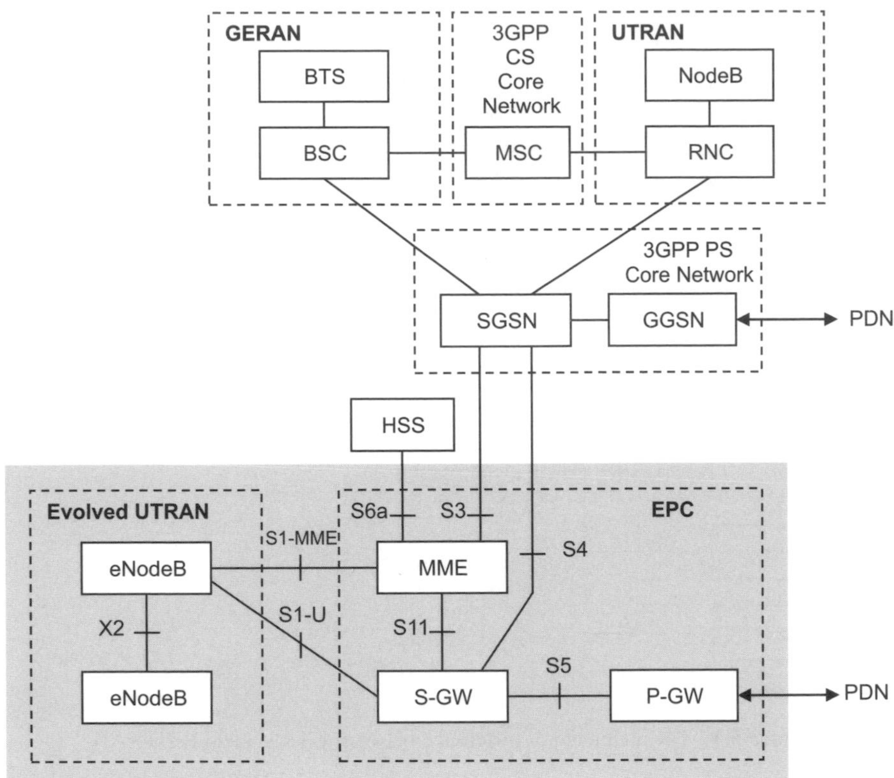

Figure 5.4 LTE/SAE elements and interfaces.

- UE scheduling (DL and UL);
- security in Access Stratum (AS);
- measurements as a basis for the scheduling and mobility management;
- IP header compression;
- encryption of the user data;
- routing of the user data between eNB and S-GW;
- handling of the paging that originates from MME;
- handling of the broadcast messaging that is originated from MME and Operations and Management System (OMS);
- selection of the MME element in case UE does not provide this information;
- handling of PWS messages, including ETWS and CMAS.

It is possible to use an additional element set, which is called Home eNB and Home eNB Gateway. Specific aspects for the Home eNB, that is, HeNB, are the following:

- HeNB are equipment that can be utilized in the customer's premises and that uses the licensed operator's spectrum;
- HeNB is meant for the enhancing of the network coverage and/or capacity;
- includes all the eNB functionalities, added by the HeNB-specific functions that are related to the configuration and security.

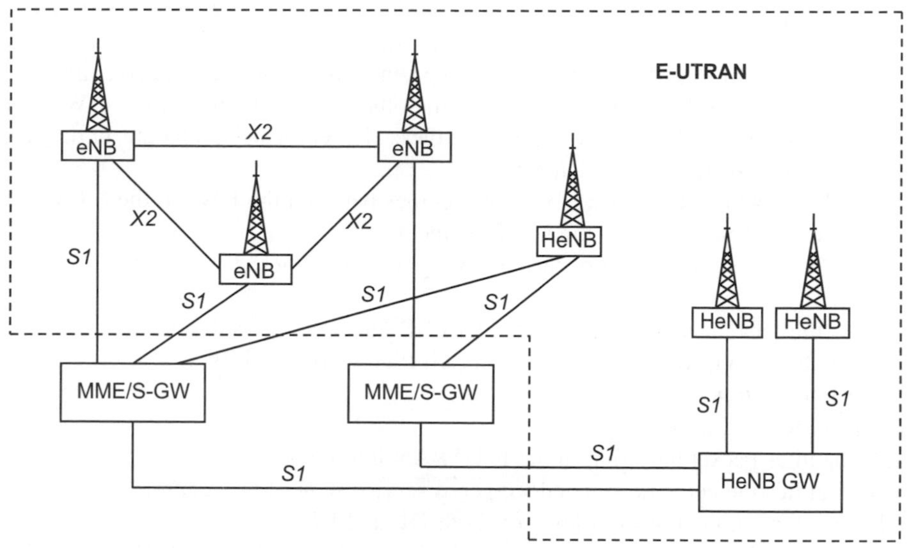

Figure 5.5 TheHome eNB concept.

The Home eNB Gateway (HeNB GW), which is related to the HeNB, solves the problem of support for a possibly very large number of S1 interfaces. It is thus an additional element that can be used for the balancing of the interfaces.

Figure 5.5 shows the principle of the HeNB concept, together with the HeNB GW elements, as interpreted from [5].

Furthermore, the HeNB concept can be utilized in the following access scenarios:

- in the closed access mode, only predefined Closed Subscriber Group (CSG) members can access the respective HeNB;
- in the hybrid access mode, both the members and nonmembers of the closed subscriber group can access the HeNB, but the members are prioritized over the nonmembers, for example, if there is congestion;
- in the open access mode, HeNB is seen by the members and nonmembers exactly as a normal eNB.

The correct functioning of the closed and hybrid modes requires additional parameters in order to support the identification and search of the HeNB by UE. The mobility management should also be aware of the HeNBs in order to perform the handovers.

Release 9 of LTE includes enhancements for the HeNB concept. Some of the most important radio access additions are [5]:

- Inbound Mobility;
- Access Control;
- new Hybrid Cell concept;
- management of out-of-date CSG info;
- Operation, Administration and Maintenance for HeNB elements;
- operator controlled CSG list;
- RF Requirements for TDD and FDD HeNBs.

5.2.2 S-GW

The Serving Gateway (S-GW) provides user plane connectivity, the UE being on one side, and the Packet Data Network Gateway (P-GW) on the other side of the physical S-GW element. Depending on the network provider's approach, these elements can be separate, or they can be combined physically as a single element.

It should be noted that no control messaging goes between the UE and the S-GW, as the control plane is taken care of by the MME element.

The S-GW element takes care of the following functionalities:

- S-GW is the local anchor point for the inter-eNB handover procedure;
- S-GW is also an anchor point for the inter-3GPP network mobility;
- Lawful Interception (LI);
- packet routing and forwarding;
- S-GW handles packet buffering in the E-UTRAN idle mode;
- S-GW handles the network initiated/triggered service request procedure;
- packet marking at the transport level for both DL and UL;
- Charging Data Record (CDR) collection, which can identify the UE, PDN and QCI;
- accounting on user and QCI granularity for the interoperator charging processes.

5.2.3 P-GW

The PDN Gateway (P-GW) provides, in the same manner as the S-GW, the user plane connectivity between UE, S-GW and P-GW. The P-GW element interfaces with the S-GW, and on the other side, with the external packet data network (PDN). In addition, P-GW includes GGSN (GPRS Gateway Support Node) functionality.

More specifically, P-GW includes the following functionalities:

- UE IP address allocation;
- Packet filtering that can be done at a user-based level. The other term for this functionality is a deep packet inspection;
- Lawful Interception (LI);
- Packet marking in the transport level, in the DL;
- Service level charging in DL and UP, as well as gating and rate enforcement;
- Rate enforcement in DL based on APN-AMBR;
- Online charging credit control.

5.2.4 MME

The Mobility Management Element (MME) is meant for the control plane signaling between the UE and other network elements like HSS. Equally, as the user plane LTE/SAE messaging does not go through MME, the control plane signaling does not go through S-GW or the P-GW of LTE/SAE.

MME handles the following functionalities:

- signaling in the Non Access Stratum (NAS);
- security of the NAS signaling;

- AS security control;
- the selection of the P-GW and S-GW elements;
- the selection of other MMEs in the handover;
- the selection of SGSN in handovers between LTE and 3GPP 2G/3G access networks;
- inter-CN node mobility signaling between different 3GPP 2G/3G access networks;
- the management of the Tracking Area (TA) lists;
- international and national roaming;
- user authentication;
- the establishment and management of bearers;
- the support of PWS message transmission, including ETWS and CMAS;
- the management of the paging retransmission of UE and other functions related to the finding the UE in the idle state.

5.2.5 GSM and UMTS Domain

As Figure 5.4 indicates, the SGSN can be used as the central point for connecting the PS domains of GERAN, UTRAN and LTE. LTE creates the packet switched connections to the external packet data networks via P-GW, while the GERAN and UTRAN uses the traditional GGSN. These legacy network elements that are relevant to LTE are the following:

- Gateway GPRS Support Node (GGSN) is responsible for terminating Gi interface towards the PDN for legacy 2G/3G access networks. For the LTE/SAE network, this node is only of interest if provided as parts of P-GW functionality and from the perspective of intersystem mobility management.
- Serving GPRS Support Node (SGSN) is responsible for the transfer of packet data between the core network and the legacy 2G/3G Radio Access Network (RAN). For the LTE/SAE network, this node is only of interest from the perspective of intersystem mobility management.
- Home Subscriber Server (HSS) is the IMS Core Network entity that is responsible for the management of the user profiles, and performs the authentication and authorization of the users, including the new LTE subscribers. The user profiles managed by HSS consist of subscription and security information as well as details about the physical location of the user.
- Policy Charging and Rules Function (PCRF) is responsible for brokering QoS Policy and Charging Policy on a per-flow basis. Figure 5.6 shows the position of PCRF in the LTE/SAE architecture.
- Authentication, Authorization and Accounting function (AAA) is responsible for relaying authentication and authorization information to and from non-3GPP access network connected to EPC.

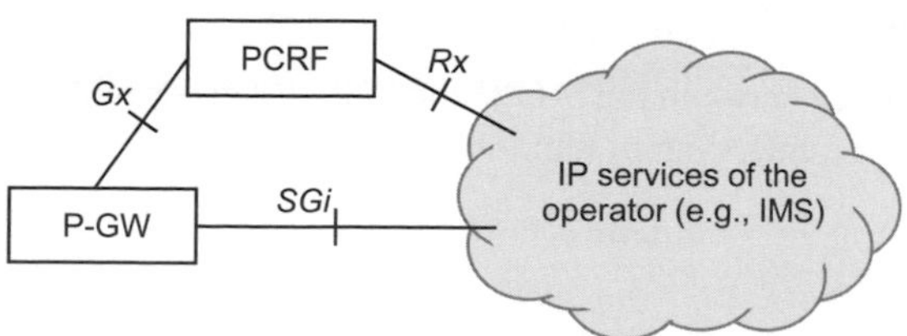

Figure 5.6 The location of the PCRF element in the LTE/SAE network.

5.2.6 Packet Data Network

The packet data network (PDN) is the IP-based network to which LTE/SAE connects via the P-GW element. It can be, for example, the Internet or the operator's IP Multimedia Subsystem (IMS).

5.3 Interfaces

The LTE/SAE system consists of several interfaces internally and between the other 3GPP 2G/3G networks, as is shown in Figure 5.4.

5.3.1 Uu Interface

The LTE radio interface LTE-Uu is defined between the eNodeB and UE. The eNodeB provides PS connectivity in such a way that previous 3G RNC functionality is integrated into the eNodeB. This flat architecture approach makes separate RNC equipment unnecessary.

5.3.2 X2 Interface

The X2 interface defines the connection between the eNodeB elements. It is meant for the inter-eNodeB handover procedures, as well as for the intercell radio resource management signaling and interface management signaling.

Physically, the interface can be, for example, fiber optics or any other solutions while the required capacity can be delivered with the required maximum delay and jitter.

5.3.3 S1 Interface

The S1 interface is divided into S1-MME and S1-U. S1-MME connects eNodeB and MME elements, whereas S1-U is used between the eNodeB and S-GW elements.

The S1-MME interface is defined for the control plane signaling between eNodeB and MME. The S1-U interface is defined between the eNodeB and S-GW in order to carry the user plane data. The data is transferred over GTP.

5.3.4 S3 Interface

The S3 interface is used for signaling between the MME and SGSN.

5.3.5 S4 Interface

The S4 interface is defined between the S-GW and SGSN. This provides a GTP-based tunnel in the user plane during the intersystem handover.

5.3.6 S5 Interface

The S5 interface is located between the S-GW and P-GW elements. Depending on the vendor solution, these elements can be integrated in the same physical element.

5.3.7 S6a Interface

The S6a interface carries the subscription and authentication information between the HSS (Home Subscriber Server) and MME.

5.3.8 S11 Interface

The S11 interface handles the signaling messages between the S-GW and MME.

5.3.9 SGi

The SGi interface is defined between the P-GW and packet data network (PDN), which can be an external public or private IP packet network. It also can be internal IP network, like IP Multimedia Sub-system.

5.3.10 Gn/Gp

As an alternative to the SGi interface, legacy Gn/Gp interface is also supported in EPS in order to create the connection to the packet data networks.

5.4 Protocol Stacks

LTE protocol stacks between different elements are divided into user and control plane. In general, the protocol stacks are reminiscent of those used in the WCDMA of UMTS.

Figure 5.7 presents the overall idea of the roles of each LTE protocol layer.

The LTE/SAE radio protocol stacks are defined in layers of 1, 2 and 3, which overlap partially with the definitions of the OSI layer structure of ISO. LTE/SAE layer 1 is related to the physical realization of the interface (like radio interface or optical fiber), whereas layer 2 is related to the data link and access, and layer 3 is related to the hosting of the access stratum protocols or nonaccess stratum signaling protocols. In LTE/SAE, the application level is included in this third layer.

The channel structure of Figure 5.7 is explained in more detail in Chapter 7.

5.4.1 User Plane

Figure 5.8 shows the complete protocol stack structure for the user plane between the UE, eNB, S-GW and P-GW.

The user plane is also defined as shown in Figure 5.9 in direct communications between two eNBs.

The functionalities of the essential user plane entities are the following:

MAC takes care of the mapping between the logical and transport channels, multiplexing and demultiplexing, reporting of the scheduling information, HARQ functions, priority handling, and transport format selection.

Among other tasks, the RLC takes care of the ARQ functions, segmentation concatenation, re-segmentation concatenation, in-sequence delivery, duplicate packet detection and re-establishment.

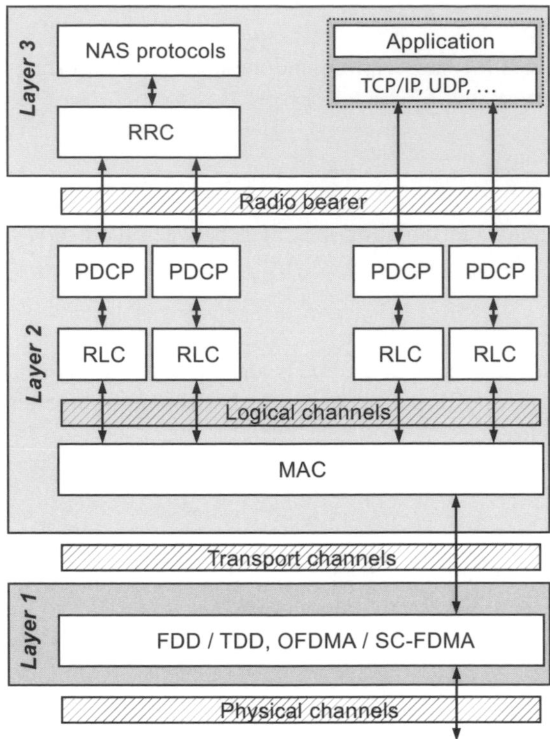

Figure 5.7 The role of LTE layers.

The PDCP layer takes care of the ciphering of the user and control plane, header compression (ROCH), and in-sequence delivery of the upper layer packet data units (PDU), duplicate elimination in lower layer SDUs, integrity protection for the control plane, and timer-based discarding.

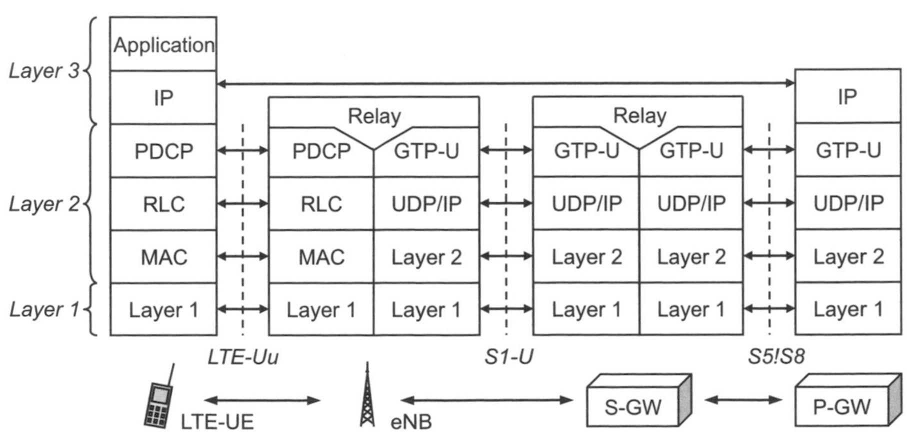

Figure 5.8 The LTE/SAE protocol layers for the user plane.

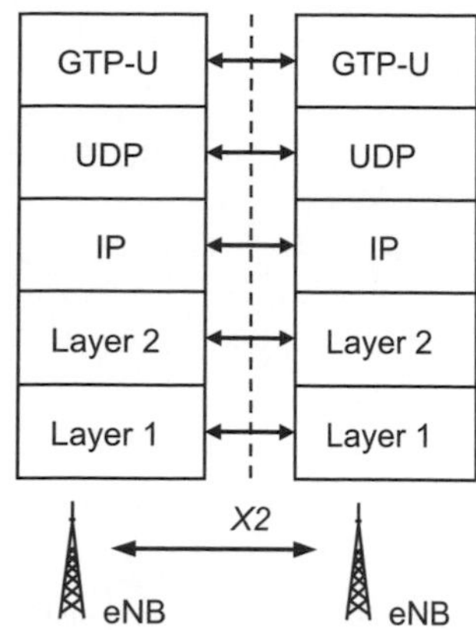

Figure 5.9 The user-plane protocol stack structure in communication between two eNBs.

5.4.2 Control Plane

The control plane protocol stack structure is shown in Figure 5.10.

Again, in the case of the direct communication between two eNBs, the protocol stack is as presented in Figure 5.11. This communication can happen, for example, when the handover procedure takes place.

5.4.3 Layer 1

LTE/SAE radio protocol layer 1 describes the physical layer. In general, it provides the means and basic functionality to deliver the bits over the air interface in both the downlink and the uplink directions.

The radio interface of LTE is based on two separate access techniques: OFDMA (Orthogonal Frequency Division Multiple Access) in the downlink direction, and SC-FDMA (Single Carrier Frequency Division Multiple Access) in the uplink direction. The functionality of OFDMA and SC-FDMA is described in more detail in Chapter 7.

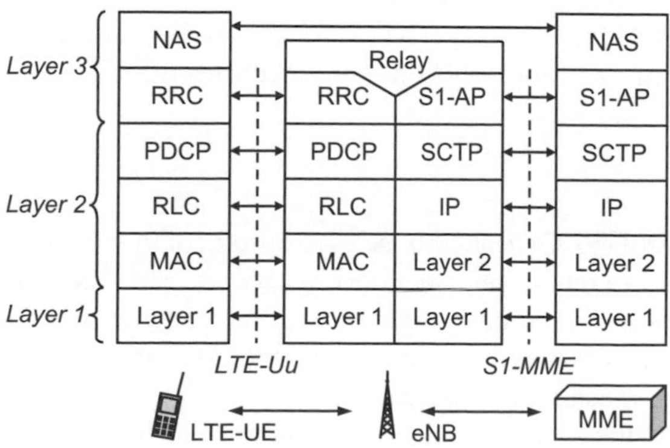

Figure 5.10 The control-plane protocol layer structure of LTE/SAE.

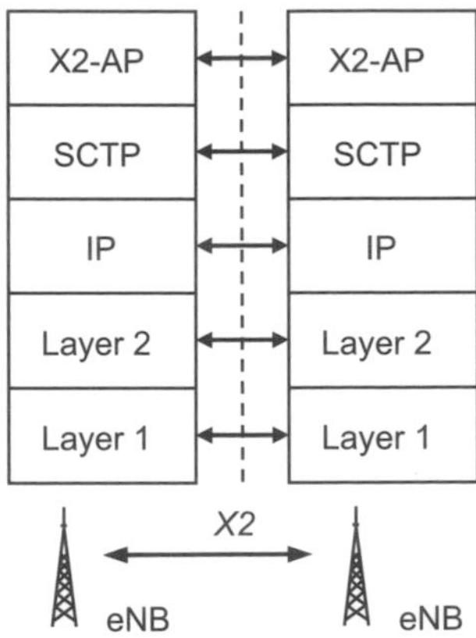

Figure 5.11 The control-plane protocol stack in direct communication between two eNBs.

A set of LTE channels is defined for signaling and data delivery. The channel definitions have been simplified compared to the previous 3G solutions of UMTS and HSPA, including the removal of dedicated channels. Instead, the shared channels are used for signaling and data delivery. In the LTE solution, the physical channels are mapped dynamically to the resources (meaning the physical resource blocks and antenna ports) that are available at the moment. This is done with help from the scheduler.

The physical layer handles the data transmission with the higher LTE/SAE layers via the transport channels. The transport channels are based on block-oriented service that takes into account the bit rate, delays, collisions and reliability of the transmission.

5.4.4 Layer 2

5.4.4.1 MAC

The MAC (Medium Access Control) protocol is the lowest protocol in layer 2. The main functionality of MAC is related to the management of the transport channels. On the other hand, MAC is communicating with the higher layers via the logical channels. MAC multiplexes the data of the logical channels onto the transmission of the transport channels, and demultiplexes it in the reception, according to the priority level of the logical channels.

MAC includes HARQ functionality (Hybrid Automatic Retransmission on reQuest). It also takes care of the handling of the collisions, and identifies the UEs.

5.4.4.2 RLC

The Radio Link Control (RLC) is next to the MAC protocol in the second LTE/SAE protocol layer structure. There is a one-to-one relationship between each Radio Bearer and each RLC.

The RLC enhances the radio bearer quality via the ARQ (Automatic Retransmission on reQuest) by using the data frames that contains sequence identities, and via the status reports, in order to trigger the retransmission mechanism.

RLC also performs the segmentation and reassembly of the data for the higher protocol layers. On the other hand, it concatenates the higher layer data into blocks for the data delivery over the transport channels as they allow only limited transport block sizes.

5.4.5 Layer 3

The layer 3 radio protocols consist of the following:

- PDCP (Packet Data Convergence Protocol);
- RRC (Radio Resource Control);
- NAS Protocols.

5.4.5.1 PDCP

Each radio bearer always uses a PDCP (Packet Data Convergence Protocol). The PDCP manages header compression, which is called ROHC (Robust Header Compression) according to the RFC 3095. PDCP also manages the ciphering and deciphering functionalities.

It should be noted that header compression is useful for IP datagram delivery, but the effect is not so significant for signaling. This means that, for signaling, the PDCP will usually only do the ciphering and deciphering without header compression.

5.4.5.2 RRC

Radio Resource Control (RRC) is an access stratum-specific control protocol for E-UTRAN. It provides the required messages for the channel management, measurement control and reporting. The control plane of RRC is a multitask entity that takes care of, for example, broadcast and paging procedures, RRC connection setup, radio bearer control, mobility functions, and LTE-UE measurement control.

The RRC functionality is discussed in more depth in Chapter 7.

5.4.5.3 NAS Protocols

The NAS protocol runs between UE and MME. It is located on top of RRC, which provides the required carrier messages for the NAS transfer. Some of the most important tasks of NAS are the authentication procedure, security control, EPS bearer management, EMC_Idle mobility handling, and paging origination in the EMC_Idle state.

5.5 Layer 2 Structure

As discussed in previous chapters, the MAC layer delivers information towards the radio interface via the transport channels, and to the RLC layer above via logical channels. The RLS in turn, delivers information for the PDCP functionalities above, which transfers the information within radio bearers.

When looking in more detail at the tasks within MAC, RLC and PDCP, we can see that there are several *service access points* between the MAC and the physical layers, referring to the individual transport channels. The service access points between the MAC and RLC layers refer to the logical channels. It should be noted that various logical channels can be multiplexed to the same transport channels. Figure 5.12 shows the downlink structure of layer 2, and Figure 5.13 shows the uplink structure.

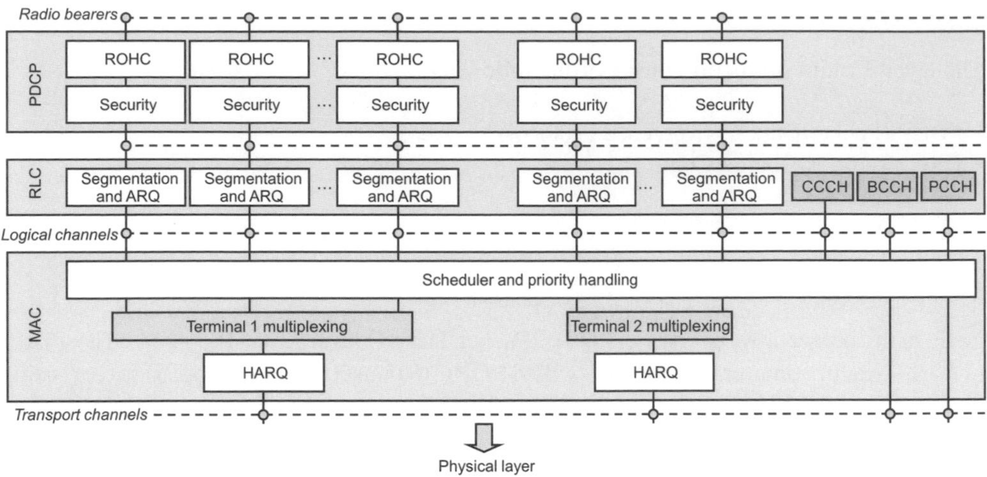

Figure 5.12 Layer 2 structure for the downlink.

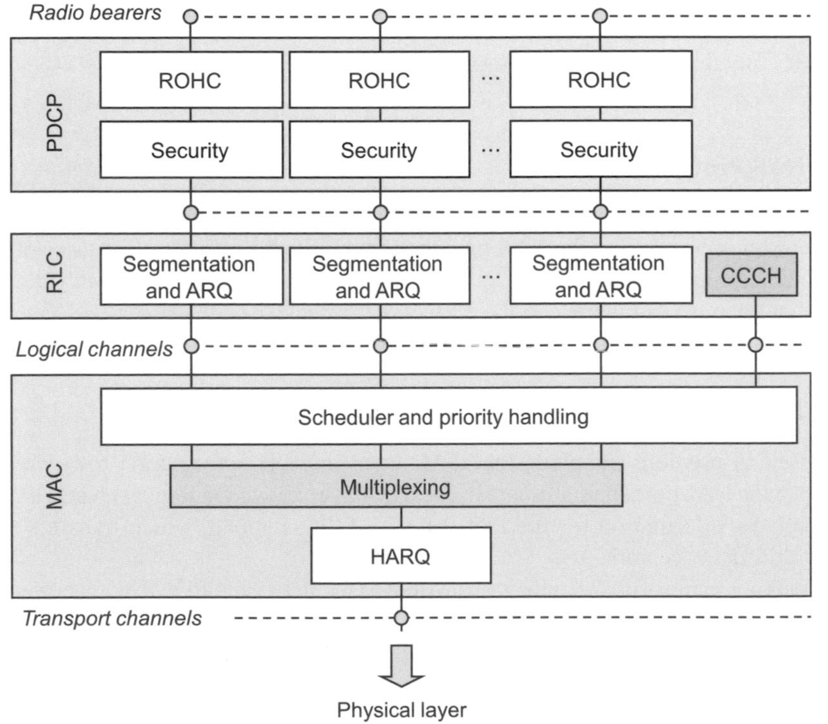

Figure 5.13 Layer 2 uplink structure.

References

[1] 3GPP TR 25.814. (2006) *Physical Layer Aspects for Evolved Universal Terrestrial Radio Access (UTRA) (Release 7)*, V7.1.0, October 2006, 3rd Generation Partnership Project, Sophia-Antipolis.
[2] 3GPP TS 36.213. (2009) *E-UTRA Physical Layer Procedures*, Section 7.1, V. 8.8.0, September 2009, 3rd Generation Partnership Project, Sophia-Antipolis.
[3] 3GPP TS 36.213. (2009) *E-UTRA Physical Layer Procedures,* Section 7.2, V. 8.8.0, 3rd Generation Partnership Project, Sophia-Antipolis.
[4] 3GPP TS 36.331. (2011) *Evolved Universal Terrestrial Radio Access (E-UTRA); Radio Resource Control (RRC); Protocol Specification,* V. 8.14.0, 2011-06-24, 3rd Generation Partnership Project, Sophia-Antipolis.
[5] Seidel, E. and Saad, E. (2010) LTE Home Node Bs and its enhancements in Release 9. Nomor Research GmbH, Munich, Germany.

6

Transport and Core Network

Juha Kallio, Jyrki T. J. Penttinen, and Olli Ramula

6.1 Introduction

This chapter presents the functional blocks and interfaces of LTE. The new architecture is illustrated using comparisons with solutions in earlier mobile communications systems. The protocol layer structure and functioning of the LTE protocols is described, and examples are given in order to clarify the principles of each protocol.

6.2 Functionality of Transport Elements

The sections that follow give a practical description of the MME, S-GW, P-GW and the respective transport modules. The information below is a snapshot of typical functionalities that can be applied in the core network of LTE/SAE. The complete list of functionalities depends on the vendor and the commercialization time schedules, so more specific information should be obtained from each vendor directly.

6.2.1 Transport Modules

Figure 6.1 shows an example of a transport module by Nokia Siemens Networks. It is characterized by the following:

- x E1/T1/JT1;
- IPSec support;
- Ethernet switching;
- ToP (IEEE1588-2008), Sync Ethernet.

In this specific case of an NSN transport element solution, there are two options. The FTLB contains 3 × GE, divided to 2 × GE electrical + 1 × GE optical via SFP module, while the FTIB contains 2 × GE.

The LTE/SAE Deployment Handbook, First Edition. Edited by Jyrki T. J. Penttinen.

Figure 6.1 An example of Flexi Multiradio BTS System Module with Flexi Transport submodule (FTLB/FTIB) by Nokia Siemens Networks. Printed by courtesy of Nokia Siemens Networks.

6.2.2 LTE Transport Protocol Stack

This basic item of the LTE/SAE network solution contains the IPv4-based protocol stacks for the user, control and management planes, and should be available from the first day of LTE deployment. LTE could also be a logical base for the support of IPv6 in order to drive the evolution path towards the evolved IP solutions.

6.2.3 Ethernet Transport

The basic LTE/SAE solution includes electrical and optical Ethernet interfaces, which provide the operator with the lowest transport cost and high transport capacity [1]. More specifically, the physical solution can be a Gigabit Ethernet 100/1000Base-T with electrical connectivity via the RJ-45 standard, and 1000Base-SX/LX/ZX with optical connectivity. Furthermore, the logical functionality includes the automatic negotiation of the mode and data rate. Figure 6.2 presents the protocol layers of the LTE/SAE Ethernet solution.

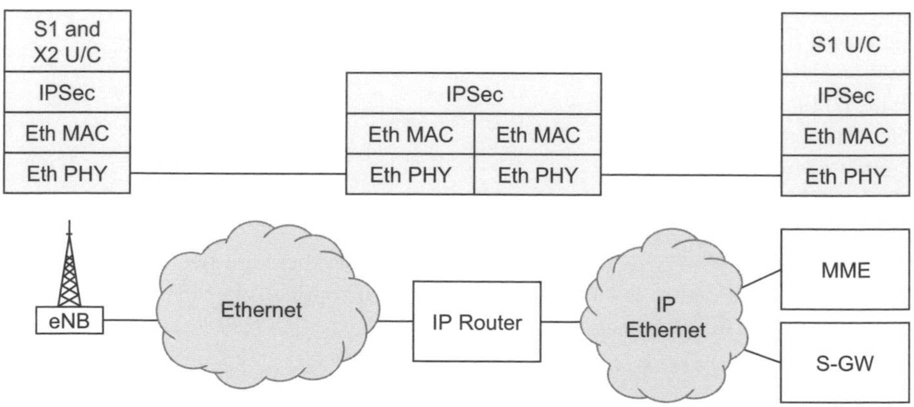

Figure 6.2 The Ethernet solution for LTE/SAE transport.

6.2.4 IP Address Differentiation

This solution provides different IP addresses for each of the LTE/SAE planes—that is, user, control, management and synchronization (U, C, M, S). The eNodeB applications can use either interface addresses or virtual addresses. In the address-sharing option, the single address is shared between all the planes while in multiple interface address solutions each plane uses a separate addresses. In the virtual address allocation, the applications are bound to the separate, virtual addresses of each plane.

6.2.5 Traffic Prioritization on the IP Layer

This functionality ensures reliable system control in such a way that it supports different user service classes. More specifically, the DiffServ Code Points (DSCP) can be configured, and the user plane DSCPs are configurable based on the QCI of the associated EPS bearer.

6.2.6 Traffic Prioritization on Ethernet Layer

This functionality ensures QoS if the transport network is not QoS-aware in the IP domain. One way of making this functionality is to use Ethernet priority bits in the Ethernet layer.

6.2.7 VLAN Based Traffic Differentiation

This functionality, shown in Figure 6.3, supports virtually separated networks for all the planes—that is, U, C, M and S planes. This is based on the ability to configure VLAN identities via the IEEE 802.1q definitions.

6.2.8 IPsec

This functionality, shown in Figure 6.4, is related to the security of the transport. Typically, IPSec can be supported in all the planes over the transport network. As an example, the

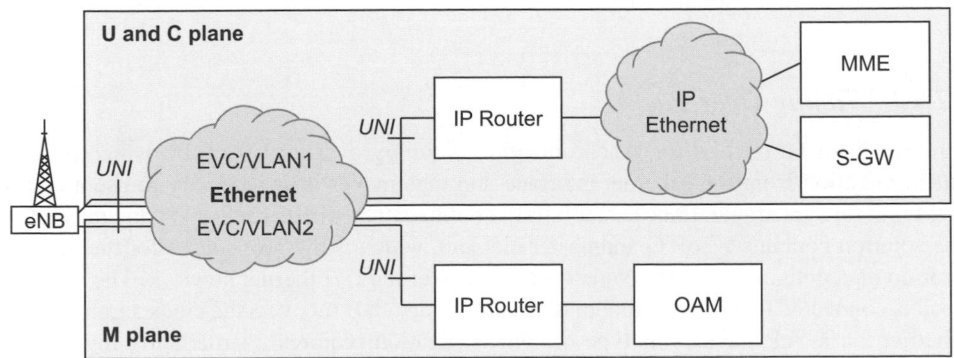

Figure 6.3 The VLAN ID can be defined separately for different planes.

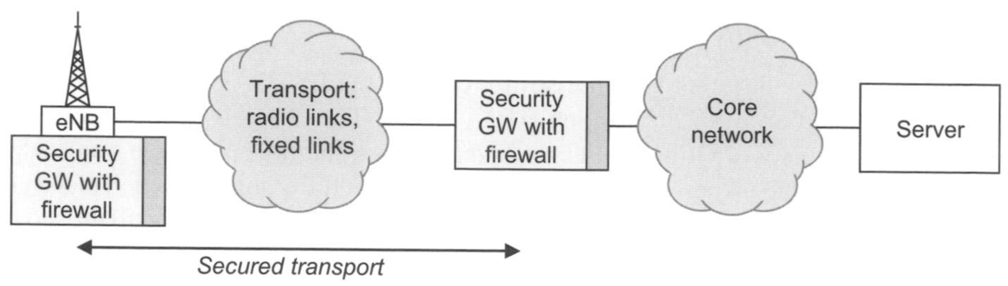

Figure 6.4 The IPSec can be utilized in the transport network.

eNodeB of Nokia Siemens Networks contains the security gateway and firewall integrated into the same element.

For more information, please refer to Chapter 11, which deals with security, and the specifications TS 33.210 (network domain security), TS 33.310 (authentication framework), and TS 401 (security architecture) [2]–[4].

6.2.9 Synchronization

A straightforward and practical solution for synchronization is the introduction of GPS. It is a functional solution for synchronization that operates in such a way that it is not necessary to take additional requirements into account on the transport network side. GPS supports both frequency and phase synchronization. Practical limitations arise from the maximum length of the data and power cable. If the GPS receiver is integrated into the antenna element, the synchronization interface down to the eNodeB element—for example, the system module of the Flexi of Nokia Siemens Networks solution—can be based on optical fiber that does not have transmission losses. A surge protector can be used between the GPS antenna and receiver and the system module, which minimizes damage in the case of, for example, thunderstorms.

As an alternative to GPS synchronization, it is also possible to carry out synchronization from the 2.048 MHz signal of the TDM infrastructure provided by equipment such as Base Transceiver Station of 2G, or NodeB of 3G. Another alternative is synchronization with the Plesiocronic Digital Hierarchy (PDH) interfaces.

6.2.10 Timing Over Packet

A more advanced method for synchronization is timing over packet (ToP), which provides synchronization from the Ethernet interface and thus makes it unnecessary to use a GPS or TDM link for synchronization. This solution is defined in the IEEE 1588-2008 documentation. The solution contains a ToP Grandmaster element, which is the root source for the synchronization data delivered for the eNodeB elements over the IP/Ethernet network. The reference clock is connected to the ToP Grandmaster, and the eNodeB recovers the clock signal over the Ethernet via a ToP slave. This type of synchronization requires a sufficiently high-quality packet data network.

6.2.11 Synchronous Ethernet

Another synchronization method is based on the synchronous Ethernet concept. It provides an accurate frequency synchronization over Ethernet links in such a way that accuracy does not depend on the network load. This functionality is based on the G.8261, G.8262 and G.8264 definitions of ITU. It distributes the frequency via layer 1 by applying an SDH-type of mechanisms. The challenge of this solution is that it must be implemented in all of the nodes that are found in the synchronization path.

6.3 Transport Network

The deployment of LTE means that the transport network should be designed to support the increased maximum radio interface data rates. This means that the existing backhaul, aggregation and backbone networks might need considerable redimensioning—that is, new hardware for enhanced capacity to guarantee the data delivery from the radio network to SAE elements, and further to the external packet-data networks.

The traditional backhaul of the operators, which is based on the TDM connectivity, can be updated to support packet data via Ethernet connectivity. This type of hybrid backhaul network is a logical option for the smooth enhancement of infrastructure that already exists. The Ethernet provides connectivity from eNB elements towards the MME and S-GW of EPC while the combination of the TDM and Ethernet provides connectivity from 2G BTS and 3G NodeB elements to the BSC and RNC, respectively.

If there is no full IP transport in the interface between the base station site and controller (for 2G and 3G) or S-GW/MME (for LTE), there is an alternative solution for base station connectivity. LTE traffic, together with previous traffic types of 2G BTS and NodeB of WCDMA and HSPA, can be delivered over the IP packet infrastructure using Carrier Ethernet Transport together with a pseudo wire transport concept. This means that if, for example, the Iub interface of 3G RAN is based on ATM transport, traffic is carried over ATM pseudowire connections, for example, in such manner that one connection is reserved for circuit-switched traffic and the other is reserved for packet-switched traffic. Similarly, 2G traffic can be delivered over a TDM pseudo wire connection between BTS and BSC in such a way that TDM signals are carried transparently over the radio access network.

6.3.1 Carrier Ethernet Transport

Carrier Ethernet Transport (CET) technology can be used for the deployment of new backhaul networks. For connectivity, so-called pseudowire solutions can be applied to emulate TDM and ATM, if native solutions are not available. Figure 6.5 shows the principle of the CET solution.

The CET concept is a cost-effective solution that can replace the traditional time-division multiplex (TDM) transport solutions such as SDH/PDH. It is possible to deploy CET for access and aggregation networks. Basically, LTE, 3G and 2G traffic types can all be delivered over packet-based backhaul infrastructure. The main benefits of the CET solution are support for standardized services of a variety of physical infrastructures, wide scalability of the bandwidth (from 1 mbps to over 10 gbps), high reliability, and support

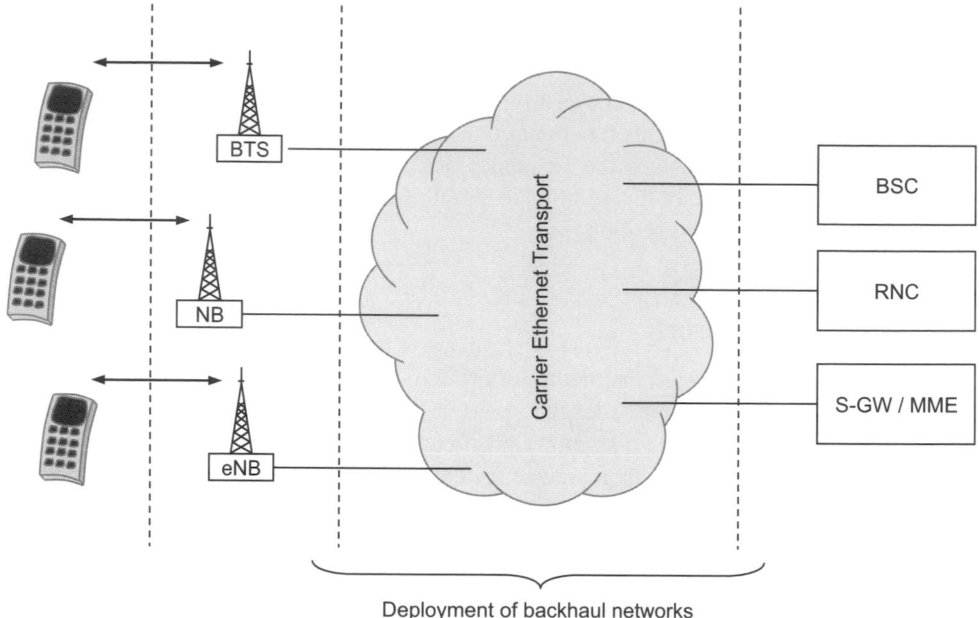

Figure 6.5 The principle of CET.

for quality of service options. It also offers the possibility to monitor, diagnose, and manage the network in a centralized way. Carrier Ethernet Transport has been standardized by the Metro Ethernet Forum, so it provides vendor-independent implementations.

6.3.2 *Transport for S1-U Interface*

The network between eNodeB elements, as well as between the eNodeB and S-GW elements, typically includes the access network itself, aggregation networks and the MPLS (Multi-Protocol Label Switching) backbone network. There can be microwave radio links within the access network in order to provide wireless interconnections, especially in areas where fiber optics are not available. In this way, handovers between eNodeB elements can be designed offering sufficiently high capacity and low delays.

The LTE *access network* can consist of partitions of several Virtual Local Area Networks (VLANs) in such a way that each partition contains one or more eNodeB elements, as shown in Figure 6.6.

The LTE *aggregation network* can be designed by applying, for example, ring topology combined with a Virtual Private LAN Service transport (VPLS). This in turn is based on the MPLS backbone concept. The aggregation network reserves a single Label Switch Path (LSP) for the respective single VLAN connection towards the LTE access network as shown in Figure 6.6. The figure also shows the option for the MPLS backbone network, based on the layer 3 routers installed in a mesh-topology, and is connected to S-GW. As in an aggregation network, a corresponding LSP is also used in this MPLS network in order to deliver the IP data traffic between the aggregation network and S-GW.

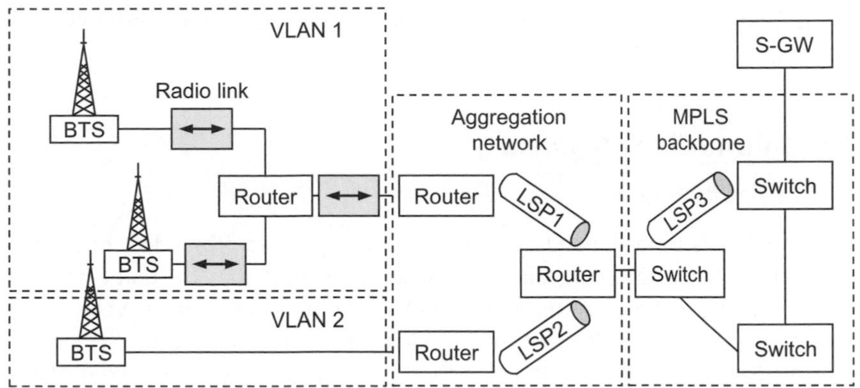

Figure 6.6 An example of traffic delivery between LTE eNB elements and S-GW.

6.4 Core Network

The logical solution for multiple radio access technologies is to use a common packet core concept. This is possible as far as the S4 interface is defined between the SGSN and Serving Gateway network entities.

The common core provides an optimized interworking functionality and Quality of Service handling between the LTE network and non-LTE access networks defined by 3GPP. It handles both LTE and 2G/3G bearers in a similar manner. Furthermore, it provides a common interface with the Home Subscriber Server (HSS). Figure 6.7 illustrates the idea of the common core concept. The QoS of each radio access network can be handled via the common Policy and Charging Rules Function (PCRF).

The class-based Quality of Service (QoS) concept specified for LTE networks in 3GPP Release 8 provides network operators with effective techniques to enable service or

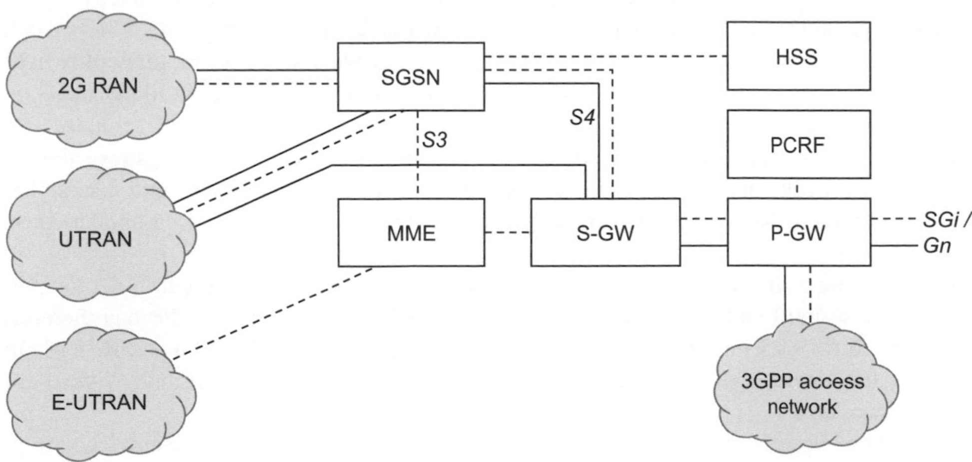

Figure 6.7 The idea of a common core concept, which can be shared between different 3GPP radio-access networks.

subscriber differentiation at the application level, and to maintain the required QoS level across the system.

6.5 IP Multimedia Subsystem

6.5.1 IMS Architecture

IMS architecture has been defined according following basic principles. First of all it is home-network centric in the sense that all services are executed by the home network. IMS architecture does support a roaming model with the familiar concepts of visited and home networks but the visited network mainly provides a local access point for SIP connectivity as Proxy Call State Control Function (P-CSCF) as well as local policy control functionalities in the form of Policy and Charging Rules Function (PCRF) as defined by 3GPP. In the home network, actual services are provided by individual Application Servers (AS), which are typically defined as logical functionalities based on their nature of service such as Telephony Application Server (TAS) for telephony supplementary services, Push to Talk Application Server (PoC AS) for Push-to-Talk services as well as the Presence Server (PS). Application servers can have different physical implementations, which can either reside in standalone hardware or can be integrated as part of some other functionality, depending on the vendor.

Core IMS architecture in the home network is built from the Interrogating Call State Control Function (I-CSCF) as well as the Serving Call State Control Function (S-CSCF). The I-CSCF is responsible for resolving a suitable S-CSCF for an IMS-registered subscriber whereas the S-CSCF is responsible for orchestration of service execution by selecting proper application servers for the session and authenticating and performing IMS registration procedure jointly with other CSCF roles and, naturally, the terminal. The Home Subscriber Server (HSS) function, defined by 3GPP, contains subscription data related to the use of the network services, including IMS and also circuit-switched and packet-switched subscription profiles. However, in practical deployments HSS does not contain both CS/PS and IMS-related subscription data but instead HSS products are introduced beside standalone HLR network elements in order to provide support for IMS and, optionally, for LTE-related subscription data. The issue that HSS products may not support legacy packet-switched (GERAN/UTRAN) or circuit-switched data means that, in some cases, communication service providers are keen to deploy LTE subscription data into HLR instead of new HSS product. The end result is likely to be the same in any case.

CSCF and S-CSCF are always located in the home network of given subscriber. If two IMS networks are involved in a communication session between end users then the standardized Network-Network Interface (NNI) will be used to connect these IMS networks.

Interworking with circuit-switched networks is a basic requirement for any IMS deployment today because most end users are still using circuit-switched services. 3GPP has therefore standardized IMS-CS interworking via the Media Gateway Control function and IMS Media Gateway functions. Figure 6.8 presents the high-level IMS architecture in both visited and home networks.

The following paragraphs will describe in more detail the IMS architecture and technology needed for native voice and video telephony over IP and Short Message Service (SMS) over IP. More details about IMS architecture can be found in [5].

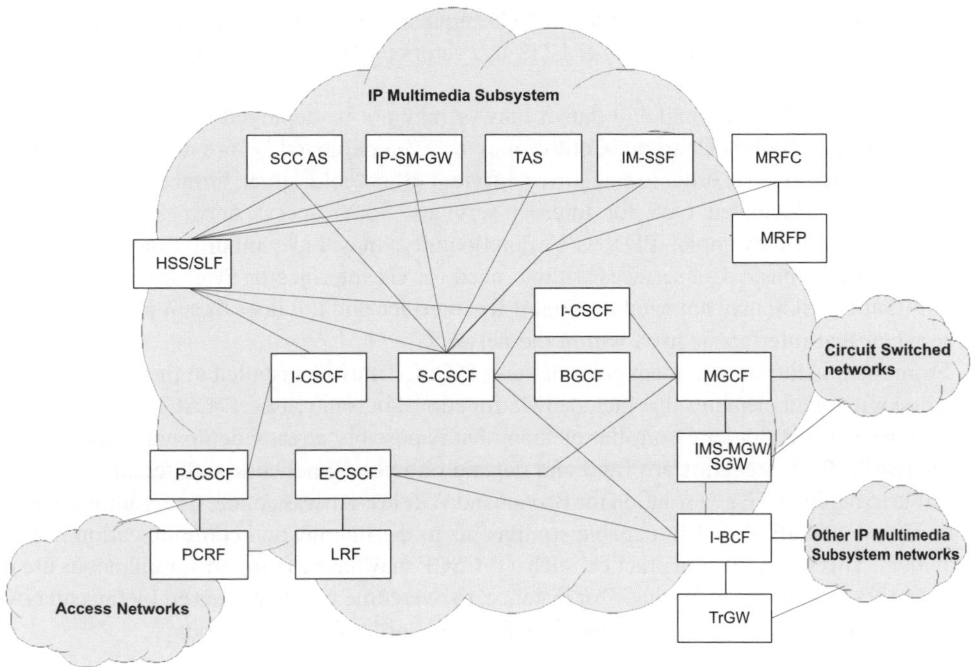

Figure 6.8 The IMS architecture.

6.5.1.1 P-CSCF

The Proxy Call State Control Function (P-CSCF) acts in a similar manner to the SIP proxy and is the first point in either the home or the visited network that the user equipment contacts in order to obtain access to IMS services. The P-CSCF will select a suitable Interrogating Call State Control Function (I-CSCF) within the home IMS network (which can be in different country) in addition to following tasks.

The P-CSCF is responsible for providing security measures to maintain the integrity and security of SIP signaling between the terminal and itself as well as asserting the subscriber's identity towards other IMS network elements such as S-CSCF. It will not be changed in a typical case during active IMS registration and it needs to be able to handle both its own subscribers and inbound roaming subscribers from other networks. Security and integrity protection is achieved through the use of IPSec but in the past IPSec has not been widely supported by SIP-capable endpoints. This is now expected to change due to the adoption of the GSMA "IMS profile for voice and SMS."

The P-CSCF is responsible for handling resource reservations via Policy Charging Control (PCC) architecture optionally deployed in the access network (such as LTE). It does this via a Diameter-based Rx interface to the Policy and Charging Rules Function (PCRF). The P-CSCF, in this respect, implements the Application Function (AF) as defined in 3GPP standardized PCC architecture. When the PCRF is used in the network it will communicate via Diameter-based Gx interface to the Policy and Charging Enforcement Point (PCEF), which in case of LTE resides in the Packet Data Network Gateway (PDN GW). The responsibilities of P-CSCF include the codec negotiation (media negotiation) between the intended SIP endpoints.

Then, based on the negotiated result, P-CSCF requests needed resources from PCRF. In the voice or video telephony service over LTE, this refers to the resources for both voice and/or video codec.

The use of PCC is optional and thus it may or may not be deployed by a communication service provider. Policy Charging Control may be even deployed before any voice or video telephony is deployed in order to categorize different users (gold, silver, bronze) or to enable IP flow-based differentiated QoS for Internet services. Some access network side products implementing, for example, PDN GW functionality, may have inbuilt functionalities to provide QoS for basic data services without need for Gx interface or PCC architecture. This means that P-CSCF may not need to support Rx interface and if it does then it may be the first time when that interface is used within the network.

From the practical deployment point of view, P-CSCF may be installed in the same physical location with other elements that include IMS functions. In some cases, P-CSCF also can reside within the Session Border Controller element that is possibly already deployed at the edge of the network. Both scenarios are valid and depend on existing network architecture. It is also possible to deploy LTE as a solution for Broadband Wireless Access connecting, for instance, an entire home and all related IP-capable equipments to the Internet and communication service providers. This means that, in practice, such a P-CSCF may have to support simultaneous use of PCRF as well as media anchoring—for instance, to overcome problems caused by far-end NAT in similar fashion to the support offered by Session Border Controllers today.

6.5.1.2 I-CSCF

The Interrogating Call State Control Function (I-CSCF) acts as the SIP proxy and typically acts as the first point of contact in the home IMS network. However in some cases the visited network may also have I-CSCF functionality in order to hide the topology of the network. The I-CSCF is contacted by the P-CSCF during IMS registration in order to obtain access to the IMS service as well as to hide the topology of the home IMS network (Topology Hiding, THIG). The interface between P-CSCF and I-CSCF is based on the 3GPP standardized Session Initiation Protocol (SIP) and routing of SIP messages is based on the Domain Name System as defined in [6].

The I-CSCF is also the first node that interfaces with the HSS of the IMS subscriber. The I-CSCF uses the Diameter-based Cx interface to fetch subscriber information from the HSS during IMS registration and decides which Serving Call State Control Function (S-CSCF) will be suitable for a given IMS subscription. If the home IMS network supports multiple HSS instances (network elements) then I-CSCF may use a Diameter-based Dx interface towards the Subscription Locator Function (SLF), which may be either colocated with other IMS functionalities (such as HSS) or deployed in standalone fashion. The SLF re-directs the Diameter request to the appropriate HSS, which contains the subscription data of the IMS subscriber.

From a voice-and-SMS-over-LTE point of view, there are no specific new requirements for the I-CSCF and it is therefore not described in more detail in this book.

6.5.1.3 S-CSCF

The Serving Call State Control Function (S-CSCF) acts as the SIP registrar for IMS subscribers by serving as an end point in the IMS network for IMS authentication (AKA) as well as coordinating which IMS services will be applied for, and in which order, for a given IMS subscriber. The S-CSCF will perform authentication and will inform the HSS about the registration status of the IMS subscriber via the Diameter-based Cx and/or Dx interfaces. The HSS needs to be aware of the identity of the S-CSCF in order, for instance, to handle the routing of terminating SIP sessions correctly when interrogated by the I-CSCF. This routing of terminating requests loosely resembles the behavior of the HLR and gateway MSC in traditional Circuit-Switched mobile networks. The interface between the I-CSCF, the P-CSCF and the S-CSCF is based on 3GPP standardized SIP.

Beyond the functionalities listed above, the S-CSCF is also responsible for deciding whether a given IMS subscription is entitled to a certain type of communication based on media type (voice, video) as well as translating used identities in the SIP signaling to SIP Uniform Resource Identity (URI) format if Telephony Uniform Resource Locator (URL) has been used by the terminal.

S-CSCF involves the required Application Servers (AS) as a part of the SIP session in order to provide actual services for the IMS subscriber. These services are based on the user profile retrieved from HSS during the registration. Alternatively, in the case the user profile has changed, the services are based on the communication service provider. For voice, video and SMS over LTE this means that the Telephony Application Server (TAS) and the IP-Short Message-Gateway (IP-SM-GW) functionalities are notified during the IMS registration by using 3rd party registration procedure. The SIP messages related to these particular services will be routed via the respective application server instances in order to invoke the service execution.

In practical deployments, the capabilities of the S-CSCF vary between network vendors. In some cases the S-CSCF may even have inbuilt application server functionalities in order to achieve higher flexibility to route SIP sessions and to manipulate SIP headers within SIP messages. Similarly, the same functionalities may be used in some cases to develop more advanced Service Control Interaction Management (SCIM). In this way, higher service interaction control can be achieved in those cases where more than one service is applied for a single SIP session. For voice, video and SMS over LTE, the S-CSCF represents an important building block in IMS architecture as described in this chapter. Nevertheless, these cases as such do not pose any significant requirements that are different from other IMS cases.

6.5.1.4 E-CSCF/LRF

The Emergency Call State Control Function (E-CSCF) is functionality that is required to complete an emergency IMS session either in visited or in home networks. The E-CSCFs is invoked by P-CSCF after P-CSCF detects that nature of session is an emergency, for instance based on the value of received the Request URI parameter. After this, the E-CSCF will resolve the required location information with the help of the Location Retrieval Function (LRF), which again is able to either use received signaling level information (P-Access-Network-Info header of SIP message) from the terminal or use the Location Service framework (LCS) possibly existing in the network. Location information is typically used, at a minimum, to select the Public Safety Answering Point (PSAP) that is responsible for emergency calls from given location. Conversion of the location to a PSAP address is done at the LRF and this PSAP's routable address (for instance SIP URI or TEL URL) is returned to the E-CSCF in order to route the call either via MGCF to circuit-switched networks in case PSAP does not have native SIP connectivity or by using SIP.

6.5.1.5 Home Subscriber Server and Subscriber Locator Function

The Home Subscriber Server (HSS) acts as the main subscriber data repository for the IMS user profile. This contains information about the identities and services related to a given subscription. The Subscriber Locator Function (SLF) is required if the IMS network has multiple HSS entities and the requesting function (e.g., I-CSCF or AS) requires information about which individual IMS user profile is located in which HSS entity.

From the voice-and-video-telephony-over-LTE point of view, data that is stored within HSS may contain, in addition to the identity of the Telephony Application Server (TAS) entity, the information about supplementary services provided. This information may be stored in XML document format as defined originally by 3GPP for the XCAP-based Ut interface but it may also be stored optionally as a binary-based format that resembles the way in which the HLR stores information in the Circuit Switched network. If XML document format is used then the HSS may not understand the actual content of this document as it is stored in the general-purpose Application Server-specific data container in the IMS user profile.

From the Short Message Service point of view, the HSS need to be provided with the identity of the IP-Short Message-Gateway (IP-SM-GW) AS entity that is responsible for handling that particular IMS subscriber. This information is then used by the S-CSCF to select the IP-SM-GW when the IMS subscriber performs IMS registration.

From the IM-SSF point of view, the HSS need to be provided with the identity of IP Multimedia—Service Switching Function (IM-SSF) entity that is responsible for handling that particular IMS subscriber. This information is then used by the S-CSCF to select the IM-SSF when the IMS subscriber performs IMS registration.

6.5.1.6 Application Servers

The Application Servers (AS) can be considered to be the work horses of IMS architecture, to provide business critical services for IMS subscribers. The underlying architecture of various Call State Control Functions are important too but have less significance when considering the service logic itself.

3GPP has standardized logical AS entities, which can be produced in a vendor-specific manner. However some groupings of functionalities can be found in the market in which voice- and video-telephony-related services are supported by a single product but then more advanced, programmable services on top of frameworks such as JAIN Service Logic Execution Environments (JSLEE) may be applied as part of other products.

From a voice-and-video-telephony point of view, one of the most important aspects of 3GPP standardized functionality is the Telephony Application Server (TAS), which is responsible for providing 3GPP-defined Multimedia Telephony (MMTel) services for IMS subscribers. From a Short Message Service over IP (SMS) point of view, the most important 3GPP standardized functionality is the IP-Short Message-Gateway (IP-SM-GW). It provides business logic for the handling of Short Message Service as well as interworking with legacy circuit-switched networks, when required. Interworking with legacy Intelligent Network (IN) services may be needed. In that case IMS architecture has dedicated AS functionality for IM-SSF, which is able to translate a SIP session to appropriate CAMEL or INAP service control protocol towards an existing Service Control Point (SCP). If there is a need to support service continuity and the network in question supports the IMS Centralized Services architecture, the so-called Service Centralization and Continuity Application Server (SCC AS) is used as part of the IMS session.

SCC AS is responsible for important tasks such as anchoring the session for possible forthcoming domain transfers due to use of Single Radio Voice Call Continuity (SRVCC). It is also used to perform Terminating Access Domain Selection (T-ADS) in order to select either a circuit-switched or an IP-based access network for the mobile-terminating (MT) call, if the terminal may be reached via both access types. Chapter 9 provides more information about SCC AS.

The Rich Communication Suite is a separate IMS application suite, which uses its own specific application server functionalities such as the Presence Server, the Instant Messaging application server and the XML Document Management Server (XDMS). In practice these functionalities are not mandatory for the implementation of "IMS profile for voice and SMS," with the exception of XDMS if it is used in the context of the Ut interface for Multimedia Telephony, but these functionalities may also be deployed in parallel if the communication service provider wishes.

If the AS requires access to an IMS user profile stored within the HSS, this access is possible via the Diameter-based Sh interface. If the network has multiple HSSs, then the Diameter-based Dx interface needs to be used towards the SLF to redirect the AS to the HSS containing the required IMS user profile. 3GPP has also defined the Diameter-based Si interface that could be used by the IM-SSF to fetch IN-related subscription data from the IMS user profile in the HSS. However, this Si-interface may not be required if the IM-SSF is able to use other mechanisms to fetch required data from the subscriber data repository, which can be the case if the IM-SSF is colocated with some other product, such as MSC. In any case, various different kinds of AS implementations may exist in practice. Not all the interfaces are necessarily supported by all the variants if the end-to-end functionality can be achieved in some alternative way, especially if there is no significant impact for the terminals or other IMS entities.

6.5.1.7 MGCF, IMS-MGW, I-BCF and TrGW

The Media Gateway Control Function (MGCF) and the IMS-Media Gateway (IMS-MGW) are functionalities that are typically involved when the SIP session is routed between the IMS subscriber and the Circuit Switched endpoint. In this case the MGCF is responsible for signaling related tasks such as conversion between SIP and SDP signaling used in the IMS network as well as signaling protocols used in Circuit-Switched networks such as ISDN User Part (ISUP), Bearer Independent Call Control (BICC) and even a specific variant of SIP protocol that tunnels the ISUP messages, SIP-I. The MGCF also controls the user plane resources required for such interworking and is located in the IMS-MGW via H.248 protocol-based 3GPP Mn interface. In a typical case multiple IMS-MGWs can be controlled by a single MGCF and vice-versa, thus maximizing the flexibility of network planning.

IMS-MGWs, at a minimum, need to be able to handle transport-level interworking, for instance between TDM and IP-based transport but also codec level interworking, which is usually called transcoding. Transcoding may support both voice and video codecs or only voice codecs, depending on the capabilities of the IMS-MGW product used.

In practical deployments in mobile networks the MGCF and IMS-MGW are typically colocated in a mobile soft-switching solution consisting of the MSC Server (MGCF) and MGW (IMS-MGW). This way it is possible to optimize the media plane routing in calls that require use of MGCF and IMS-MGW and either originate or terminate with a Circuit-Switched mobile terminal as no separate transit MGW may be required.

Voice and video telephony over LTE requires the MGCF and MGW to be able to support codecs mandated by 3GPP specifications as well as GSMA "IMS profile for voice and SMS." Support for High Definition voice with the Wideband Adaptive Multi Rate (WB-AMR) speech codec requires additional capability from MGCF and IMS-MGW to support interworking of SIP sessions with Circuit-Switched calls by using Transcoder Free Operation (TrFO) or Tandem Free Operation (TFO) depending on the call scenario. These two technologies are mandatory in order to support WB-AMR in Circuit-Switched networks.

In order to support interworking between SIP-based video telephony and 3G-324M as defined by 3GPP, depending on the capabilities of the products used for IMS-CS interworking, either an integrated or standalone video gateway should be used. If a standalone video gateway installation is used that is different from the MGW used for audio-only calls, then routing of calls need to be done in such way that voice calls and video calls towards IMS are routed, for instance, with a different prefix in front of the called party number in order to use different gateways correctly.

When interworking via IMS Network-Network Interface (IMS-NNI) to other IP-based networks it is possible to deploy the Interconnection Bearer Control Function (I-BCF) together with Transition Gateway (TrGW) functionality. I-BCF and TrGW may be used to provide security functionalities to prevent Denial of Service (DoS) attacks from unsecure IP interconnections but also to perform user-plane related functionalities such as transcoding, if required for IMS sessions that break-in or break-out from IMS. It may also be possible, depending on the product that offers I-BCF and TrGW functionalities, to use the same product for SIP-I interworking between the Circuit-Switched core networks as defined in [7]. In this way it is possible to achieve synergies between these different domains.

6.5.1.8 Media Resource Function Controller and Processor

Media Resource Function Controller (MRFC) and Media Resource Function Processor (MRFP) provide media plane related functionalities in case they are needed from the IMS network. These capabilities typically mean injection of in-band tones and announcements and collecting in-band information such as DTMFs. These functions may also provide support for network-based conferencing similar to that which exists in Circuit-Switched mobile networks today as a Multiparty supplementary service. Typical commercial MRFC/MRFP products are flexible and support multimedia for various purposes including conferencing.

Despite the fact that 3GPP originally standardized two separate functionalities for the Media Resource Function (MRFC/MRFP), in commercial products these are typically sold as standalone entities with the ability to deploy functionalities separately, if required. In addition to the standalone element it is possible that some vendors may have colocated the relevant functionality for certain service (e.g., voice conferencing or capability to deliver in-band voice announcements for voice telephony) in some existing products, thus providing more value for their communication service provider customers.

Voice and video telephony over LTE is considered to require support for in-band interaction similar to that which exists in today's Circuit-Switched mobile telephony. This means that announcements similar to those given by the network and tones need to be available when voice is deployed over the IMS network. Similarly, but less often, consumer ad-hoc conferencing functionality is also required from Circuit-Switched networks, which also imposes similar requirements on the IMS network.

6.5.1.9 SRVCC and ICS Enhanced MSC Server

Current modern mobile networks have a MSC Server system that enables communication service providers to use packet-switched transport for Circuit-Switched calls as well as for signaling. Similarly, Media Gateway platforms may have additional capabilities to support other use cases beyond Circuit-Switched calls.

3GPP Release 8 has defined new functionality for MSC Server to assist in a service continuity procedure known as Single Radio Voice Call Continuity (SRVCC) as part of 3GPP TS 23.216 [z]. SRVCC handles the continuation of a voice call when the terminal moves from LTE to a Circuit-Switched network. A SRVCC-enhanced MSC Server has a specific GTP-based Sv interface towards the MME function. This Sv interface is defined in 3GPP TS 29.280. It is used by MME for requesting MSC Server to reserve required radio access resources from the target Circuit Switched radio access (GERAN/UTRAN) for SRVCC. SRVCC compatible MSC Server prepares resources for IuCS or A interface, which may be locally or remotely connected. If target radio access is controlled by another MSC Server (MSC-B) then the SRVCC enhanced MSC Server will perform normal Inter-MSC relocation as defined in [8]. After the target Circuit-Switched radio access resources have been committed then SRVCC-enhanced MSC Server will establish a call on behalf of terminal to a specific address given by MME via Sv interface. This address is related to the current SCC AS of that particular subscription and involved in the original call establishment.

The SRVCC procedure has been gradually improved between 3GPP Release 8 and Release 10 to support more functionality, such as the capability to support multiple simultaneous calls (active and held) as well as the capability to perform reverse SRVCC from a Circuit-Switched network to LTE. In order to support functionalities beyond 3GPP, Release 8 additional requirements set by 3GPP IMS Centralized Services (ICS) architecture need to be taken into use. This can occur in a phased manner if IMS-based voice over LTE has been deployed commercially by using the 3GPP Release 8 standardization baseline. 3GPP Release 9 introduces specific "MSC Server assisted mid-call" functionality in [9], which is based on ICS-enhanced MSC Server functionality defined in [10]. This functionality is required to support SRVCC for multiple ongoing calls (active and held) if the terminal is not ICS enabled—that is, if it does not have the capability to use a Circuit-Switched network as the bearer for the session that is established by using SIP as defined in [10]. If reverse SRVCC is required then Circuit-Switched calls originated by the terminal need to be anchored in the IMS (SCC AS), which means that the IMS Centralized Service architecture needs to be fully deployed into the network.

References

[1] Metro Ethernet (n.d.) Metro Ethernet Forum, www.metroethernetforum.org (accessed August 29, 2011).

[2] 3GPP TS 33.210. (2009) *3G Security; Network Domain Security (NDS); IP Network Layer Security.* V. 8.3.0, 2009-06-12, 3rd Generation Partnership Project, Sophia-Antipolis.

[3] 3GPP 33.310. (2010) *Network Domain Security (NDS); Authentication Framework (AF).* 8.4.0, 2010-06-18, 3rd Generation Partnership Project, Sophia-Antipolis.

[4] 3GPP TS 33.401. (2011) *3GPP System Architecture Evolution (SAE); Security Architecture.* 8.8.0, 2011-06-24, 3rd Generation Partnership Project, Sophia-Antipolis.

[5] Poikselka, M., Mayer, G., Khartabil, H., Niemi, A. (2004) *The IMS: IP Multimedia Concepts and Services in the Mobile Domain.* John Wiley & Sons, Ltd, Chichester.

[6] IETF RFC 3263. (2002) *Session Initiation Protocol (SIP): Locating SIP Servers.*

[7] GSMA PRD IR.83. (2009) *SIP-I Interworking Description.* Version 1.0, 17, GSMA, London.

[8] 3GPP TS 23.009. (2009) *Handover Procedures*. V. 8.2.0, 2009-09-28, 3rd Generation Partnership Project, Sophia-Antipolis.
[9] 3GPP TS 23.237. (2010) *IP Multimedia Subsystem (IMS) Service Continuity; Stage 2*. V. 8.7.0, 2010-03-26, 3rd Generation Partnership Project, Sophia-Antipolis.
[10] 3GPP TS 23.292. (2010) *IP Multimedia Subsystem (IMS) Centralized Services; Stage 2*. V. 8.8.0, 2010-06-14, 3rd Generation Partnership Project, Sophia-Antipolis.

7

LTE Radio Network

Francesco D. Calabrese, Guillaume Monghal, Mohmmad Anas, Luis Maestro, and Jyrki T. J. Penttinen

7.1 Introduction

This chapter presents the LTE radio interface and related topics. First, an LTE spectrum is described in theory and practice. Then, the LTE multiplex in the downlink and the uplink—that is, OFDM and SC-FDMA—is described. Frequency Division and Time Division variants of LTE are both shown. The LTE link budget is explained with practical examples. The hardware solutions of eNodeB and the antenna systems are explained in detail, as well as practical issues concerning hardware reuse in LTE deployment.

7.2 LTE Radio Interface

The LTE radio interface is based on the frequency division multiplexing technique. Orthogonal Frequency Division Multiplex (OFDMA) is used in the downlink direction whereas Single Carrier Frequency Division Multiple Access (SC-FDMA) is used in the uplink direction. OFDMA provides good protection against the rapidly varying radio conditions, including fast-fading and multi-path propagated radio components. It is not a very efficient solution, though, as the peak-to-average power ratio PAPR behavior causes difficulties in the equipment's circuit design. For that reason, SC-FDMA is selected in the uplink as the terminal can handle these challenges easier [1].

Long-Term Evolution supports both FDD (Frequency Division Duplex) and TDD (Time Division Duplex). In the FDD mode, the uplink and downlink transmission happens in separate frequency bands, whereas the TDD mode uses timeslots of the same frequency band for downlink and uplink transmission. Both of these modes can be used efficiently in such a way that the total bands are the same, varying between 1.4 and 20 MHz. Depending on the bandwidth and other functionalities such as MIMO variants and the modulation scheme, the maximum data speed that LTE provides is up to about 300 mbps in the downlink and 75 mbps in the uplink.

Due to the flexible bandwidth definitions of LTE, it can be deployed in many mobile operator scenarios. The smallest bands are applicable in situations where the operator does not have too

The LTE/SAE Deployment Handbook, First Edition. Edited by Jyrki T. J. Penttinen.

many extra frequencies due to the operator's other systems. Although the smallest LTE band can provide the lowest data speeds and capacity, it is justified as an interim solution in frequency refarming. In this way, the same band can be shared between GSM/UMTS and LTE. With the growth of LTE subscriber penetration, the 2G/3G band can be reduced while LTE takes a larger share of the band. For example, there are LTE terminals available for the 900 MHz band at the beginning of the service, and as LTE matures, there will also be terminals for other bands that offers multi-system functionalities. LTE provides a more flexible arrangement of the frequencies than UMTS, which is fixed to a figure of 5 MHz (or, in some cases 3.8 or 4.2 MHz, depending on the vendor solutions).

The LTE bandwidth dictates how many subcarriers can be used in that band. This, on the other hand, dictates the number of the radio resource blocks—that is, how much capacity can be offered in that area. A single radio resource block corresponds to 12 consecutive subcarriers.

7.3 LTE Spectrum

There are various frequency band options available for the LTE system, depending on the country and continent. LTE networks can be deployed in the existing and new frequency bands as:

- the 900 MHz and 1800 MHz bands that are currently widely utilized for the GSM system;
- the 850 MHz and 1900 MHz bands that are widely utilized for GSM in the northern America region;
- the 700 MHz band that has typically been used for the analog TV broadcast networks and that are being refarmed due to the digitalization of the TV systems;
- the 2100 MHz band outside of the Americas, and the combined 1700 MHz and 2100 MHz bands in the Americas, which were widely used for the previous 3G systems—that is, UMTS/WCDMA and HSPA;
- a new 2600 MHz band is becoming available in various parts of the globe.

It is probable that the initial LTE deployment will happen in the 2100 MHz and in the combined 2100 MHz/1700 MHz bands, as well as in the 900 MHz band.

Figures 7.1 and 7.2 show the graphical presentation of the FDD and TDD frequency bands of LTE.

7.4 OFDM and OFDMA

7.4.1 General Principle

LTE uses OFDM (Orthogonal Frequency Division Multi-Carrier) in the downlink—that is, in the direction from the eNodeB to the UE. This direction is sometimes also referred as a forward link. OFDM complies with the LTE requirements for the spectrum flexibility, and enables a cost-efficient base for wide frequency bands that provide high peak data rates. The LTE downlink physical resource can be seen as a time-frequency grid. In the frequency domain, the spacing between the adjacent subcarriers, Δf, is 15 kHz. In addition, the OFDM symbol duration time is $1/\Delta f +$ cyclic prefix. The cyclic prefix is used to maintain orthogonality between the subcarriers even for a time-dispersive radio channel. One resource element carries QPSK, 16QAM or 64QAM.

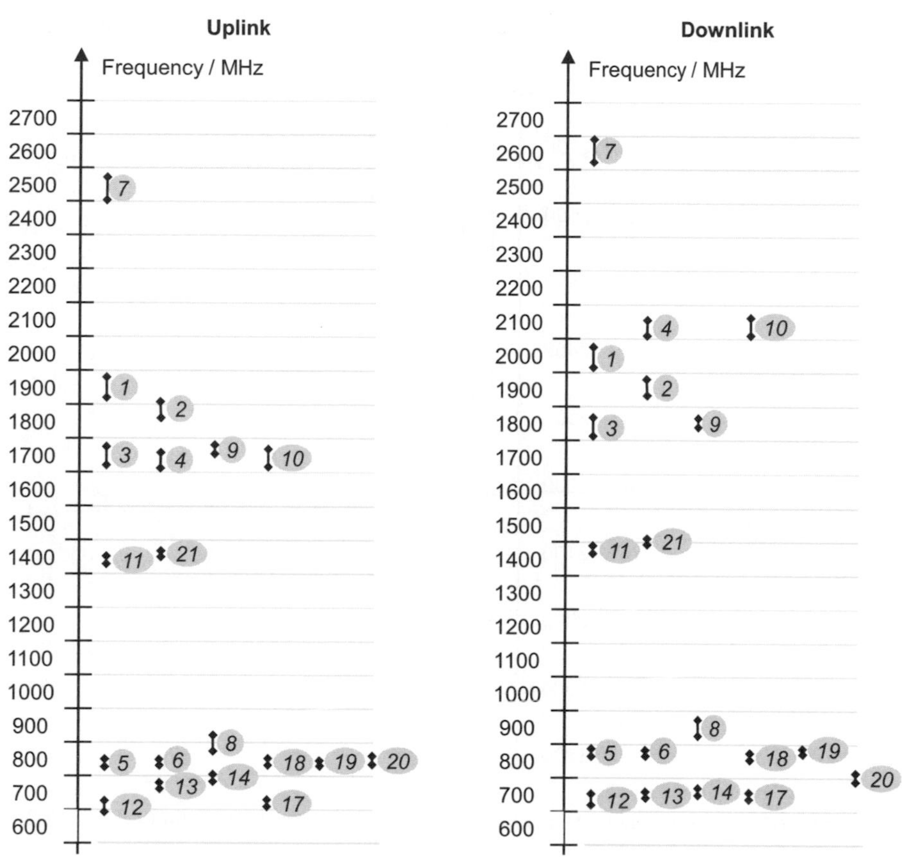

Figure 7.1 The LTE spectrum for FDD downlink and uplink. The references are found in Table 7.1.

Figure 7.3 shows the basic principle of the difference between earlier 3G bandwidth, which is fixed to 5 MHz, and the flexible LTE bandwidth. The dynamic definition of the bandwidth is actually one of the main benefits of LTE over WCDMA and HSPA. A smaller band allows efficient frequency band refarming between LTE and other systems—for example, WCDMA and GSM, which is especially beneficial in cases when there is not too much band in use. On the other hand, the largest LTE bandwidths provide the highest data rates, which is the main factor differentiating them from the WCDMA and HSPA data rates within their fixed 5 MHz band.

Orthogonal Frequency-Division Multiplexing (OFDM) is a modulation technique for data transmission that has been known since the 1960s [1]. Nowadays, OFDM is used in many standards such as European Digital Audio Broadcasting (DAB), Terrestrial Digital Video Broadcasting (DVB-T), Asynchronous Digital Subscriber Line (ADSL) and High-bit-rate Digital Subscriber Line (HDSL). It can also be found in the IEEE 802.11a Local Area Network (Wi-Fi) and the IEEE 802.16 Metropolitan Area Network (WiMAX) [2]. It has been defined as the medium access technology for the downlink in Long Term Evolution (LTE), chosen among other candidates such as Multi-Carrier Wide-band Code-Division Multiple Access (MC-WCDMA) and Multi-Carrier Time-Division Synchronous-Code-Division Multiple Access (MC-TD-SCDMA) [3].

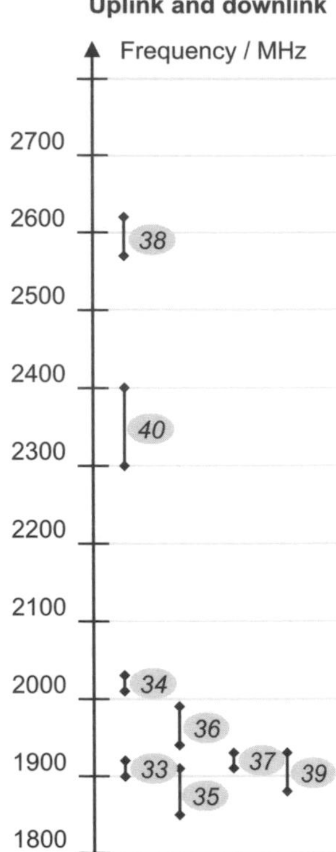

Figure 7.2 The frequency bands of TDD LTE. The references are found in Table 7.2.

Table 7.1 FDD frequency bands for LTE

Band nr	Uplink frequency (MHz)	Downlink frequency (MHz)	DL and UL Bandwidth (MHz)
1	1920–1980	2110–2170	60
2	1850–1910	1930–1990	60
3	1710–1785	1805–1880	75
4	1710–1755	2110–2155	45
5	824–849	869–894	25
6	830–840	875–885	10
7	2500–2570	2620–2690	70
8	880–915	925–960	35
9	1749.9–1784.9	1844.9–1879.9	35
10	1710–1770	2110–2170	60
11	1427.9–1447.9	1475.9–1495.9	20
12	698–716	728–746	18
13	777–787	746–756	10
14	788–798	758–768	10
15	N/A	N/A	N/A
16	N/A	N/A	N/A
17	704–716	734–746	12
18	815–830	860–875	15
19	830–845	875–890	15
20	832–862	791–821	30
21	1447.9–1462.9	1495.9–1510.9	15

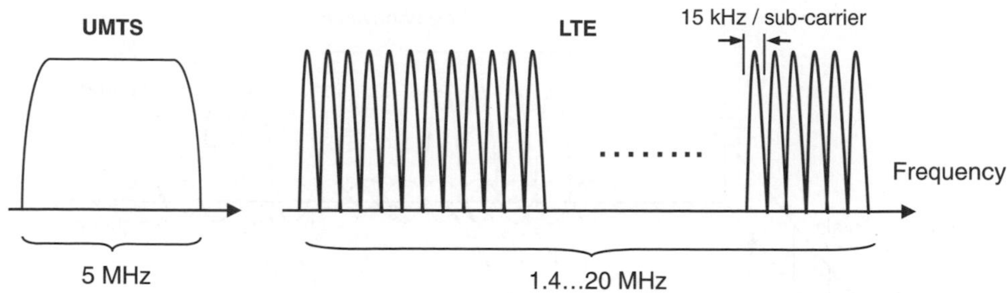

Figure 7.3 The frequency band of LTE consists of several subcarriers while UMTS uses one complete carrier for all the traffic of a single cell.

OFDM is based on splitting the data stream to be transmitted onto several orthogonal subcarriers, allowing an increased symbol period. Since low-rate modulation schemes are more robust to multipath, it is more effective to transmit many low-rate streams in parallel than one single high-rate stream. The goal of using these subcarriers is to obtain a channel that is roughly constant (flat) over each given sub-band, making equalization much simpler at the receiver. Furthermore, OFDM allows the use of low-complexity Multiple-Input Multiple-Output (MIMO) techniques. Finally, OFDM provides a flexible use of the bandwidth and can achieve high peak data rates [4]–[6].

OFDM is based on the well-known Frequency Division Multiplexing (FDM) technique. In FDM, different streams of information are mapped onto separate parallel frequency channels. OFDM differs from traditional FDM in the following ways [7]–[10]:

- The same information stream is mapped onto a large number of narrowband subcarriers increasing the symbol period compared to single carrier schemes.
- The subcarriers are orthogonal to each other in order to reduce the Inter-Carrier Interference (ICI). Moreover, overlap between subcarriers is allowed to provide high spectral efficiency.
- A guard interval, often called Cyclic Prefix (CP), is added at the beginning of each OFDM symbol to preserve orthogonality between subcarriers and eliminate Inter-Symbol Interference (ISI) and ICI.

Figure 7.4 depicts the concepts introduced above. In the frequency domain, the overlap between subcarriers is easily seen as well as the fact that they are orthogonal to each other. On the other side, in the time domain, a guard interval is present at the beginning of each OFDM symbol.

Table 7.2 TDD frequency bands for LTE

Band nr	Uplink frequency (MHz)	Downlink frequency (MHz)	DL and UL Bandwidth (MHz)
33	1900–1920	1900–1920	20
34	2010–2025	2010–2025	20
35	1850–1910	1850–1910	60
36	1930–1990	1930–1990	60
37	1910–1930	1910–1930	20
38	2570–2620	2570–2620	50
39	1880–1920	1880–1920	40
40	2300–2400	2300–2400	1000

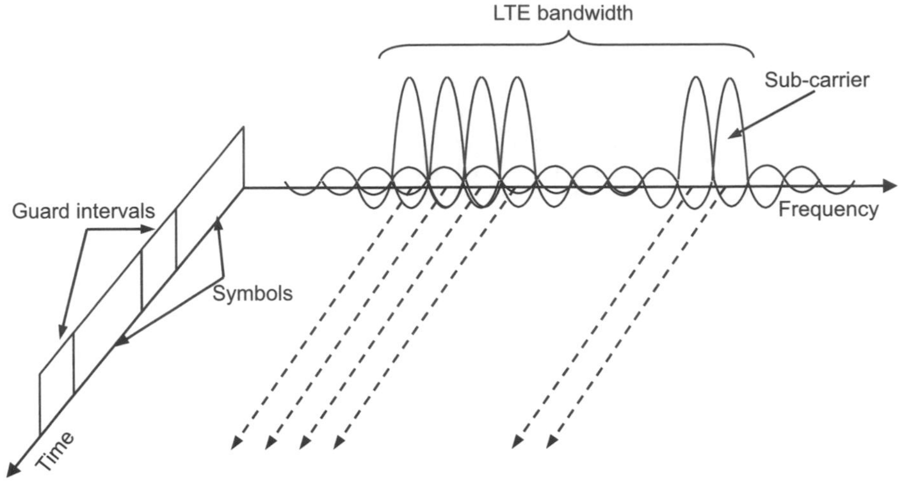

Figure 7.4 Frequency-time interpretation of an OFDM signal.

7.4.2 OFDM Transceiver Chain

Figure 7.5 presents a simplified block diagram of a single-input single-output (SISO) OFDM system. On the transmitter side, the modulated (QAM/PSK) symbols d are mapped onto N orthogonal subcarriers. This is accomplished by means of an Inverse Discrete Fourier Transform (IDFT) operation. Most commonly, the IDFT is performed with an Inverse Fast Fourier Transform (IFFT) algorithm, which is computationally efficient. Next, the CP is inserted and a parallel-to-serial conversion is performed prior to transmission over the air.

At the receiver end, the reversal operations are performed. Once the received signal reaches the receiver, the CP, which could have experienced interference from previous OFDM symbols, is removed. Then, a Fast Fourier Transform (FFT) operation brings the data to the frequency

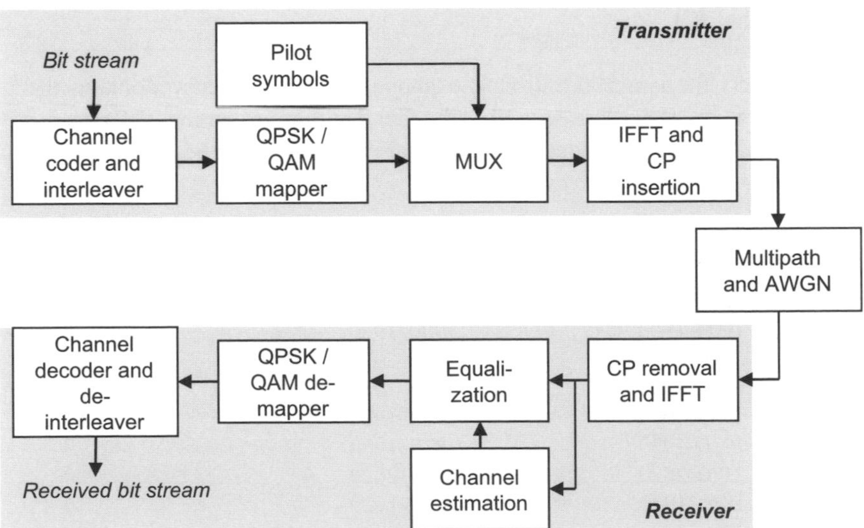

Figure 7.5 SISO OFDM simplified block diagram.

domain. This way, channel estimation and equalization is simplified. Note that in order to be able to carry out the latter operations, known symbols called pilots are inserted in certain frequency positions/subcarriers on the transmitter side. At the end of the chain, the equalized data symbols are demodulated yielding the received bit stream.

7.4.3 Cyclic Prefix

As mentioned before, a guard interval is added at the beginning of each OFDM symbol to mitigate some of the negative effects of the multipath channel. If the duration of the guard interval T_g is larger than the maximum delay of the channel $\tau_{\max}$, all multipath components will arrive within this guard time and the useful symbol will not be affected, avoiding Inter-Symbol Interference (ISI) as can be seen in Figure 7.6.

One particular instance of the guard interval is the so-called cyclic prefix. In this case the last N_g samples of the useful OFDM symbol with N samples in total are copied to the beginning of the same symbol. Since the number of cycles of each orthogonality function per OFDM symbol will be maintained as an integer, this strategy also allows the orthogonality properties of the transmitted subcarriers to be kept, avoiding ICI. Figure 7.7 shows the cyclic prefix concept where

$$T_u = N \times T_0 \tag{7.1}$$

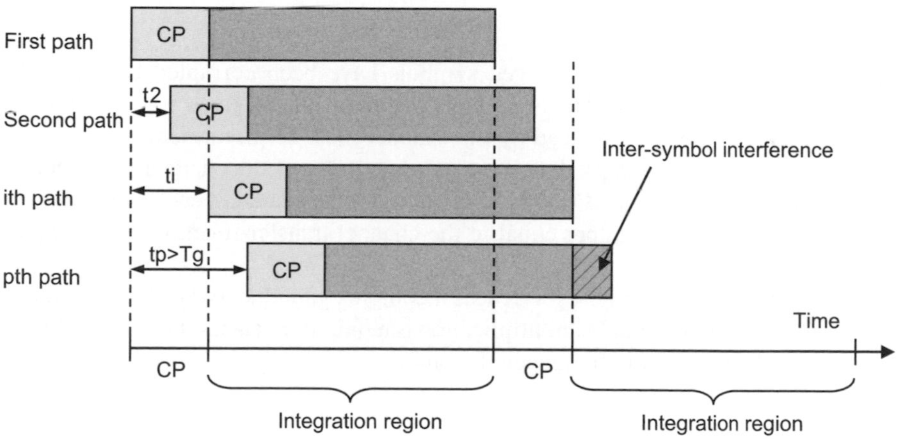

Figure 7.6 Cyclic prefix (CP) avoiding ISI.

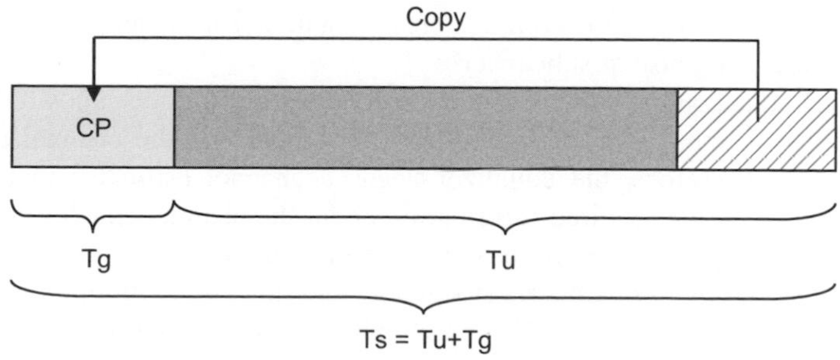

Figure 7.7 Cyclic prefix as a copy of the last part of an OFDM symbol.

$$T_g = N_g \times T_0 \tag{7.2}$$

$$T_s = \left(N + N_g\right) \times T_0 \tag{7.3}$$

The symbols mean the following: T_u is the useful OFDM symbol where data symbols are allocated, T_g is the duration of the cyclic prefix, and T_s is the total duration of the OFDM symbol.

The insertion of the CP results in a Spectral Efficiency Loss (SEL), which is not so important compared to the benefits that it provides in terms of ISI and ICI robustness. The SEL can be interpreted as the loss of throughput that the OFDM transmission system will suffer by the addition of the cyclic prefix. It can be written as [11]:

$$SEL = \frac{T_g}{T_g + T_u} \tag{7.4}$$

It can be seen that the loss of spectral efficiency is directly related to the ratio between the duration of the CP and the total duration of an OFDM symbol.

7.4.4 Channel Estimation and Equalization

In wireless OFDM systems, the received symbols have been corrupted by the multipath channel. In order to undo these effects, an equalization of the received signal that somehow compensates the variations introduced by the channel must be performed.

Assuming that the CP is longer than the maximum delay of the channel and a nonvariant channel over the duration of an OFDM symbol (slow-fading channel), each subcarrier symbol is multiplied by a complex number equal to the channel transfer function coefficient at this subcarrier frequency.

In other words, each subcarrier experiences a complex gain due to the channel. In order to undo these effects a single complex multiplication is required for each subcarrier yielding low-complexity equalization in the frequency domain:

$$y[k] = \frac{z[k]}{h[k]} = d[k] + \frac{w[k]}{h[k]} \tag{7.5}$$

where $y[k]$ is the equalized symbol in the kth subcarrier, $z[k]$ is the received symbol at the kth subcarrier after FFT and $h[k]$ is the complex channel gain at subcarrier k. $w[k]$ represents the additive white Gaussian noise at subcarrier k.

Note that this equalization has been performed assuming perfect knowledge about the channel. However, in most of systems that employ equalizers, the channel properties are unknown *a priori*. Therefore, the equalizer needs a channel estimator that provides the equalization block with the required information about the channel characteristics.

Different approaches have been proposed to estimate the channel in OFDM systems but *pilot-aided channel estimation* is the most suitable solution for the mobile radio channel. In LTE it is the proposed solution [3]. This technique consists of transmitting symbols, often called pilot symbols, known by both the transmitter and the receiver, in order to estimate the channel at the receiver. This approach presents an important tradeoff between the number of pilots used to perform the estimation and the transmission efficiency. The more pilots are used, the more accurate the estimation will be, but also the more overhead will be transmitted reducing the data rate.

As an example, the following examples depict the mapping of cell-specific reference signals in LTE [12] for different numbers of antenna ports and with normal CP. These pilot symbols are distributed in both frequency and time domains and they are orthogonal to each other in order to allow for accurate channel estimation.

Figure 7.8 shows the idea of the LTE radio resource block, and Figures 7.9, 7.10 and 7.11 show the mapping of the reference signals.

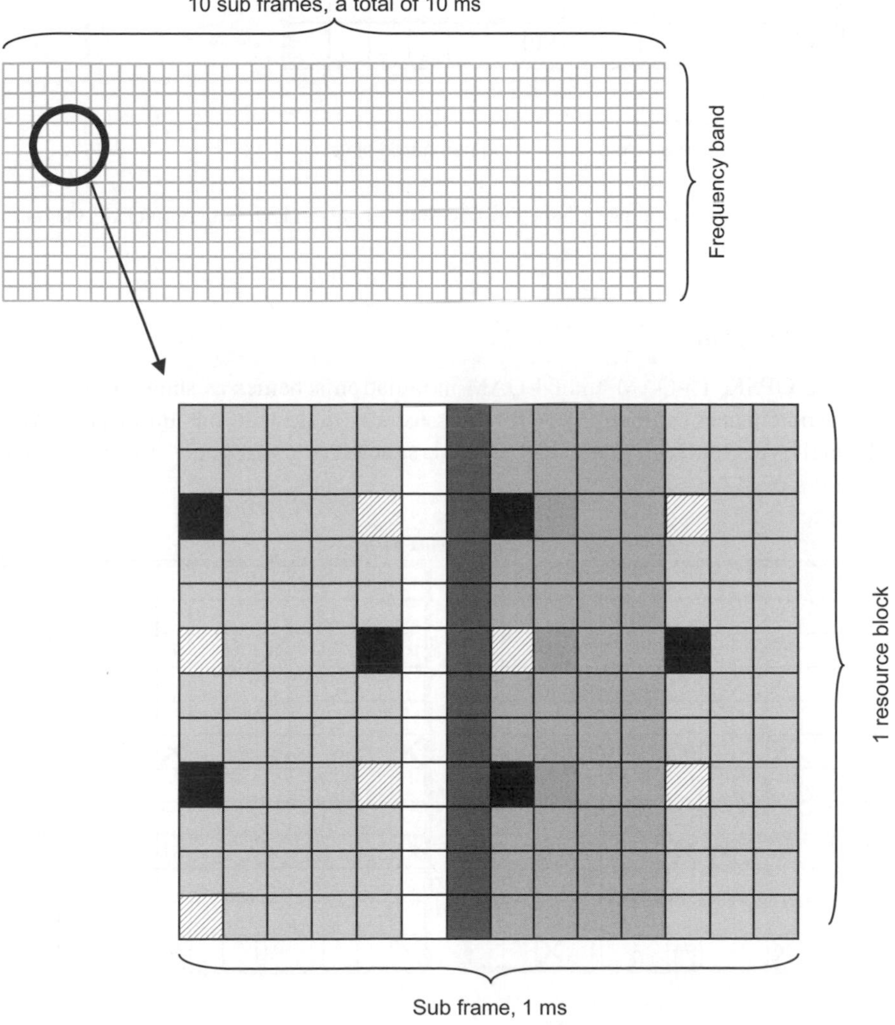

Figure 7.8 The forming of the LTE radio resource block.

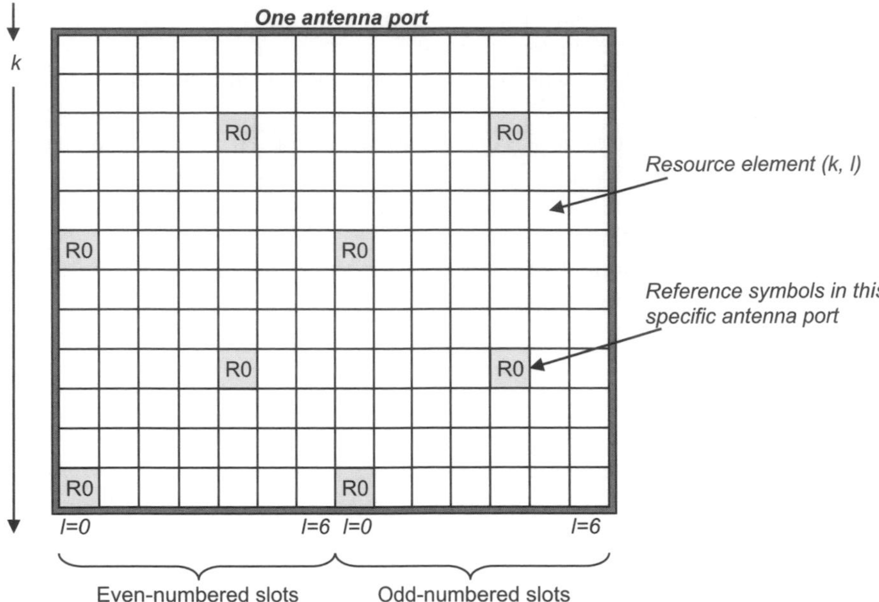

Figure 7.9 Mapping of downlink cell-specific reference signals in LTE with normal CP, that is, in one antenna port setup of LTE.

7.4.5 Modulation

LTE can use QPSK, 16-QAM and 64-QAM modulation schemes as shown in Figures 7.12 and 7.13. The channel estimation of OFDM is usually done with the aid of pilot symbols. The channel type for each individual OFDM subcarrier corresponds to the flat fading.

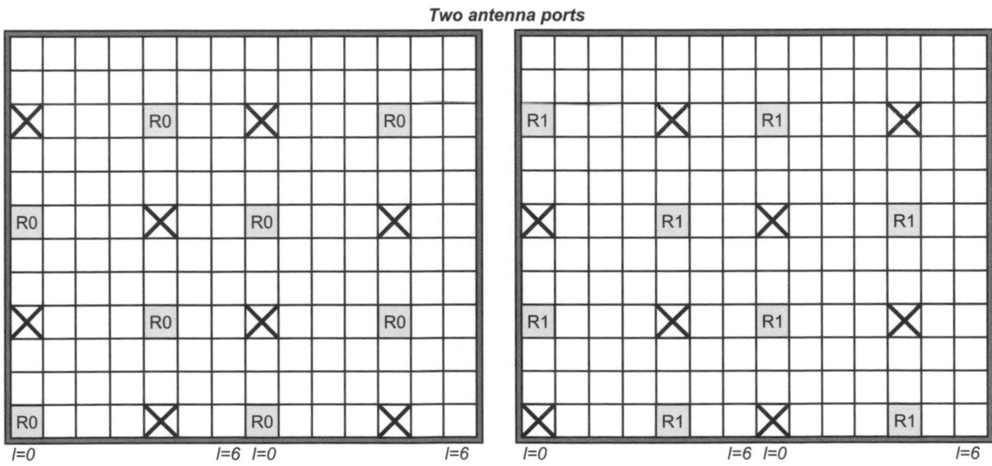

Figure 7.10 Two-port MIMO in LTE. The cross indicates the resource elements that are not used in the respective antenna port.

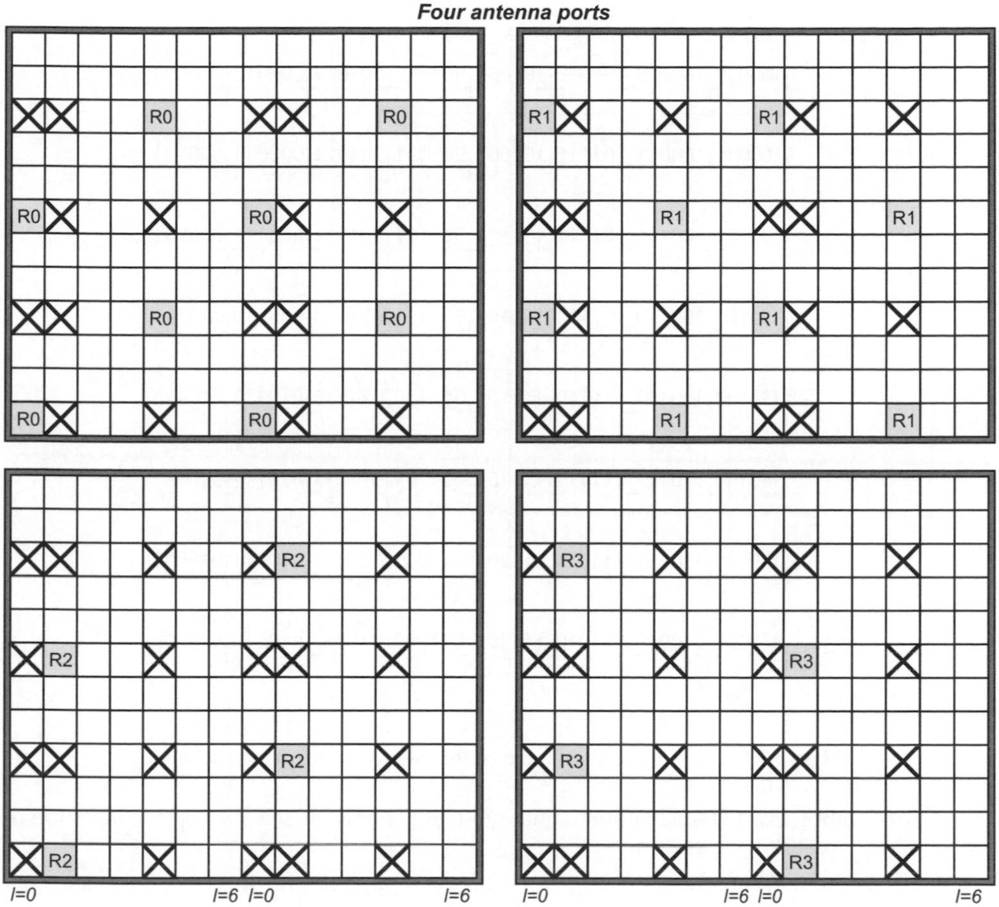

Four antenna ports

l=0 l=6 l=0 l=6 l=0 l=6 l=0 l=6

Figure 7.11 Four antenna port setup.

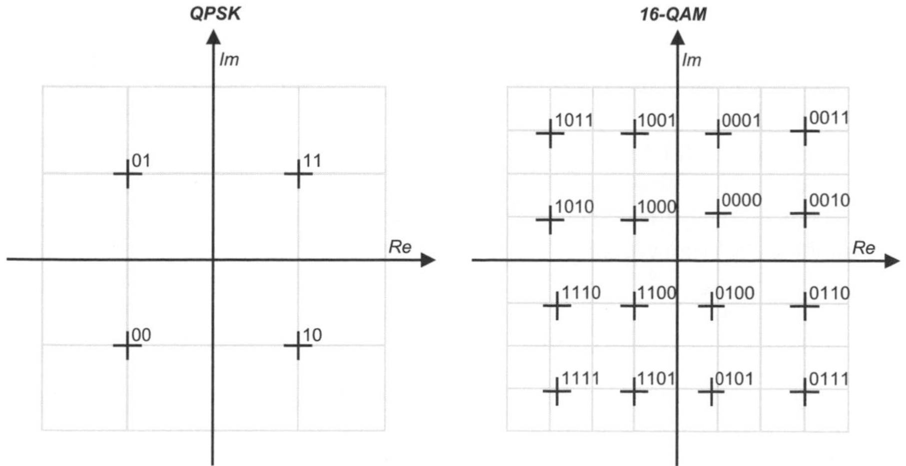

Figure 7.12 The I/Q constellation of the QPSK and 16-QAM modulation schemes.

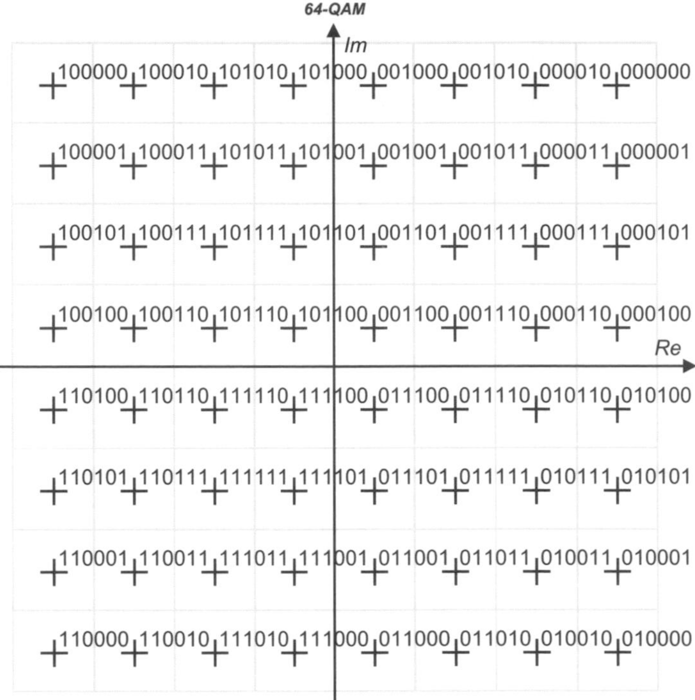

Figure 7.13 The I/Q constellation of the 64-QAM modulation scheme.

The pilot-symbol-assisted modulation on flat fading channels involves the sparse insertion of known pilot symbols in a stream of data symbols.

QPSK modulation provides the largest coverage areas but with the lowest capacity per bandwidth. 64-QAM results in a smaller coverage but it offers more capacity.

7.4.6 Coding

LTE uses turbo coding or convolutional coding, the former being more modern providing, in general, about a 3 dB gain over the older and less effective, but at the same time more robust, convolutional coding.

The creation of the OFDM signal is based on the Inverse Fast Fourier Transform (IFFT), which is the practical version of the Discrete Fourier Transform (DTF) and relatively easy to deploy as there are standard components for the transform calculation. The reception uses the FFT to combine the original signal.

7.4.7 Signal Processing Chain

After the coding and modulation of the user data, its OFDM signal is formed by applying serial-to-parallel conversion. This is an essential step in order to feed the IFFT process. Before bringing the parallel subcarriers of the user data, the subcarrier mapping also takes the required amount of parallel subcarriers from the other users—that is, ODFMA is applied. All these streams are fed into the IFFT input in order to carry out the Inversed Discrete Fourier Transform in a practical way. It is important to note that the process from the serial symbol stream to S/P conversion, the subcarrier mapping process, and the N-point IFFT process happen in the frequency domain, whereas the process from the IFFT conversion happens in the time domain.

The OFDM symbols are formed by adding the cyclic prefix at the beginning of the symbols in order to protect the signal against multipath propagated components. Then, the windowing, digital-to-analog conversion, frequency up-conversion, RF processing and finally the actual radio transmission are performed in the transmitter of the eNodeB. The OFDM transmission is only used in the downlink, so the LTE-UE does have the OFDM receiver and the SC-FDMA transmitter.

7.5 SC-FDM and SC-FDMA

Single-carrier frequency-division multiplexing (SC-FDM), sometimes referred to as discrete Fourier transform (DFT)-spread OFDM, is a modulation technique that, as its name indicates, shares the same principles as OFDM. The same benefits in terms of multipath mitigation and low-complexity equalization are therefore achievable [13].

The difference, though, is that a discrete Fourier transform (DFT) is performed prior to the IFFT operation at the transmitter side, which spreads the data symbols over all the subcarriers carrying information and produces a virtual single-carrier structure. Figure 7.14 shows the principle of the SC-FDMA transmission.

As a consequence, SC-FDM presents a lower peak-to-average-power ratio (PAPR) than OFDM [14]. This property makes SC-FDM attractive for uplink transmissions, as the UE benefits in terms of transmitted power efficiency.

Furthermore, DFT spreading allows the frequency selectivity of the channel to be exploited because all symbols are transmitted in all the subcarriers. Therefore, if some subcarriers are in deep fade, the information can still be recovered from other subcarriers experiencing better channel conditions. On the other hand, when DFT despreading is performed at the receiver, the noise is spread over all the subcarriers and generates an effect called noise enhancement, which degrades the SC-FDM performance and requires the use of a more complex equalization based on a minimum mean square error (MMSE) receiver [13].

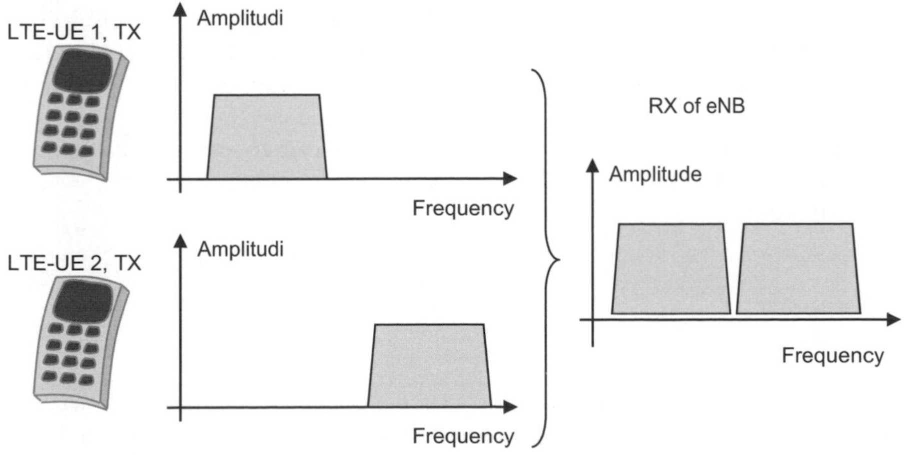

Figure 7.14 The principle of the SC-FDMA transmission.

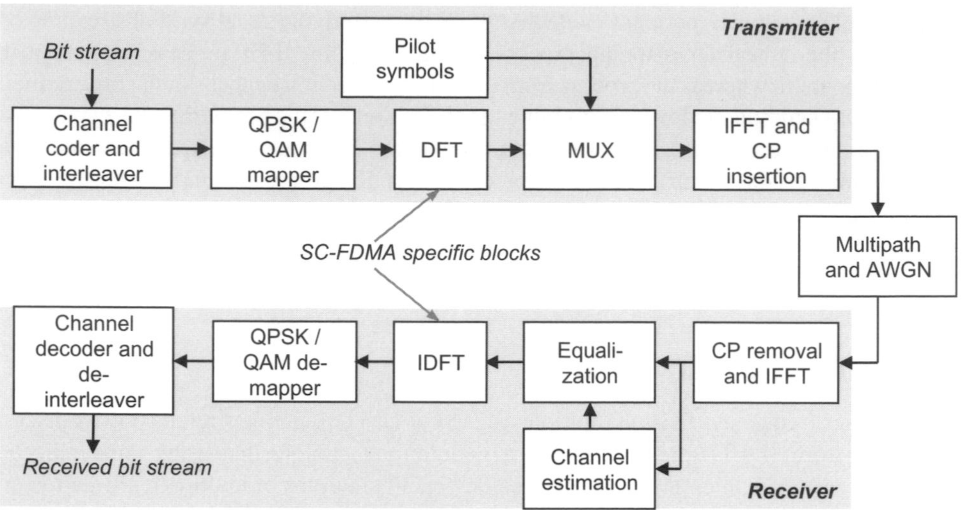

Figure 7.15 SISO SC-FDM simplified block diagram.

7.5.1 SC-FDM Transceiver Chain

Figure 7.15 presents the block diagram of a SISO SC-FDM system. It can be seen that the main difference compared to the OFDM diagram in Figure 7.5 is the FFT/IFFT block, which spreads the data symbols over all the subcarriers prior to the IFFT operation. The rest of the blocks remain the same as in OFDM.

7.5.2 PAPR Benefits

As mentioned before, OFDM shows large envelope variations in the transmitted signal. The different subcarriers carrying parallel data could constructively add in phase leading to instantaneous peak power compared to the average power level. Signals with a high PAPR require highly linear power amplifiers to avoid excessive intermodulation distortion. The power amplifiers therefore have to be operated with a large backoff from their peak value. This eventually translates into low power efficiency. This effect is particularly critical for uplink transmissions on the UE side.

As SC-FDM is spreading the data symbols over all the subcarriers, then an averaging effect is achieved and thus transmission peaks are diminished, resulting in a lower PAPR (see Figure 7.16) [15].

7.6 Reporting

7.6.1 CSI

In LTE, the LTE-UE reports to the network via UE Channel State Information (CSI). Some of the key feedback types in LTE are CQI, RI and PMI. The CSI feedback is meant to deliver information for the eNodeB about the DL channel state. This allows the eNodeB to decide the scheduling. The principle of the channel feedback of LTE is quite similar to the WCDMA/

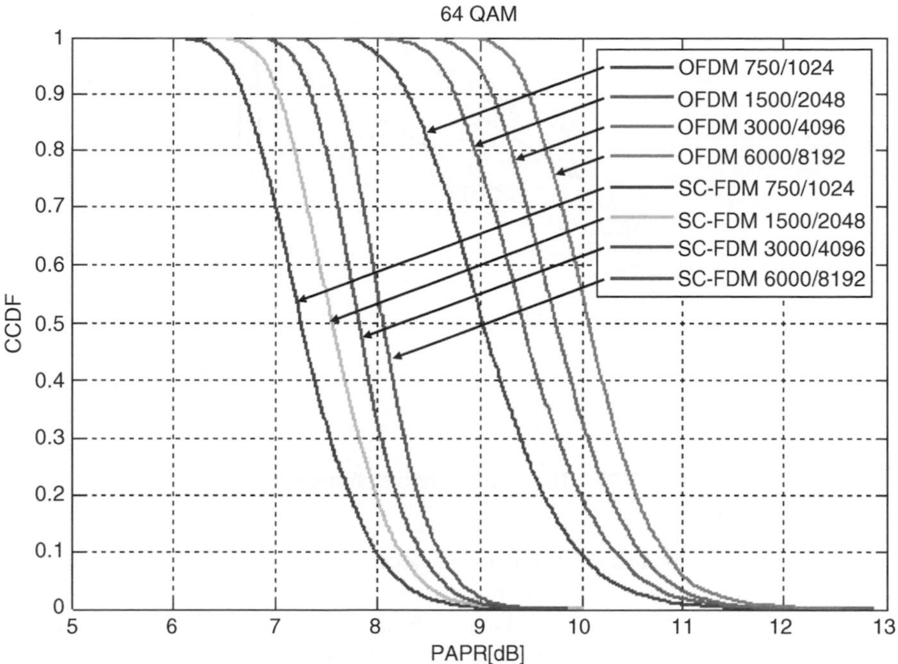

Figure 7.16 PAPR for 64QAM modulation and different OFDM/SC-FDM bandwidths.

HSPA, the most important difference being the frequency selectivity in the case of LTE reporting

LTE-UE measures the CSI during the call, and sends it for eNodeB via PUCCH or PUSCH channel depending on the situation. The three types of channel state information are the following: CQI (Channel Quality Indicator), RI (Rank Indicator) and PMI (Precoding Matrix Indicator). It should be noted, though, that the CSI that LTE-UE sends to the eNodeB is meant as general information for decision making. The eNodeB is not obligated to follow it.

In the uplink direction, there is a procedure called channel sounding that delivers information about the UL channel state. The information is carried with a Sounding Reference Symbols (SRS).

Figure 7.17 shows the principle of the method.

7.6.2 CQI

The most intuitive channel feedback is the CQI. This contains 16 levels (0–15), from which level 0 is out of the range. The CQI value—that is, the index—indicates the modulation and coding scheme (MCS) used at the time as indicated in Table 7.3. During the LTE data call, LTE-UE reports to eNodeB the highest CQI index corresponding to the MCS for which the transport block BLER does not exceed 10%. This, in turn, can be interpreted directly as the quality of the connection at a given time. The CQI value can vary as fast as the TTI interval. In practice, when measuring the CQI values—for example, via the radio field test equipment—the statistics might be shown in such a way that during the selected period, for example, 1 second, the statistics

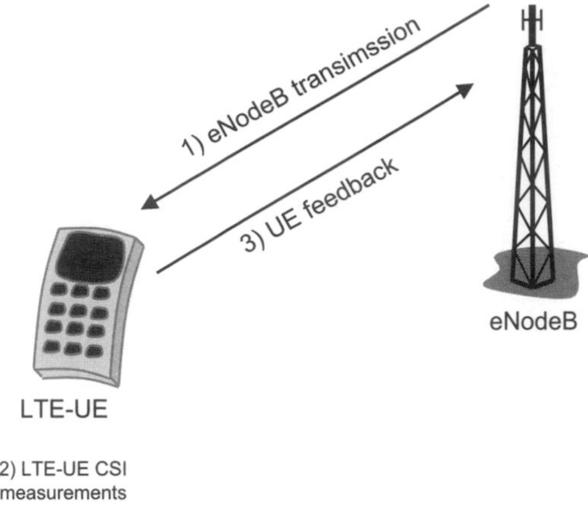

Figure 7.17 The principle of the UE measurements.

show all the CQI values that occurred and their respective percentages. This information can be further post-processed in order to create histograms over the investigated area.

LTE-UE always has a minimum of 2.33 ms for the processing of the CQI measurement. This is due to the synchronization of the downlink and the uplink in such a way that the CQI report transmitted in the uplink subframe $n + 4$ corresponds to the reference period of the downlink subframe n for FDD. Figure 7.18 clarifies the synchronization of the reporting.

7.6.3 RI

The Rank Indicator (RI) is used as a reporting method when LTE-UE is operating in MIMO modes with spatial multiplexing. It is not used for a single antenna operation or TX diversity.

Table 7.3 The CQI values

Index	Modulation scheme	Code rate (x 1024)	Efficiency
1	QPSK	78	0.15
2	QPSK	120	0.23
3	QPSK	193	0.38
4	QPSK	308	0.60
5	QPSK	449	0.88
6	QPSK	602	1.2
7	16-QAM	378	1.5
8	16-QAM	490	1.9
9	16-QAM	616	2.4
10	64-QAM	466	2.7
11	64-QAM	567	3.3
12	64-QAM	666	3.9
13	64-QAM	772	4.5
14	64-QAM	873	5.1
15	64-QAM	948	5.6

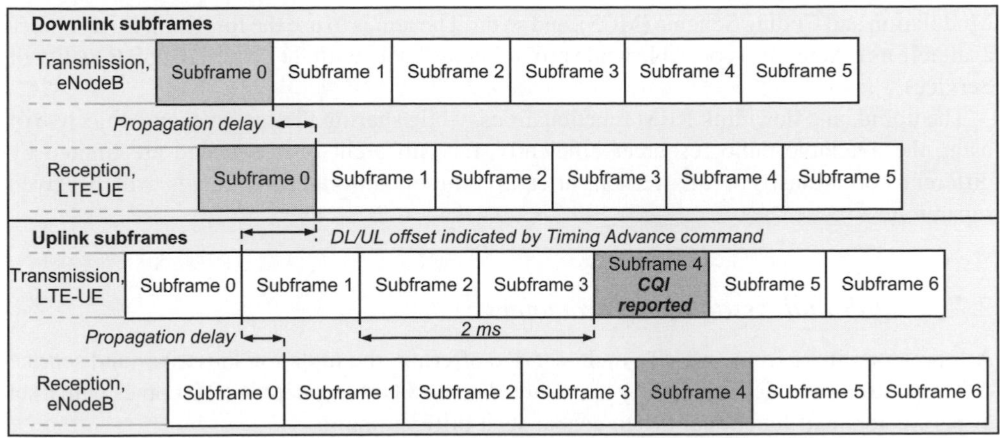

Figure 7.18 The synchronization LTE-UE reporting.

The RI is, in reality, a recommendation of LTE-UE for the number of layers to be used in spatial multiplexing. The RI can have a value of 1 or 2 in the case of 2-by-2 antenna configuration, and value of 1, 2, 3, or 4 in the case of a 4-by-4 antenna configuration.The RI is always associated with one or more CQI reports

7.6.4 PMI

The Pre-coding Matrix Indicator (PMI) based on Table 7.4, gives information about the preferred Precoding Matrix. It should be noted, though, that, as in the RI, the PMI is relevant only when the MIMO operation is active. The MIMO operation combined with the PMI feedback forms a closed-loop MIMO.

7.7 LTE Radio Resource Management

7.7.1 Introduction

The term Radio Resource Management (RRM) generally refers to the set of strategies and algorithms used to control parameters like transmit power, bandwidth allocation, the

Table 7.4 The MIMO pre-coding matrix indicator table (PMI)

Codebook	One layer	Two layers
0	$\frac{1}{\sqrt{2}}\begin{bmatrix}1\\1\end{bmatrix}$	N/A
1	$\frac{1}{\sqrt{2}}\begin{bmatrix}1\\-1\end{bmatrix}$	$\frac{1}{2}\begin{bmatrix}1 & 1\\1 & -1\end{bmatrix}$
2	$\frac{1}{\sqrt{2}}\begin{bmatrix}1\\j\end{bmatrix}$	$\frac{1}{2}\begin{bmatrix}1 & 1\\1 & -j\end{bmatrix}$
3	$\frac{1}{\sqrt{2}}\begin{bmatrix}1\\-j\end{bmatrix}$	N/A

Modulation and Coding Scheme (MCS), and so on. The aim is to use the limited radio resources available as efficiently as possible while providing the users with the required QoS (Quality of Service).

The uplink and downlink RRM functionalities, while sharing the same general objective of using the available radio resources efficiently, face different problems and are limited by different conditions. For this reason, after a common introduction, details will be given separately.

7.7.2 QoS and Associated Parameters

As operators move from single to multi-service offering, the tools for subscriber and service differentiation become increasingly important. The EPS QoS concept in LTE comes with a set of parameters and functionalities to enable such differentiation.

The lowest level for QoS control in LTE is represented by the bearer. A bearer uniquely identifies a set of packet flows receiving a common forwarding treatment in the nodes encountered from the terminal to the gateway. A packet flow is uniquely identified by the 5-tuple: source IP address and port number, destination IP address and port number, protocol ID.

Bearers can be classified as GBR or non-GBR and as default or dedicated. Table 7.5 shows some examples of bearers and their classification. It is worth noting that a dedicated bearer can be GBR or non-GBR whereas a default bearer can only be non-GBR.

There exists one default bearer per terminal IP address. The default bearer is set up when the terminal attaches to the network and a serving GW is selected for it. Dedicated bearers are required to provide a different QoS to different flows belonging to the same IP address of a terminal.

GBR bearers require transmission resources to be reserved when the user is admitted by an admission control function. Such bearers are chosen, based on operator polices, for services for which it is preferred to block a service request rather than degrading the performance of a service request that has already been admitted. Non-GBR bearers, instead, may experience congestion-related packet loss, which occurs where there are resource limitations.

Each EPS bearer (GBR and non-GBR) is associated with the following bearer-level QoS parameters, signaled from the Access Gateway (aGW) (where they are generated) to the eNode-B (where they are used):

- **Quality Class Identifier (QCI)**: a scalar that is used as a reference to access node-specific parameters that control bearer-level packet- forwarding treatment (e.g., bearer priority, packet delay budget and packet loss rate), and that have been preconfigured by the operator owning the eNode-B. A one-to-one mapping of standardized QCI values to standardized characteristics is captured in [16].

Table 7.5 Bearers classification

GBR type	Default bearers	Dedicated bearers
Non-GBR bearers	Bearer setup at terminal attachment	For example, Internet browsing, chat, e-mail
GBR bearers	N/A	For example, VoIP, streaming

Table 7.6 QCI mapping table and typical services

QCI	Resource Type	Priority	L2 packet delay budget	L2 packet loss rate	Example services
1		2	100 ms	10-2	Conversational Voice
2	GBR	4	150 ms	10-3	Conversational Video (Live Streaming)
3		3	50 ms	10-3	Real Time Gaming
4		5	300 ms	10-6	Non-Conversational Video (Buffered Streaming)
5		1	100 ms	10-6	IMS Signaling
6		6	300 ms	10-6	Video (Buffered Streaming) TCP-based (e.g., www, e-mail, chat, ftp, p2p file sharing, etc.)
7	Non-GBR	7	100 ms	10-3	Voice, Video (Live Streaming) Interactive Gaming
8		8	300 ms	10-6	Video (Buffered Streaming) TCP-based (e.g., www, e-mail, chat, ftp, p2p file sharing, etc.)
9		9			

- **Allocation Retention Priority (ARP)**: the primary purpose of ARP is to decide whether a bearer establishment/modification request can be accepted or needs to be rejected if there are resource limitations. In addition, the ARP can be used by the eNode-B to decide which bearer (s) to drop during exceptional resource limitations (e.g., at handover).

Additionally, for GBR bearers, the maximum bit rate (MBR) and GBR are defined. These parameters define the MBR—that is, the bit rate that the traffic on the bearer may not exceed—and the GBR—that is, the bit rate that the network guarantees (e.g., via admission control) it can sustain for that bearer. There also exists an aggregate MBR (AMBR), which sets a limit on the maximum bit rate that can be consumed by a group of non-GBR bearers belonging to the same user.

3GPP specifications [16], [17] define a mapping table for nine different QCIs, as shown in Table 7.6.

7.8 RRM Principles and Algorithms Common to UL and DL

7.8.1 Connection Mobility Control

Connection mobility control is concerned with the management of radio resources in connection with idle (RRC_IDLE) or connected (RRC_CONNECTED) mode mobility. In idle mode, the cell reselection algorithms are controlled by setting parameters (thresholds and hysteresis values) that define the best cell and/or determine when the UE should select a new cell. Further, LTE broadcasts parameters that configure the UE measurement and reporting procedures. In connected mode, the mobility of radio connections has to be supported. Handover decisions may be based on UE and eNode-B measurements. Such decisions may also take other inputs, such as neighbor cell load, traffic distribution, transport and hardware resources, and operator-defined policies into account [18]. Connection mobility control is located at L3 in the eNode-B.

7.8.1.1 Handover

The intra-LTE handover in the RRC_CONNECTED state is UE assisted and network controlled. One of the goals of LTE is to provide seamless access to voice and multimedia services with strict delay requirements, which is achieved by supporting a handover from one cell – the source cell—to another – the target cell. The decentralized system architecture of LTE facilitates the use of a hard handover. A hard handover (break-before-make type) is standardized for LTE, whereas a soft handover (make-before-break type) is not included, which makes the problem of providing seamless access even more critical.

The handover procedure in LTE can be divided into three phases: initialization, preparation, and execution, as shown in Figure 7.19. In the initialization phase, the UE carries out the channel measurements from both source and target eNode-Bs, followed by the processing and reporting of the measured value to the source eNode-B. The channel measurements for

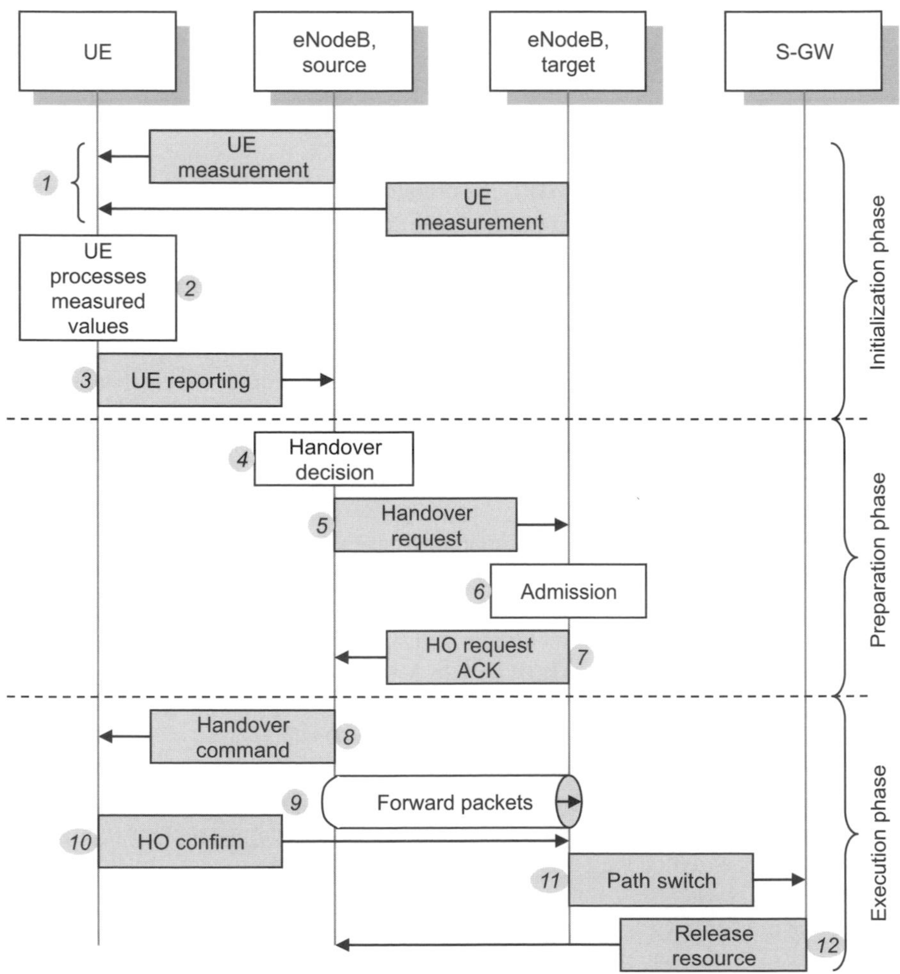

Figure 7.19 Intra-LTE handover procedure [18].

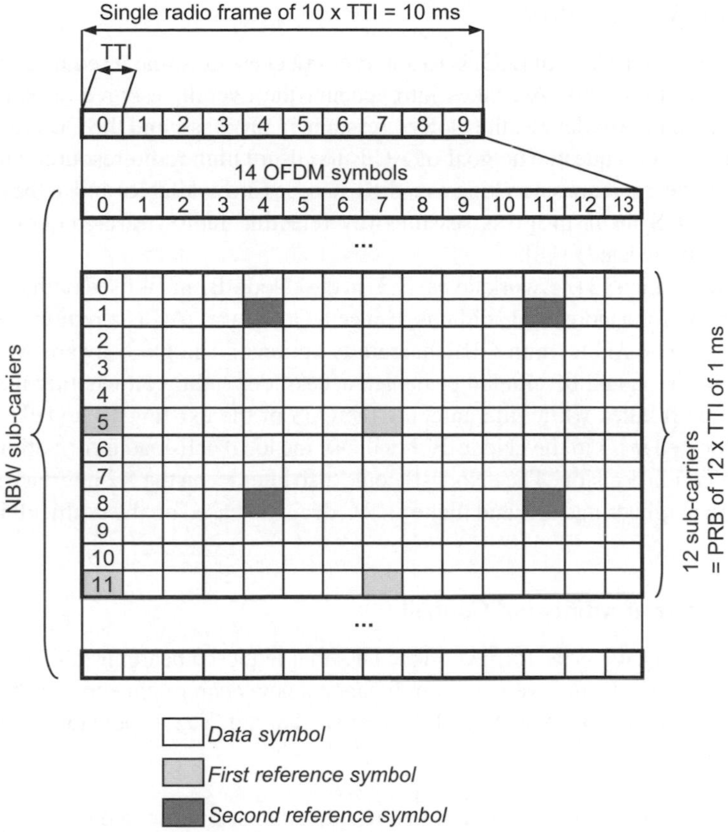

Figure 7.20 Frame structure of E-UTRA FDD containing 14 OFDM symbols per TTI including downlink subcarrier structure with reference signal (pilot) structure for one eNode-B transmit antenna port [3].

handover are done at the downlink and/or uplink reference symbols (pilots). The downlink reference symbols structure in an E-UTRA FDD frame is illustrated in Figure 7.20.

In the preparation phase the source eNode-B makes a handover decision, and it requests a handover with the target eNode-B. The Admission Control (AC) unit in the target eNode-B makes the decision to admit or reject the user, which is sent to the source eNode-B using a handover request ACK or NACK. Finally, in the execution phase, the source eNode-B generates the handover command towards the UE. The source eNode-B then forwards the packet to the target eNode-B. After this, the UE performs synchronization with the target eNode-B and accesses the target cell via the Random Access Channel (RACH). When the UE has successfully accessed the target cell, the UE sends the handover confirmation message to the target eNode-B to indicate that the handover procedure is complete. The target eNode-B sends a path-switch message to the aGW to inform it that the UE has changed the cell, followed by a release-resource message. The source eNode-B is informed of the success of handover. After receiving the release-resource message the source eNode-B releases radio as well as user-plane- and control-plane-related resources associated with the UE context [18].

7.8.2 Admission Control

The task of Admission Control (AC) is to admit or reject establishment requests for new radio bearers. In order to do this, AC takes into account the overall resource situation, the QoS requirements, the priority levels, the QoS of sessions in progress, and the QoS requirement of the new radio bearer request. The goal of AC is to ensure high radio resource utilization (by accepting radio bearer requests as long as radio resources are available) and at the same time to ensure proper QoS for in-progress sessions (by rejecting radio bearer requests when they cannot be accommodated) [18].

AC is located at Layer 3 (network layer) (L3) in the eNode-B, and is used both for the setup of a new bearer and for handover candidates. Hence a QoS-aware AC is a requirement for GBR bearers in LTE. The AC for non-GBR bearers is optional. The QoS- aware AC determines whether a new UE should be granted or denied access, depending on whether the QoS of the new UE will be fulfilled while guaranteeing the QoS of the existing UEs [19].

As AC is located in L3 in the eNode-B, it will use the local cell-load information to make an admission/rejection decision. The eNode-B could also interact on the X2 interface sharing load information in neighboring cells and make an AC decision based on the multicell information.

7.8.2.1 The Role of Admission Control

Admission Control (AC) is performed on the UEs that request a bearer establishment with the eNodeB. This occurs at handovers, or simply when a new bearer connection is being created. The AC decides whether the bearer can be established or not. The responsibility of the AC can be seen as twofold:

- to ensure that the eNodeB has enough free resourced to accommodate the incoming bearer;
- to ensure that the eNodeB will be able to maintain the overall expected level of QoS with the introduction of the new bearer.

In order to evaluate these two conditions, the eNodeB can take into account the QoS parameters of all the UEs and of the incoming bearer. It can also take into account the channel conditions of the connected UEs via their CQIs. However, no CQI is available for the incoming UE as it is not connected yet. Therefore in order to take into account the channel conditions of the incoming user, the eNodeB has to rely mostly on the following layer-3 measurements:

- the RSRP, which indicates the wideband received pilot power;
- the RSSI, which indicates the wideband received power including interference.

7.8.2.2 Examples of Algorithms

In this section, we give an overall list of standard algorithms.

Number of Connections

The simplest AC algorithm is simply to accept up to an arbitrary number of bearers. This method is, of course, overly simple and the main drawback is obviously the complete disregard for the QoS constraints of the users. This type of algorithm could, however, be found in early LTE eNodeBs as the early rollout of LTE is focused on best effort services. Best effort services only concern AC because there should not be so many that only a low throughput can be provided to them.

Fixed Capacity Based

Capacity-based algorithms assume a certain capacity for the system and ensure that the sum of all the GBRs does not exceed the capacity. The main drawback of such algorithms is that they do not take into account the channel conditions of any user. Instead, they simply assume that the cell can accommodate a certain throughput.

Average Required Resource Based

Average required resource based algorithms calculate:

- The average percentage of resource needed for each UE to fulfill their GBR. In order to calculate this value, the CQI can be used as well as the resource allocation history of the UE.
- The expected average percentage of resource needed by the incoming UE to fulfill its GBR.

The sum of these numbers should not exceed the total available resource for the incoming bearer to be admitted.

7.8.3 HARQ

In LTE, both retransmission functionalities Automatic Repeat reQuest (ARQ) and HARQ are provided. ARQ provides error correction by retransmissions in acknowledged mode at the Radio Link Control (RLC) sublayer of Layer 2. HARQ is located in the MAC sublayer of Layer 2 and ensures delivery between peer entities at Layer 1 [20].

If a data packet is not correctly received, the HARQ ensures a fast Layer 1 retransmission from the transmitter (UE). In this way the HARQ provides robustness against LA errors (due, for example, to errors in CSI estimation and reporting) and it improves the reliability of the channel [21], [22], [23].

In case of HARQ retransmission failure, the ARQ in the RLC sublayer can handle further retransmissions using knowledge gained from the HARQ in the MAC sublayer.

7.8.4 Link Adaptation

7.8.4.1 The Role of Link Adaptation

The Link Adaptation (LA) is a fundamental functionality for a radio channel. It is the mechanism that chooses the appropriate Modulation and Coding Scheme (MCS) for a transmission in order to maximize the data transmitted over the channel. In LTE, Link Adaptation is also referred to as fast Adaptive Modulation and Coding (AMC) as the MCS can be changed every TTI (every 1 ms).

In LTE, the Physical Uplink Shared CHannel (PUSCH) supports BPSK, QPSK and 16 QAM at various coding rates, while the Physical Downlink Shared CHannel (PDSCH) supports QPSK, 16QAM and 64QAM with various coding rates.

In order to optimize resource use, AMC usually aims at maintaining a BLock Error Rate (BLER) on the order of 10%, while relying on HARQ to provide a packet error rate significantly smaller than 1% to the RLC sublayer. This relatively high BLER target allows the system to use high MCS thus taking full advantage of the link capacity.

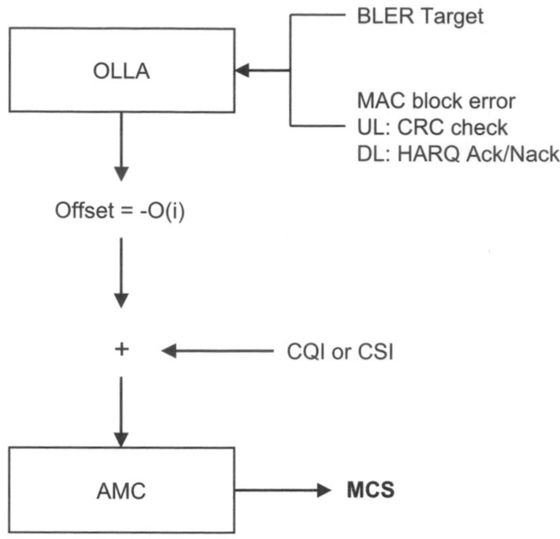

Figure 7.21 Interaction of OLLA and AMC.

7.8.4.2 Outer Loop Link Adaptation

AMC can use various channel state information (CSI in UL and CQI reports in DL) in order to determine the MCS with an appropriate block error probability. However, due to the various possible channel evaluation errors, it is unlikely that the expected block error rate occurs.

In order to keep the BLER at first transmission as close as possible to the target, an OLLA algorithm is needed to offset the channel measurements, as shown in Figure 7.21, for a user *i* and a bandwidth *bw*.

The offset $O(i)$ is adjusted following the same rules of outer loop PC in WCDMA [24]:

- if a first transmission on PUSCH or DSCH is correctly received, $O(i)$ is decreased by $OD = S \cdot BLERT$
- if a first transmission on PUSCH or DSCH is not correctly received, $O(i)$ is increased by $OU = S \cdot (1 - BLERT)$

where S represents the step size and BLERT the BLER to which the algorithm will converge if the offset $O(i)$ remains within a specified range $Omin <= O(i) <= Omax$.

7.8.5 Packet Scheduling

The PS is an entity located in the Medium Access Control (MAC) sublayer, which aims to utilize efficiently the downlink and uplink shared channel resources. The main role of the PS is to multiplex the users in the time and frequency domains. Such multiplexing takes place via mapping of users to the available physical resources. The PS is able to perform mapping of users to the Physical Resource Block (PRB)s on a Transmission Time Interval (TTI) (1 ms) basis and is therefore referred to as fast scheduling.

7.8.5.1 Multi User Diversity

One of the reasons for using fast scheduling in LTE is the possibility of using Multi-User Diversity. Indeed, if the system is affected by time- and frequency-selective fading, the PS entity can exploit multi-user diversity by allocating the users to the portions of the bandwidth that exhibit favorable channel conditions. In this way, radio channel fading, which used to be a limitation to the performance of wireless systems, is turned into an advantage.

7.8.5.2 The Packet Scheduling Problematic in Short

The packet scheduler can operate using the following information:

- inputs from link adaptation regarding achievable MCS;
- the Quality of Service (QoS) parameters associated with each UE.

The QoS parameters of a UE explicitly constrain the eNodeB with regard to how much throughput should be delivered. On the other hand, the channel quality reports of a UE indicate to the eNode-B how resources should be allocated in order to provide a certain throughput.

The packet scheduler should allocate resources with the aim of satisfying as well as possible the QoS constraints of the UEs. Meanwhile it should also maximize the throughput available for best effort services (services with no specific QoS constraints). In reality, the scheduling decisions need to take into account a large set of factors including payloads buffered in the UE, HARQ retransmissions, UE sleep cycles, and so on.

7.8.5.3 The Role of the PDCCH

In LTE downlink, one of the physical resources division units is the Time Transmission Intervals (TTI). From a time-domain perspective, a TTI consists of 11 or 14 OFDM symbols (depending on the cyclic prefix settings), which corresponds to 1 ms. The first OFDM symbols of the TTI (up to three) consist of the Physical Downlink Control CHannel (PDCCH), while the remaining OFDM symbols can be used for data channels. Every TTI, the PDCCH is read by all connected UEs. It contains both UL and DL allocation information: sets of allocated PRBs, MCS, and antenna diversity modes.

Each UE will in turn read the DL TTI and send data over the UL TTI accordingly. Complete and detailed information on the procedure for reading and decoding the PDCCH can be found in the 3GPP specification [25]. The PDCCH is designed to be a robust, low-error channel. As a pure signaling channel, it is also limited in capacity, thus limiting the number of users that can be scheduled per TTI, both in the DL and the UL.

7.8.5.4 Some Packet Schedulers

It is important to note that the packet scheduler is not standardized by 3GPP. The main consequences are that:

- Different eNodeB products are very likely to use different packet schedulers, which can be tuned and set in very different ways, using very different parameters.

- Packet scheduling is a way for an eNodeB manufacturer to differentiate itself from its competitors. Therefore, Packet scheduling is an extensive source of research.

This section presents important principles for packet scheduling in LTE based on the current state of research. The aim of the present section is to provide a deeper understanding of the packet-scheduling problematic. Those principles generally apply to both the UL and the DL. The specificities and extra constraints of packet scheduling in DL and UL are detailed in later chapters.

PDCCH and Complexity Limitations

For this reason, it is proposed that in case of multiple users, a decoupled time/frequency domain packet scheduling structure would be adopted where [1], [3]:

- in the first step, the time domain scheduler selects a certain number of UEs to be scheduled out of all the connected UEs;
- in the second step, the frequency domain scheduler determines which PRBs will be allocated to which UE.

Figure 7.22 clarifies this issue. Another important limitation of the packet scheduler is that it operates on a per TTI basis: every 1 ms. In such a short time, the computing capacity of even an eNodeB is limited so simplicity must be preferred for a packet scheduler. For example, complex iterative algorithms may be prohibitively demanding. In order to cope with the complexity issues, the literature is mainly centered around proposing simple metric-based algorithms, as in [1], [3]. Metric-based algorithms consist simply of:

- for the time domain packet scheduler: selecting the N_TD with the highest metrics;
- for the frequency domain scheduler: allocating the UE with the highest metric of each PRB.

Diversity Algorithms

One of the key features of packet scheduling in LTE is that it enables multiuser diversity gain. Usually, fading caused by multipath propagation is seen as a drawback, but the LTE flexible resource-allocation capabilities turn multipath propagation fading into a potential gain in a very simple way. Indeed, multiuser diversity gain consists simply of scheduling only users at an SINR peak for each PRB. The following frequency domain packet schedulers are example of diversity-based packet schedulers:

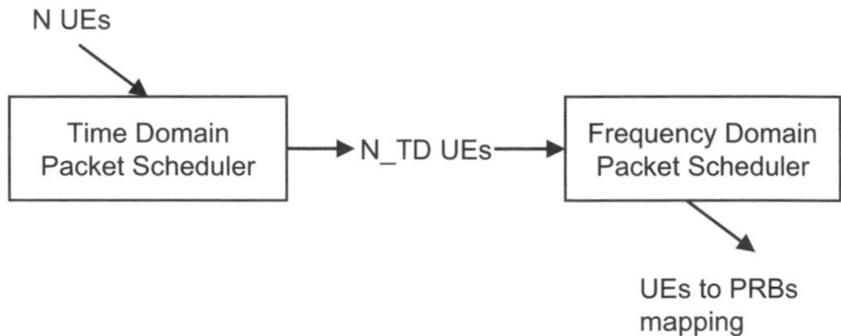

Figure 7.22 Split time and frequency packet scheduling.

- The Maximum Throughput algorithm [3] schedules, on each PRB, the UE with the highest CQI, thus maximizing the overall received throughput. This scheduling strategy is very unfair as it prioritizes mainly the UEs that are close to the base station.
- The Proportional Fair algorithm [3] schedules, on each PRB, the UE with the highest ratio: CQI to average CQI. Assuming that all the UEs undergo the same fading statistics, the Proportional Fair algorithm schedules, on average, the same amount of resource to each UE. This scheduling strategy is interesting as it introduces the notion of fairness (each user gets the same amount of resource), while each UE gets the best of its channel conditions. It has been shown that the PF algorithm provides a system throughput gain of up to 40% depending on the transmission diversity scheme used [3], [4].

GBR Aware Algorithms

Simple diversity algorithms may be appropriate mechanisms to cope with best effort traffic, which, by definition, does not require any specific delivery quality. However, QoS users require more sophisticated scheduling to comply with their constraints.

Still within the decoupled frequency/time domain scheduling framework, the algorithms proposed in the literature generally focus on:

- enforcing the GBR through the time domain scheduler;
- provide multiuser diversity benefits—for example, with a scheduler like proportional fair, via the frequency domain packet scheduler.

Following that method, a lot of packet schedulers inherited from time domain division multiplexing systems can simply be reapplied to LTE. An example of a time-domain packet scheduler that enforces GBR constraints is barrier function-based packet scheduling [6], which consists in prioritizing:

- First the GBR UEs which don't comply with their GBR constraints. Within that set, the UEs which are furthest from their GBR target have higher priority.
- Then the GBR UEs that comply with their GBR constraint and the best effort users. Within that set, a time domain proportional fair metric is used.

This overall GBR control principle is further refined in [5] where two principles are shown. Firstly, in order to provide an appropriate control to the time domain scheduler, the frequency domain scheduler should be independent from the time domain scheduler in the sense that its average behavior should not be influenced by the nature of the time domain scheduler. Secondly, it has been shown that the time domain scheduler cannot necessary ensure the GBR if the bit rate is too high. In order to prevent this, it is necessary to include a throughput control functionality in the frequency domain scheduler.

Delay Aware Algorithms

Many applications like video, sound streaming or VoIP may have a tight delay budget in their QoS constraints. Video and sound-streaming UEs, can deliver their requested QoS using a barrier function based scheduling in the very same way than for GBR UEs. It consists simply of prioritizing, with the time domain scheduler, the UEs that are getting close to their delay budget compared to other users.

In practice, one of the challenges with delay budget UEs is that the packet scheduler must ensure the transmission of the full amount of data that is close to expiry before expiry. For example in a traffic type with a large packet, if a large packet is getting close to expiry, the packet scheduling must ensure that the full packet will have been delivered in time. For this reason, the time domain scheduler must be aware of the packet size and predict how much in advance a UE must be handled to deliver the full packet in time.

VoIP UEs present some further challenges associated with their typically small packet size and very tight delay constraints (typically 50 ms). With VoIP, small packets need to be scheduled very often, which tends to overload the PDCCH when too many VoIP users are present in the cell at the same time. If VoIP UEs are overprioritized, this can result in the shared channel being underused by lack of PDCCH resources.

In order to avoid PDCCH overloading, it is necessary, as described in [7], to enforce VoIP packet bundling. This simply consists of scheduling several consecutive VoIP packets from one user in a single TTI. From the time-domain packet scheduler perspective it simply means that only UEs with several packets in their buffer should be scheduled. The consequence is that VoIP users are scheduled less often but with more data at a time, thus releasing PDCCH resources for other UEs, which enables better use of the shared channel.

Persistent and Semi-Persistent Scheduling: A Little Help for VoIP

Another way to cope with VoIP traffic is to use persistent or semi-persistent scheduling. Those two modes allow (with various degrees of flexibility) the eNodeB to schedule VoIP UEs on predetermined PRBs in predetermined TTIs without having to specify it in the PDCCH every time. The main advantage of these modes is that they save PDCCH resources and therefore allow other traffic types to use the shared channel more efficiently. However, the downside is that it cannot benefit from diversity gain.

7.8.6 Load Balancing

Load balancing (or load control) has the task of handling uneven distributions of the traffic load over multiple cells. The purpose of load balancing is thus to influence the load distribution in such a manner that radio resources remain highly utilized and the QoS of in-progress sessions is maintained as far as possible and the probability of call dropping is kept sufficiently small. Load balancing algorithms may result in the handover or cell reselection decisions with the purpose of redistributing traffic from highly loaded cells to underutilized cells. Load-balancing functionality is located in the eNode-B.

Figure 7.23 shows that the AC, handover, and load control are closely coupled RRM functionalities. Handover is made when an active user in the source cell could be best served in the target cell. The AC, with feedback from the load-control functionality, decides whether an incoming call (new or handover call) should be accepted or blocked. The AC then informs the load control about the change in load conditions due to the admission of a new or handover call. If an incoming call cannot be served in the originating cell, and if the call can be served by an adjacent cell, the call is immediately handed over to the adjacent cell. This is called directed retry – a well known concept used in Global System for Mobile Communication (GSM) [26], and could potentially be used for LTE as well.

The load control keeps track of the load condition in a cell and if it is overloaded it drops a Best Effort (BE) call to maintain the QoS of the active calls in the cell. One way to reduce the probability of call dropping is to make a handover to an adjacent cell if this call could be served in the adjacent cell with the required QoS. This is called load-based handover. Beside call dropping or handover of lower priority calls, the QoS of lower priority calls can be degraded to

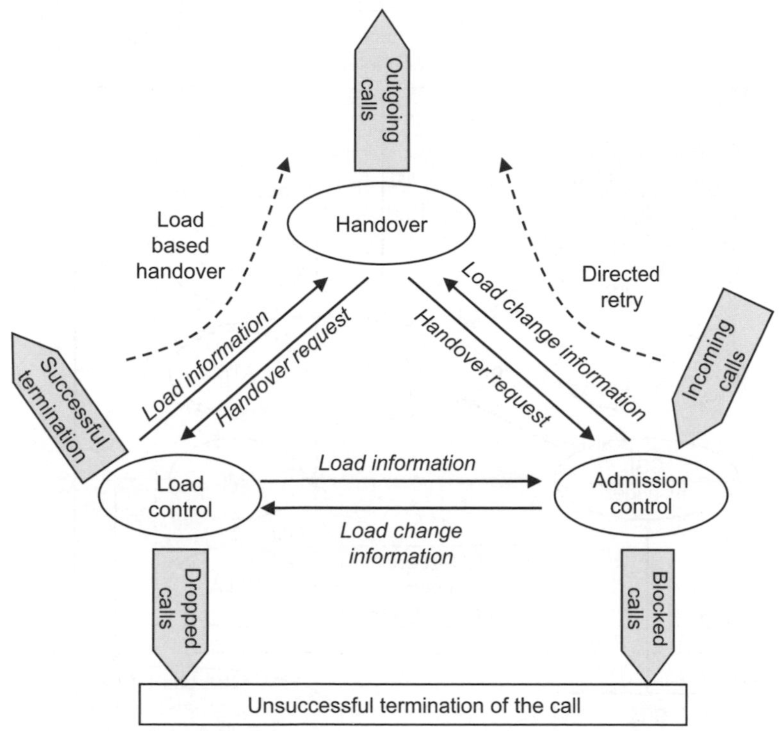

Figure 7.23 Interaction between handover, load control, and admission control.

free resources. This is especially useful in situations where no appropriate adjacent cell is available and therefore call drops can be avoided at the expense of degraded quality of lower priority calls.

7.9 Uplink RRM

An overview of the functionalities of UL RRM, their interaction and location in the protocol stack is shown in Figure 7.24.

After a description of some uplink-specific scheduling and link-adaptation aspects, a description of the signaling mechanisms needed to support them will be given.

7.9.1 Packet Scheduling: Specific UL Constraints

7.9.1.1 PRB Contiguity Constraint

In the uplink the complexity of packet scheduling mainly arises from the fact that the PRBs allocated to the same user have to be adjacent in frequency. This constraint is connected to the physical access transmission multiplexing mode used in LTE UL: SC-FDMA. The PRB contiguity constraint greatly limits the flexibility of the scheduling and therefore negatively affects the multi-user frequency and diversity gain that can be derived from it. On the other hand, this puts less stress on the PDCCH as the PDCCH UL allocation field is shorter: it only

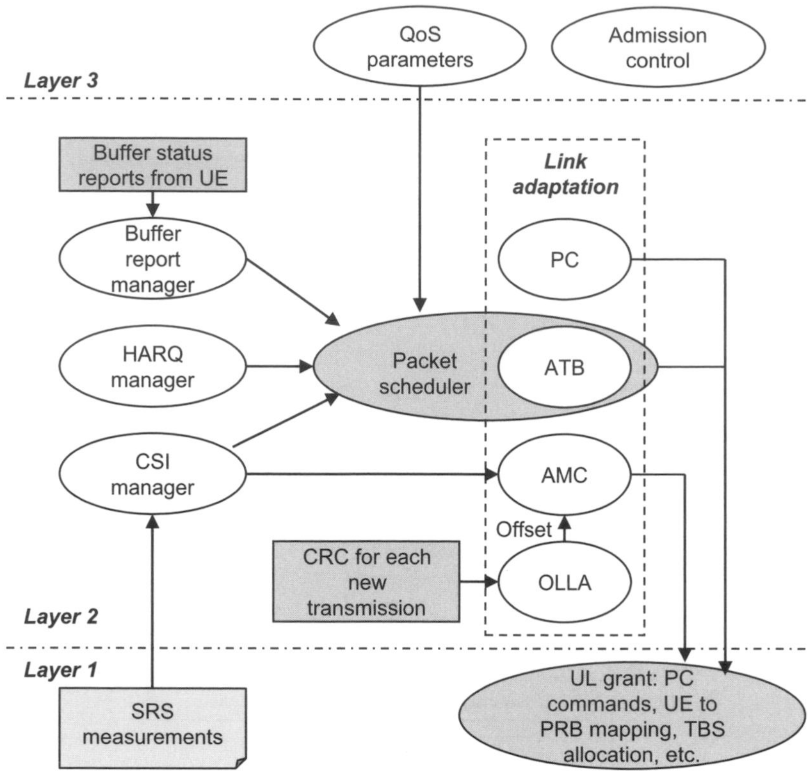

Figure 7.24 Interaction between RRM functionalities with focus on scheduling and adaptation.

consists of the first allocated PRB (PRBs are sorted by ascending frequency) and the number of PRBs allocated.

7.9.2 Link Adaptation

In thc following, the mechanisms that control the adaptation of the main transmitting parameters, that is MCS, bandwidth and power, are described.

7.9.2.1 Adaptive Modulation and Coding

It is well known that Adaptive Modulation and Coding (AMC) can significantly improve the spectral efficiency of a wireless system [27]. The MCS selection algorithm is based on mapping tables that return an MCS format (and hence a Transport Block Size (TBS)) after having received an SINR value and, optionally, the BLock Error Rate (BLER) target at first transmission as input. In the LTE uplink the supported data modulation schemes are QPSK, 16-QAM and 64-Quadrature Amplitude Modulation (64-QAM) [28].

The expected instantaneous throughput per TTI for a given MCS and SINR can be defined as:

$$T(MCS, SINR) = TBS(MCS) \cdot (1 - BLEP(MCS, SINR)) \tag{7.6}$$

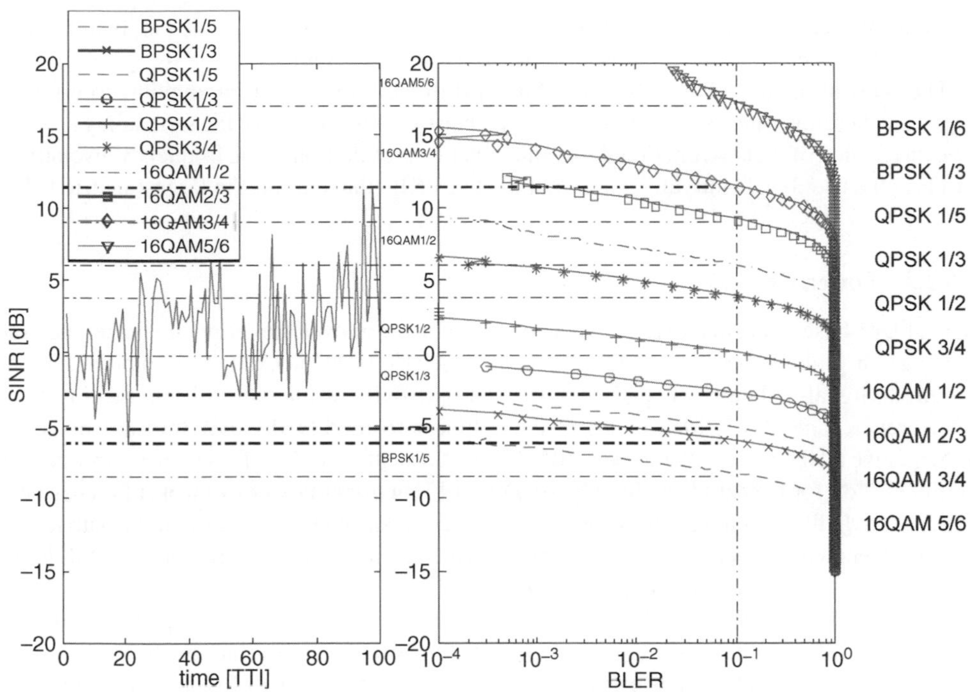

Figure 7.25 AMC mechanism: MCS selection based on estimated SINR.

where the Block Error Probability (BLEP) represents the probability that the transmitted block is going to be in error.

A possible algorithm for the selection of the MCS consists in selecting the MCS that maximizes the throughput under the constraint that the estimated BLEP is smaller than or equal to the BLER target at first transmission.

The AMC can be performed on a slow basis, for example with the same rate of the power control commands to exploit the slowly changing channel variations, or on a faster basis—for example every TTI—to exploit the high instantaneous SNR conditions. A detailed performance analysis of the AMC functionality is carried out in [29] where the fast AMC is shown to exhibit a gain above 20% compared to the slow AMC, as indicated in Figure 7.25.

7.9.2.2 Adaptive Transmission Bandwidth

In a SC-FDMA system, which enables bandwidth scalability, the adaptability of the transmission bandwidth represents a fundamental feature given the variety of services that an LTE system is called to provide. The ATB, therefore, becomes a necessary technique to cope with different traffic types, varying cell load and power limitation in the UE.

Some services, for example, VoIP, require a limited amount of bandwidth while a user with BE type of traffic may receive as much bandwidth as is available as long as there is data in the buffer and power available at the UE. Power limitations also represent a constraint that highlights the importance of the ATB. Due to adverse channel conditions, the PSD that a user is required to transmit with may be so high as to limit the user to a limited bandwidth. A varying cell load also calls for the adaptability of the transmission bandwidth as the bandwidth a user can receive depends on the number of other users in the system.

The ATB ultimately allows the allocation of different portions of bandwidths to different users and therefore offers significant flexibility when exploited as part of the scheduling process. The integration of the two functionalities allows better exploitation of the frequency diversity by limiting the user bandwidth allocation to the set of PRBs that exhibit the largest metric value.

7.9.2.3 Power Control

In a OFDM-based system like LTE, where orthogonality removes the intracell interference and the near-far problem typical of CDMA systems, the role of PC is changed into providing the required Signal-to-Interference-plus-Noise Ratio (SINR) while controlling the inter-cell interference. The classic idea of PC in the uplink is to modify the user transmit power so as to receive all the users with the same SINR at the Base Station (BS). This idea is known as full compensation for the path loss. In 3GPP the idea of Fractional Power Control (FPC) has been introduced. In this scheme the users are allowed to compensate for a fraction of the path loss so that the users with higher path loss will operate with a lower SINR requirement and will likely generate less interference to neighboring cells.

The agreed FPC scheme to set the power on PUSCH is based on an Open Loop Power Control (OLPC) algorithm, aiming to compensate for slow channel variations. In order to adapt to changes in the intercell interference situation or to correct the path-loss measurements and power amplifier errors, aperiodic close-loop adjustments can also be applied. The user transmit power is set according to formula (7.7) expressed in dBm [25].

$$P = \min\{P_{\max}, P_0 + 10 \cdot \log_{10} M + \alpha \cdot L + \Delta_{MCS} + f(\Delta_i)\} \quad [dBm] \tag{7.7}$$

where P_{max} is the maximum user transmit power, P_0 is a user-specific (optionally cell-specific) parameter, M is the number of PRBs allocated to a certain user, α is the cell-specific path-loss compensation factor that can be set to 0.0 and from 0.4 to 1.0 in steps of 0.1, L is the downlink path loss measured in the UE based on the transmit power P_{DL} of the reference symbols [20], Δ_{mcs} is a user-specific parameter (optionally cell specific) signaled by upper-layers; Δ_i is a user-specific close-loop correction value, and the $f(\cdot)$ function performs an absolute or cumulative increase depending on the value of the UE-specific parameter *Accumulation-enabled*.

If the absolute approach is used, the user applies the offset given in the PC command using the latest OLPC command as reference. If the cumulative approach is used, the user applies the offset given in the PC command using the latest transmission power value as reference. In the latter case Δ_i can take one of four possible values: -1, 0, 1, 3 dB.

In case the Closed Loop Power Control (CLPC) term is not used, the formula is simplified to include only the open-loop terms as indicated in (7.7).

$$P = \min\{P_{\max}, P_0 + 10 \cdot \log_{10} M + \alpha \cdot L\} \quad [dBm] \tag{7.8}$$

The exchange of the different signals related to PC is exemplified in Figure 7.26.

7.9.3 Uplink Signaling for Scheduling and Link Adaptation Support

The PS and LA entities rely on the CSI gathered via Sounding Reference Signals (SRSs) to perform channel-aware scheduling and AMC. Similarly, the allocation of time-frequency

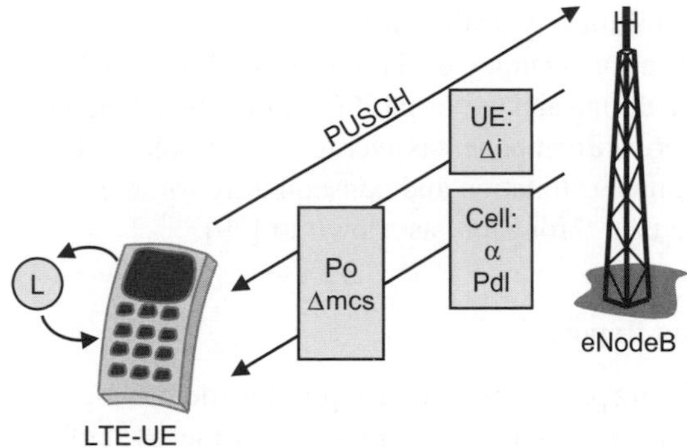

Figure 7.26 Power control signaling.

resources to users requires knowledge of their buffer status to avoid allocating more resources than are needed. Finally, the knowledge of how close the user is to its maximum transmit power is especially relevant for ATB operations. For this reason it is worth describing in more detail the signaling needed to support such operations as simplified in Figure 7.27.

7.9.3.1 Channel State Information

The uplink CSI can be described as the SINR measurement of the SRS. The CSI measurements are used to gain knowledge of the channel and perform fast AMC and Frequency-Domain Packet Scheduling (FDPS).

The SRS is transmitted over all or part of the scheduling bandwidth. Users in the same cell can transmit in the same bandwidth without interfering with each other thanks to the orthogonality provided by Constant Amplitude Zero AutoCorrelation (CAZAC) sequences and the uplink synchronous transmission. In reality there exists a constraint on the number of users in one cell that can simultaneously sound the same bandwidth without interfering with each other. The PSD on the pilot channel is the same as the one used on the data channel. User

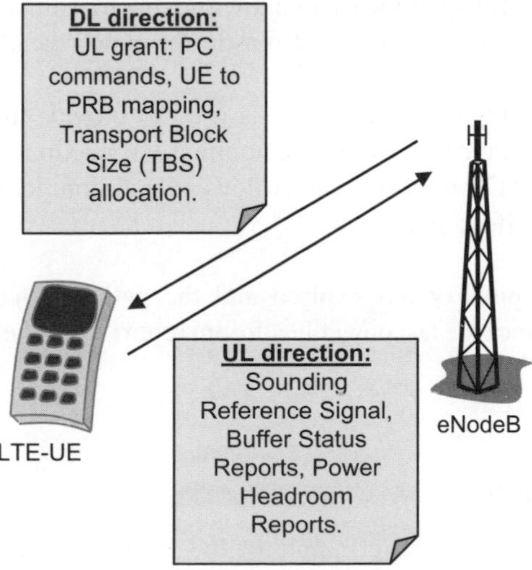

Figure 7.27 Signaling exchange between UE and eNode-B.

equipment power capabilities typically impose a limit on the sounding bandwidth, or, alternatively, on the level of accuracy of the corresponding SINR measurements. Typically, due to the dynamic scheduling and the variability of the instantaneous interference conditions in the uplink, the interference component is averaged over a certain time window. This is shown to be beneficial for channel estimation and consequently for an improvement of average cell throughput and outage user throughput as shown in [29].

7.9.3.2 Buffer Status Reports

The Buffer Status reporting procedure is used to provide the serving eNode-B with information about the amount of data available for transmission in the UL buffers of the UE.

A BSR can be triggered in one of three forms:

- **"Regular BSR"**: the UE buffer has to transmit data belonging to a radio bearer (logical channel) group with higher priority than those for which data already existed in the buffer (this includes as special case the situation in which the new data arrive in an empty buffer) or in case of a serving cell change.
- **"Padding BSR"**: UL resources are allocated and the number of padding bits is equal to or larger than the size of the BSR MAC control element.
- **"Periodic BSR"**: issued when the periodic BSR timer expires.

BSR are reported on a per Radio Bearer Group (RBG) basis as result of a compromise between the need to differentiate data flows based on QoS requirements and the need to minimize the resources allocated for signaling. Each RBG groups radio bearers with similar QoS requirements.

7.9.3.3 Power Headroom Reports

Due to the open-loop component of the standardized PC formula (see (2.5)) the eNode-B cannot always know the PSD transmitted by the UE. Such information is important for different RRM operations including the allocation of bandwidth, modulation and the coding scheme. Assuming that the eNode-B knows the user bandwidth, the transmission power can be derived from the information on the PSD. For this reason the power headroom reports have been standardized in [30].

The Power Headroom reporting procedure is used to provide the serving eNode-B with information about the difference between the nominal UE maximum transmit power and the estimated power for UL-SCH transmission. A Power Headroom Report (PHR) is triggered if either of the following criteria is met:

- a predefined timer expires or has expired and the path-loss has changed more than a predefined threshold since the last power headroom report when the UE has UL resources for new transmission;
- the predefined timer expires.

7.10 Downlink RRM

Figure 7.28 presents the high-level principle of the RRM systems. The following sections present the functionality in more detail.

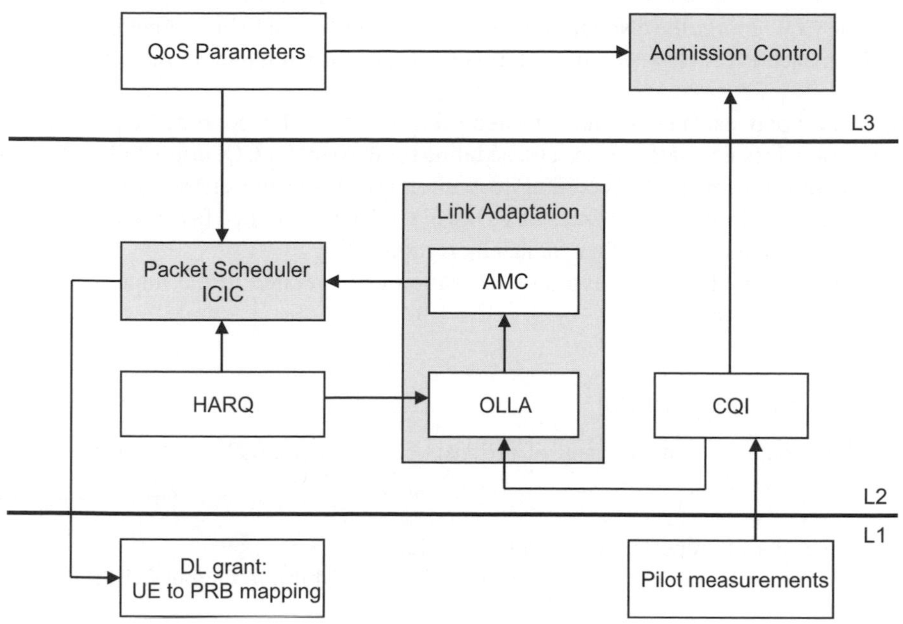

Figure 7.28 Downlink RRM system overview.

7.10.1 Channel Quality, Feedback and Link Adaptation

The power levels received by each UE from the signaling eNodeB (the eNodeB to which the UE is connected) and the various interfering eNodeBs of a network directly affect the Signal to Interference and Noise Ratio (SINR) variations in both time and frequency domain. Each UE provides a CQI to the eNodeB that indicates the channel state along these two dimensions. This allows for the DL packet-scheduling decisions to be made with channel information.

7.10.1.1 Channel Quality Indicator

In order to allow UEs to provide CQIs to the eNodeB, the 3GPP standard specifies a Reference Signal (RS) scheme. Every TTI, the eNodeB broadcasts reference pilots spread over the whole frequency bandwidth. The pilot schemes are described in [31].

The CQI for a given sub-band is the index of the MCS supported with a block error probability not exceeding 0.1 [32]. The sub-band size depends on the transmission bandwidth. In that sense, the link adaptation in the DL is very simple as the UE directly provides a MCS. The eNodeB is, of course, free to use a value different from the CQI. This can happen, for example, if the OLLA scheme described in Section 7.8.4.2 is used.

According to the 3GPP standard, the CQI can be reported in a periodic or aperiodic fashion. For periodic reporting, the reporting frequency can be set by the eNodeB. Several CQI transmission modes exist that offer various tradeoffs between CQI codeword size and CQI frequency accuracy. All the modes are described in [32]. The choice of the CQI transmission periodicity and mode affect directly the performance of the system [9].

With the CQI reports, the eNodeB has information on how much throughput is supported for each UE on each PRB. However, this information is not fully reliable for mainly two reasons:

- The CQI is not instantaneous: due to the reporting delay and the reporting frequency, the CQI information does not match the exact scheduling time. An older CQI is more likely to provide less accurate information. According to the remark in the present section regarding the change of the radio channel in time, the CQI will tend to be outdated faster if the UE moves at a higher speed and if the carrier frequency is higher.
- The CQI is prone to various types of estimation error related to the imperfect nature of receivers in general.

7.10.2 Packet Scheduling

The challenges that exist in the design of a DL packet scheduler mostly arise from the following elements:

- the channel quality reports are not fully reliable;
- advanced transmission techniques like Multiple Input Multiple Output (MIMO).

7.10.2.1 Resource Allocation Types: A Downlink PDCCH Related Limitation

The LTE DL TTI structure comes with an important constraint. The limited size of the PDCCH allows for only a limited number of UEs to be scheduled every TTI. Besides, in order to keep the number of UEs that can be scheduled every TTI as high as possible, LTE specifies PDCCH resource-allocation code words that allow only a limited flexibility as to the PRB configurations that can be allocated to any UE. There are three types of resource-allocation field referred to as [25]:

- Resource Allocation type 0, where the allocation granularity is the Resource Block Group, a set of consecutive PRBs the size of which depends on the transmission bandwidth.
- Resource Allocation type 1, which enables distributed allocations but where the minimum distance between two allocated PRBs is of the Resource Block Group size.
- Resource Allocation type 2, where the allocation simply consists of a set of consecutive "Virtual Resource Blocks," where Virtual Resource Blocks to PRB map changes in a pseudo-random fashion on a time slot basis. The concept of Virtual Resource Blocks is explained in [31].

7.10.2.2 Asynchronous HARQ

In the DL, HARQ is asynchronous. In that case, the packet scheduler must take care of scheduling the retransmissions in the same way as the rest of the data because the UE does not know, ahead of time, when the retransmissions are scheduled. A retransmission scheduling strategy is described in [3]. In this example, the time domain scheduler treats retransmission UEs equally to others, however the frequency domain scheduler:

- first schedules the non-retransmission UEs and leave enough PRBs for the retransmissions UEs;
- then schedules the retransmission UEs.

7.10.2.3 Packet Scheduling and MIMO Spatial Multiplexing

One of the major features of LTE is MIMO spatial multiplexing. Spatial multiplexing allows for transmitting several streams of data over the same PRBs. The number of streams that a UE can support at the same time depends on the antenna configuration at the receiver and transmitter and, to a great extent, on the instantaneous channel state. Together with the CQI, the UE feeds back a "rank indicator," which informs the eNodeB of how many data streams can be supported by the UE. When spatial multiplexing is enabled in the eNodeB, it is the role of the Packet Scheduler to decide whether spatial multiplexing is used or not for a UE. For close-loop spatial multiplexing, the UE feeds back a Pre-coding Matrix Indicator.

More complex is MU-MIMO, which sends several data streams on the same set of PRBs but to various UEs. Research is still largely ongoing on MU-MIMO packet scheduling.

7.10.3 Inter Cell Interference Control

7.10.3.1 The Role of ICIC

LTE offers the possibility to work as a reuse 1 system: a cellular system where all the frequency resource is fully used in all cells. The main problem of a reuse 1 cellular system is the poor radio conditions at the cell edge due to high interference. The role of the ICIC is to control the downlink power allocation in the frequency domain in order to provide frequency resources with limited interference.

7.10.3.2 Fixed Power Frequency Allocation Schemes

One simple way to perform ICIC is by allocating a fixed-frequency power profile to each eNodeB so that:

- The part of the frequency band that has higher transmit power corresponds to low power transmission in adjacent base stations to create a frequency allocation zone for this eNodeB with higher SINR, mostly destined for cell-edge users.
- The part of the frequency band that has lower transmit power corresponds to high power in adjacent base stations in order to avoid creating interference in the high SINR frequency-allocation zones of adjacent eNodeBs. The low power frequency band is destined mainly for UEs closer to the eNodeB.

Figure 7.29 shows an illustration of this type of scheme.

Such schemes can be easily be mapped in a hexagonal cellular grid, for example. There is almost no evidence in the literature that these techniques can achieve a significant cell-edge throughput gain as, most of the time, the packet scheduler can compensate the low SINR of cell edge users by allocating them more resources.

7.10.3.3 Dynamic ICIC

Dynamic ICIC changes the power allocation based on the current power allocation schemes of neighboring cells. Dynamic power allocation schemes can use the following information:

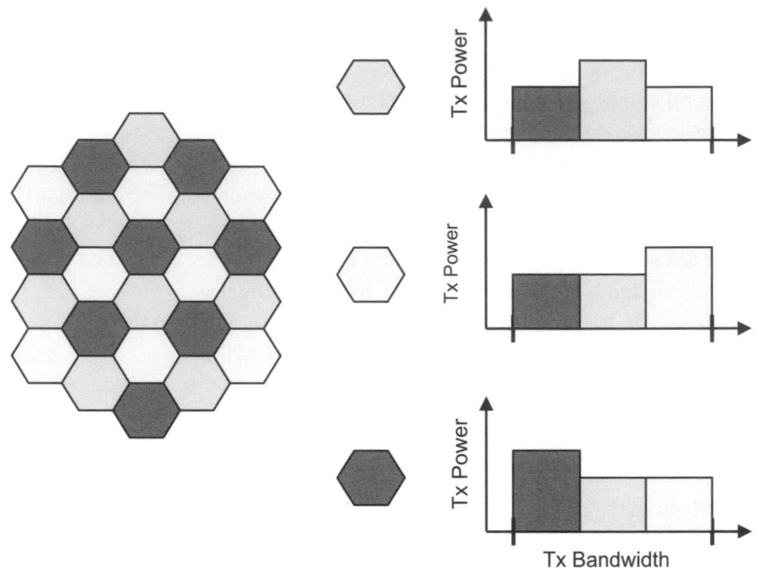

Figure 7.29 On the left-hand side: an example of cell layout with three types of cell. On the right-hand side: the transmitting Power patterns associated to each cell type. Each cell has a high power/low interference frequency zone.

- the resource usage status from neighboring eNodeBs;
- the CQIs indirectly contain information about the power allocation in neighboring cells as they contain some information about interference on the various PRBs of each UE.

Again, there is no evidence in the literature that Dynamic ICIC can bring any significant gain for cell-edge UE. Note that Dynamic ICIC must be used cautiously as fast power variations can diminish the relevance of the CQIs and therefore create large block error rates in the system [8].

7.11 Intra-LTE Handover

The LTE uses scalable bandwidth up to 20 MHz (1.4, 3, 5, 10, 15, 20 MHz) based on the number of used subcarriers. The use of scalable bandwidth in LTE allows the handover measurement to be carried out on different bandwidths. Hence, measurement bandwidth is a parameter of L1 filtering and should be optimized for different environments—for example, user speeds. The frequency-selective multipath fading will have an impact on handover performance, depending on the measurement bandwidth. Handover decisions are typically based on the downlink channel measurements standardized in 3GPP, which consist of RSRP, RSRQ, and so on [33]. As shown in Figure 7.30, these measurements are filtered using an L3 filter [34], before using the measurement results for the evaluation of the reporting criteria or for the measurement reporting, as:

$$F_n = (1 - a) \cdot F_{n-1} + a \cdot M_n \tag{7.9}$$

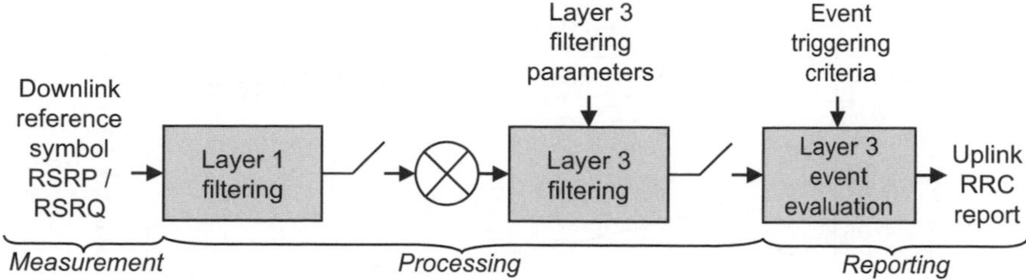

Figure 7.30 Handover initialization phase including handover measurement, filtering and reporting in the UE.

where

- $\boldsymbol{M_n}$ is the latest received measurement result from the physical layer;
- $\boldsymbol{F_n}$ is the updated filtered measurement result, that is used for evaluation of reporting criteria or for measurement reporting;
- $\boldsymbol{F_{n-1}}$ is the old filtered measurement result, where $\boldsymbol{F_0}$ is set to $\boldsymbol{M_1}$ when the first measurement result from the physical layer is received; and
- $\boldsymbol{a} = 1/2^{(\boldsymbol{k}/4)}$, where $\boldsymbol{k}$ is the *filterCoefficent* for the corresponding measurement quantity received by the *quantityConfig*;
- the relative influence on the updated filtered measurement result of the latest received measurement and the old filtered measurement result is controlled by the factor $\boldsymbol{a}$.

The handover measurement report triggering is standardized in 3GPP as events A1, A2 . . . A5. The handover decision is based on the filtered measurement result, $\overline{F_n}$, and is executed if, for example, the condition in Equation (7.9) is satisfied, where *Hys* is the hysteresis margin. The time-to-trigger (TTT) window is introduced as the time to wait before making a handover decision during which the same cell remains the potential target cell.

$$\overline{F_n}_{\text{Target Cell}}[n] \geq \overline{F_n}_{\text{Source Cell}}[n] + Hys \quad [\text{dB}] \tag{7.10}$$

Introducing the TTT window is one way to suppress the number of unnecessary handovers called ping-pong handovers. The ping-pong handover is defined as a handover to a neighboring cell that returns to the original cell after a short time. Each handover requires network resources to reroute the call to the new eNode-B. Thus, minimizing the expected number of handovers minimizes the signaling overhead. Another solution to reduce the number of handovers is to introduce a handover avoidance timer, which allows handover only after the timer expires.

One of the key performance indicators (KPIs) of handover performance is the reduced number of handovers, which would mean a reduced signaling overhead. The use of larger measurement bandwidth makes a significant improvement in the performance in terms of number of handovers in low Doppler environments. For example at 3 km/h by increasing the measurement bandwidth from 1.25 to 5 MHz a decrease of 30% in average number of handovers is noticed [35].

Although a higher measurement bandwidth provides performance, in situations when different cells are operating at different transmission bandwidth an idea is to limit the measurement bandwidth to a well-defined fixed value. With the high Doppler shift, larger measurement bandwidth does not provide any significant performance gain in terms of number of handovers. Further, for an adaptive choice of filtering period, depending on the user speed, the gain for using larger measurement bandwidth can be made negligible for a penalty in terms

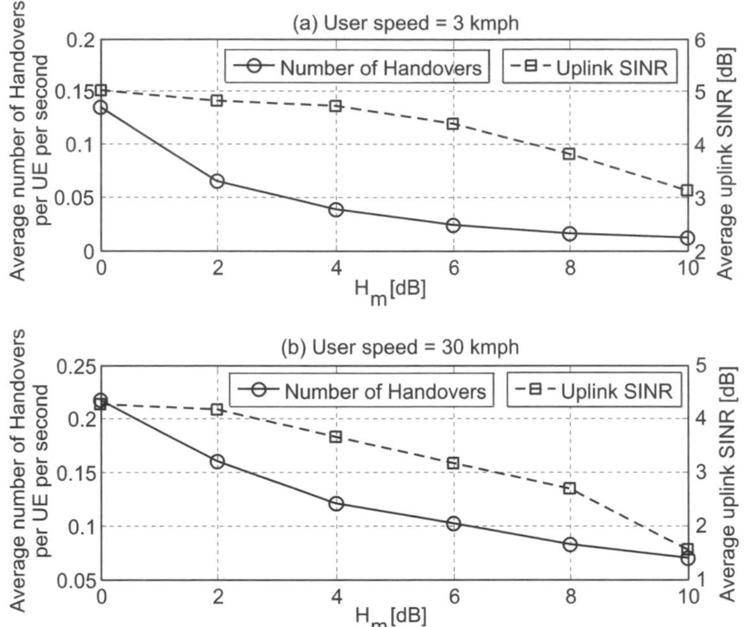

Figure 7.31 Effect of varying Hys for the intra-LTE handover on average number of handovers and average uplink SINR at the user speeds of 3 and 30 kmh.

of signal quality. Hence, it is recommended that 1.25 MHz of measurement bandwidth be used for a good choice of L3 filtering period.

The average number of handovers decreases with an increase in the hysteresis margin as seen in Figure 7.31. The reduction in the average number of handovers is a desired criteria but it also leads to a reduction in uplink quality, which is not desired as has been concluded in [36].

References

[1] 3GPP TS 36.300. (2010) *E-UTRA overall description*, V. 8.12.0, 3rd Generation Partnership Project, Sophia-Antipolis.

[2] Hanzo, L., Münster, M., Choi, B. and Keller, T. (2003) *OFDM and MC-CDMA for Broadband MultiUser Communications, WLANs and Broadcasting*. Wiley-IEEE Press, Hoboken, NJ.

[3] 3GPP TR 25.814. (2006) *Physical Layer Aspects for Evolved Universal Terrestrial Radio Access (UTRA)* (Release 7), V. 7.1.0, 3rd Generation Partnership Project, Sophia-Antipolis.

[4] Wengerter, C., Ohlhorst, J. and Elbwart, A.G.E. von. (2005) Fairness and throughput analysis for generalized proportional fair frequency scheduling in OFDMA. *IEEE Proceedings of the Vehicular Technology Conference (VTC)*, **3**, 1903–1907.

[5] Lopez, D., Ubeda, C., Kovacs, I., Frederiksen, F. and Pedersen, K. (2008) Performance of downlink UTRAN LTE under control channel constraints. IEEE Proceedings of the Vehicular Technology Conference (VTC), May, Singapore, pp. 2512–16.

[6] Pedersen, K., Kolding, T., Kovacs, I., Monghal, G., Frederiksen, F., Mogensen, P. (2009) Performance analysis of simple channel feedback schemes for a practical OFDMA system. *IEEE Transactions on Vehicular Technology*, **58** (9), 5309–5315.

[7] Monghal, G., Laselva, D., Michaelsen, P.-H., Wigard, J. (2010) Dynamic packet scheduling for traffic mixes of best effort and VoIP users in E-UTRAN downlink. IEEE Proceedings of the Vehicular Technology Conference (VTC), May, Taipei, Taiwan.

[8] Monghal, G., Kumar, S., Pedersen, K.I. and Mogensen, P.E. (2009) Integrated fractional load and packet scheduling for OFDMA systems. Proceedings of the IEEE International Conference on Communications (ICC), June, Dresden.

[9] Anas, M., Rosa, C., Calabrese, F.D., Pedersen, K.I. and Mogensen, P.E. (2008) Combined admission control and scheduling for QoS differentiation in LTE uplink. Proceedings of the 68th IEEE Vehicular Technology Conference (VTC), September, Calgary, Canada.

[10] 3GPP TR 25.892. (2004) *Feasibility study of orthogonal frequency division multiplexing (OFDM) for UTRAN enhancement* (Release 6), V6.0.0, 3rd Generation Partnership Project, Sophia-Antipolis.

[11] Olives Vidal, P. (2003) Analysis of OFDM for UTRAN enhancement. Master's thesis, Aalborg University, Institute of Electronic Systems.

[12] 3GPP TS 36.2011. (2007) *Evolved Universal Terrestrial Radio Access (E-UTRA); physical channels and modulation*, V8.0.0, 3rd Generation Partnership Project, Sophia-Antipolis.

[13] Falconer, S.L., Ariyavisitakul, S.L., Benyamin-Seeyar, A. and Eidson, B. (2002) Frequency domain equalization for single-carrier broadband wireless systems. *IEEE Communications Magazine*, **40** (4), 58–66.

[14] Lim, M. and Goodmand, D.J. (2006) Single carrier FDMA for uplink wireless transmission. *IEEE Vehicular Technology Magazine*, **1**, 30–38.

[15] Berardinelli, G., Ruiz De Temino, L.A., Frattasi, S., Rahman, M. and Mogensen, P. (2008) OFDMA vs. SC-FDMA: performance comparison in local area IMT-A scenarios. *IEEE Wireless Communications*, (October), 64–72.

[16] 3GPP 23.401. (2011) *General Packet Radio Service (GPRS) enhancements for Evolved Universal Terrestrial Radio Access Network (E-UTRAN) access*. V. 8.14.0, 2011-06-12, 3rd Generation Partnership Project, Sophia-Antipolis.

[17] 3GPP 23.203. (2011) *Policy and charging control architecture*, V. 8.12.0, 2011-06-12, 3rd Generation Partnership Project, Sophia-Antipolis.

[18] 3GPP TS 36.300. (2010) *Evolved Universal Terrestrial Radio Access (EUTRA) and Evolved Universal Terrestrial Radio Access (E-UTRAN); Overall Description; Stage 2*, V. 8.12.0. 2010-04-21, 3rd Generation Partnership Project, Sophia-Antipolis.

[19] Hosein, P. (2003) A class-based admission control algorithm for shared wireless channels supporting QoS services. Proceedings of the 5th IFIP TC6 International Conference on Mobile and Wireless Communications Networks (MWCN), October, Singapore.

[20] 3GPP 25.813. (2006) *E-UTRA and E-UTRAN; radio interface protocol aspects*, V. 7.1.0, 2006-10-18, 3rd Generation Partnership Project, Sophia-Antipolis..

[21] Pokhariyal, A., Pedersen, K.I., Monghal, G., Kovacs, I.Z., Rosa, C., Kolding, T.E. and Mogensen, P.E. (2007) HARQ aware frequency domain packet scheduler with different degrees of fairness. Proceedings of the IEEE Vehicular Technology Conference (VTC), April, Dublin, Ireland, pp. 2761–2765.

[22] Pokhariyal, A., Kolding, T.E., Mogensen, P.E. (2006) Performance of downlink frequency domain packet scheduling for the UTRAN Long Term Evolution. IEEE Proceedings of Personal, Indoor and Mobile Radio Communications, September, Helsinki, Finland.

[23] Monghal, G., Pedersen, K.I., Kovacs, I.Z. and Mogensen, P.E. (2008) QoS oriented time and frequency domain packet schedulers for the UTRAN long term evolution. Proceedings of the IEEE Vehicular Technology Conference (VTC), May, Singapore.

[24] Sampath, A., Kumar, P.S. and Holtzman, J.M. (1997) On setting reverse link target SIR in a CDMA system. Proceedings of IEEE Vehicular Technology Conference (VTC), May, Phoenix, AZ.

[25] 3GPP TS 36.213. (2009) *E-UTRA Physical Layer procedures, Section 7.1*, V. 8.8.0, 3rd Generation Partnership Project, Sophia-Antipolis.

[26] Eklundh, B. (1986) Channel utilization and blocking probability in a cellular mobile telephone system with directed retry. *IEEE Transactions on Communications*, **34** (4), 329–337.

[27] Goldsmith, A.J. and Chua, S.-G. (1998) Adaptive coded modulation for fading channels. *IEEE Transactions on Communications*, **46** (5), 595–602.

[28] 3GPP TS 36.101. (2008) *Evolved Universal Terrestrial Radio Access (E-UTRA); User Equipment (UE) radio transmission and reception*, V. 8.2.0, 2008-06-06, 3rd Generation Partnership Project, Sophia-Antipolis.

[29] Rosa, C., Villa, D.L., Castellanos, C.U., Calabrese, F.D., Michaelsen, P.H., Pedersen, P.I. and Skov, P. (2008) Performance of Fast AMC in E-UTRAN Uplink. Proceedings of the IEEE International Conference on Communications (ICC), May, Beijing, China, pp. 4973–4977.

[30] 3GPP TS 36.321. (2008) *E-UTRA Medium Access Control (MAC) protocol specification*, V. 8.4.0, 3rd Generation Partnership Project, Sophia-Antipolis.

[31] 3GPP TS 36.211. (2009) *E-UTRA physical channels and modulations*, V. 8.9.0, 3rd Generation Partnership Project, Sophia-Antipolis.

[32] 3GPP TS 36.213. (2009) *E-UTRA Physical Layer procedures, Section 7.2*, V. 8.8.0, 3rd Generation Partnership Project, Sophia-Antipolis.

[33] 3GPP TS 36.214. (2009) *Evolved Universal Terrestrial Radio Access (EUTRA); Physical Layer; Measurements*, V. 8.7.0, 2009-09-29, 3rd Generation Partnership Project, Sophia-Antipolis.

[34] 3GPP TS 36.331. (2011) *Evolved Universal Terrestrial Radio Access (E-UTRA); Radio Resource Control (RRC); Protocol specification*, V. 8.14.0, 2011-06-24, 3rd Generation Partnership Project, Sophia-Antipolis.

[35] Anas, M., Calabrese, F.D., Östling, P.E., Pedersen, K.I. and Mogensen, P.E. (2007) Performance analysis of handover measurements and Layer 3 Filtering for UTRAN LTE. Proceedings of the IEEE International Symposium on Personal, Indoor and Mobile Radio Communications (PIMRC), September, Athens, Greece.

[36] Anas, M., Calabrese, F.D., Mogensen, P.E., Rosa, C. and Pedersen, K.I. (2007) Performance. evaluation of received signal strength based hard handover for UTRAN LTE. Proceedings of the 65th IEEE Vehicular Technology Conference (VTC), April, Dublin, Ireland.

8

Terminals and Applications

Tero Jalkanen, Jyrki T. J. Penttinen, Juha Kallio, and Adnan Basir

8.1 Introduction

LTE brings mobile communications to a new level. The LTE network supports much higher data rates than any of the previous wide-area mobile network solutions of 3GPP. This has an important impact on the terminals.

8.2 Effect of Smartphones on LTE

8.2.1 General

The last few years have seen tremendous increase in the number of smartphones on the market. The number of smartphones users is expected to grow further in coming years as cheaper and better smartphones are appearing. This growth has resulted in high revenues for operators but, at the same time, high congestion in the networks has been observed. Recent lab studies (in the UMTS test network) have suggested that it is not data utilization that is to be blamed but rather high Layer 3 (L3) signaling consumption of smartphones that has bogged down the networks.

Most Smartphone applications (e-mail, Facebook, etc.) are always on, running in the background and frequently obtaining updates from application servers. To make the situation even worst, smartphones keep connecting and disconnecting from the network, to save battery power. A single connection and disconnection from the network results in a large number of signaling messages, which can reduce the performance of the RRC state machine [1], [2].

For example the Google Android phone simply drops the RRC connection (Fast Dormancy) when it has no data to send, and if there is new data a new request is made for an RRC connection. A lab study showed that an online multiplayer board game played for 30 minutes resulted in 3201 signaling messages on Google Android while a single voice call only requires on average 50 L3 signaling messages. A 30-minute online game played on Google Android in the UMTS test network used control plane resources of around 64 voice call users, which is significant amount.

Some operators have tried to cope with this problem by changing the frequency spectrum and also upgrading HSPA+ protocols, but the results are still not satisfying in the larger urban areas.

The LTE/SAE Deployment Handbook, First Edition. Edited by Jyrki T. J. Penttinen.

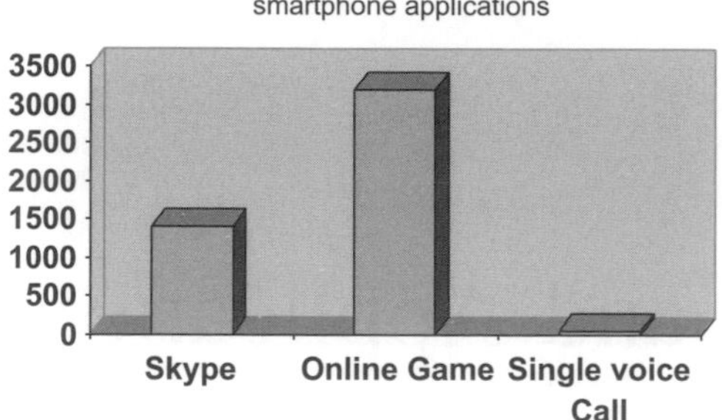

Figure 8.1 The effect of smartphones on signaling.

Such high signaling results in frequent dropped calls and users are unable to make data connections.

Figure 8.1 below shows the comparison between different applications and their network signaling utilization in a 3G test network.

The following statistics are related to Figure 8.1:

- **Skype**—Skype tested on iPhone 3.1 for one hour while it was running idle in the background: 1400 signaling messages were generated in the network.
- **Online Game**—Online Poker played for 30 minutes on Google Android with other online players: 3201 signaling messages were generated.
- **UMTS Voice Call**—a single voice call consumed only 50 signaling messages.

8.2.2 Is LTE Capable Enough to Handle the Challenge?

LTE can bring multiple challenges and improvements to signaling congestion. It is understood that LTE devices will generate more signaling traffic, as the applications will always be running in the background and VoIP is going to be used extensively over LTE. From a core network perspective, as the network is flattened now, the Mobility Management entity (MME) and Serving Gateway (SGW) are going to face a large amount of data and control traffic flow. As the eNodeBs are directly connected to the MME, this can increase the signaling load on the MME as compared to SGSN as MME now handles the paging requests to all eNodeBs, exposure to all inter-eNodeB mobility events, idle/active state transitions, and lastly NAS signaling ciphering and integrity protection.

LTE RRC architecture is much simplified now, however. There are only two RRC states—RRC_Idle and RRC_Connected, unlike in UMTS where RRC has five states: IDLE, CELL_FACH, CELL_DCH, CELL_PCH and URA_PCH. The main reason for only having two RRC states in LTE is that there is no concept of dedicated and common transport channels in LTE. Data is transferred through a shared transport channel. This will greatly improve the performance of RRC state machine and reduce the signaling overhead. Secondly the availability of DRX period in RRC_Connected state will give the freedom to UEs to remain in connected state for longer time as the power consumption will no longer be an issue for UEs in the connected state unlike in UMTS (fast dormancy used by UEs in UMTS).

8.2.3 LTE RRC States

The RRC connection establishment procedure for the LTE is relatively simple compared to the UMTS RRC connection-establishment procedure. In UMTS, NBAP and ALCAP, signaling procedures are required across the Iub interface between Node B and RNC, used to set up new radio link and new transport connection. But as the LTE architecture does not involve RNC, it removes the requirement for these signaling procedures. With LTE, the initial (Non-Access Stratum (NAS) message is transferred as part of the RRC connection establishment procedure while, in UMTS, NAS messages are always transferred after the RRC connection is established successfully. At the same time, the amount of information sent in signaling messages has been reduced, resulting in the substantial reduction of layer 3 signaling messages.

All the improvements in LTE RRC architecture will have a significant effect on the signaling load generated by the ALWAYS ON nature of smartphones applications. Currently LTE smartphones are not available on the market, so it is too early to predict the real behavior of smartphones in the commercial LTE network.

8.3 Interworking

8.3.1 Simultaneous Support for LTE/SAE and 2G/3G

One interesting question related to the LTE/SAE deployment interaction with the existing 2G and 3G networks is the device capability—especially the issue of whether it is possible to have a device that includes both LTE/SAE and 2G/3G support simultaneously. It is clear that support for 2G/3G is built in for many, if not all, LTE/SAE devices so that they are able to switch to 2G/3G whenever LTE/SAE is not available. However, it is not very straightforward to build a device that would be connected at the same time both to LTE/SAE and 2G/3G networks. This would require two separate radios to run simultaneously, both connected to the core network. It would also probably require two SIM cards.

This kind of device would solve a number of issues related to handover problems between LTE/SAE and 2G/3G. For example, voice-related functionality such as CS Fallback or SRVCC would not be needed because the LTE/SAE device would always be connected to the 2G/3G network, meaning that the native CS voice service of 2G/3G would be available without any need for handover. However, the complexity of the solution involving two radios, related to crucial mobile phone design features such as size, cost, battery consumption, and interference, means that in the near future it is unlikely that we will see such devices being deployed.

The 3GPP specifications define the states of the LTE user equipment (in this book referred as LTE-UE) and state transitions, including inter-RAT procedures. The states have been divided into the RRC_CONNECTED state (when an RRC connection has been established) and the RRC_IDLE state (when no RRC connection is established). The RRC states of the LTE-UE are characterized as follows.

8.3.1.1 RRC Idle State

The RRC Idle state means that the LTE-UE controls mobility, and takes care of monitoring the paging channel in order to act when there is an incoming call waiting for the customer. In this state, the LTE-UE monitors the system information change. For the LTE terminal models that support the Earthquake and Tsunami Warning System (ETWS), the terminal monitors the notifications that can be delivered via the paging channel. In addition to monitoring the paging channel, the LTE-UE performs neighboring cell measurements, cell selection and the cell reselection procedure, and in general is able to acquire system information from the LTE/SAE network. In this stage, the discontinuous reception (DRX) of the LTE-UE in question can be configured by the upper layers.

8.3.1.2 RRC Connected State

The LTE-UE is only capable of delivering unicast data in the downlink and uplink in the RRC Connected state.

Mobility is controlled by the LTE/SAE network, meaning that the network takes care of the handover and cell-change procedures order, possibly with an additional network assistance (NACC) to the 2G radio access network (GERAN). In this state, the LTE-UE still monitors the paging channel and/or System Information Block Type 1 contents in order to detect system information changes. As in the Idle state, the LTE-UE also monitors the ETWS notifications if the terminal is able to support the ETWS system. The LTE-UE monitors control channels that are associated with the shared data channel to determine if data is scheduled for it. The important task of LTE-UE is to provide channel quality and feedback information for the LTE/SAE network, which makes decisions related to radio resource and mobility management, including handovers when the neighboring cells are more attractive for the connection. For this, the LTE-UE also performs neighboring cell measurements and reports the measurement results to the LTE/SAE network for comparison of the best cells and as a basis for decisions about the handovers. In addition, the LTE-UE acquires system information. At lower layers, the UE may be configured with a UE-specific DRX.

8.3.1.3 Mobility Support

The mobility of the LTE-UE between LTE and 2G is illustrated in Figure 8.2, and the same idea is presented for 3G and CDMA2000 in Figures 8.3 and 8.4, respectively. In the latter, HRPD refers to High Rate Packet Data.

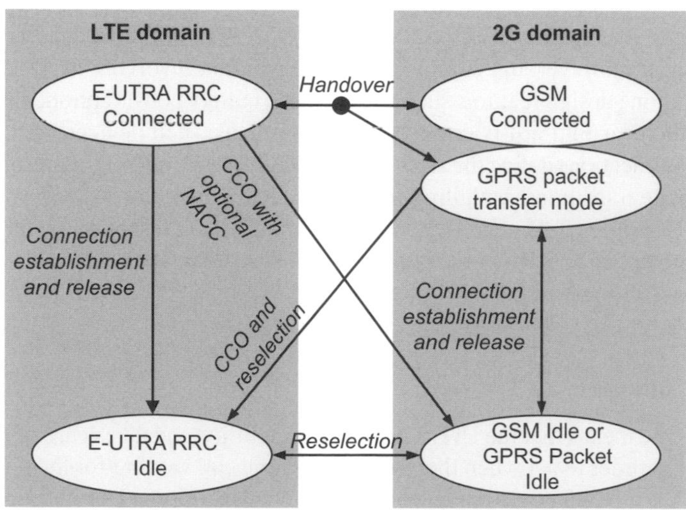

Figure 8.2 The LTE-UE states and the inter-RAT mobility procedures with GSM network as interpreted in [3].

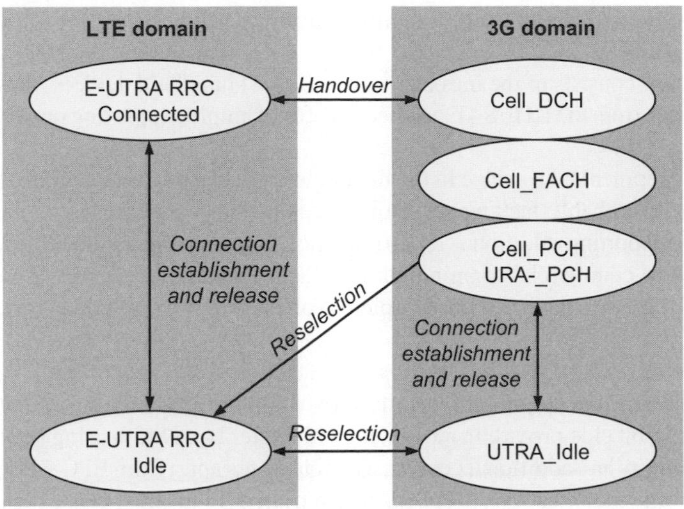

Figure 8.3 The LTE-UE states and the inter-RAT mobility procedures with UMTS network as interpreted in [3].

8.3.2 Support for CS Fallback and VoLTE

One way to look at the device capability evolution is listed below:

1. In the first phase the LTE/SAE device is a pure data-only dongle supporting perhaps only one LTE frequency band.

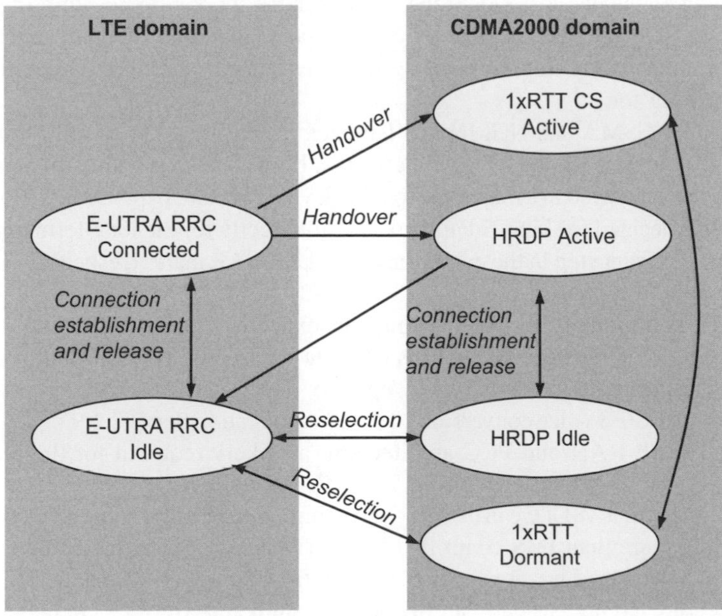

Figure 8.4 Mobility procedures between E-UTRA and CDMA2000 as interpreted in [3].

2. In the second phase, dongles become more advanced, adding support for additional LTE frequency bands.
3. The next phase consists of the introduction of the first handheld devices, these being rather simple devices from the LTE/SAE perspective, for example supporting only CS Fallback for the voice service.
4. Further development takes place in the handheld device area—for example the first VoLTE support is still probably lacking some of the features.
5. A need for supporting additional LTE frequency bands in the handheld devices becomes obvious due to commercial roaming taking place.
6. Full-scale LTE/SAE device with complete VoLTE support including functions such as SRVCC.

On the network side, development typically moves along the same lines. First, LTE/SAE is used only as a big bit pipe providing customers with better bandwidth using their data dongles. This does not require any additional core network elements apart from EPC—IMS or AS are not needed, nor are any modifications to the CS core required. In the next phase the operator needs to deploy additional core network elements and update the CS core to offer functions such as CS Fallback, VoLTE and SRVCC to its customers.

This issue is directly related to the type of devices being supported—as long as all or the majority of LTE/SAE devices are data dongles there is not really a great hurry to offer anything other than just the bit-pipe service by the operator. This is due to customers typically using their data dongles as the name would imply—just for pure data traffic such as accessing Internet services. The penetration of handheld devices, which are very much more likely to be used for communication purposes (i.e., voice and messaging), basically drives the commercial demand for the CS Fallback and/or VoLTE deployment.

There is very little public information regarding support for CS Fallback in the forthcoming LTE/SAE devices, thus it remains to be seen how widely this functionality can be deployed on the network side if the devices in the market do not support it. However, as normal it is expected that device vendors, or at least some of them, are listening to operators' demands and roadmaps because it makes sense to produce something that your customer is asking for.

As indicated by GSMA and NGMN, CS Fallback is considered to be the common interim solution for the voice service in LTE/SAE networks. If this becomes reality then we can expect CS Fallback to be supported in a major way on the device side too. In practice this means that an operator that has decided to skip CS Fallback and go directly for the long-term goal of VoLTE might miss an important step in the evolution of the LTE/SAE market, especially if competing operators are able to support it.

Since VoLTE is the long-term target it cannot be expected that it will be widely available as "commercial-grade" in the near future. This is due to IMS-based IP voice deployment being a major piece of implementation work for any operator due to functionality such as the IMS core system (including CS/PS voice conversion), related Application Servers, SRVCC support, QoS support in LTE/SAE RAN and PCC architecture, all likely required for the real CS voice replacement service.

It should be noted that VoLTE can be deployed using a much leaner approach too, in the sense that it is nothing more than IMS controlled P2P service exchanging media between devices using RTP/RTCP stream. Thus it is possible to run the first tests/trials using VoLTE via very simple architecture having basically an IMS core system and SIP/RTP clients in the devices connected to each other via IP, but this approach is not really feasible for the full commercial operator voice launch, which more-or-less requires that all the bells and whistles of the existing CS voice should be supported even though there is very little use for some of them. Naturally, regardless of the network support, an operator must also wait for devices supporting VoLTE to be widely available.

A related issue is the question of supporting CS Fallback and/or VoLTE for the inbound roamers. In theory this could impose some requirements for the operator to support functions that might not be used by the operator itself. Consider, for example, Operator A, which does not support CS Fallback, having an LTE/SAE roaming agreement with Operator B using CS Fallback. Customers of Operator B are used to their LTE/SAE devices supporting voice in the home network via CS Fallback. Now, when they roam into Operator A's network, what happens? It is possible for the device to detect that CS Fallback is not supported and if this device is a "voice-centric" device it likely will not use the LTE/SAE network at all but stick to 2G/3G where voice is available. Thus the Operator A LTE/SAE network would not be used at all by this kind of inbound roamer, potentially leading to unhappy customers and less roaming revenues for Operator A.

8.4 LTE Terminal Requirements

8.4.1 Performance

The Release 8 specifications define the maximum output power level of +23 dBm for the class 3 LTE terminal, with a tolerance of +/−2 dB. This applies to all the E-UTRA bands 1–14, 17 and 33–40 as shown in Chapter 7. It should be noted that for E-UTRA bands 2, 3, 7, 8 and 12, an additional lower tolerance limit of 1.5 dB can be applied. Taking into account the upper limit, the definitions basically allow a maximum of +25 dBm power level for the class 3 LTE terminal. Due to higher order modulation and transmit bandwidth, the maximum power can be reduced by 1. . .2 dB depending on the modulation (QPSK/16-QAM) and allocated resource blocks.

The minimum output power of the LTE terminal is defined as the mean power in one subframe—that is, within a 1 ms time period. The power level is −40 dBm for all the bandwidth values of LTE. The power level of off-mode is a maximum of −50 dBm.

The LTE terminal switch time from the off state to the on state is defined in such a way that the transient time period is 20 us, as shown in Figure 8.5.

The specifications [1] define criteria for the transmit modulation quality for the terminal's RF transmissions. The quality is specified in terms of Error Vector Magnitude (EVM) for the allocated resource blocks (RBs), which should not exceed 17.5 dN for QPSK, and 12.5 dB for 16-QAM. Other quality criteria are EVM equalizer spectrum flatness derived from the equalizer coefficients generated by the EVM measurement process, carrier leakage that is caused by IQ offset, and in-band emissions for the nonallocated resource blocks. Validation of the quality can be done via the conformance testing.

There are also extensive list of other requirements for the LTE terminal, including the limits for the spurious emissions, adjacent channel leakage ratio and so on. For radio link designing purposes, it is sufficient to assume the LTE-UE transmitter ideal with the expected maximum output power of +23 dBm as indicated in the class 3 terminal definitions.

The source [1] also defines the requirements for the LTE-UE receiver for both single and multi-antenna cases. One of the essential items is the minimum sensitivity figure of the terminal. This value depends on the E-UTRA band and channel bandwidth in such a way that the value oscillates between −102.2. . . −106.2 dBm in the 1.4 MHz bandwidth over all

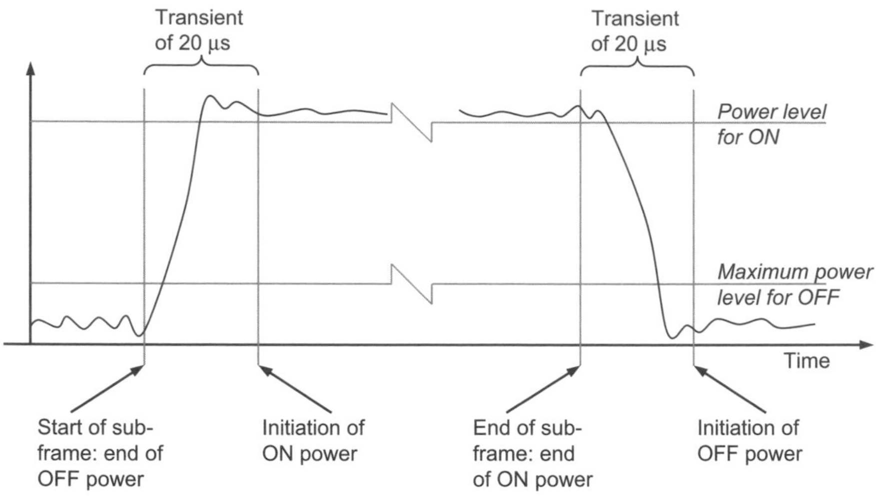

Figure 8.5 The off and on modes of the LTE terminal.

the E-UTRA bands for the QPSK modulation, assuming a single antenna reception, while the value is −91... −94 in the 20 MHz bandwidth. For radio link budget calculation purposes, the complete table is found in [1]. Various other receiver requirements are listed, including the reporting accuracy of the channel quality indicator (CQI) under different radio conditions, and the rank indicator (RI) reporting accuracy.

8.4.2 LTE-UE Categories

LTE-UE contains a field called ue-Category, which defines the combined uplink and downlink capability of the equipment. There are five terminal categories. Table 8.1 shows the downlink and uplink physical layer parameter values for the UE Category.

8.4.3 HW Architecture

LTE-UE contains functional blocks for the transmission and reception of signaling and data. The processing of the bit streams in the reception of the LTE-UE is based on the OFDM as

Table 8.1 The values of the transport blocks of the LTE-UE categories

LTE-UE Category	Maximum number of DL-SCH transport block bits received within a TTI in downlink	Maximum number of bits of an UL-SCH transport block transmitted within a TTI in uplink
Category 1	10 296	5160
Category 2	51 024	25 456
Category 3	102 048	51 024
Category 4	150 752	51 024
Category 5	299 552	75 376

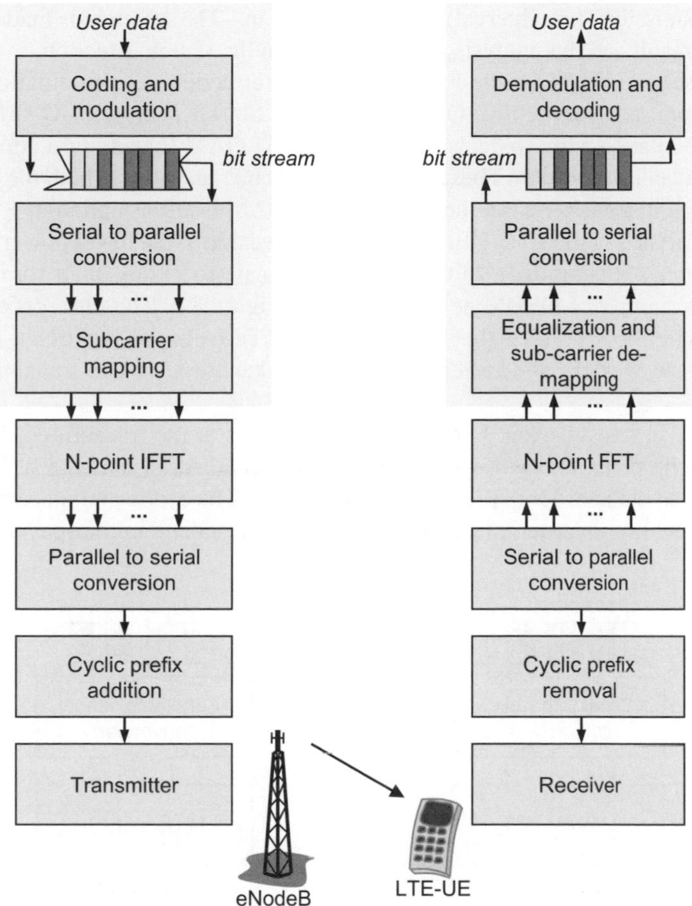

Figure 8.6 The reception of LTE-UE.

shown in Figure 8.6. The figure shows the case for a single user. For multiple users, each one feeds the user data to the serial/parallel conversion, and the subcarrier mapping is done before the N-point IFFT takes place. Equally, on the receiver side, the stream is handled separately for each user by taking the data flow of each user from the N-point FFT to the equalization and subcarrier demapping.

OFDMA is a feasible multiplexing scheme, especially for the LTE downlink. Despite the fact that it results in additional complexity due to resource scheduling, it is much more efficient and its latency performance is better compared to packet-oriented solutions. In the OFDMA system, users are allocated a certain number of subcarriers for a predetermined amount of time via the physical resource blocks (PRB). The resource blocks have time and frequency dimension which can be compared to the "extended" concept of the GSM timeslots and transceivers, which varies over time. The eNodeB is in charge of scheduling the radio blocks.

First, before the OFDM signal processing of the LTE-UE, the user data is modulated and coded. The selection of the modulation and coding scheme depends on the radio conditions and is dynamic. It is thus called adaptive modulation and coding (AMC), and follows the same principles as in the HSPA specifications. The minimum interval of the scheme modification is one TTI (Transmission Time Interval) frame—that is, 1 ms. The selection is made based on the measurement result and the analysis of the eNodeB in the uplink direction.

The coding scheme of LTE can be conventional or Turbo coding. The modulation scheme can be selected from a set of QPSK (Quadrature Phase Shift Keying), 16-QAM (Quadrature Amplitude Modulation with 16 decision points) and 64-QAM (Quadrature Amplitude Modulation with 64 decision points). These modulation schemes are valid in both the downlink and uplink, although in the latter case, the support of 64-QAM is left as optional.

In the transmission of the LTE-UE, SC-FDMA is used due to the lower power consumption of the transmitter's RF amplifier. In the high-level signal processing chart format, it is quite similar to the principle of OFDM as can be seen in Figure 8.6. The difference between the downlink direction's OFDMA is that the uplink's SC-FDMA has an additional IFFT block in the receiver of the eNodeB, and an additional FFT block in the LTE-UEs transmitter as shown in Figure 8.7.

The last task in transmission, before the signal arrives at the transmitter element, is the addition of cyclic prefix and pulse shaping into the signal. As in the case of OFDM, pulse shaping is carried out in order to prevent spectral growth. The cyclic prefix is an essential part of LTE functionality in order to enhance performance under multipropagated radio path

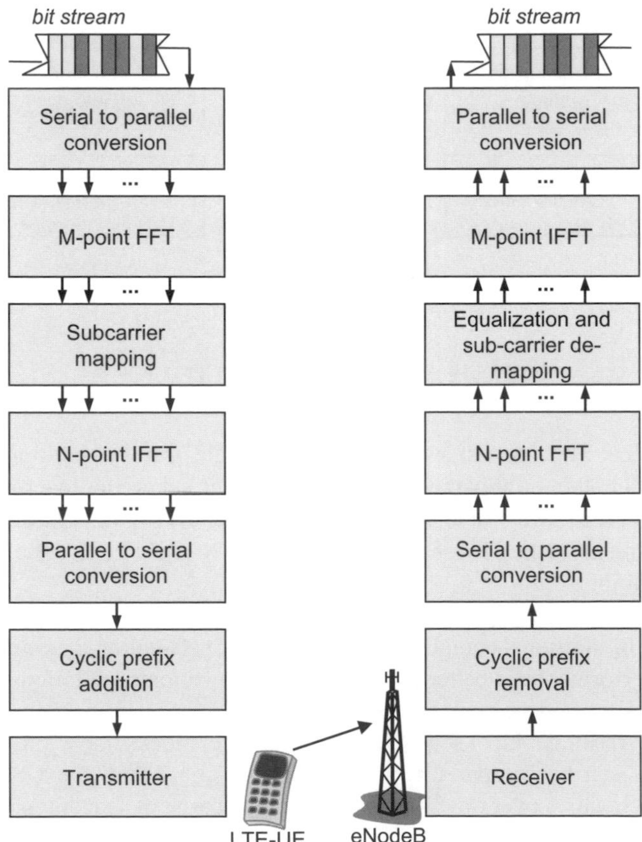

Figure 8.7 The block diagram of the signal processing of the LTE-UE transmitter and eNodeB receiver based on the SC-FDMA.

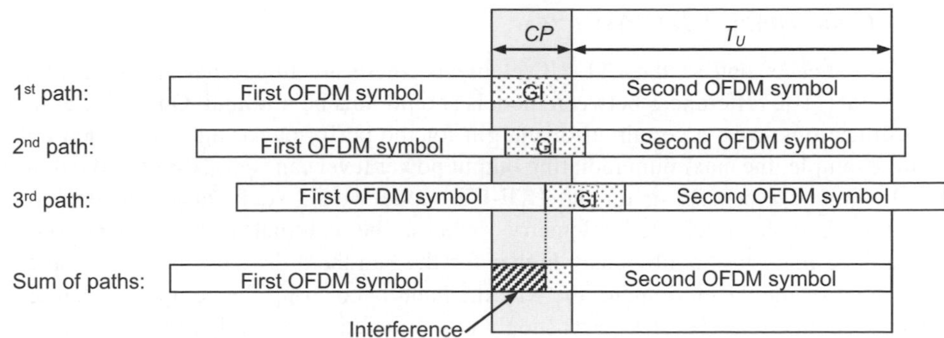

Figure 8.8 The Cyclic Prefix edits the intersymbol interference in LTE.

conditions. In the transmitter, the symbol is extended in the IFFT block. This extension is called the cyclic prefix. It should be long enough to cope with the multipath components—that is, to eliminate significant interfering paths. On the other hand, if the cyclic prefix is too long, it wastes capacity. Thus one of the many optimization tasks of the operator is to select the value for the cyclic prefix duration. Figure 8.8 shows the basic principle of the cyclic prefix.

The MIMO functionality adds the complexity of the transmitter and receiver as the number of antennas increases. In fact, the receiver of the LTE-UE is required to have at least two receiving antennas, and plain OFDM SISO transmission is not supported at the eNodeB side.

The SC-FDMA that is used in the uplink of LTE differs from the downlink requirements. The most important issue on the UE side is the power consumption. The high Peak to Average Power Ratio (PAPR) and the resulting efficiency degradation that is found in OFDM signaling is also an important reason why OFDM has not been selected for the downlink. SC-FDMA was selected instead because its radio performance is much more suitable for the transmission of the LTE-UE. One of the key benefits of SC-FDMA is that the transmitter and receiver architecture are almost identical to those used in OFDMA, with the exception that the SC-FDMA provides lower PAPR.

Figures 8.6 and 8.7 indicate that the OFDMA and SC-FDMA are quite close to each other. More specific differences are:

- a constellation mapper that converts the incoming bit stream into the single carrier symbols (BPSK, QPSK, or 16QAM depending on channel conditions);
- a serial/parallel converter formats the time domain symbols into blocks that are fed as an input to the FFT process;
- M-point DFT converts the time domain symbol blocks into M discrete tones;
- subcarrier mapping allocates the DFT output tones to the specified subcarriers;
- N-point IDFT converts the mapped subcarriers into the time domain for the transmission
- cyclic prefix and pulse shaping are carried out;
- RFE converts the digital signal to the analog form and up-converts it to RF for the transmission.

The underlying SC-FDMA signal represented by the discrete subcarriers is a single carrier, unlike in the case of OFDM. This is a result of the fact that the SC-FDMA subcarriers are not independently modulated. This results in lower PAPR than in OFDM.

8.4.4 Conformance Test Aspects

In conformance testing of the LTE-UE elements, there are functional and performance requirements. The differences between models should thus be minimal. On the other hand, the specifications leave a certain error margin for the performance figures in such a way that, for example, the maximum radiating output power level can vary 2 dB over the defined level. In practice, depending on the LTE-UE model (USB stick, handheld device, or other—for example, integrated telemetry device), the external losses depend on the position. The body loss is obviously higher for the handheld devices, and if the antenna is integrated inside the equipment, the way the hand is covering the equipment can cause additional attenuations. These aspects should be estimated accordingly in order to design a realistic radio link budget.

In practice, the LTE-UE receiver conformance tests include performance measurements for the complete receiver. Prior to performance testing, the receiver's sub-blocks must be verified, and the possible source of specific distortions quantified and reduced. These investigations are executed during the receiver's design phase. If various receivers are included in the same device, the measurements must be carried out for each receive chain separately before the verification tests of the MIMO performance [4].

A simplified block diagram of a typical LTE transceiver is shown in Figure 8.9.

The components of the LTE-UE have a higher degree of integration, with single components performing multiple functions. This is a positive aspect from the customer's point of view as the physical size of the handset can be considerably smaller than in the early days of the 3G specifications. On the other hand, this also generates challenges for testing. In addition to the integrated multifunctional components, the other practical issue can simply be the reduced space in the printed board for the signal measurements.

In the case of LTE, the signal from a receiver's RF block cannot be demodulated into I and Q components using analog techniques. Instead, the down-converted IF signal is digitized via analog/digital conversion (ADC) as seen in Figure 8.9, and then fed to baseband processing in order to execute the demodulation and decoding. The measurement of the output of the ADC is not as straightforward as it is in the digital form. The LTE measurement equipment manufacturers thus typically provide the option to analyze the digital form of the ADC output directly with the measurement equipment. The analysis may contain, for example, the ADC performance via RF measurements directly on digital data, and numerical error vector magnitude (EVM) measurements with sufficiently in-depth graphical presentation.

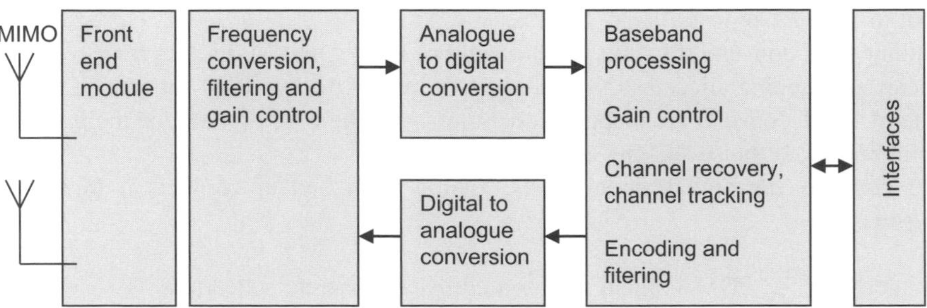

Figure 8.9 A general LTE transceiver block diagram.

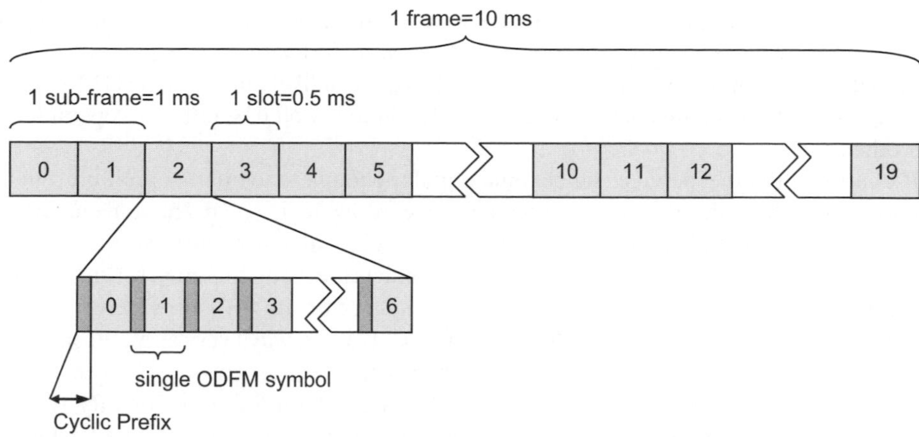

Figure 8.10 The generic LTE frame structure.

Figure 8.10 illustrates the FDD frame structure of LTE. As can be seen, the LTE frame duration is 10 ms. The LTE frame is divided into 10 subframes. Furthermore, each subframe contains two 0.5 ms long slots. The slots include six or seven ODFM symbols, for the normal and extended cyclic prefixes, respectively.

8.5 LTE Applications

8.5.1 Non-Operator Applications

Over The Top (OTT) solutions have been probably the most talked about subject in the mobile industry for some time now. In practice this refers to the threat caused by various Internet-based communication service providers to the mobile voice and messaging environment traditionally held by operators. Major VoIP and IM providers such as Skype, Google and Microsoft, along with a huge number of smaller innovative players, have been gradually entering the multi-billion dollar market of mobile communication. In practice this is handled via a client in the mobile device connecting to the service in Internet via IP access offered by the mobile operator—thus, in the typical OTT scenario, operator services are not required apart from the role of offering that "bit pipe."

Lack of major investment by the operator to supply customers with the OTT solution as well as the short period of time normally required to implement it have been generally described as the major advantages of the OTT solution. The biggest disadvantage for the operator is the financial loss related to the "bit pipe" role, which more-or-less means outsourcing the operator communication services to the OTT provider. This kind of "Access Network Holding Company" direction would mean a major change, even though one could argue that, for example, the FNO/ISP parts of many mobile operators have been doing exactly that for many years already.

One aspect of deploying a nonstandardized OTT application is that the operator doing this cannot expect other operators to select the same application. In practice, this could lead to a situation where only the customers of a particular operator are able to contact each other and, for example, a friend using another operator cannot be invited to a session. Traditionally, one of the key aspects of an operator service has been that it is interoperable across different operators and devices. So, for example, you can make a voice call or send SMS to basically anybody without any knowledge of what device and which network the recipient is using. With the OTT solution this situation changes quite a lot because it is quite unlikely that a single OTT would be deployed across the operators or that different OTT solutions would be fully interoperable with each other.

Of course an Internet service can become popular enough to avoid this problem, such as Facebook, which at the time of writing has (according to its own announcement) 500 million users. Skype and Google also have major reach, which for some users or use cases means that there is no problem because all your contacts are using the same service anyway. However, if Ylva is a customer of Operator A having a deal with OTT Provider A and Pelle uses Operator B utilizing OTT Provider B, it is probably not always very straightforward for Ylva and Pelle to communicate, especially when using services such as Presence or IM which normally lack the "interoperability via PSTN break-out/in" functionality typically existing for voice. For a normal Internet user this might not be a major issue as such because they are more or less used to, for example, IM service providers not being fully interoperable with each other, but in the mobile phone environment this might not be that obvious.

One of the key disadvantages of the OTT solution is generally though to be the lack of SRVCC (or similar) functionality offering LTE/SAE IP-based voice with more-or-less seamlessly handover to 2G/3G CS voice whenever the user moves out of the LTE network's coverage for whatever reason. Depending on the operator, this can be a big issue, especially if the LTE coverage is sparse. On the other hand, in a geographically limited location such as Hong Kong it is possible to have a full LTE/SAE coverage rather easily—thus interaction with 2G/3G is a less important aspect.

A rather brutal way around this problem is for the operator simply to state that there is no handover functionality between LTE and 2G/3G networks. That is, users should understand when they, for example, start a voice call in the LTE cell and then move to a location lacking LTE coverage, the voice call is simply dropped. The same goes for a data session. From the deployment perspective this is an extremely straightforward and easy way because an operator does not have to deploy any kind of SRVCC, PS Handover functionality, and so on, at all. However, from the marketing point of view it certainly poses a major challenge—how can one make customers understand and agree to this clear lack of functionality when compared for example to the current situation where handovers between 2G and 3G networks are more-or-less seamless?

One could argue, though, that the existing 3G network can be used to provide the OTT solution with necessary fallback functionality—that is, to use the IP-based voice over the data connection in the same way regardless of whether there is LTE or 3G coverage. In some sense this approach makes sense because the data connections offered by the typical 3G network are probably good enough to support most OTT requirements. This requires that there should be a countrywide 3G network, because in the 2G network the OTT solution typically does not work so well. One downside of the always-on OTT solution is battery consumption. Most devices on the market today are unable to keep up the data connection required by the OTT solution for a whole working day, which can be considered to be a kind of minimum requirement for a regular phone used by a typical office worker.

From the network point of view, the OTT solution is seen to be more-or-less the same as pure data using best effort mechanisms—that is, it might have an impact compared to the more traditional operator solution that is more or less completely separated. Depending on the implementation there are some ways around this, involving the operator setting specific QoS/QCI parameters to the OTT communication traffic to "promote" it above the basic data, but they are not necessarily straightforward to set up, especially when thinking about roaming scenarios involving support required from the other operators. VoLTE, on the other hand, benefits from using, for example, dedicated QoS parameters and a dedicated APN.

Anyway, it is obvious that an operator needs to consider this complex area carefully, from the technical point of view and especially from the commercial perspective.

8.5.2 Rich Communication Suite

Rich Communication Suite (RCS) is a GSMA program aiming to develop an interoperable communication package using IMS. The IMS core system is used as the underlying service platform, taking care of issues such as authentication, authorization, registration, charging, and routing. RCS presents a new IMS based operator service, which gives alternatives for OTT solutions. For further information see [5].

From the end-user point of view RCS would enable communication such as instant messaging, video sharing and buddy lists. These capabilities would be available on any type of device using an open communication between devices and networks. In a nutshell, RCS can be considered to be a set of services based around the address book in the phone or other device. The main features of RCS are:

- enhanced phonebook, with service capabilities and presence enhanced contacts information;
- enhanced messaging, which enables a large variety of messaging options including chat and messaging history;
- Enriched Call, which enables multimedia content sharing during a voice call.

One major theme for the RCS is the service capability, showing the services that the user can use with friends.

Wider and large-scale IMS deployment, interoperability between different terminal vendor RCS clients, and RCS service interworking between operators are listed as the key aims of the RCS Initiative. RCS reuses existing services and components—for example SMS and MMS are used as a part of RCS as they are today: no modifications are needed.

RCS is developed in a series of releases, in a somewhat similar fashion to 3GPP. The first release was, rather logically, called RCS Release 1 and it was finalized in December 2008. Functions included Social Presence and capability exchange, IM, Rich Call (Image and Video Share) and a basic Network Address Book offering synchronization functionality for the address book in the device with the network. Release 1 is a pure mobile-only service. In essence this means that the native CS voice functionality of 2G/3G networks is used as a part of RCS.

Release 2 was finalized in June 2009. The main part of this release is the introduction of Broadband Access Support—for example a RCS PC client using ADSL access network is now a part of RCS environment. Release 2 adds PS Voice for those Broadband Access clients using access networks which do not offer native CS voice support, like WLAN, ADSL or IPoAC. This PS Voice is an IP-based solution using a subset of the 3GPP-defined MMTel (Multimedia Telephony) standard. The reason why PS Voice was defined instead of using VoLTE, offering an identical mechanism, was simply that in the year 2009 VoLTE did not yet exist.

Release 3 was finalized in December 2009 with new features of NVAS (Network Value Added Services), which is a network-based service supporting media processing during the share experience. It can be used, for example, to convert images sent via Image Share automatically. GSMA PRD IR.84 based Video Share has also been introduced, allowing Video Share to be used without an underlying voice call. Location information is a new part of the presence information exchanged between the users. Release 2 didn't have a full functionality of send and receive of SMS/MMS on the BA client due to missing 3GPP specification but Release 3 fixes that. Finally, it is worth noting that it is now possible to use the BA client as primary client—that is, the BA client can act as a standalone device whereas in Release 2 the user always had a mobile client as the primary contact point, with the BA client a secondary device. This delivers the complete multidevice environment for RCS.

Latest Release 4 was finalized in Dec 2010, including functionality:

- LTE Market;
- replacement of PS Voice with IR.92 VoLTE;
- voice + supplementary services;
- video quality enhancement;
- addition of 384 and 768 kbit/s bearers and corresponding modes of H.264 codec;
- messaging evolution;
- alignment IR.92 with text messaging;
- introduction of OMA CPM (Converged IP Messaging) as "Messaging Service Evolution";
- IP based Messaging;
- SMS service enhancement overcoming the current SMS service limitations;
- MMS service enhancement overcoming the current MMS service limitations;
- SMS/MMS service enhancement in multi-device configuration;
- inclusion of Legacy SMS/MMS users in an RCS chat session;
- network-based Common Message Storage;
- content sharing enhancement;
- image share—real time synchronization of interaction;
- video share—real time synchronization of interactions;
- web screen sharing;
- network address book synchronization enhancement;
- synchronization service discovery;
- synchronization process triggered from network;
- profile and contact management enhancement;
- group of contacts;
- VIP groups.

Full implementation of Release 4, including, for example, CPM functionality means a major change to the operator core network. So from the deployment point of view, for example, Release 3 is easier to handle. It is possible to select a feature subset if the operator so requires—that is, it is possible to launch, for example, Release 4 without Content Sharing Enhancement just as it is possible to extend the basic Release with some operator-specific application/service.

Figure 8.11 shows an example of RCS Release 4 architecture. The PS/CS GW is used for interworking between CS (Circuit -Switched) and PS (Packet -Switched) voice—that is, VoLTE. SUPL indicates a Secure User Plane Location element as documented in [6]. Msg

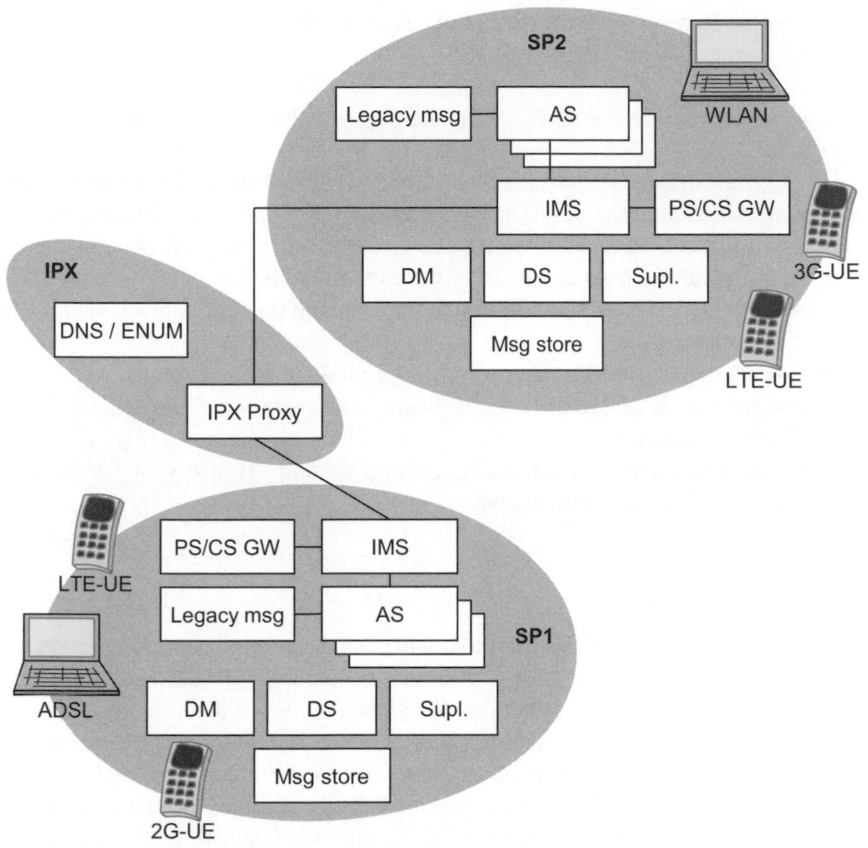

Figure 8.11 Simplified example of RCS architecture.

Store relates to the CPM (Converged IP Messaging) Message Storage Server as illustrated in OMA CPM specifications. Legacy Msg refers to SMS/MMS service used via IWF (Interworking Functions) located in the ASs (Application Servers) group, which in addition to IWF node(s) includes various other nodes used by the RCS services, for example:

- Presence Server;
- Messaging Server;
- XDM Server;
- Multimedia Telephony Application Server;
- Video Share Application Server, as utilized in [7].

The latest development of RCS is called RCS-e, which basically takes RCS Release 2 and enhances it by making many features optional and adding some new functionality. RCS-e has been developed as a "go-to-market" initiative by big European operators such as Deutsche Telecom, Orange-FT, Telecom Italia, Telefonica and Vodafone, which are working towards commercial launches of RCS-e starting in late 2011. In essence, RCS-e provides a "simple interoperable extension to voice and text today." RCS-e was announced at the Mobile World Congress in February 2011 [8].

The following differences between RCS Release 2 and RCS-e have to be taken into account when thinking about the RCS deployment:

- OMA Presence is optional in RCS-e; The support is mandatory in clients but not on the network side.
- Use of SIP OPTIONS messages is increased and enhanced in RCS-e to give a "dynamic capability exchange" functionality; (A) RCS uses SIP OPTIONS typically only when setting up a voice call (to find out the Sharing capability of the other end), otherwise the capability exchange is handled via Presence; (B) RCS-e uses SIP OPTIONS for all the contacts for first time discovery + periodic poll (in addition to On Demand poll per contact, which is similar to the RCS mechanism).
- Video and Image Share and File Transfer are optional for RCS-e clients.
- Details of Broadband Access (e.g., PC client using ADSL) are out of scope for the first release of the RCS-e specification.
- RCS-e also modifies aspects related to provisioning and discovery of friends and adds specific UX (User Experience) guidance.

8.5.3 LTE/SAE and RCS

"Why is RCS relevant to the LTE/SAE deployment considerations?" the reader might now ask. The answer is that by deploying RCS Release 4 the operator gets not only the "rich communication service package" offered by RCS but also the "basic voice package" offered by VoLTE. This is due to RCS fully endorsing IR.92 (VoLTE) in Release 4 as the solution for providing the necessary voice functionality. RCS and VoLTE are aligned and interoperable. This includes functionality such as Supplementary Services. Some discussion is still ongoing related to issues such as how to reuse the VoLTE principles in non-LTE access networks but, in practice, this alignment allows, for example, a RCS Broadband Access client using an ADSL access network to talk with a VoLTE terminal using a LTE access network.

As a part of Release 4 a specific IR.92 Endorsement Document has been produced. This shows what parts of IR.92 RCS clients using non-LTE access networks must implement. For example, LTE RAN-related aspects such as RoHC or LTE DRX are not applicable for devices that do not use the LTE access network but rather some other IP-based broadband access network such as WLAN [9].

A more general RCS and LTE market alignment has also taken place with the GSMA RCS Program. This is documented in a separate, more high-level positioning paper "Considerations for an RCS operator when introducing LTE" [10].

RCS does not require an LTE/SAE network as such but it can certainly benefit from it. The increased bandwidth offered by LTE/SAE can be seen by the end user as a better quality video stream when using Video Share, for instance. Also other type of services like Social Presence are able to exchange information such as presence updates and profile images between the users in a smoother way because they are able to take the advantage of lower latency and increased bandwidth of LTE/SAE. Use of LTE/SAE should also somewhat increase the battery life of the RCS device. On the other hand, LTE/SAE can benefit from RCS as it offers a way to provide customers with something other than just the basic voice service.

It is worth noting that, specifically concerning RCS-e, the details of LTE/SAE can be considered to be mostly out of scope, at least for the current release of RCS-e available at the time of writing. That is due to RCS-e initiative concentrating on the go-to-market aspects—the very near future—rather than thinking about VoLTE and other more long-term goals. Thus the details of how RCS-e works with LTE/SAE remain to be seen. In any case, for the basic RCS that is defined in the Releases 1 to 4, this relationship is clear based on the description above.

Generally speaking the conclusion seems to be that VoLTE and RCS are very much complementary services: one concentrates on providing the voice service to end users whereas other offers a variety of multimedia applications to enrich that basic voice functionality. Both reuse the same basic standardized components such as SIP, XDM and IMS. Thus, from the deployment perspective, it would appear logical to think about implementing RCS when deploying VoLTE and vice versa. This would likely ease the financial burden of deploying the IMS core system, application servers and integration with the existing network because an operator would get more-or-less "two birds with one stone" instead of trying to justify that implementation cost just for RCS or just for VoLTE. The roaming and interconnection aspects of these two interoperable IMS based services are also very similar—that is, a model deployed for VoLTE is also very feasible for RCS (and vice versa).

One related aspect of this is that the IMS core system deployed for the needs of RCS/VoLTE can obviously be reused for other purposes, such as replacing the existing CS core network of FNO with IMS, which is taking place at the moment as the first real commercial use case for IMS. Naturally, this is more relevant for those operators with both mobile and fixed arms.

Finally, one must remember that the RCS track is not the only option for deployment of "rich communication clients" for LTE/SAE—there are plenty of alternatives, especially in the OTT world. Those are described in Section 8.5.1.

References

[1] 3GPP TS 36.306. (2010) *Evolved Universal Terrestrial Radio Access (E-UTRA); User Equipment (UE) radio access capabilities (Release 8)*. 14 p. V. 8.7.0, 2010-06, 3rd Generation Partnership Project, Sophia-Antipolis.

[2] 3GPP TS 36.101. (2010) *User Equipment (UE) radio transmission and reception*, V. 8.12.0, 3rd Generation Partnership Project, Sophia-Antipolis.

[3] 3GPP TS 36.331. (2011) *Evolved Universal Terrestrial Radio Access (E-UTRA); Radio Resource Control (RRC); Protocol specification*, V. 8.14.0, 2011-06-24, 3rd Generation Partnership Project, Sophia-Antipolis.

[4] Rumney, M. (2010) New challenges for LTE and MIMO receiver test challenges, www.eetimes.com/design/test-and-measurement/4210806/New-challenges-for-LTE-and-MIMO-receiver-test?pageNumber=1 (accessed August 29, 2011).

[5] Holma, H. and Toskala, A. (2011) LTE for UMTS. *Evolution to LTE-Advanced*. 2nd edition, John Wiley & Sons, Ltd, Chichester.

[6] GSM (n.d.) Description of Rich Communications, www.gsmworld.com/rcs (accessed August 29, 2011).

[7] GSMA IR.84. (2009) *Video Share Phase 2 Interoperability Specification 2.0*, November 12.

[8] GSM (2011) Rich Communications news, www.gsmworld.com/newsroom/press-releases/2011/6047.htm (accessed 29 August 2011).

[9] GSM (2011) Rich Communication Suite Release 4. Endorsement of GSMA IR.92 GSMA VoLTE—"IMS Profile for Voice and SMS" Version 1.0, 14 February 2011, www.gsmworld.com/documents/rcs_rel4_end_gsma_ir92_v1.pdf (accessed August 29, 2011).

[10] GSM (n.d.) RCS over LTE. Considerations for an RCS operator when introducing LTE, www.gsmworld.com/documents/RCS/RCS_over_LTE_deployment_considerations_v1.0.pdf (accessed August 29, 2011).

9

Voice Over LTE

Juha Kallio, Tero Jalkanen, and Jyrki T. J. Penttinen

9.1 Introduction

Voice is perceived to be the most important service supplied by today's communication service providers for their end users. In this sense, voice also represents an important service when considering the commercial LTE service portfolio despite the fact that early deployments of LTE may be more related to datacentric services such as browsing or video. In the early stage of LTE standardization it was decided that LTE architecture would no longer have Circuit-Switched technology [1], similar to that used as the backbone of today's mobile networks. This brought number of new challenges for communication service providers who have to provide voice and Short Message Service in a new manner without increasing the time schedule and complexity of the overall LTE deployment project. Initially, this issue was considered in 3GPP to be handled with the introduction of IP Multimedia Subsystem (IMS) [2] but it turned out that deployment of this technology was not yet mature enough to be used in some networks.

This issue was enough to drive industry to push a number of different rather fragmented alternatives for voice over LTE. These alternatives ranged from use of the CS network to provide voice services for LTE attached subscribers with the Circuit-Switched call-control protocol stack all the way to more radical alternatives that use third-party VoIP services, such as those that are currently already used a lot as "Over-the-Top"-communication services on the Internet. Despite these alternatives the outcome of discussion within 3GPP was that two ways were selected to provide voice for LTE attached subscriber, which therefore can be considered to be the most important ones.

The first architectural model being standardized as part of 3GPP Release 8 is called CS Fallback for Evolved Packet System (EPS) or CSFB [3], [4], which was defined to be a gap-filler solution before the complete IMS is available end-to-end, but LTE has been launched in order to fulfill needs of increasing mobile broadband services. However, now CS Fallback for EPS is also understood to complement deployment of a communication service provider voice and messaging service even when native IP-based voice has been deployed. The original idea of being merely a gap-filler type of solution, CS Fallback for EPS can therefore currently be considered as a typical procedure in LTE.

The LTE/SAE Deployment Handbook, First Edition. Edited by Jyrki T. J. Penttinen.

Since its introduction, CS Fallback for EPS has been divided into two different functionalities, which can be deployed either separately or together depending on business requirements. The first functionality, which is also considered important in the early stage of LTE deployment, is the ability to transfer Short Message (SM) over LTE as signaling information, and use the existing Circuit Switched (CS) core network to assist in this process. This is called as SMS over SGs, which was given this name because of the new interface that is required to connect Evolved Packet Core and the Circuit Switched core networks in order to enable this functionality [5]. The second functionality, which again has more uncertainties from the call-performance point of view compared with today's mobile voice service, is the ability to complete voice calls, transfer Unstructured Supplementary Service Data (USSD), perform Location Services (LCS), and manipulate supplementary service settings via call unrelated signaling procedures by performing, at first, a fallback to legacy GERAN/UTRAN radio access for the duration of this procedure.

The second architectural model, which is commonly considered as the target architecture, is based on the use of the IP multimedia subsystem. GSMA has defined is as "IMS Profile for Voice over SMS" [4]. In this profile, voice and Short Message Service have been defined by referring to the 3GPP Release 8 standards.

Readers should not consider these two mechanisms (CS Fallback for EPS and IMS based voice over LTE) as competitive but as complementary. Functionality-wise, CS Fallback for EPS can be deployed simultaneously with IMS-based Voice over LTE and the logic to select appropriate mechanism for call delivery or attempt will be performed automatically either by the network or the terminal.

CS Fallback for EPS may also need to be used to complement a lack of native VoIP capabilities in the network, for instance when sufficient IMS roaming has not yet been established or when the underlying Evolved Packet System does not support emergency services or location services as defined by 3GPP.

Therefore, despite the number of different architectural alternatives available to solve the complete voice-over LTE challenge, this chapter will provide an insight into the details and technologies behind the architectures.

9.2 CS Fallback for Evolved Packet System

CS Fallback for Evolved Packet System (EPS) is defined in 3GPP Release 8 to enable service providers to continue use of Circuit-Switched networks to provide voice and short-messaging services for EPS subscribers. Use of this mechanism requires support from LTE radio access, the evolved packet core as well as from the Circuit-Switched network. The Stage 2 specification defining overall functionality of CS Fallback for EPS is 3GPP TS 23.272 [4].

CS Fallback for EPS has been defined to cover two separate functionalities, which can be either used simultaneously or independently. The first functionality provides the capability to transfer Short Message (SM) over LTE radio access and the second functionality allows use of voice or video telephony as well as other services such as Unstructured Supplementary Service Data (USSD), and Location Services (LCS) via the traditional Circuit Switched core network. The latter functionalities are achieved by performing fallback from LTE radio access into legacy GERAN/UTRAN radio access for the duration of the procedure—in practice, this means, for example, for the duration of the voice or video call. During the time the terminal is engaged with the ongoing call and attached to GERAN or UTRAN radio access then possible active packet-switched connections are downgraded to a level that is available from that legacy

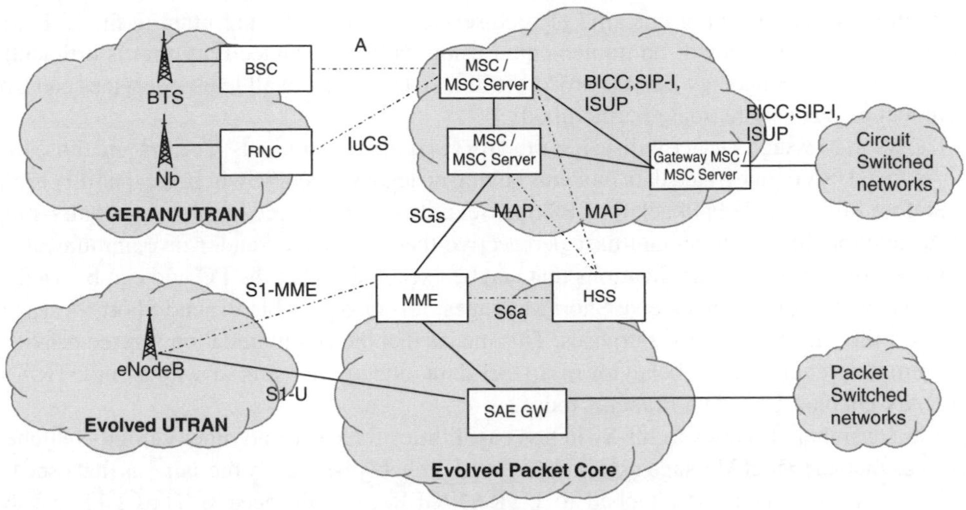

Figure 9.1 The architecture of CS Fallback for EPS.

access. This could mean that it would still be possible to use simultaneous packet and voice connectivity in case of UTRAN or complete suspension of packet connection in case of GERAN without Dual Transfer Mode (DTM) functionality. Therefore it is possible to continue use of packet-switched connection (e.g., registration to Internet social network services) despite the fact that fallback has been performed, depending on network architecture and capabilities.

From an architectural perspective, CS Fallback for EPS requires that the Evolved Packet Core, namely MME, and the corresponding Circuit-Switched core network, namely MSC or MSC Server, is connected using SGs interface that is based on Stream Control Transfer Protocol (SCTP). 3GPP selected a very similar Gs interface as basis of SGs interface but extended that with new requirements caused by new use cases required for CS Fallback for EPS. Additionally SGs interface no longer require use of traditional SS7 infrastructure but use preconfigured SCTP associations (IP connections) between network elements identified by pairs of IP addresses and port numbers.

CS Fallback for EPS does not require any specific provisioning for the HSS. It reuses existing teleservices for voice service, short messaging services and synchronous transparent data services (i.e., video telephony). Communication service providers, after enabling CS Fallback for EPS support in the network, will automatically allow all supporting terminals to use it based on configuration within Evolved Packet Core, namely in MME. MME is able to prevent use of CS Fallback for EPS or even SMS over SGs but these mechanisms are not defined in any 3GPP standard and are not subscription-specific.

Figure 9.1 represents the architecture of CS Fallback for EPS.

Instances of the use of this technology will be described in more detail below.

9.3 SMS Over SGs

Short Message Service has been, and continues to be, tremendously successful. Many different services have been developed across the globe that use Short Message Service as enabling technology. Continuation of this kind of such service is extremely important in the LTE era. Naturally, new services will be implemented using native IP connectivity that is efficiently enabled by LTE technology but this process will take time and not all legacy services can (for technical or business reasons) be modified.

The Short Message Service will be required for two kinds of terminals. The first are those that are, in fact, USB or integrated data modems having no legacy Circuit-Switched capability at all. Therefore full-scale CS Fallback for EPS functionality is not required by such terminals—only SMS functionality. From an end-user perspective there are some vendor or communication service-provider branded applications that can be executed within the PC and can be used to either send or, more often, receive Short Messages. It is also possible to send Short Messages for Over the Air configuration purposes. This means that the communication service provider can remotely configure the behavior of a USB data modem in a similar way as in UTRAN/ GERAN-enabled USB data modems today.

If the terminal also has Circuit-Switched capabilities for voice and other Circuit-Switched services then the Short Message user interface will probably be exactly the same as that used by end user when terminal is attached to UTRAN/GERAN radio access. Thus LTE can be considered as yet another form of transport for the old Short Message Service.

It is also likely that, in the first LTE deployments, when LTE roaming capability is introduced, communication service providers will face the same kind of requirements (possible regulatory ones) to provide a Short Message Service in order to inform roaming customers about cost of roaming data service and also inform them when the cost of service exceeds a certain threshold. This is known as "roaming bill shock prevention" and is currently mandated by the European Union (EU).

9.3.1 Functionality

In order to support delivery of Short Messages over LTE in a similar fashion to the way in which Short Messages are delivered today in existing mobile networks, changes have been introduced to all relevant signaling interfaces between the terminal and the serving MSC or MSC Server network element.

In a LTE-Uu interface between the terminal and the eNb, a Short Message is transferred transparently within a Non-Access Stratum (NAS) protocol defined in 3GPP TS 24.301 [6]. This interface is between the terminal and the Mobility Management Entity (MME). The MME then forwards the content to the serving MSC or the MSC Server, which supports an SGs interface as defined in 3GPP TS 29.118 [7]. Short Message content can be either a single or a concatenated payload and both mobile-originated and mobile-terminating Short Message are supported.

Support for SMS over SGs does not require any specific functionality from an Evolved Nb (eNb) element because the actual payload is transferred as NAS signaling transparently between the LTE-attached terminal equipment and the serving MME.

The current understanding in the industry is that 3GPP Release 8 provides a sufficient standardization baseline for SMS delivery within LTE signaling (NAS), and no improvements are required from later 3GPP releases.

9.3.2 Combined EPS/IMSI Attachment

In order to be able to access the services provided by the circuit-switched network via the SGs interface, the terminal should initiate a combined EPS/IMSI attachment procedure.

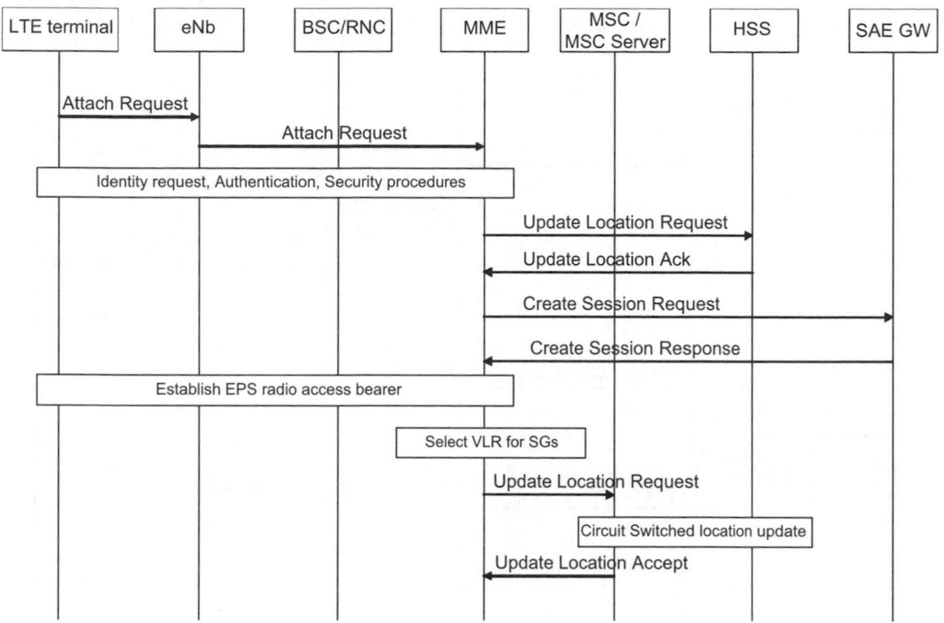

Figure 9.2 The combined EPS/IMSI attachment procedure.

The condition for the functioning of this is that the procedure is supported and allowed by MME for the service in question so that the Location Update via SGs interface can be performed in the MSC or MSC Server element, and that these elements support the SGs interface.

The terminal may perform the attachment with a more detailed indication of the nature of attachment/capabilities of terminal. If the terminal supports only SMS over SGs, then it may indicate this with an "SMS-only" indication in the Combined EPS/IMSI attachment to MME.

The MSC or MSC Server will not perform any authentication because the subscriber has already been authenticated by EPS. However the MSC or MSC Server will perform Update Location towards the HSS (HLR) in the normal fashion, as the location update would have been received, for example, via Gs interface, and will retrieve the user profile from HSS in return.

After this procedure has been completed then MSC or MSC Server will consider the subscriber to be reachable via SGs interface, and mobile-terminating transactions such as Short Message will be routed via that interface.

Figure 9.2 represents the combined EPS/IMSI attachment procedure.

After this procedure has been completed successfully then the terminal can either originate or terminate service requests that use procedures defined for CS Fallback for EPS and SMS over SGs.

9.3.3 *Mobile Originated Short Message*

Mobile originating (MO) Short Message delivery is initiated by a terminal by performing UE-triggered Service Request towards Evolved Packet Core. After this event, the terminal is able to send an Uplink NAS Transport message to the MME, which is mapped into the Uplink Unitdata message of SGs interface towards serving MSC or MSC Server. These uplink messages contain the actual payload of the Short Message.

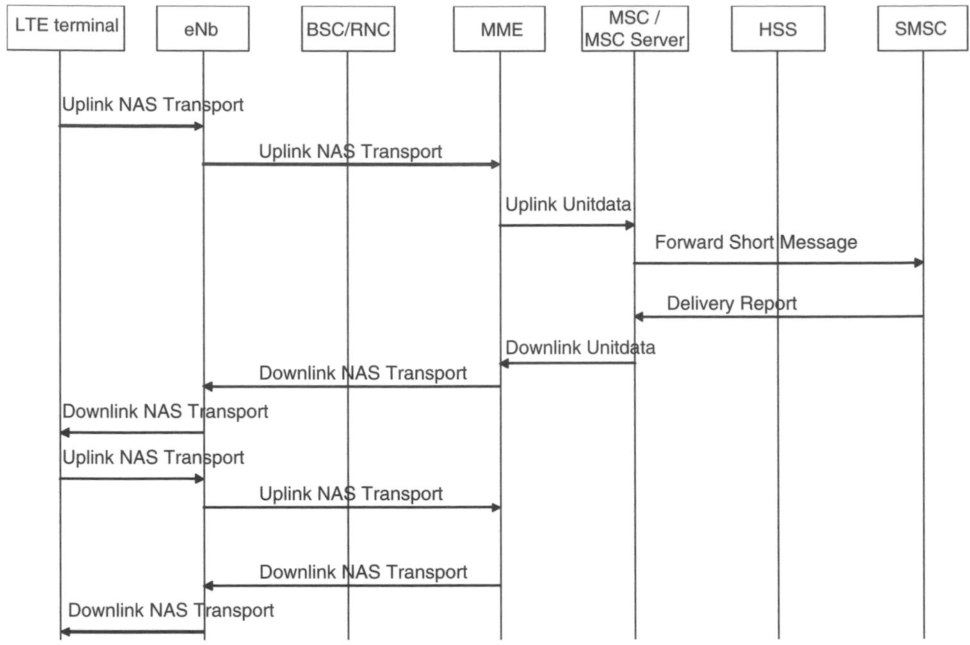

Figure 9.3 The signaling flow for the mobile originated short message delivery.

The MSC or the MSC Server will perform the necessary tasks for the mobile-originated Short Message in a similar fashion to that which would have been used if the message had been received from terminal via GERAN/UTRAN radio access. This means that, for example, the required Intelligent Network services and charging can be performed at this point.

Eventually, the message is forwarded to the Short Message Service Center (SMSC) of the served subscriber via MAP interface, which then complete the Short Message delivery in the normal fashion towards the originating terminal.

The message flow in Figure 9.3 represents the originating Short Message service delivery via LTE radio access using SMS over SGs.

9.3.4 Mobile Terminating Short Message

Mobile-terminating (MT) Short Message delivery is a slightly more complicated procedure than mobile-originating delivery. In this case, as the terminal is indirectly attached to the serving MSC or MSC Server (Circuit-Switched core network) by MME, then the SMSC of the sender will route the Short Message by using the MAP interface to the particular MSC or MSC Server having SGs association for served recipient. This procedure is handled in exactly the same fashion as Short Message delivery in mobile networks today. An MSC or MSC Server with an SGs association will perform the necessary tasks for a mobile-terminating Short Message in a similar fashion as is performed today if the subscriber is registered via GERAN/ UTRAN radio access. After these tasks, MSC or MSC Server will perform paging via SGs interface. If paging is successful, the Short Message will be delivered via SGs interface within the Downlink Unitdata message to MME, which again will forward the payload transparently to the terminal with the Downlink NAS transport message. Finally, the delivery of the Short

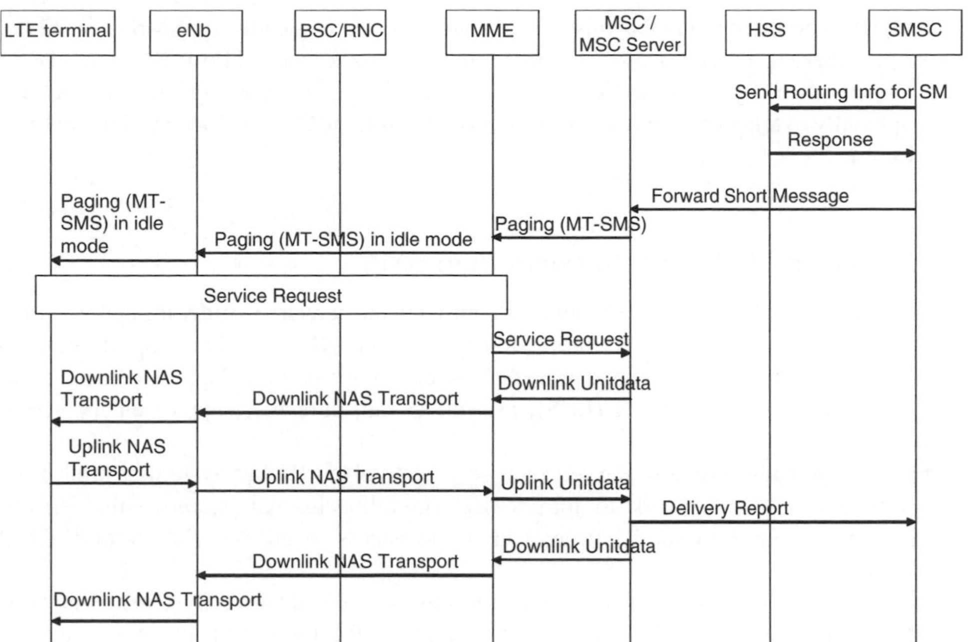

Figure 9.4 The signaling flow for the mobile terminating short message delivery.

Message will be completed by exchanging the delivery report from the terminal to SMSC of the sender, and by sending an acknowledgement of the completion of the procedure to the terminal of the recipient.

The message flow in Figure 9.4 represents the terminating Short Message delivery via LTE radio access using SMS over SGs.

9.3.5 Deployment View

When a communication service provider is willing to deploy SMS over SGs functionality in order to support the Short Message Service for LTE-attached subscribers as defined in 3GPP TS 23.272 then this can be achieved by implementing support for this functionality into the serving Evolved Packet Core (EPC) as well as the serving MSC or MSC Server network elements.

The SGs interface needs to be designed taking into account the Short Message Service traffic profile (dynamic capacity) as well as the number of subscribers (static capacity) required from VLR, which then provides information about the required amount of processing capacity from the MSC or MSC Server point of view. It can be expected that, overall, the Short Message Service traffic profile is likely to be similar to the traffic profile that is used today with mobile broadband subscriptions.

The deployment of SMS over SGs can be designed in such way that the SGs interface from MMEs to MSC or MSC Server network elements does not have to be aligned with the network architecture of the GERAN or UTRAN radio access network. This means that there is no need to implement SGs interface from MME to the MSC or MSC Server that controls BSC or RNC of the same geographical area (Location Areas, LA). Instead it is possible to nominate a few MSC or MSC Server network elements and in this way centralize the configuration.

The reader should be aware that this kind of solution is recommended for SMS over SGs use cases but if there is a need for use cases that require fallback from LTE to GERAN/UTRAN radio access then, again, network architecture should be designed in such a way that geographically overlapping areas are designed jointly with the Circuit-Switched core network SGs design.

9.4 Voice and Other CS Services than SMS

The complete CS Fallback for EPS feature is required in the network in order to support Circuit Switched (CS) services other than Short Message Service (SMS), and to support voice and video telephony, and non-call-related procedures such as Unstructured Supplementary Service Data (USSD), Location Services (LCS), as well as call-unrelated supplementary service control procedures.

These functionalities require the terminal serving Evolved Packet System as well as the Circuit Switched core network to jointly perform fallback—that is, move the terminal for duration of requested service from LTE radio access to either GERAN or UTRAN radio access.

These functionalities require that terminal, serving the Evolved Packet System as well as the circuit-switched core network, will jointly perform fallback—that is, move terminal for duration of requested service from LTE radio access to either GERAN or UTRAN radio access. In case of voice and video telephony call, establishment will be suspended for the duration of fallback procedure, including possible handover of Packet Switched connections to target radio access. The perceived delay differs depending on the technology used for radio access and in some cases it can be considered to be unacceptable from the end user's point of view. For this reason, and because service providers are extremely concerned about the end user experience, there have been efforts in the industry in 3GPP Release 9 to improve the radio access side of LTE access network in order to reduce this delay. Also these technologies depend on the target access technology (GERAN or UTRAN).

Such functionalities include for instance RRC redirection procedure that is triggered by eNb towards the terminal with or without supporting information to assist terminal to select appropriate target radio access for fallback. Similarly, it is possible that Packet Switched connections are handed over from LTE to UTRAN based on PS-PS relocation during fallback procedure. Whatever mechanism will be applied and agreed in the industry to be used with the terminal will impact on the duration of call establishment and therefore also on the end user's experience. However, the current understanding is that end-user experience can be kept at a reasonable level and thus CS Fallback for EPS functionality for voice and video calls is still a valid proposition for networks.

For USSD, LCS and call unrelated supplementary service procedures, similarly, fallback is performed for the duration of procedure. This way no changes were required for 3GPP specifications in order to transfer related signaling procedures via the Evolved Packet System and the serving MSC or MSC Server in the Circuit-Switched network.

It is currently possible that CS Fallback for EPS will be further enhanced in forthcoming releases beyond 3GPP Release 8 in order to improve the end-user experience by reducing the call establishment time. Some of these mechanisms are already being standardized as part of 3GPP Release 9, which is therefore considered by some as the most suitable standardization baseline for commercial deployment. Nevertheless, it is not clear yet if all the required items in the end-to-end deployment are able to support this baseline. Functionality-wise it is also possible to have mixed configurations in cases where, for example, the Circuit-Switched core network is only based on 3GPP Release 8 but the Evolved Packet System is based on newer 3GPP release 9. Differences in SGs interface are not so great that they would prevent this kind of configuration altogether. However it is naturally feasible to have the same baseline in all related entities end-to-end.

The following paragraphs describe more details use cases other than Short Message Service made possible through use of CS Fallback for EPS.

9.4.1 Voice and Video Call

Voice and video telephony is currently supported by mobile Circuit-Switched networks and therefore support is expected to continue in LTE as well. CS Fallback for EPS functionality can support this by performing the fallback procedure that is, moving an LTE-attached terminal for the duration of a procedure to GERAN/UTRAN access and then returning the terminal back to LTE after the procedure is completed.

In case of mobile originated voice and video call this means that terminal jointly with eNb and MME decides to perform fallback. Such fallback procedure can be either for normal voice/video call or emergency voice call. In case of emergency call MME may indicate nature of call via S1-AP request message via S1 interface to the eNb, which then can take this information into account in order to secure that fallback can be successfully performed in any condition.

MSC or MSC Server having SGs association does not have any visibility for mobile originating calls performed by terminal. Actual call establishment procedure is performed by MSC or MSC Server which serves the GERAN/UTRAN radio access to which terminal performs fallback.

For mobile-terminating voice and video calls, since the terminal is attached to the serving MSC or MSC Server via SGs interface by MME, the routing of mobile terminating calls is performed by using an HLR enquiry from the gateway MSC or MSC Server. The HLR will request the VLR from the serving MSC or MSC Server, to allocate a mobile station roaming number (MSRN), which then is used by gateway MSC or MSC Server to route the call. Eventually, when the call is received by the serving MSC or MSC Server that has the SGs association for the called subscriber's terminal, this MSC or MSC Server performs a paging procedure via the SGs interface by routing the message to MME and finally to the terminal. This paging message may contain an optional calling-line identity parameter, which can be shown to the called subscriber together with information about the incoming call. If the called subscriber is not willing to perform the fallback, for instance if there is some critical data application connected via LTE, then it is possible to reject incoming call at this stage without causing fallback to occur. However, if the terminal accepts the incoming paging it replies with an Extended Service Request message to MME, which again replies to the serving MSC or MSC Server with a Service Request message via SGs interface. This message, when received by the MSC or MSC Server, can be compared to the successful acceptance of a CS Fallback for EPS invocation. After this, the terminal starts to perform PS-PS relocations to move the active PS connections to GERAN/UTRAN as well as moving itself to GERAN/UTRAN radio access. In case this procedure is completed successfully then terminal will send a paging response to MSC or MSC Server that controls the GERAN/UTRAN radio access in order to complete call is completed normally.

Figure 9.5 represents a basic mobile-originating call from an LTE-attached terminal using procedures defined in CS Fallback for EPS.

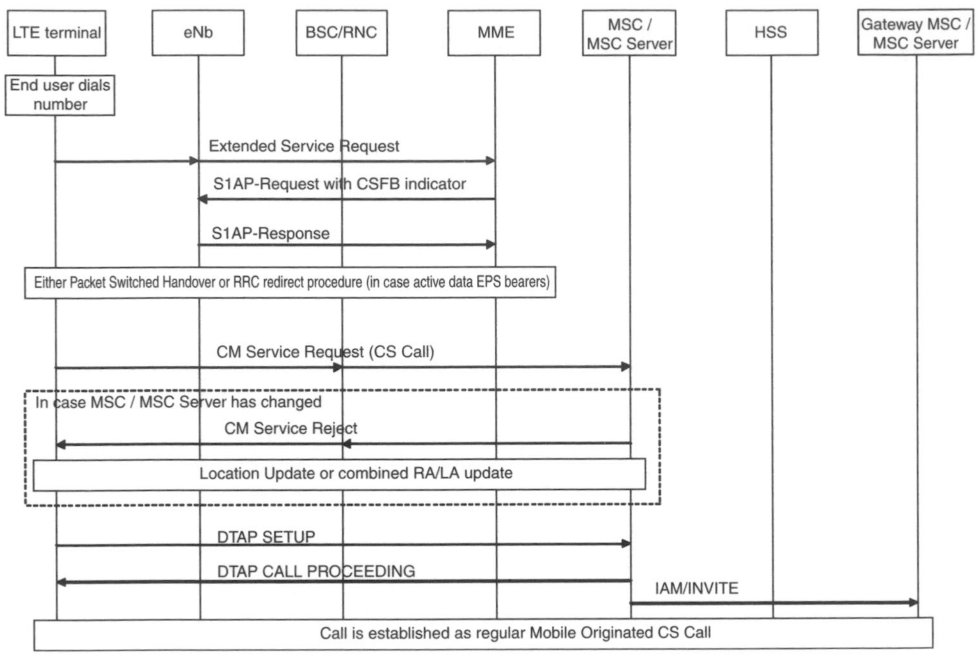

Figure 9.5 The mobile originating call from LTE attached terminal via the CS Fallback functionality.

A mobile-terminating call attempt when the called subscriber's terminal is attached to LTE is a more complicated one than the mobile originating call or other call-unrelated procedures. There is greater complexity involved because Tracking Areas and Location Areas may not be perfectly aligned and therefore the MSC or MSC Server that will be contacted by terminal after fallback has been performed may be different from the MSC or MSC Server that was originally paging the terminal via SGs interface. 3GPP has addressed this problem by reusing the existing Mobile Terminating Roaming Retry (MTRR) procedure in order to reroute the call within the Circuit Switched core network via a new MSC or MSC Server, different from one that was originally used for paging the terminal via SGs. This procedure is defined in 3GPP TS 23.272 and originally in 3GPP TS 23.018 [8].

Figure 9.6 represents the mobile-terminating call to an LTE attached terminal using procedures defined in CS Fallback for EPS without involvement of MTRR.

If MTRR is needed in conjunction with a mobile-terminating call then the call flow in Figure 9.7 occurs. The use of the MTRR requires end-to-end support for the procedure in the Circuit-Switched core network, and this needs to be indicated using a specific flag from the gateway MSC or MSC Server via HSS to the serving visited MSC Server. In this way the serving visited MSC or MSC Server is able to detect that the MTRR procedure can be performed.

9.4.2 Call Unrelated to Supplementary and Location Services

Call-unrelated procedures for Unstructured Supplementary Service Data (USSD), Location Services (LCS) as well as call-unrelated supplementary service control also uses similar fallback to that used for voice/video calls—that is, they move the terminal to GERAN/UTRAN

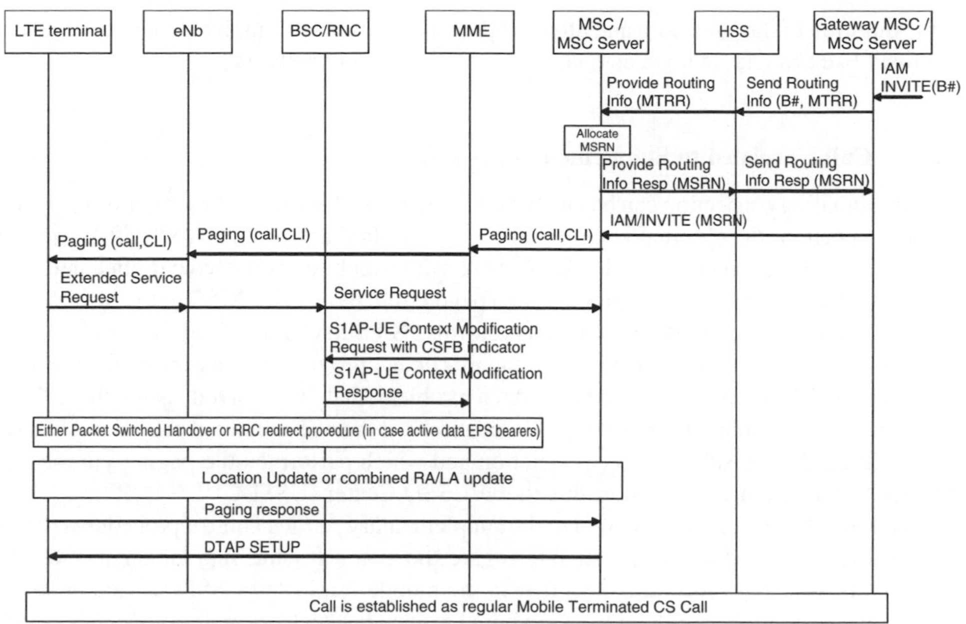

Figure 9.6 The mobile terminating call to LTE attached terminal via the CS Fallback functionality without the involvement of MTRR.

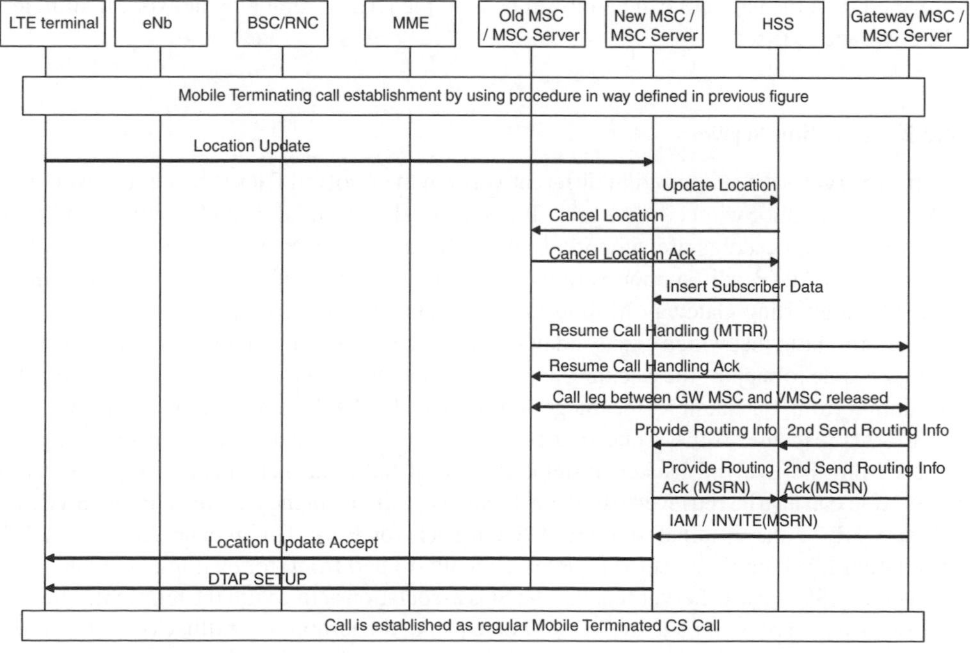

Figure 9.7 The mobile terminating call to LTE attached terminal via the CS Fallback functionality with the involvement of MTRR.

for the duration of the procedure and after the procedure has been completed then terminal can be moved back to LTE radio access coverage by the serving network.

9.4.2.1 Call Unrelated to Supplementary Services

A USSD signaling connection can be either mobile or network initiated. In a mobile-originated USSD connection, the terminal is expected to perform fallback to GERAN/UTRAN jointly with eNb and the serving Evolved Packet Core as with a mobile-originated call. In a network-initiated USSD connection, the terminal is first paged from the serving MSC or the MSC Server via the SGs interface to the MME and then performs PS-PS relocation if required and fallback to GERAN/UTRAN. In this case current standards do not provide any possibility for handling the misaligned Tracking Area/Location Area problem that was solved using the MTRR procedure with the mobile-terminating call. Therefore, if the call-unrelated procedure fails the MSC or MSC Server with the SGs association and which performed the paging via the SGs interface to the terminal will report this to the USSD Center (USSDC).

In the case of a call that is unrelated to the supplementary service control procedures, and if the end user is willing to modify or interrogate the state of some supplementary service, functionality-wise this procedure is similar to the mobile originated USSD service or call in such manner that terminal will, jointly with both Evolved Packet Core and serving eNb, decide to perform fallback—that is, to move the terminal from LTE radio access to GERAN/UTRAN. This is only valid in the mobile-originating direction and is never requested by the network. Functionality-wise, the terminal will perform exchange of call-unrelated signaling as defined by 3GPP TS 24.010 [9] by signaling a connection to the MSC or MSC Server that controls the GERAN/UTRAN to which the terminal performed fallback. After the procedure has been performed then the terminal can be moved back to the LTE coverage by network if sufficient LTE coverage exists.

9.4.2.2 Location Services

Location Services are supported in different ways in the Evolved Packet System as well as in Packet- and Circuit-Switched networks. It is expected that, in LTE, both control- and user-plane location procedures are supported. In the former can it means that both the Evolved Packet Core and LTE radio support various positioning methods and can deliver the outcome of this to the requesting Gateway Mobile Location Center (GMLC) when so requested. This model is similar to that which exists today in mobile networks. In latter case it means that the terminal is able to support the Secure User Plane Location (SUPL) method to deliver location information from, for example, the integrated Assisted Global Positioning System (A-GPS) as a SUPL 2.0 response to the requesting server via the IP connection without requiring any support from the Evolved Packet System. If none of these mechanisms exists, the location information can also be retrieved using mechanisms available in the existing Circuit-Switched core network, if the terminal supports CS Fallback for EPS. This means that the Mobile Terminating Location Request (MT-LR) is typically routed from, for example, the GMLC to the serving MSC or MSC Server that has the SGs association active with the terminal. Then the MSC or MSC Server will page the subscriber, causing the terminal to fallback—that is, move from LTE access to GERAN/UTRAN for the duration of the positioning. The terminal will reply to the paging request via GERAN/UTRAN. As with the USSD, with LCS services current standards do not describe any solution to cope with the misaligned Tracking Area/Location

Area problem, which was solved using the MTRR procedure with mobile-terminating calls. Again, if LCS fails, the MSC or MSC Server associated with the SGs and that performed paging via an SGs interface to the terminal will report negative a acknowledgment to the MT-LR towards the GMLC.

9.4.3 Deployment View

When CS Fallback for EPS for voice or call-unrelated procedures other than SMS over SGs is deployed into network then this typically requires joint network planning between the Evolved Packet System and Circuit-Switched network domains.

The SGs interface should be designed in such manner that it is built between those MMEs that handle the LTE radio access coverage of same geographical area as GERAN/UTRAN radio access handled by MSCs or MSC Servers. It is necessary to take into account both the traffic profile (dynamic capacity) as well as the increased number of subscribers in VLR (static capacity) in order to dimension the MSC or MSC Server appropriately to fit possible increased signaling traffic caused by SGs interface. In addition, the probability of the usage of MTRR needs to be estimated, which also depends on the initial design of the SGs interface and how perfect the Tracking Area/Location Area alignment is in practice. Absolute alignment is probably not a feasible target and therefore MTRR may be required in all networks that support CS Fallback for EPS for voice and video telephony calls.

Use of USSD or LCS services without parallel voice call needs to be estimated. If LCS services are only used in parallel with voice calls (for instance a pizza deliverer may request the location of a subscriber during a call in order to provide better service) then possible problems caused by a lack of a similar MTRR procedure for call-unrelated procedures may not cause any issues at all. However, if such new procedures that support these services do not exist in the current network, it is important to take this lack of the network support into account in designing the services in order to avoid increasingly failing MT-LR or NI-USSD requests rates.

9.5 Voice and SMS Over IP

LTE is able to provide an enhanced end-user experience for mobile broadband by reducing overall end-to-end latencies as well as increasing the throughput of data connectivity in way that the end user hasn't experienced with older 3GPP technologies. For this very same reason it is logical to consider that the LTE bearer is able to transfer voice and other kind of communication between end users more efficiently. Finally, the lack of native Circuit-Switched technology in LTE removes the possibility of directly reusing existing Circuit-Switched mobile networks to provide these services.

In future, when deciding on the model to use, communication service providers will have to choose how to continue important voice and messaging services in LTE. It is assumed that—whatever technologies are deployed at first as gap fillers—the eventual target will be to deploy native IP-based technologies in order to lower the complexity and costs related to the maintenance of the network.

The selection process for suitable technology to implement voice and SMS over IP depends on the service requirements and the overall business model of the service provider. It is possible that solutions could range from the use of third-party Over the Top VoIP services such as those widely used on the Internet today to more traditional regulated voice service models is based on the IMS and 3GPP-defined supplementary services. Most likely, traditional mobile service providers will continue with this more traditional path because the overall business model to provide voice would otherwise be more radically changed.

Another issue that also affects the selection of technology is the service continuation requirement. Let us consider a situation in which a communication service provider has provided both 2G and 3G voice and short message services and considers using LTE technology. This can be either considered as completely new kind of service in which service parity with legacy Circuit-Switched network is not important or there has to be certain level of service parity with the existing network. In the former case, voice and SMS services can probably be provided with rather conservative costs and by using completely new equipment. However in latter case the costs involved to create service parity can be high, depending on the technology available in the market. This can also mean that there is a need to interface with backend systems that have already been deployed such as the Intelligent Network Service Control Point (SCP) via CAMEL or INAP protocols, as well as having traditional MSC- or MSC Server-based services available for voice and SMS over IP communication. Naturally this needs to be analyzed when considering the end-to-end service portfolio for voice and messaging and varies a lot between different communication service providers. No solution can be given that is suitable for them all.

3GPP has made a significant effort to define a single standardized baseline for voice and SMS over IP by using Session Initiation Protocol (SIP) technologies. In this way, voice and Short Message Services can be deployed using IMS, which ensures easier interoperability among products from multiple vendors conforming to 3GPP specifications as well as targets to create an ecosystem similar to that which was created in the early days of GSM and 3G [10]–[12]. The use of IMS architecture for LTE also creates the possibility of achieving true Fixed Mobile Convergence (FMC) in which LTE is considered as another technology to access the IMS, among other fixed and mobile broadband and narrowband access technologies. The IMS can therefore be considered as access agnostic but aware architecture.

In addition to the specifications defining the detailed architecture as well as protocol-level requirements for these services, GSMA has decided to publish a more specific profile for "IMS profile for voice and SMS" as an IR.92 document [13] the purpose of which was to select the most important features from all relevant 3GPP specifications in order to achieve faster time-to-market and successful interoperability in a multivendor environment. A similar process was performed earlier by GSMA as a Rich Communication Suite initiative for multimedia services such as video sharing and capability enhanced presence in order to ensure important benefits for those services. Therefore it can be considered natural for GSMA to lead this process.

9.5.1 IP Multimedia Subsystem

The IP Multimedia Subsystem is 3GPP-defined architecture to enable controlled use of multimedia services. It was initially standardized as part of 3GPP Release 5 but its inception in networks around the world has taken more time than was originally expected.

Initially IMS was used as a platform for services such as Push-to-Talk, Presence and Instant Messaging, but these services were not very successful for various reasons and use of them was very marginal. The introduction of IMS in mobile networks was slow for a number of reasons. The first and possibly the main reason was the lack of killer services that can be only deployed based on IMS architecture. IMS has been considered to be a successful platform for voice services but today mobile networks support voice in an extremely efficient way after the introduction of mobile soft-switch technology in 3GPP Release 4—that is, the MSC Server system. Voice over IP using GPRS or WCDMA technologies has not been considered to be feasible due to lack of practical QoS capabilities as well as less efficient use of such transport compared to native Circuit-Switched radio bearers. This situation may change when sufficient 3GPP Release 7 capabilities for High Speed Packet Access (HSPA) are deployed but then there are similar possibilities for enhancing Circuit-Switched-based technology to use CS over HSPA, which again increases radio efficiency to nearly the same level as VoIP over HSPA. Uptake of VoIP over HSPA from ecosystem point of view is thus not considered likely to take off before Voice and SMS over LTE as native IP-based services will be completed, if at all.

However, despite challenges in the mobile network segment, IMS has been more successful in the fixed network segment where it has successfully helped communication service providers to replace aging fixed-network equipment with modern VoIP-capable, multivendor-proof, IMS-compliant technology. Similarly, IMS has been also in the public view related to enterprise services, namely services like hosted IP Centrex, which some service providers have deployed commercially.

Reader should not consider that lack of IMS deployments are related to the technological immaturity of IMS or specifications defining the IMS architecture. Difficulties with IMS have been mainly related to lack of such killer services beyond voice that would have required IMS.

In the future, LTE-related use cases are expected to drive the introduction of IMS into mobile networks as IMS is the only globally recognized technology to implement this kind of service. Another recognized business opportunity for IMS in mobile networks will be the Rich Communication Suite (RCS), which, from current indications, may be deployed before LTE is deployed, thus creating a basis for the later introduction of Voice and SMS over LTE. In the future it is also foreseen that IMS will be used as a centralized service architecture to enable the same services but in a highly efficient manner for both fixed and mobile based terminals—that is, true Fixed Mobile Convergence. In any case migration from Circuit-Switched-based telephony to Voice over IMS will happen in an evolutionary manner and it may take a long period of time to have both technologies in use within the same network.

9.5.2 Voice and Video Telephony Over IP

The basic telephony service that is implemented using IMS architecture requires at a minimum that the terminal and the IMS network support 3GPP-defined protocols such as Session Initiation Protocol (SIP), Session Description Protocol (SDP) as well as relevant service logic for required application servers such as Telephony Application Server (TAS) for Multimedia Telephony.

More advanced services can be applied using either Service Delivery Framework (SDF) or by using a traditional Intelligent Network, either integrated by using traditional CAMEL or INAP protocols or with SIP protocol (IMS Service Control, ISC). These capabilities likely will vary a lot based on network and business requirements set by the service provider.

Before any service can be provided, terminals have to register with IMS as defined by 3GPP TS 23.228 [14] and based on protocol-level implementation defined in 3GPP TS 24.229 [15].

During this registration phase the terminal will be authenticated by using Authentication and Key Agreement (AKA) based on 3GPP TS 33.203. Authentication can be either based on Universal Subscriber Identity Module (USIM) or IMS Subscriber Identity Module (ISIM), which can be embedded within the Universal Integrated Circuit Card (UICC) as a software module. The UICC is commonly known by end users as "SIM"-cards whereas USIM or ISIM applications are not directly visible for end users. It should be noted that the use of LTE services is not possible with the traditional SIM card, and therefore scenarios that are based on the use of the SIM for IMS AKA in the voice and SMS over LTE are not relevant in this context. Therefore, at a minimum UICC equipped with USIM is sufficient for voice and SMS over LTE using IMS architecture without a requirement to upgrade ISIM support for UICCs that have already been deployed.

When the terminal has been successfully registered with the IMS network then relevant application server instances are notified about this event by third-party registration with the IMS Service Control (ISC) interface. The necessary application servers are configured in the static IMS subscription profile of the subscriber and stored within HSS. The S-CSCF of the subscriber will fetch the data via a Cx interface and perform the necessary third-party registration on behalf of the terminal for these application servers. Application servers may then fetch subscriber-related data from HSS via the Sh interface, or in some other way ensure that the required subscription profile is available for service execution.

Typical application servers involved in voice and SMS handling are likely to include the following:

- Telephony Application Server (TAS) as defined in 3GPP TS 23.002 to provide Multimedia Telephony supplementary services.
- IP-Short Message-Gateway (IP-SM-GW) as defined in 3GPP TS 23.204, for instance in order to provide interworking with Circuit Switched Short Message Service architecture.
- IP Multimedia—Serving Switching Function (IM-SSF) as defined in 3GPP TS 23.002 to provide interworking towards an IMS-capable Intelligent Network Service Control Point (SCP) with IMS specific implementation of CAMEL protocol.
- Service Centralization and Continuity Application Server as defined in 3GPP TS 23.292 to provide Terminating Access Domain Selection (T-ADS) and to assist in service continuity between LTE and Circuit-Switched networks.

One of the most important application servers is TAS, which has a key role in service execution for supplementary services defined by 3GPP as part of Multimedia Telephony and taken into use within GSMA "IMS profile for voice and SMS" (IR.92 document).

The SCC AS will be involved as part of every IMS session in order to anchor the session in case service continuity will be invoked at later point in time by the underlying access network. Similarly, it will be used to select an appropriate access network for terminating IMS session if the terminal is simultaneously attached into both a Circuit-Switched and an IP (LTE) network.

IM-SSF capability may actually use the traditional Circuit-Switched based CAMEL protocol or even the INAP protocol that is already deployed and used by SCP and service applications. This results in a situation where IM-SSF does not comply with the original 3GPP specification but instead represents a more practical implementation that fulfills the business needs of service providers.

Relevant application servers are described in more detail below.

9.5.2.1 Session Establishment

A voice or video telephony session is established by using capabilities offered by a Session Initiation Protocol (SIP) as well as an embedded Session Description Protocol (SDP), which is responsible for the actual codec negotiation.

SIP protocol is responsible for establishing, maintaining and terminating sessions in the IMS network. It was originally defined by IETF but was then adopted and enhanced by 3GPP to fulfill requirements set by the IMS architecture. The SIP is a multipurpose protocol that can be used as part of different applications whereas protocols typically used in Circuit-Switched networks are intended only for single purposes (e.g., establishing calls, transferring short messages). This flexibility has also caused interoperability problems in the past, which 3GPP and later GSMA as part of the "IMS profile for voice and SMS" have tried to overcome by defining strict guidelines about how SIP and related protocols should be used by terminals and networks within IMS architecture. This possibility for multivendor interoperability has been called one of the strongest benefits of IMS.

The SDP protocol is a protocol that is used to describe nature of an IMS session in such a manner that a media connection between the endpoints involved can be established and modified, if needed. The SDP is encapsulated within the message body of SIP signaling when required. In a nutshell, this process to agree common media connection is called SDP negotiation. SDP negotiation is formed from one or more offer and answer transactions, which may be required to negotiate all the required parameters such as codecs, individual codec types, codec attributes as well as IP addresses and associated port numbers for both the local and remote sides of the media connection. Similarly, as with the SIP, the SDP has also been defined to be very flexible and therefore 3GPP and later GSMA "IMS profile for voice and SMS" have narrowed the required functionalities in order to achieve early time-to-market and better interoperability within a multivendor environment.

This book will not describe previously mentioned protocols in more detail. If the reader is interested about these protocols, there exist books that give the protocols in greater detail.

9.5.2.2 Multimedia Telephony

3GPP defined Multimedia Telephony within Release 7 to define a standardized set of supplementary services as well as a generic telephony service over the IMS architecture. The basis of this work was derived from ETSI TISPAN (Telecommunications and Internet converged Services and Protocols for Advanced Networking) work, which was completed earlier, to provide a similar capability for fixed IMS deployments. In practice, 3GPP took a subset of mobile network features as a basis for this work. 3GPP also modified some of the existing services in order to match the Multimedia Telephony requirements with the mobile networks. 3GPP enriched the way in which some services can be triggered (for example the presence-condition-based call diversion service) but also introduced a new kind of call diversion service, which was similar to Call Forwarding Not Reachable (CFNRc) in today's Circuit-Switched mobile networks.

Multimedia Telephony was implemented in such a way that same the services can be applied to any terminal regardless of the access network technology used. This will make it possible, in the future, to deploy centralized IMS architecture with same set of supplementary services and thus lower operating expenses than would otherwise be incurred by maintaining multiple different service domains for each access technology. The use of Multimedia Telephony is therefore expected to be the basis for true Fixed Mobile Convergence (FMC), merging fixed, cable and mobile accesses to same unified service architecture.

GSMA "IMS profile for Voice and SMS" was based on a mindset that, at a minimum, a similar level of services needs to be made available from IMS network as is today available from Circuit-Switched mobile networks. Multimedia Telephony has a degree of flexibility that is beyond the capabilities of currently available technology in Circuit-Switched networks. It was considered to be an improvement but one that was likely to be deployed after the basic voice service was successfully deployed.

The profile that GSMA defined aimed to be a subset of the services and capabilities that are related to the original Multimedia Telephony of 3GPP. This subset is aligned with the capabilities of the current circuit-switched mobile networks. This provides a service that is independent of the access technology used (circuit-switched or IMS) regardless of the location where the subscribers are roaming. The independence is also valid for the synchronization between the service settings between different domains that handle the service executions (TAS and MSC).

The following sections will detail how Multimedia Telephony has been adopted at a high level by GSMA in "IMS profile for Voice and SMS." It should be noted that this profile only defines services for voice and video telephony but not for SMS or other IMS services. This is despite the fact that some networks today have, for example, call-forwarding services for SMS. This naturally does not mean that such services, which provide additional value for service providers, should not be used but instead it opens the door for competition between service providers.

Supplementary Services

Multimedia Telephony, as defined by 3GPP and endorsed by GSMA under the item "IMS profile for Voice and SMS," covers a set of services and conditions for situations where services need to be invoked with the existing circuit-switched mobile networks.

The supplementary service settings are maintained as a single XML document, which has been defined in 3GPP TS 24.623 and enhanced in 3GPP Release 9 with an optional capability negotiation mechanism that allows endpoints (terminal and network) to negotiate which services and capabilities are supported. This XML document, which can be stored within the HSS and the XML Document Management Server (XDMS), contains both operator-defined content such as provisioned services but also data that reflects the configuration settings that have been configured by the end user. In other words, it is not possible for the end user to add new services (e.g., Call Diversion Unconditional) to an XML document unless it has been provided to the user beforehand by the service provider. The network will use the data stored in this document to invoke services in Telephony Application Server (TAS) functionality located in serving IMS network. The TAS is typically always involved in an IMS session if the session is subject for Multimedia Telephony services.

Table 9.1 represents the defined subset of services in "IMS profile for Voice and SMS."

Supplementary Service Control

The Multimedia Telephony-related supplementary service settings are held within XML documents stored on the network side.

This is a structured XML document that can be manipulated either by the service provider or the end user. The end user may possibly have some non-standard means to modify the XML document—for example, by using the Web portal of the service provider or even using facility

Table 9.1 Services in IMS profile for voice and SMS

Supplementary service	Description of service
Originating Identification Presentation 3GPP TS 24.607	Provides identity of calling subscriber to called subscriber. Same as Calling Line Presentation (CLIP) in CS networks.
Terminating Identification Presentation 3GPP TS 24.608	Provides identity of called subscriber to calling subscriber. Same as Connected Line Presentation (COLP) in CS networks.
Originating Identification Restriction 3GPP TS 24.607	Restrict presentation of calling subscriber's identity to called subscriber. Same as Calling Line identity Restriction (CLIR) in CS networks.
Terminating Identification Restriction 3GPP TS 24.608	Restrict presentation of called subscriber's identity to calling subscriber. Same as Connected Line identity Restriction (COLR) in CS networks.
Communication Diversion Unconditional 3GPP TS 24.604	Forward session unconditionally to new destination. Same as Call Forwarding Unconditional (CFU) in CS networks.
Communication Diversion on not Logged in 3GPP TS 24.604	Forward session in case called subscriber is not registered into network.
Communication Diversion on Busy 3GPP TS 24.604	Forward session in case called subscriber is busy or indicates user busy situation by pressing "red button." Similar to Call Forwarding Busy (CFB) in CS networks.
Communication Diversion on not Reachable 3GPP TS 24.604	Forward session in case called subscriber's terminal is not reachable. Similar to Call Forwarding Not Reachable (CFNRc) in CS networks.
Communication Diversion on No Reply 3GPP TS 24.604	Forward session in case called subscriber has not answered to incoming call. Similar to Call Forwarding No Reply (CFNRy) in CS networks.
Barring of All Incoming Calls 3GPP TS 24.611	Prevent incoming session to particular subscriber.
Barring of All Outgoing Calls 3GPP TS 24.611	Prevent outgoing session from particular subscriber.
Barring of Outgoing International Calls 3GPP TS 24.611	Prevent outgoing international session from particular subscriber.
Barring of Incoming Calls - When Roaming 3GPP TS 24.611	Prevent incoming sessions while subscriber is roaming outside home network.
Communication Hold 3GPP TS 24.610	Enables subscriber to place ongoing session on hold and retrieve it afterwards.
Message Waiting Indication (MWI) 3GPP TS 24.606	Subscriber is able to receive indication about left messages in for example, Voice Mail System. In case of Circuit Switched networks this is typically handled via SMS service, which is also possible in this case as well unless network supports MWI.
Communication Waiting 3GPP TS 24.615	Enables subscriber to have indication about incoming session while having active session.
Ad-Hoc Multi Party Conference 3GPP TS 24.605	Subscriber is able to establish ad-hoc conference (audio & video). This is similar than Multiparty supplementary service of Circuit Switched networks but can be enriched to support video conferencing, if required.

codes (for instance use "*21*C number#" to activate Call Diversion Unconditional service) but also in the 3GPP standardized way by using XML Configuration Access Protocol (XCAP) via the 3GPP Ut interface.

XCAP was originally defined by IETF RFC 4825 and later on taken into use within the IMS architecture by both Open Mobile Alliance and 3GPP to manipulate XML-based documents related to various IMS-based services. For instance OMA specified that XCAP be used to manipulate authorization rules associated with Presence service. Similarly Resource Lists could also be manipulated with the help of XCAP.

The XCAP client is located within the terminal and will be able to render and manipulate XML documents associated with specific services, according to 3GPP TS 24.623. XCAP maps the specific XML elements in document as HTTP URIs, which can then be accessed directly by the client in order to add, modify or remove actual data in an XML document. XCAP server functionality, which may reside in different physical network element such as one that also acts as Telephony Application Server (TAS), is responsible for acting as a repository of the XML documents associated with a particular service. The XCAP server will also terminate XCAP requests from the XCAP client and manipulate XML documents based on those requests but also based on limitations (validation) of the schema related to the XML document.

Before XCAP requests can be accepted from the XCAP client there is a need to perform authentication either by using HTTP Digest or Generic Authentication Algorithm (GAA) as defined in 3GPP TS 24.623. This authentication may be performed either by a specific Aggregation Proxy (AP) functionality or by the XCAP Server (TAS) which is the target for the XCAP request. The AP is optional functionality but can be used to harmonize network architecture if other XCAP-based applications have also been deployed, such as Presence, according to OMA specifications. If the AP authenticates the XCAP request then it will insert a specific HTTP X 3GPP Asserted Identity header into the HTTP request before sending the request to the targeted XCAP server. Routing of XCAP requests to correct XCAP servers by the AP is based on the Application Usage ID (AuID) field inside the XCAP request, which identifies that the request be related to, for instance, Multimedia Telephony. The AuID defined by 3GPP for Multimedia Telephony is "simservs.ngn.etsi.org."

9.5.2.3 Text Telephony

Text telephony has been deployed in some Circuit-Switched mobile networks today to assist hearing-impaired subscribers to communicate over the Circuit Switched voice channel with in-band messaging. This is an especially important feature in conjunction with emergency calls in, for instance, the United States. This 3GPP defined method is named Global Text Telephony (GTT) and defined in 3GPP TS 23.226 [16]. 3GPP defined methods to transfer Text Telephony over both voice (GTT-Voice) and 3G-324M video telephony (GTT-CS) but in fact GTT-Voice has been the one that is mostly used today.

GTT provides the interworking of Cellular Text Telephony modem (CTM) technology, used in radio access, with V.18, used in fixed networks. This way, compatibility with existing deployed Text Telephony equipment can be achieved from a mobile terminal. CTM has been specified in [16].

The situation is different with IMS-based voice and video telephony since it was decided that CTM technology should not be used in order to transfer Text Telephony from the IMS terminal. Instead, ITU-T T.140 [17] was selected to be used. T.140 protocol is text based and it uses the IETF RFC 4103 [18] payload format as defined in 3GPP TS 26.235 [19].

If the terminal is capable for both Circuit-Switched and IMS-based voice and video telephony then it will need to support both CTM and T.140-based Text Telephony depending on the access network used.

The network needs to support interworking between CTM and T.140 for IMS sessions that are established between the IMS terminal and the Circuit-Switched network. This interworking is the responsibility of MGCF and IMS-MGW, which represents the SIP User Agent towards the IMS terminal and therefore acts as an active endpoint for SDP negotiation.

9.5.2.4 Policy Control

The 3GPP-defined Policy and Charging Control (PCC) is also reused in case of IMS-based voice and video telephony to ensure that the appropriate QoS can be guaranteed from the EPS [20].

The PCC framework provides service- and flow-based policy control in both roaming and nonroaming scenarios. PCC functionality has not been widely deployed in GERAN- or UTRAN-based Packet-Switched networks but a clear trend has been to improve the service awareness of the network especially due to the increased data traffic caused by mobile broadband uptake globally. Similarly, another trend has been to minimize the number of Access Point Names (APN) that need to be used by the terminal to obtain access to different services. These two requirements together mean that modern mobile broadband networks will eventually perform, for example, Deep Packet Inspection (DPI), in order to detect service use from all IP traffic exchanged between the terminal and the network. Additionally, policy control can be performed based on location, network, time of day, category of subscriber and other relevant criteria, which can be taken into account within the Policy and Charging Rules Function (PCRF).

PCC can use a stored user profile and, if available, service-based information received from the Application Function (AF) via a Diameter-based Rx interface. Readers should note that the DPI can also be the AF in some cases. In case of IMS-based voice and video telephony, AF resides within P-CSCF. P-CSCF participates in codec negotiation that is performed by using Session Description Protocol (SDP) between end points and therefore is able to request sufficient resources from the underlying access network such as the EPS, based on the results of this negotiation. The PCRF that receives a request from the P-CSCF will eventually take local policies into account as well as the PCC-related user profile and request needed resources from the Policy Control Enforcement Point (PCEF) via a Diameter-based Gx interface. The PCEF is colocated, in the EPS, in the Packet Data Network Gateway (PDN-GW). It is also possible to request the PCC framework to inform the AF/P-CSCF about various events that could occur on the access network side such as loss of IP connectivity, or handover from one radio access technology to another.

In the nonroaming scenario, all PCC-related functionalities are located in the home network as well as the Rx and Gx interfaces. However, when subscriber is roaming then it is possible that PCRF is actually split into two functions: visited-PCRF and Home-PCRF. A Diameter-based S9 interface can be used between these two functions to deliver user profile information and tunnel Rx interface messages between the home and visited networks. The user profile is always stored within the home network of the subscriber. In IMS-based voice and video telephony, when the subscriber is roaming, then PDN-GW and P-CSCF are located in the visited network as defined in IMS roaming architecture. This means, then, that PCC decisions are performed in the visited network possibly with information received from Home-PCRF via the S9 interface.

Table 9.2 CQI mapping

QCI value	Purpose
1	Conversational voice (GBR)
2	Conversational video (GBR)
5	IMS signaling (non-GBR) and possibly also Ut/XCAP interface

Policy control with EPS architecture is simpler than with legacy GERAN/UTRAN-based Packet-Switched architecture. EPS uses specific QoS Class Identifier (QCI) values defined in 3GPP TS 23.203 to associate certain kinds of traffic with certain type of EPS bearer. The EPS bearer is similar to PDP in legacy technologies. EPS has two kind of bearers: default and dedicated ones. The default bearer is always established by the network when the terminal attaches to EPS but default EPS bearers cannot support Guaranteed Bit Rate (GBR) traffic, which makes it unsuitable for conversational traffic such as voice sessions. On the other hand, a dedicated EPS bearer is able to support voice and other non-best-effort traffic when required. Dedicated EPS bearers need to be always established by the network when required, which is also a different kind of behavior from that in existing legacy GERAN/UTRAN-based PS networks today. The idea of the dedicated EPS bearers can be compared with the "secondary PDP context" defined in the previous packet data domain functionality of 3G.

Table 9.2 represents QCI mapping for voice and video telephony as well as for IMS signaling as defined in 3GPP TS 23.203.

Other values are intended for other applications beyond the scope of this chapter. It is possible that the service provider will define its own mapping for QCI values but the use of standardized mapping is highly recommended (at least related to previously listed QCI values) in order to harmonize functionality across different networks.

The QCI functionality is used in order to provide QoS in the access network, which is EPS in this case. However, when IMS-based voice and video telephony is transported within the IP backbone then it is also important to ensure that adequate QoS can also be achieved within those IP connections. This also means IP-based interconnections between different operators either using SIP-I or native IMS-NNI. For this reason, GSMA has defined mapping from the QoS used in the access network to the DiffServ Code Point (DSCP) value, which can be used within IP backbone. This mapping, which is documented in GSMA PRD IR.34 [21], is based on six different DiffServ Per Hop Behavior (PHB) ranging from Expedited Forwarding (EF) to the Best Effort (BE) categories. In the case of IMS voice and video telephony, this means that QCI = 1/2 is mapped to the EF PHB category whereas IMS signaling may be mapped into AF31 (Assured Forwarding for Interactive traffic) category.

The GSMA-defined IP eXchange (IPX) is expected to be used for voice and video telephony traffic exchanged between networks by using IP. The IPX is a logical continuation of work done for GRX, which has been successfully used as an interoperator backbone for GPRS traffic since 2000. IPX enhances the GRX model by providing QoS awareness as well as a similar payment model to that which exists in today's Circuit-Switched networks. The IPX, when used for IMS based NNI for voice and video telephony among other IMS services, will use similar DiffServ Code Point-based mapping of IP traffic in order to ensure the Per Hop Behavior expected by the service. For further information about IMS NNI, see GSMA PRD IR.65 [22] and for further information about the IPX, see GSMA PRD IR.34 [21].

In addition to these, GSMA has also recently defined roaming guidelines for LTE over the GRX infrastructure within GSMA PRD IR.88 document [23]. This document details the access part of EPS access but naturally in case of roaming scenarios IPX and GRX are applicable

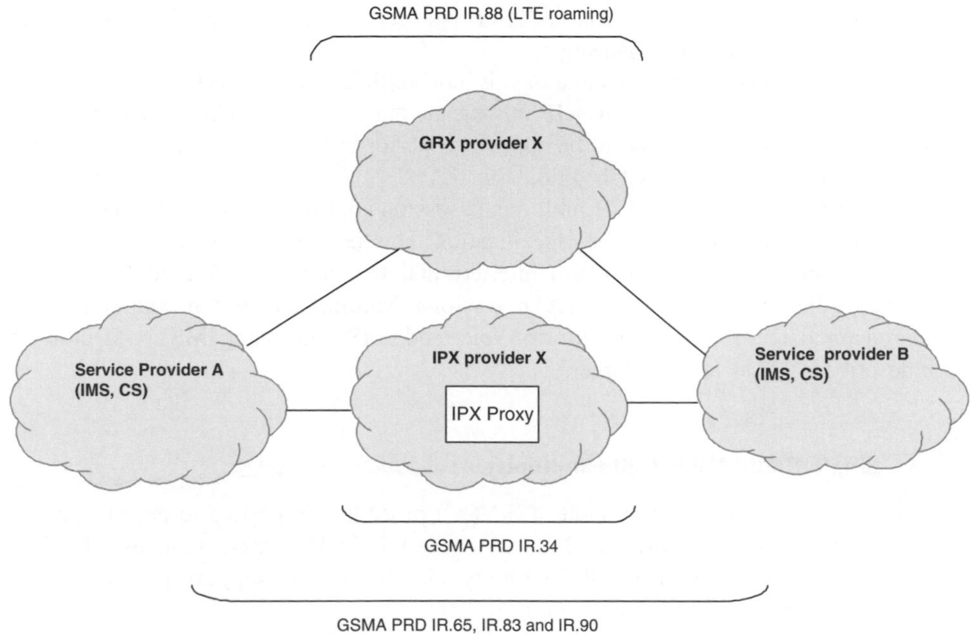

Figure 9.8 Voice over LTE, LTE roaming and CS interworking using IPX/GRX.

depending on services that are being used. Additionally IMS Network Network Interfaces (NNI) have been documented as part of GSMA PRD IR.65 [22] and recently updated to cover new requirements caused by Voice over LTE as well as interworking to Circuit Switched IP interconnect by using profile defined in GSMA PRD IR.83 [24]. Rich Communication Suite related recommendation is described in GSMA PRD IR.90 [25].

Figure 9.8 represents the scope of these different GSMA guidelines from Voice over LTE, LTE roaming, and the CS interworking point of view using IPX/GRX.

9.5.2.5 Service Continuity

Service continuity can be considered to be built from multiple layers each of which performs its own role in order to secure continuation of service. The lowest layer is provided by the access network and the higher layer can be provided by service application. Both layers are used in IMS to provide maximal end-user experience.

LTE is based on 3GPP technology and provides mobility as an inbuilt functionality between different types of access technologies. This enables the terminal to move without changing the IP address allocated to the terminal during the attachment phase between LTE, UTRAN, GERAN and Wireless LAN (WLAN) access networks. In the beginning of the commercial phase of LTE, it is most likely that service continuity can occur between the 3GPP based access networks—between LTE, GERAN and UTRAN. The support of service continuity between LTE and non-3GPP access networks, like WLAN and WiMAX, may be introduced later, if the market requires this functionality.

This access-domain service continuity is sufficient if the target access network technology is able to fulfill requirements set by the service in question. For Voice over IP it means different requirements from those set by non-real-time packet services such as Presence, Instant Messaging or even Signaling Connection.

If the target access network technology is not sufficient then service-domain service continuity is required. For Voice over IP this means that Voice Call Continuity procedures defined in 3GPP TS 23.347 need to be applied to perform domain transfer from a packet-switched domain to a circuit-switched domain.

In addition to basic Voice over IP, multimedia session continuity has also been defined by 3GPP as part of a previously described specification. This session continuity enriches domain transfer with the capability to perform interterminal transfers of multimedia sessions or individual media components of particular sessions. Multimedia session continuity is not considered mandatory for the introduction of Voice and SMS when using IMS architecture and is not described further in this chapter.

9.5.2.6 Single Radio Voice Call Continuity

The Single Radio Voice Call Continuity (SRVCC) procedure is needed to perform domain transfers for voice sessions between Circuit-Switched and LTE access domains. SRVCC is based on 3GPP Release 7 Voice Call Continuity and enhanced to support network-initiated behavior triggered by the access network [26], [27].

SRVCC has been defined in 3GPP Release 8 for both UTRAN and EPS but it can only be implemented for LTE if required and if VoIP/HSPA is not required by the service provider. Figure 9.9 represents the architecture required to perform the SRVCC procedure.

In order to initiate the SRVCC procedure from LTE to a Circuit-Switched network, eNb will receive measurements from the LTE terminal. If the terminal has an ongoing IMS session and a

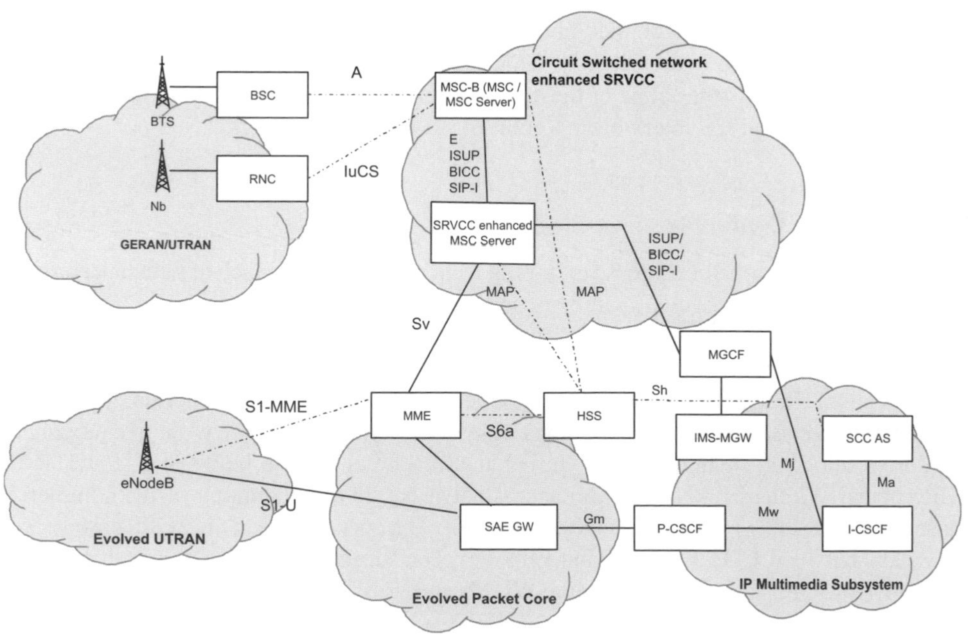

Figure 9.9 The architecture that is required to perform SRVCC procedure.

related dedicated EPS bearer with Guaranteed Bit Rate (GBR) QoS property, then eNb will decide whether SRVCC needs to be performed based on the preconfigured triggering condition. If SRVCC is to be performed then it indicates the need for this to the serving MME, which will start to prepare resources for the target networks. The target network in this case means the target circuit-switched network, which requires that the MME select a suitable SRVCC-enhanced MSC Server with a Sv interface to the MME but also the target legacy packet-switched network when the MME needs to contact the SGSN via a S3 interface.

After the MME has selected and approached the SRVCC-enhanced MSC Server it will trigger the reservation of target circuit-switched network resources needed to perform SRVCC. The SRVCC-enhanced MSC Server will contact the target RNC or the BSC if the target circuit-switched radio access is controlled by the same network element. However, if inter-MSC relocation is needed whilst SRVCC procedure takes place, MME will also contact MSC or MSC Server that controls the target BSC or RNC via the MAP based E-interface (i.e., MSC-B).

SRVCC-enhanced MSC Server will perform normal the preparations needed on the Circuit-Switched side and after this make contact with the anchor functionality that resides within IMS architecture—that is, the SCC AS that has initially anchored the IMS session during the session establishment phase. The SCC AS address which is called Session Transfer Number for Single Radio (STN-SR) is obtained from HSS by MME during the LTE attachment phase and is given via Sv interface to the SRVCC-enhanced MSC Server. The SCC AS associates the incoming session (via MGCF) from SRVCC enhanced MSC Server with the ongoing IMS session and updates the IMS session with the new local descriptor, which has the new IP address and port number that should be used for RTP and RTCP sessions from this point onwards. This new information is, in practice, the IP address of the MGW that provides the IMS-MGW functionality and was allocated for the incoming session from the SRVCC-enhanced MSC Server to the SCC AS.

It should be highlighted that the SRVCC procedure consists of preparation and the actual domain transfer phase. It is possible that the preparation phase happens and resources for the target Circuit-Switched network are reserved but then the actual domain-transfer phase is not performed because the terminal has moved back to LTE coverage again. However, if the domain transfer phase has been started and the SCC AS is contacted by the SRVCC-enhanced MSC Server then the terminal is moved to Circuit-Switched radio access. The reverse SRVCC needed to perform domain transfer from Circuit Switched to LTE radio access is not supported until 3GPP Release 10-based equipment is available in the network.

Figure 9.10 represents the SRVCC procedure as defined in 3GPP Release 8 timeframe.

In parallel with the SRVCC procedure, the MME will also perform a packet-switched handover (not shown in Figure 9.10) to GERAN/UTRAN if required. In order to recognize which EPS bearers need to be transferred as SRVCC handovers and which as packet-switched handovers, the MME needs to be aware which EPS bearer has the voice component. This is made possible by the use of PCC architecture in such a way that the P-CSCF will request the PCRF to provide QoS in the form of an EPS dedicated bearer with the Guaranteed Bit Rate (GBR) and QoS Class Identifier value 1 (Voice). The MME will be informed by the PDN GW about which EPS bearer has the previously mentioned properties and based on that knowledge the MME will pick the correct EPS bearer from other active EPS bearers.

If the target radio access technology is GERAN then the continuation of the data connection after the SRVCC requires the use of Dual Transfer Mode (DTM) feature in GERAN. If the target radio access technology is UTRAN then the Multi-RAB functionality of UTRAN ensures continuation of data connection.

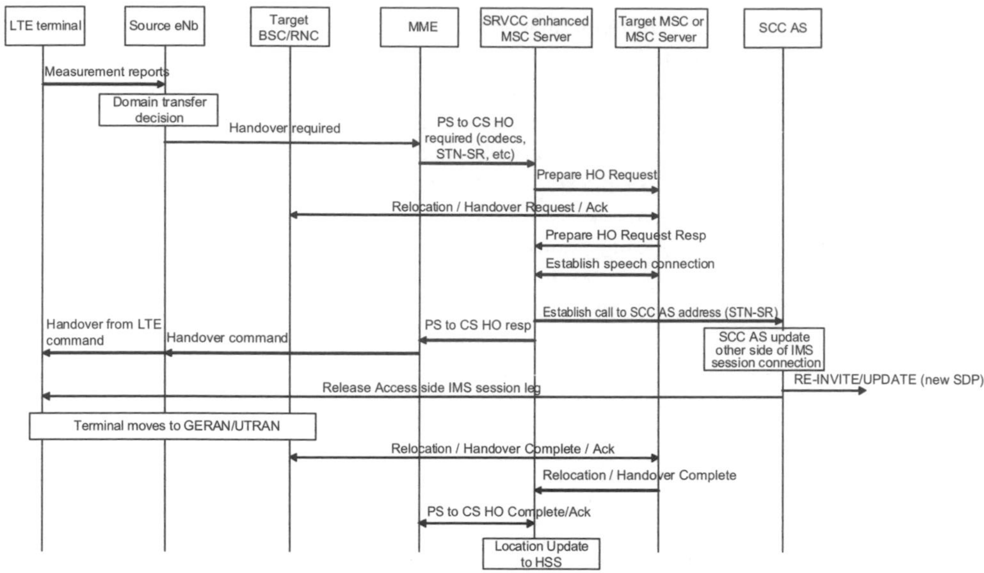

Figure 9.10 The SRVCC procedure.

9.5.2.7 Codecs

Voice and video telephony requires the use of speech and video codecs in order to transfer speech and video frames between endpoints.

In Circuit-Switched mobile networks speech codecs have typically been defined by 3GPP or related organizations, which have ensured the proper interworking in a multivendor environment as well as establishing a good basis to cover rules related to Intellectual Property Rightss (IPR). Table 9.3 represents a list of commonly used speech codecs in 3GPP networks today across the globe.

More detailed information about speech codecs can be found in 3GPP TS 26.103.

3GPP has also ensured that end-to-end codec transparency can be achieved in call scenarios involving Circuit-Switched networks through the use of such features as Out of Band Transcoder Control (OoBTC), also known as Transcoder Free Operation (TrFO), as well as Tandem Free Operation (TFO). These mechanisms enable the network to transparently transfer received speech frames transferred from one terminal to another. If the network needs to apply any in-band information (announcements or tones) for the speech path or perform some reconfiguration for speech connections—for example, due to the invocation of supplementary services such as Explicit Call Transfer—then it is possible to reinvoke TrFO and TFO automatically. These features are in key a position to ensure that the best voice quality can be obtained when the call is made between the Circuit-Switched and VoIP-capable IMS terminal.

The following sections will provide more details about the use of these codecs in IMS architecture as part of "IMS profile for voice and SMS."

Speech Codecs

3GPP and GSMA in "IMS profile for voice and SMS" have mandated the use of the narrowband UMTS AMR codec in LTE in order to ensure backward compatibility to Circuit-Switched networks and to ensure sufficient voice quality provided by that codec.

Table 9.3 Codec definitions by 3GPP

Speech codec	Technology	Additional information
Full Rate (FR)	GERAN	Speech codec consuming full traffic channel in GERAN. Moderate speech quality compared to newer codecs. Fixed mode available in this speech codec is 13.0 kbit/s.
Half Rate (HR)	GERAN	Speech codec for higher channel density in GERAN. Enables two speech calls to use same traffic channel in GERAN. Reduced speech quality compared to full rate codecs. Fixed mode available in this speech codec is 5.60 kbit/s.
Enhanced Full Rate (EFR)	GERAN	Improved variant of full rate speech codec for GERAN. Speech quality close to AMR codec in good traffic channel conditions. Fixed mode available in this speech codec is 12.20 kbit/s.
AMR Full Rate (FR-AMR)	GERAN	Adaptive Multi Rate (AMR) speech codec with capability to change bit rate and robustness of codec based on traffic channel condition (link adaptation). Compatible with UMTS AMR2 speech codec. Available modes in this speech codec range from 4.75 kbit/s to 12.20 kbit/s. The highest mode is same than with Enhanced Full Rate.
AMR Half Rate (HR-AMR)	GERAN	Adaptive Multi Rate (AMR) speech codec with capability to have two speech calls within full rate traffic channel in GERAN. Compatible with UMTS AMR2 speech codec. Available modes in this speech codec range from 4.75 kbit/s to 7.95 kbit/s.
UMTS AMR and UMTS AMR2	UTRAN	Adaptive Multi Rate (AMR) speech codec with possibility to change bit rate and robustness of codec based on demand. Codec named as UMTS AMR2 when used in dual-mode terminals and rate adaptation is possible only with every second speech frame (i.e., compatible with AMR in GERAN). Rate control is typically performed in UTRAN access only to achieve higher capacity from radio cell. Quality-related rate adaptation is not triggered from terminal or radio access. Available modes in this speech codec range from 4.75 kbit/s to 12.20 kbit/s. The highest mode is the same as with Enhanced Full Rate.

(continued)

Table 9.3 (*Continued*)

Speech codec	Technology	Additional information
Wideband AMR (WB AMR-FR and UMTS WB AMR)	GERAN/UTRAN	Wideband speech codec to provide significantly better voice quality than what is available with previously listed narrowband speech codecs. Rate control is typically performed in UTRAN access only to achieve higher capacity from radio cell. Quality related rate adaptation is not triggered from terminal or radio access. Available modes in this speech codec range from 6.60 kbit/s to 23.85 kbit/s. However in practice the maximum mode available in circuit-switched mobile networks is 12.65 kbit/s, which is considered as tradeoff between voice quality and resource consumption in the air interface. Similar restrictions do not exist in case of end-to-end IMS sessions.

In practice the codec implementation is exactly same as one that was defined in the early days of UMTS.

In order to transfer AMR-encoded speech frames, according to 3GPP TS 26.114, it is required that the terminal or another endpoint in the IMS architecture should support framing as defined by IETF RFC 4867. 3GPP TS 26.114 defines in more detail which attributes of RTP profile need to be implemented as mandatory functionality in order to achieve higher multi-vendor interoperability.

These attributes define, for instance, whether an octet-aligned mode or a bandwidth-efficient mode needs to be used, what level of redundancy (if any) should be supported, and so on.

One notable difference between the use of the UMTS AMR codec in IMS and in Circuit-Switched networks is related to use of the rate-control procedure. The rate-control procedure can be invoked by the endpoint involved in the call to request the other endpoint to send either higher or lower mode of speech codec. If TrFO or TFO procedures are used within Circuit-Switched networks or if the connection is an end-to-end IMS session, then rate adaptation is performed between the terminals involved without network involvement.

In Circuit-Switched networks, rate control can be supported to either increase the robustness of the codec against bit errors (link adaptation with GERAN) or in order to achieve higher capacity from same radio bandwidth (as for example, in case of UTRAN). However, today's VoIP terminals, which are capable of supporting AMR codec, typically do not support the capability to request rate adaptation but are only capable of responding to requests from other side of the session (e.g., Media Gateway). This is due to the fact that the VoIP terminal is not typically aware of conditions within the IP access or backbone, making it impossible to perform any decisions to change applied rate from one to another. This has been recently changed in 3GPP so that terminal can request rate adaptation (for instance to a lower rate) from the other side of connection if an indication of congestion is received by the terminal as an Explicit Congestion Notification (ECN). This is an optional procedure that may not be supported by the terminal or other endpoint and thus it cannot be assumed that rate adaptation will be performed at all in such cases.

Nevertheless, even if the terminal or network are not ECN capable, it is still possible for the rate control request to be received by the IMS terminal from the circuit-switched network via

Media Gateway. Because of this, the minimum requirement for the IMS terminal is that it will need to be able to respect the received request and perform the rate adaptation.

High Definition Voice

3GPP has mandated the use of Wideband AMR codec if wideband speech is required to be supported but, overall, wideband speech service itself is not mandated by the IMS network. This way, again, backward compatibility can be ensured with Circuit-Switched mobile networks.

The wideband AMR codec in Circuit Switched networks uses a 12.65 kbit/s mode as the highest mode providing moderate bandwidth consumption on the radio access side (note that narrowband AMR uses 12.20 kbit/s, which is close to 12.65 kbit/s). However, a similar limitation does not exist in the case of VoIP and it is possible to use the highest mode, 23.85 kbit/s, if this is negotiated between endpoints.

The following list summarizes one of the biggest differences in introduction of wideband AMR as VoIP compared to the Circuit Switched networks:

- No changes are required for the access network side (e.g., EPS). In the case of Circuit-Switched networks, wideband AMR needs changes for radio access and core networks. Naturally, the introduction of wideband AMR requires changes to IMS network entities such as the MGCF/IMS-MGW as well as the MRFC/MRFP, if those functionalities are involved with wideband sessions.
- Theoretically, higher voice quality can be achieved with the capability to use higher wideband AMR modes than are available with Circuit-Switched technology today. See above.
- Wideband AMR can be used in conjunction with video codec thus providing a higher end-user experience for video telephony. Currently, 3GPP defined Circuit-Switched video telephony (3G-324M) in practice does not use wideband AMR.
- End-to-end wideband codec support between fixed VoIP and 3GPP VoIP endpoints may require support for transcoding between G.722 and wideband AMR codec (G.722.2). It is also possible in the future that other wideband fixed VoIP codecs such as ITU-T G.729.1 or Skype Silk may need to be supported with such capabilities, depending on business requirements.

Despite the differences it is strongly assumed that wideband speech will be taken into use with the introduction of voice and video telephony over LTE. In the future a new speech codec could be introduced as result of recently started standardization activity in 3GPP Release 10. This codec is currently being discussed as part of a work item named “Codec for Enhanced Voice Services,” which aims to develop a partly backward compatible, super-wideband codec with better voice quality than any known 3GPP codec. Introduction of wideband AMR codec has taken many years and therefore it is fair to assume that the introduction of the new codec into commercial use will also likely take time. On the other hand, Internet-based VoIP services that will be available and also used with LTE will have—or, at least, in the future will evolve gradually to have—wideband speech codecs (for example IETF’s ongoing program to define royalty-free Internet Wideband Audio codec also known as “Harmony”) and therefore 3GPP has to also keep up with developments in order to stay competitive enough.

Video Codecs

Video codecs are required in the context of IMS-based multimedia when possibly highly compressed video stream is transferred between endpoints. In the future, video codecs will be used to encode and decode various different kinds of video use cases (files containing movies, streamed content from various Internet-based user-generated content services, etc.). However the most important use case within the scope of this chapter is related to real-time encoding/decoding of video content, which will be described in more detail next. The actual payload format to transfer video over RTP is based on the relevant IETF RFCs and referred by 3GPP, whenever appropriate.

Since early days of 3G the biggest difference in Circuit-Switched telephony was the introduction of 3G-324M video telephony in order to enable video calls between 3G terminals and eventually to native IP endpoints in the H.323 and SIP domain. H.323 has been largely replaced with SIP-based technology today and migration to IMS architecture will lead this process in the future on a faster timescale. 3G-324M was really the first technology that made video telephony available for the large-scale consumer market. It has been integrated as feature in nearly all 3G phones sold today. Network support also exists in many parts of world where 3G has been deployed. Despite the penetration of 3G-324M technology in video telephony, in mobile networks it has not been big commercial success and video calls form a significantly low proportion of the total amount of calls across the globe.

Low interest in mobile video telephony, especially in the consumer segment, has not hindered efforts to continue development of more advanced video codecs such as ITU-T H.264/Advanced Video Codec (AVC). One issue that may prevent improvements for 3G-324M Circuit-Switched video telephony is related to available bandwidth, which is 64 kbit/s and shared between video and audio codec streams and ITU-T H.245 logical channel control protocol in a multiplexed manner. Current understanding is that this bandwidth is already quite well used and, for instance, in order to obtain better video quality, that would cause a deterioration in audio quality and vise-versa. Therefore next big step is expected to be the uptake of native IP based video telephony in IMS architecture using 3GPP defined protocols and codecs.

Video codecs mandated by 3GPP is ITU-T H.263 Profile 0 Level 45 as defined in 3GPP TS 26.114. However 3GPP encourages implementations to support video codecs that provide higher end-user experience, such as MPEG-4 (Part 2) Visual Simple Profile Level 3 as well as ITU-T H.264/MPEG-4 (Part 10) AVC. The latter, in particular, is expected to be used in many commercial implementations because of the excellent performance and quality that it provides.

In order to complete video calls between IMS and Circuit-Switched terminals, specific video-enabled Media Gateway functionality is required. This kind of Media Gateway is able to perform interworking between 3G-324M-related user plane protocols (H.223/H.245) and speech and video codecs to those used within the IMS network. Naturally, in the best situation, at least codecs are compatible on both sides of the call in order to remove the need for transcoding that may result in worse end-user experience than without transcoding. The Media Gateway also needs to be capable of performing transrating—that is, to adapt codec's bit rate between the endpoints involved, because that way it is possible to connect endpoints to each other using varying transport networks (Circuit Switched versus native IP).

Finally, the IMS network may also support value-added services for video telephony such as video announcements, video ring-back content, video conferencing and others, which requires that the required functionalities for speech and video related RTP connections are handled in Media Resource Function Processor (MRFP) in order to provide the required service.

9.6 Summary

Both CS fallback for LTE as well as the LTE-based VoIP with SR-VCC solution can be deployed either in a phased manner or simultaneously depending on the terminal's capabilities.

It is also possible to deploy IMS for non-VoIP services, such as video sharing, presence or instant messaging. Even a primary voice service is deployed by using CS fallback for LTE procedures. This kind of architecture ensures maximum flexibility for the operator to plan and deploy LTE as well as IMS solutions.

References

[1] GSMA (2010) *IMS Profile for Voice over SMS 1.0*, IR.92, GSM, London.
[2] Nokia Siemens Networks (2007) R2-074678, Stage 3 Aspects of Persistent Scheduling, Nokia/Nokia Siemens Networks, Jeju, South Korea.
[3] Nokia Siemens Networks (2007) R2-074679, Persistent scheduling for UL, Nokia/Nokia Siemens Networks, Jeju, South Korea.
[4] 3GPP TS 23.272 (2010) Circuit Switched (CS) fallback in Evolved Packet System (EPS), V. 8.1.0, 3rd Generation Partnership Project, Sophia-Antipolis.
[5] 3GPP TS 23.204. (2010) *Support of Short Message Service (SMS) over generic 3GPP Internet Protocol (IP) access*, V.8.3.0, 3rd Generation Partnership Project, Sophia-Antipolis.
[6] 3GPP TS 24.301. (2011) *Non-Access-Stratum (NAS) protocol for Evolved Packet System (EPS); Stage 3.* V. 8.10.0, 3rd Generation Partnership Project, Sophia-Antipolis.
[7] 3GPP TS 29.118. (2010) *Mobility Management Entity (MME) - Visitor Location Register (VLR) SGs interface specification.* V. 8.8.0, 3rd Generation Partnership Project, Sophia-Antipolis.
[8] 3GPP TS 23.018. (2011) *Basic call handling; Technical realization.* V. 8.4.0, 3rd Generation Partnership Project, Sophia-Antipolis.
[9] 3GPP TS 24.010. (2008) *Mobile radio interface layer 3; Supplementary services specification; General aspects*, V. 8.0.0, 3rd Generation Partnership Project, Sophia-Antipolis.
[10] Halonen, T., Romero, J. and Melero, J. (2003) *GSM, GPRS and EDGE Performance*, 2nd edn, John Wiley & Sons, Ltd, Chichester.
[11] Barreto, A., Garcia, L. and Souza, E. (2007) GERAN Evolution for Increased Speech Capacity, Vehicular Technology Conference, 2007. VTC2007-Spring, April 2007.
[12] Holma, H. and Toskala, A. (2007) *WCDMA for UMTS*, 4th edn, John Wiley & Sons, Ltd, Chichester.
[13] GSMA PRD IR.92 (2010) *IMS Profile for Voice and SMS 1.0. 18*, GSM, London.
[14] 3GPP TS 23.228. (2010) *IP Multimedia Subsystem (IMS); Stage 2.* V. 8.12.0, 3rd Generation Partnership Project, Sophia-Antipolis.
[15] 3GPP TS 24.229. (2011) *IP multimedia call control protocol based on Session Initiation Protocol (SIP) and Session Description Protocol (SDP); Stage 3.* V. 8.16.0, 3rd Generation Partnership Project, Sophia-Antipolis.
[16] 3GPP TS 23.226. (2008) *Global text telephony (GTT); Stage 2.* V. 8.0.0, 3rd Generation Partnership Project, Sophia-Antipolis.
[17] ITU-T T.140. (1998) *Protocol for multimedia application text*, International Telecommunications Union, Geneva.
[18] IETF RFC 4103. (2005) *RTP Payload for Text Conversation*, Internet Engineering Task Force, Fremont, CA.
[19] 3GPP TS 26.235. (2008) *Packet switched conversational multimedia applications; Default codecs.* V. 8.0.0, 3rd Generation Partnership Project, Sophia-Antipolis.
[20] 3GPP TS 23.203 (2008) *Policy and charging control architecture*, V.8.3.1, 3rd Generation Partnership Project, Sophia-Antipolis.
[21] GSMA PRD IR.34. (2010) *Inter-Service Provider IP Backbone Guidelines*, V. 4.9, GSM, London.
[22] GSMA PRD IR.65. (2010) *IMS Roaming and Interworking Guidelines*, V. 5.0, GSM, London.
[23] GSMA PRD IR.88. (2010) *LTE/SAE Roaming Guidelines*, V. 3.0, GSM, London.
[24] GSMA PRD IR.83. (2009) *SIP-I Interworking Guidelines*, V. 1.0, GSM, London.
[25] GSMA PRD IR.90. (2010) *RCS Interworking Guidelines*, V. 2.0, GSM, London.
[26] 3GPP TS 23.216. (2008) *Single Radio Voice Call Continuity (SRVCC)*, V. 8.1.0. V. 8.7.0, 3rd Generation Partnership Project, Sophia-Antipolis.
[27] 3GPP TS 23.206. (2007) *Voice Call Continuity (VCC) between Circuit Switched (CS) and IP Multimedia Subsystem (IMS)*, V.7.5.0, 3rd Generation Partnership Project, Sophia-Antipolis.

10

Functionality of LTE/SAE

Jyrki T. J. Penttinen and Tero Jalkanen

10.1 Introduction

This chapter presents the state of the LTE UE and network. This is explained using examples of typical procedures prior to and during an LTE call. This chapter also presents the signaling flows for different cases, together with the protocol messages. End-to-end functionality is explained with examples. Operations-related topics are presented for fault, configuration, backupandrestore, and inventory management. Principles for online and offline charging are presented. Finally, signal protection is presented.

10.2 States

There are two states in EPS and LTE UE: EMM (EPS Mobility Management) and ECM (EPS Connection Management). Figure 10.1 clarifies the transitions between the states, and how they overlap with each other. In the EMM-Deregistered state, the UE is not reachable for the LTE/SAE network. The transition from the EMM Deregistered state to the EMM-Registered state happens via the Attach procedure or via the Tracking Area Update from GERAN or UTRAN. At the same time, the UE moves to ECM-Connected state. After the release of the signaling, the UE moves to ECM-Idle state. It still keeps the EMM-Registered state. The UE can move back to the ECM-Connected/EMM-Registered stage when the signaling connection has been established. The UE can also move directly to the EMM-Deregistered state if the Detach procedure is carried out. This Detach procedure can also move UE to the EMM-Deregistered state from the EMM-Registered/ECM-Connected state.

The UE is not reachable in the EMM-Deregistered state. In the EMM-Registered/ECM-Connected state, there is an RRC connection as well as a S1-MME connection on, that is, this is the active communication mode of the terminal. In the EMM-Registered/ECM-Idle stage, the UE can still be reached by the network via the paging procedure. Figure 10.2 clarifies the idea of state transition in practice.

The LTE/SAE Deployment Handbook, First Edition. Edited by Jyrki T. J. Penttinen.

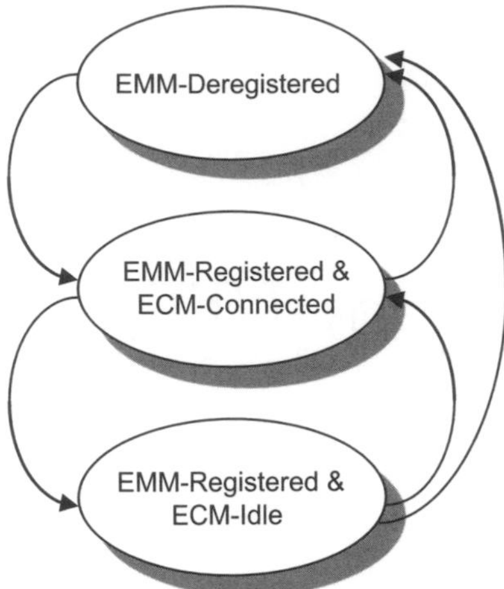

Figure 10.1 The states of the LTE UE and network.

10.2.1 Mobility Management

Mobility Management is needed in order to keep track of the location of the UE. Without this functionality, it would not be possible to page the UE. The Mobility Management procedures are Paging and Tracking Area Update (TAU).

The paging procedure provides the means for the network to send the initial paging message for the UE sufficiently accurately—that is, at the Tracking Area (TA) level. It is important for the network to know the location of the UE in an area that consists of various eNodeBs and their respective cells.

The tracking area update is performed when the UE notices that it has entered a new Tracking Area. This is noted via the best observed list, which the UE is updating constantly based on the situation.

There are two EMM states: *EMM-Deregistered* and *EMM-Registered*. They describe whether the UE is registered in the MME. If so, the network knows that the UEs can be reached by paging. Otherwise, if the UE is in the EMM-Deregistered state, the MME does not

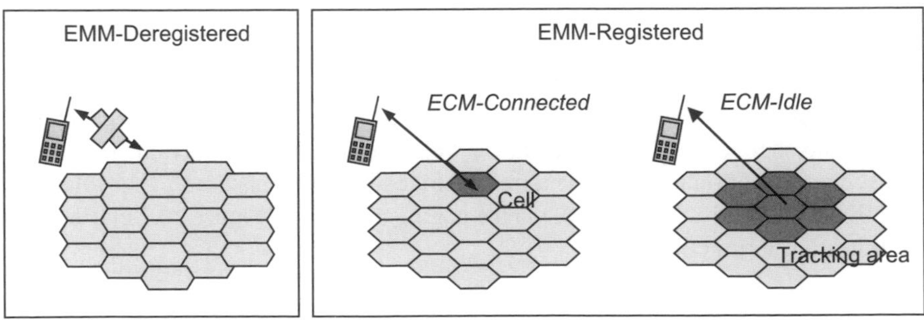

Figure 10.2 The idea of the LTE state transition. The presence of UE is not known by the network in EMM-Deregistered state. In EMM-Registered state, the location of the UE is known at cell level when the UE is connected, and at tracking area level when the UE is in idle mode.

contain any information about the location of the UE. This means that the UE is not reachable for the network.

The LTE Attach Procedure triggers the transition from the EMM-Deregistered state to the EMM-Registered state. Also a tracking area update (TAU) from a 2G (GERAN) or 3G (UTRAN) network triggers this transition. In the EMM-Registered state, the UE can be paged and it is thus reachable for the network.

10.2.2 Handover

The LTE/SAE handover can happen internally within the same LTE/SAE network, which means that the handover happens between eNodeB elements. The handover can also happen between different 3GPP networks—that is, between two different LTE/SAE networks, or between LTE/SAE and legacy 2G/3G networks (the GERAN and UTRAN networks).

3GPP has defined requirements for the handover performance as for the interruption delay form the initiation and termination of the handover procedure. For the fastest case, which is the inter-eNodeB handover, the average interruption can be a maximum of 54 ms in the downlink direction and 58 ms in the uplink direction for the user plane. The average interruption time for the signaling plane can be a maximum of 56 ms. For the handover between LTE/SAE and UTRA, the respective maximum delays in the user plane can be up to 150 ms and 300 ms in the downlink and uplink directions, respectively.

Depending on the handover type, the mobility anchor point can be fixed to the eNodeB, S-GW or P-GW. The heaviest handover procedure—that is, the handover between LTE/SAE and non-3GPP networks like CDMA2000—the P-GW element acts as a mobility anchor point. S-GW is the mobility anchor point for the inter-RAT handovers between 3GPP networks as well as for the intra-eNodeB handovers within the same LTE/SAE network. The lightest handover—that is, the inter-eNodeB handover—means that the mobility anchor point is in the original eNodeB element.

10.2.3 Connection Management

There are two EPS Connection Management (ECM) states, *ECM-Idle* and *ECM-Connected*. They describe the signaling connectivity between the UE and the EPC.

In the ECM-Idle state, there is no connection for the signaling between the UE and the MME. In the ECM-Connected state, there exists a signaling connection between the UE and the MME. In this state, the signaling can happen via an RRC connection between the UE and eNodeB, and via an S1-MME connection between eNodeB and the MME.

The connection management procedures can be divided into the following:

- random access procedure;
- LTE attach procedure;
- user data connection setup procedure;
- connection release procedure.

When the UE is in the EMC-IDLE state—that is, not in the EMC-CONNECTED state—its location is known by the network only at the level of the tracking area. At this stage, it is possible for the network to call the UE through the paging message that is sent via the cells over the tracking area. By answering to paging, or when UE initiates the connection, it moves to the

EMC-CONNECTED state. At this state, the location of the UE is known at the single cell level—the network knows explicitly the cell with which the UE is communicating. After the inactivity timer expires, the UE moves to the EMC-IDLE state. When the UE finally moves to the EMM-DEREGISTERED state, its location is not known by the network any more until EMM-REGISTERED state is reached again.

10.2.3.1 Random Access Procedure

The random access procedure is performed when the UE wants to initiate a connection with the network. This is the only moment when the overlapping radio transmissions can occur in the uplink, if more than one UE sends the random access message at the same time.

The aim of the random access procedure is to establish an RRC signaling link between the UE and eNodeB.

LTE defines two types of random access procedure: contention-based and non-contention-based random access procedure. The typical mode is the contention-based random access procedure as shown in Figure 10.3.

In this case, the UE is either in the ECM-IDLE state, wanting to connect to the network, or alternatively, the UE wants to initiate the attach procedure when it is already connected to the network. The random access procedure is necessary when there is data to be sent, either in the uplink or the downlink. The latter is indicated by the paging message of the network. A tracking area update also requires the UE to perform the random access procedure.

Non-contention based random access procedure is utilized in special cases, UE being in the ECM-CONNECTED state, as shown in Figure 10.4. This random access type is utilized when the networks makes UE to perform the handover procedure to another cell. This can also happen when the UE is not synchronized with the network but there is data to be received by UE.

10.2.3.2 LTE Attach Procedure

When the UE makes the LTE attach procedure via the RRC signaling link by sending an LTE attach request message to the eNodeB, the EMM-DEREGISTERED state is changed to EMM-REGISTERED. This can happen, for example, when the UE is switched on. The message is then forwarded to MME, which may perform initial procedures like authentication. As a result,

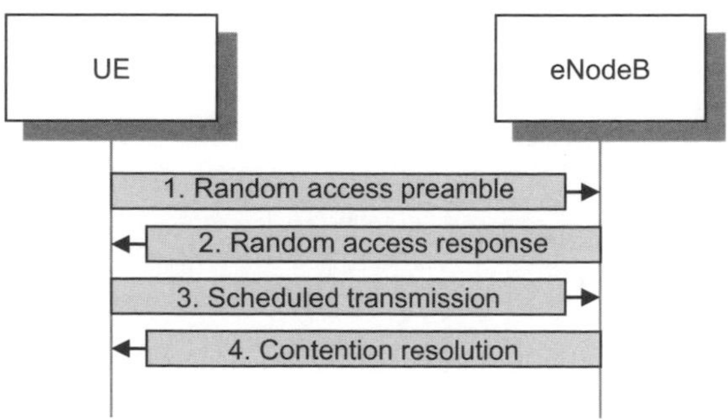

Figure 10.3 Contention-based random access procedure.

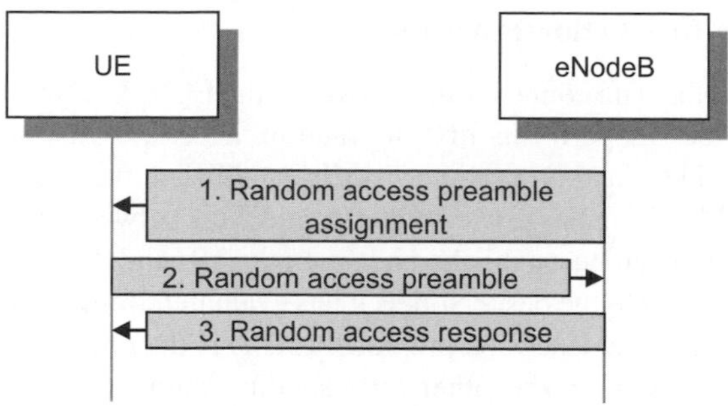

Figure 10.4 Non-contention-based random access.

the UE registers to the SAE network via the signaling between MME and HSS as shown in Figure 10.5, and directly enters the EMC-CONNECTED state, even if there is no data transmission at this very moment. Then, after a certain inactivity time, the UE and network change the state from EMC-CONNECTED to EMC-IDLE.

Following the signaling flow in Figure 10.5, the first step in the LTE attach procedure is the random access procedure as described previously (1). This establishes the Radio Resource Control (RRC) link that is used for further signaling between UE and eNodeB. Then, the UE sends the LTE attach request to eNodeB, which forwards the request to the MME (2). At this moment, before other signaling is possible, the network may perform the authentication procedure (3). This happens if no context is created elsewhere for this UE. Then, if the authentication is successful, the MME sends a tracking area update message to the HSS (4). This means that the HSS now knows that the MME is serving the UE. The HSS now sends the user's subscription data to the MME, which the MME acknowledges (5). After this, the HSS still acknowledges the tracking area update message that the MME sent previously (4). As a result, a default EPS bearer has now been created.

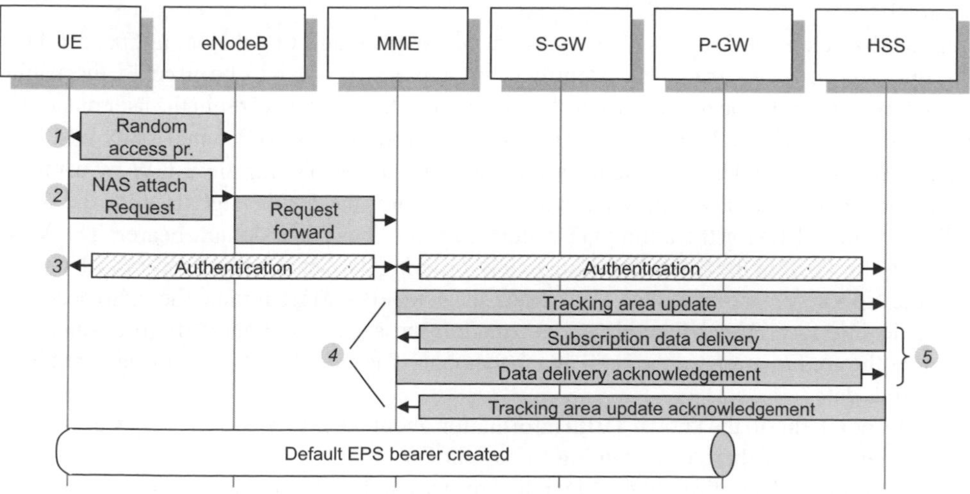

Figure 10.5 The LTE Attach procedure.

10.2.3.3 User Data Connection Procedure

When the UE initiates the data connection, it moves to the ECM-CONNECTED state. In order to do the transition, the UE performs first the random access, by using the Radio Resource Control (RRC) signaling. At this moment, MME establishes the signaling link via the S1 interface. Then, the MME creates a user-plane connection between the UE and the S-GW.

Before the connection can be established between the UE and the eNodeB, the UE sends a random access message (1) to the best eNodeB it has evaluated from the local list, according to the random access procedure described previously. This is the only moment of the signaling when collisions can occur with the other UEs sending their possible own random access messaging at the same time.

Then, the UE sends am *NAS Attach Request*, that is, Non-Access Stratum Attach Request (2) to the eNodeB, according to the LTE attach procedure described previously. Before any other signaling, authentication is performed between the UE, MME and HSS as a part of the attach procedure, unless there is already a context created between the UE and the network. When the authentication procedure is finished, the MME sends Tracking Area Update information to the HSS. This basically means that the HSS now knows that this specific MME is serving the UE from now on, which HSS acknowledges to the MME as described in the LTE attach procedure. If the secure procedure results in failure—that is, the authentication was not successful—the call initiation is terminated at this point.

In order to continue from this point, the following signaling is performed for the establishment of the default EPS bearer. A user plane connection is created between eNodeB, S-GW and P-GW. Then, an IP address is given to the UE by the P-GW, followed by the establishment of the radio bearers. The uplink data transmission can take place as of this moment, if there is data to be transmitted. There is a confirmation messaging between the eNodeB, the MME and the S-GW, which makes it possible to deliver downlink data if it exists, followed by the last confirmations of this phase. The GTP (GPRS Tunneling Protocol) links are needed in the downlink as well as in the uplink for the default bearer, the user plane being between the eNodeB and the P-GW, and the control plane being between the S-GW and the P-GW. At this point, TEID (Tunnel Endpoint Identifier) information is used in the relevant elements.

In order to create the EPS default bearer, the more specific signaling flow is as shown in Figure 10.6.

The MME element signals with the relevant S-GW (4) requesting the default bearer setup. The message includes the MME identifier and address in order to be used in the paging procedure. After this, the S-GW asks the P-GW to set up the default bearer in the user plane. The TEID information is used at this point for identifying the downlink GTP tunnel endpoints. Now, the P-GW assigns the UE an IP address (5), and sends a message for the S-GW by using the TEIDs to inform the downlink endpoints. Furthermore, the S-GW signals with the MME delivering the TEID for the uplink GTP tunnel, for the user plane default bearer. The MME forwards the TEID of the S-GW for the eNodeB, ordering it to set up radio bearers with the UE, with the NAS acceptance message. As a result, the eNodeB starts to initiate the radio bearers in the radio interface. The UE sends a confirmation message to the eNodeB (6), which also includes the confirmation for the NAS attach message. The default EPS bearer is now ready for the uplink data.

At this point, due to the known TEID information about the S-GW, the UE may start sending data in the uplink if there is something to be sent (7).

Next, the eNodeB confirms to the MME that the user and control plane bearers are ready, and that the UE is now capable of sending data in the uplink direction. At the same time, the eNode

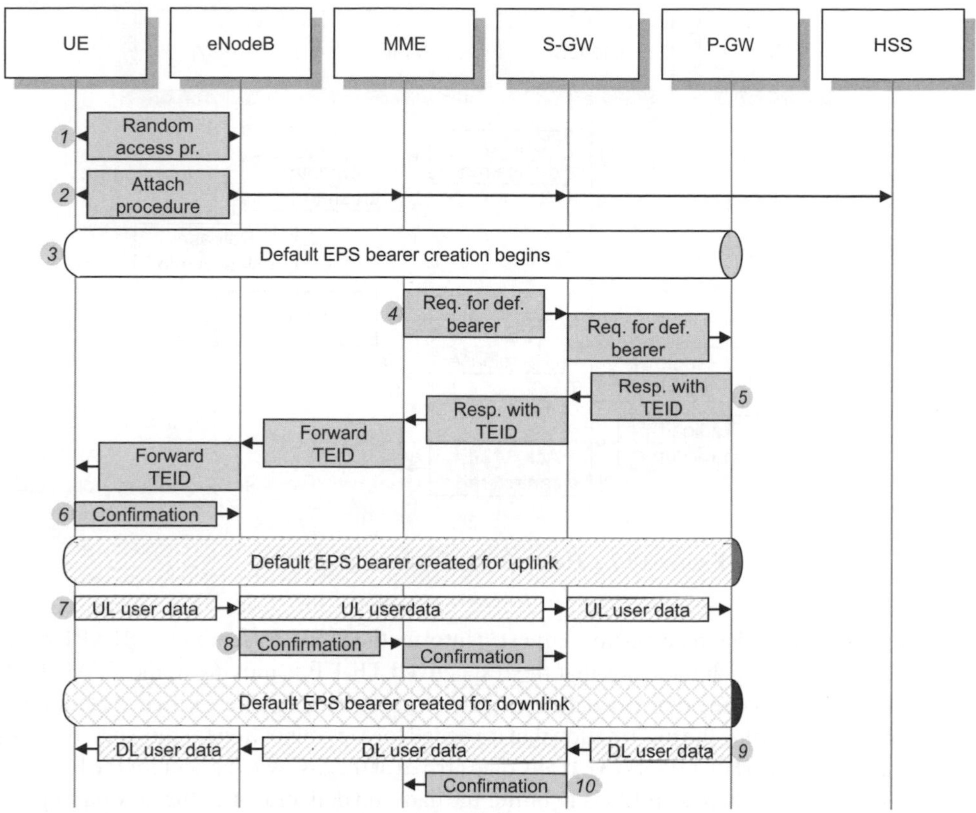

Figure 10.6 The signaling flow for the LTE data transmission. Prior to the actual flow, the random access and LTE attach procedures have been performed.

B sends information about its TEID in the downlink GTP tunnel endpoint side, forwarding the NAS attach confirmation to the MME (8). The eNodeB's TEID information is also forwarded to the S-GW.

At this point, the S-GW may now start sending data in the downlink, if there is something to be sent (9). Confirmation is still sent from S-GW to MME (10), which finalizes the signaling prior to the data transmission.

10.2.3.4 Connection Release Procedure

The connexion release procedure returns the ECM-CONNECTED state to the ECM-IDLE state. This can happen, for example, when there has been no data transmission for a certain time. Figure 10.7 shows the principle of the connection release signaling flow. When the connection release is triggered, the eNodeB starts signaling with the MME. The aim of the signaling is to release both the user and the signaling plane connections that were activated for the respective LTE-UE. First, if the eNodeB notices, for example, a sufficiently long time period without downlink or uplink traffic for a certain user, the eNodeB triggers the procedure by sending a connection release request for the MME element (1).

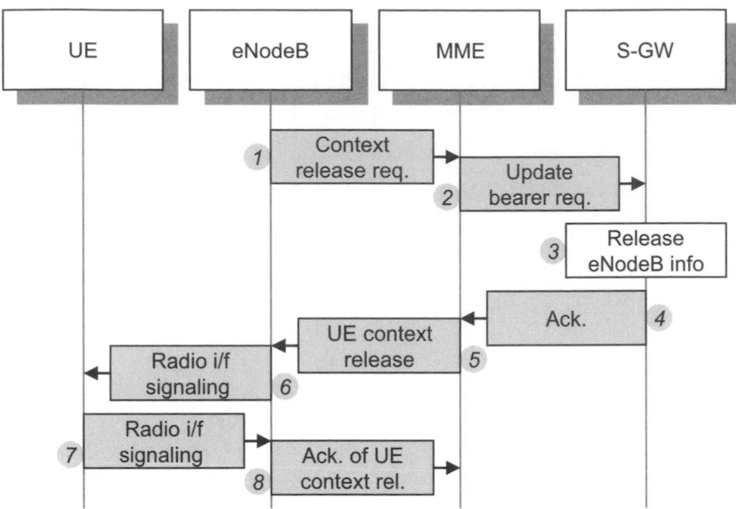

Figure 10.7 The connection release procedure.

As soon as the MME has received the request, it informs the S-GW by sending an S1AP Update Bearer Request (2), which indicates that the respective LTE-UE is moving to the ECM-IDLE state and is not available any more for data transmission unless paging is performed. At this point, the S-GW starts releasing the information that is needed for the connection for this specific LTE-UE, including the TEID identifiers (3). If any data arrives at the S-GW to be sent for the LTE-UE in the downlink, the only possibility is to buffer the data and deliver it after the new paging has been performed. The S-GW now informs the MME about the successful connection release (4), and the MME further informs the eNodeB about the release by sending an S1AP UE Context Release Command (5). The eNodeB element signals with LTE-UE (6, 7), and the connection release procedure is finalized by the acknowledgment of the eNodeB for the MME element (8).

10.2.4 Authentication

The authentication is further strengthened in the LTE/SAE system throughout the signaling compared to the previous GSM and UMTS solutions. The authentication procedure with the respective signaling flow is explained in more detail in Chapter 11.

10.2.5 Tracking Area

The Tracking Area is used for the optimal initiation of the LTE connections. The Tracking Area concept becomes relevant when the UE is in the ECM-Idle state, and when the network wishes to start the communication. At this point, the location of the UE must be known in such a way that the network can deliver the initial signaling for the UE in a certain geographical location. The signaling is then sent via all the cells that belong to the Tracking Area. Figure 10.8 clarifies the principle.

The ECM-Idle state can be reached within the cells that belong to the Tracking Area in which the UE is currently registered. Please note that the UE may be registered within various Tracking Areas.

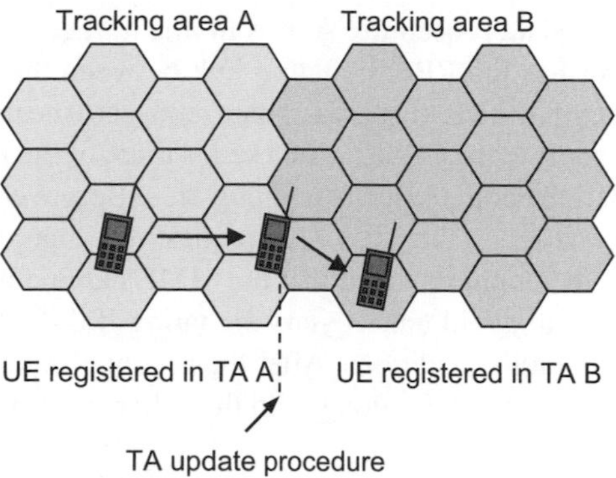

Figure 10.8 The tracking area is updated when UE drifts to the new tracking area, in order to be found by the network when it performs paging within the cells in the most actual tracking area. The ID of UE in ECM-IDLE state is S-TMSI.

The MME procedure gives a Temporary Mobile Subscriber Identity (S-TMSI) for the UE. The S-TMSI uniquely identifies a certain piece of UE within the Tracking Area—that is, two UEs can never exist with the same S-TMSI in the same Tracking Area. When the UE is in the ECM-Idle state, the MME procedure can request (within a single-cell area, or within one or more Tracking Areas) that the UE associated with a certain S-TMSI switch to the ECM-Connected state. This MME request is performed via the paging signaling.

The UE registers to a certain Tracking Area by observing the signal level of the cells that are candidates for the communication. The best cell—according to certain criteria—represents the current Tracking Area. The Tracking Area code is, in fact, delivered via the best cell. When the order of the best cells is changed, the Tracking Area code also changes at some point. In this instance, the UE is obligated to signal to the network in order to update its registration to the new Tracking Area.

Figure 10.9 clarifies the TAU procedure signal flow.

In the beginning (1), the UE starts receiving a new tracking area indicator via the broadcast channel, and the UE notices that it is not in the current list of the tracking area or tracking areas that UE has given to the network. As the UE is not registered within this new tracking area, it

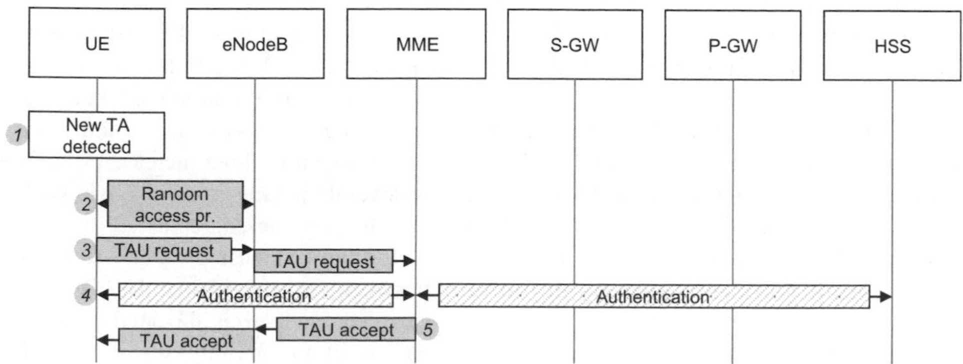

Figure 10.9 Tracking area update procedure. The periodic TAU is done in the same manner.

triggers the tracking area update procedure. At this moment, the random access procedure is initiated (2), which establishes the RRC signaling link between the UE and the eNodeB.

As a next step, the UE sends the tracking area update request to the network (3). This message is forwarded from the eNodeB to the MME, which keeps track of the mobility of the UEs. This message contains information about the latest tracking area. Before continuing, authentication may take place (4). Then, based on the UE's TAU request information, the MME generates a new list of tracking areas. It should be noted that the MME can include the previous tracking area in the new list in order to avoid unnecessary signaling if the UE is in the border of two tracking areas with varying radio conditions. After the accepted tracking area update request, the MME sends the new list to the UE, along with the other information that belongs to this message (5).

In addition to the signaling due to the physical moving between the Tracking Areas, the Tracking Area Update (with or without new Tracking Area) can be done periodically. This means that the UE regularly signals with the network, and the Tracking Area is thus revised. This helps the network to keep the registration alive. This prevents unnecessary paging signaling if the UE disappears from the network without giving information about the state transition to EMM-Deregistered—for example, if the battery suddenly drains.

The dimensioning of the Tracking Area is one of the important operator's tasks. If the area is too large, the paging message is sent over too many eNodeBs, which increases the signaling load of the network. On the other hand, if the Tracking Area is too small, the UE has to inform the network each time it enters new Tracking Area, which again produces unnecessarily high amount of signaling.

The in-depth network optimization includes the identification of the optimal Tracking Area size that produces the lowest signaling load. This obviously depends on the geographical area and how much movement there is locally. The level can be investigated via the network statistics, for example, via the Operations and Management System (OMS).

10.2.6 Paging Procedure

While the UE is in the ECM-IDLE state—that is, when the terminal is registered in the network but not in connected mode—it follows the tracking area information of the broadcast channels of the cells. Whenever there is reason to find the UE—that is, when new downlink data is waiting to be sent from the network to the UE—it must be found in order to change to the connected mode as only that allows the actual data transmission. The UE is thus paged via all the cells that belong to the tracking area or several tracking areas where the UE is last registered. It should be noted that unlike in the GPRS routing area or basic GSMs location area where only one area is possible at the time, LTE allows the registration of UE to multiple tracking areas.

More specifically, the MME follows up the individual UEs' tracking areas, and sends a paging message via the corresponding eNodeBs. As the signaling load increases when the tracking areas are greater, and when the UE registers to several tracking areas, it is an important optimization task of the operator to design the strategy for the tracking areas.

The actual paging procedure contains two phases. In the first phase eNodeB sends out a paging indication message via the paging channels, based on the request from the MME. This message indicates that a certain paging group is being paged. Each UE is allocated to a certain paging group based on its IMSI, or a temporal version of the identity, that is, S-TMSI. Whenever UE realizes that the paging group where it belongs is being paged, the UE starts receiving the complete paging message for more details.

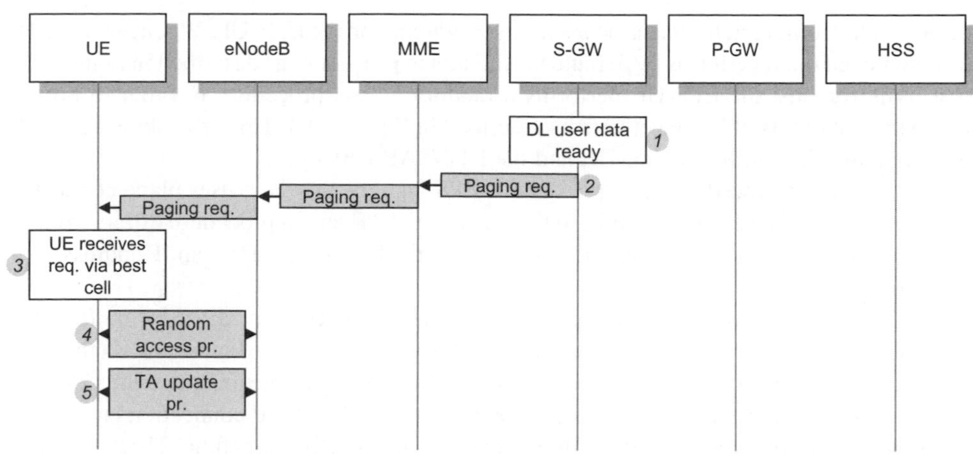

Figure 10.10 Paging signaling flow.

As a next step, the UE then starts changing from the ECM-IDLE to the ECM-CONNECTED state. As a first part of this procedure, the UE makes the random access procedure, followed by the location area update procedure.

Figure 10.10 shows the principle of paging signaling. In this example, the S-GW has received data to be sent to UE (1). As the location of UE is not known, the S-GW sends a paging request to the MME (2). The correct, serving MME is found based on the information stored in the S-GW, which was obtained from the initial LTE attach procedure of the UE, and possibly the MME relocation procedure during the period in which the UE is registered.

The MME then sends the paging request message (S1AP) for all the eNodeBs that are defined in the tracking areas in question. These eNodeBs then send the initial paging indication message, which contains information about the way to read the actual, more detailed message. This indication is repeated until the UE responds, or until the maximum defined repetition number is reached.

If the UE is still in the tracking area where it was registered, it receives the paging indication within the broadcast channel of the best cell that it is currently monitoring (3). The UE realizes that it belongs to the paging group referred to in the paging indication. In this way, it can read the more detailed message, as the instructional also reveals the physical resources for the delivery of the paging message via the eNodeBs in the area.

All the UEs belonging to the same paging group now read the more detailed message. The UEs belonging to other paging groups within the same tracking area do not need to react. The UE that detects its own IMSI or S-TMSI from the message knows now that it is paged and reacts by starting the random access procedure, whilst the other UEs in the same paging group simply discard the message and continue following the other, forthcoming paging indicators. The correct UE then continues signaling as shown in the User Data Connection Procedure signaling chart.

10.3 End-to-End Functionality

Following the signaling flows of Figure 10.6, an example of the end-to-end VoIP connection establishment and utilization can be observed.

Let us start from scratch—that is, when the user switches on the LTE-UE. Whenever the LTE-UE is switched on, it performs by default an LTE attach procedure as described in Figure 10.5. As a basis for this, the LTE-UE performs a random access procedure in order to provide information about itself to the network as described in Figure 10.3. This procedure creates the RRC signaling link between LTE-UE and the LTE/SAE network.

The LTE attach procedure sets up a default EPS bearer—that is, a user plane connection between the UE and the core network. In this sense, the LTE attach procedure differs from the previous mobile network functionalities. As a result, the UE also receives an IP address at the same time.

As a next step, the Voice over IP service is initiated by setting up SIP (Session Initiation Protocol) signaling between LTE-UE and the IMS through the LTE/SAE network. At this time, the setting up of the voice service happens in the application layer, meaning that the LTE/SAE network is simply a transparent means for the bit transfer. The VoIP connection is set up by following the normal procedures of IMS—that is, via the SIP signaling. These signaling messages are carried over the EPS and are delivered to the IMS [1]. The only interacting role of LTE/SAE could be the provision of the PCRF (Policy and Charging Rules Function) in order to offer QoS differentiation. If the LTE/SAE network is QoS aware, it is possible to dedicate a special EPS bearer which can give messages higher priority over the other services. If the LTE/SAE network is not able to provide variable QoS for different applications, the VoIP traffic is simply delivered via the default EPS bearer over the LTE/SAE part.

The VoIP service is described at a more detailed level in Chapter 9 as a part of the LTE/SAE application.

10.4 LTE/SAE Roaming

10.4.1 General

This section presents the principles related to LTE/EPC roaming, including IMS services such as VoLTE used on top of the LTE access. LTE/SAE introduces a major change to the radio network, the core network and potentially also the way roaming is arranged.

It is worth noting, though, that this area is partially undergoing work in international forums such as GSMA. Thus, some assumptions are used in this chapter reflecting the most probable solutions before the final models are agreed.

One important issue to note before going into the details: roaming and interconnection are two completely separate things. Both require some kind of connectivity to take place between two operators but they are vastly different:

- Roaming takes place when Mr. Palin hops into the plane and travels from his native Great Britain to France. His phone switches from UK Operator A to French Operator A.
- Interconnection takes place when Mr. Cleese, who is using UK Operator B, calls Mr. Jones who is served by UK Operator C.

It is possible to have roaming and interconnection taking place at the same time—for example, in the previous example when Mr. Palin, roaming in France, calls Mr. Cleese, who is a customer of another operator in UK. So, confusing roaming with interconnection is like mixing walking with swimming: both are related to moving but no other similarity exists. Luckily people are not doing this as much as they used to do some years ago. . .

Roaming is one of cornerstones of the GSM success story—being able to seamlessly use the same mobile phone at home and when abroad basically regardless of what part of the world you have ended up in has proven to be very popular among end users. LTE/SAE must therefore offer this valuable function. Sharing videos or images could be especially popular when one is roaming—for example, while sightseeing in Tokyo, hiking on an Appalachian Trail, or canoeing in Lake Saimaa.

LTE can be used just as a large bit pipe, allowing users to access whatever Internet services they wish, but it can also be used to provide operators with more specific services. As in previous generations and versions of mobile communications, the voice service is obviously one of the most important services in LTE, where it can be delivered using OTT players (the bit pipe model), Fallback to 2G/3G (CSFB), or Voice over LTE (VoLTE) as provided by the IMS core system. These methods lead to some special requirements for LTE/SAE roaming, related to issues such as QoS and service-based charging, which have to be taken into account in the design of commercial LTE/SAE roaming models. Voice can be assumed to be the first realistic service in LTE offered by the operator, thus the models deployed for the use of voice will likely have a big impact for other services.

On the other hand, the most likely scenario is that LTE/SAE roaming is not a vital issue for the operators in the very initial phase of LTE deployment due to the small number of LTE terminals. It is expected that support for LTE/SAE roaming by operators will grow only gradually when real commercial need comes into the picture and the number of operators having the commercial LTE/SAE service increases. In general, as has been the case, for example, with the GPRS Roaming Exchange (GRX), it would be beneficial to have a single common worldwide solution specified for LTE/SAE roaming that will be supported by most or all of the operators even before the concrete commercial need arises. This is in order to avoid the typical last-minute rush in which individual operators quickly try to arrange any kind of solution providing IP connectivity from the VPLMN to the HPLMN, which would lead to a number of incompatible solutions being deployed.

As a typical LTE/SAE device is capable of using the 3G access network whenever necessary, it is also capable of reusing existing 3G roaming. In practice the end-user would be able to access services such as Internet browsing or e-mail via 3G from his/her new LTE/SAE device from the VPLMN if the LTE/SAE roaming agreement/connection is not in place or if the VPLMN is not LTE/SAE capable at all. Nevertheless, it is expected that the general advantages of LTE/SAE over 3G, such as lower production costs, increased bandwidth, lower delay, and better QoS support, mean that there is interest in the operator community to launch LTE/SAE roaming despite the current 3G PS roaming in many cases offering good enough service for roamers.

In this context, the term "LTE/SAE roaming" also includes core related topics—that is, the EPC and EPS are in the scope of the roaming. The reason for the terminology is that the GSMA is using the title "LTE/SAE roaming" in such a way that also EPS related topics are combined under the same title.

10.4.2 Roaming Architecture

This section illustrates the main network elements involved in LTE/SAE roaming, including the two major architectural alternatives that are compared in Section 5.6. The general document for the technical details of LTE/SAE roaming is IR.88 [2].

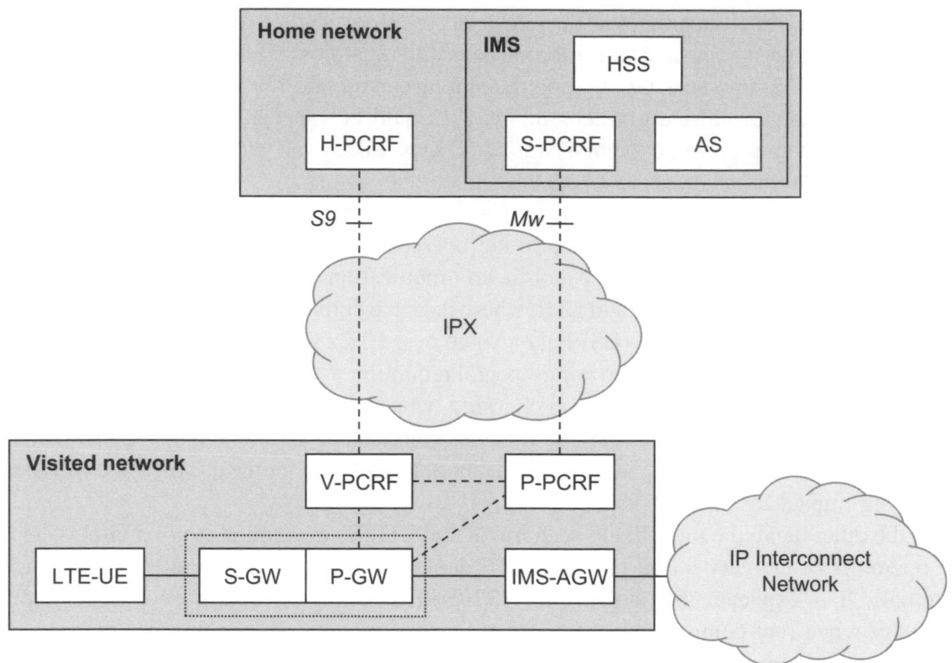

Figure 10.11 The high-level architecture of the LTE/EPS roaming. The dotted line represents the control plane and solid line a media plane.

The following nodes and interfaces are relevant for LTE/SAE roaming, as indicated in Figure 10.11:

- The Mobility Management Entity (MME), including the AAA, paging and other control plane management.
- The Serving Gateway (SGW), acting as an "LTE SGSN" when compared with the Serving GPRS Support Node in the GPRS environment.
- The Packet Data Network Gateway (PGW), acting as an "LTE GGSN" in a similar way to the Gateway GPRS Support Node in the case of GPRS.
- The Policy and Charging Rules Function (PCRF) is a QoS resource, as well as a usage and charging authorization linkage between service and transport layers.
- The Home Subscriber Server (HSS) acts as an "LTE HLR" comparable to the GPRS home location server.

In addition to these main nodes additional supporting elements are also required when deploying LTE/SAE roaming. As an example, Diameter Edge Proxies/Relays are needed by the interoperator interfaces that use Diameter connected via the S6a and S9 interfaces, as shown in Table 10.1.

If the aim is to offer, for example, a VoLTE service as specified in IR.92 [10] for the LTE customers, in addition to LTE and EPC, the deployment of the IMS core system and related AS infrastructure are also needed. Furthermore, it is important to ensure that the LTE bearers specified for VoLTE are available. PCC architecture is also needed, and the CS core must be upgraded to support SRVCC functionality.

Table 10.1 LTE/SAE roaming interface definitions

Nodes	Interface	Protocol
MME – HSS	S6a	Diameter Base Protocol (IETF RFC 3588 [3]) and 3GPP TS 29.272 [4])
SGW – PGW	S8	GTP (GTP-C 3GPP TS 29.274 [5] and GTP-U 3GPP TS 29.281 [6]) or PMIP (IETF RFC 5213 [7]) and 3GPP TS 29.275 [8])
hPCRF – vPCRF	S9	Diameter Base Protocol (IETF RFC 3588 [3]) and 3GPP TS 29.125 [9])

Compared to the situation of today, LTE/SAE roaming introduces some new protocols to be supported by the inter-PLMN infrastructure:

- Diameter (used by e.g., MME-to-HSS interface);
- SCTP (used by Diameter as the transport protocol).

It can be estimated that even if these protocols do not have a major impact as such, they need to be taken care of in the planning of, for example, service- and application-aware nodes or firewalls that are used in LTE/SAE roaming. A related issue is that the GTP-aware firewalls normally used in a 2G/3G PS roaming environment might not understand GTPv2 used by LTE/SAE in the S8 interface for the user plane (the control plane still uses GTPv1). This is something to check when deploying LTE/SAE roaming reusing the existing components of 2G/3G roaming.

10.4.3 Inter-Operator Connectivity

In the roaming situation, one of the most important functions is the connectivity between the VPLMN and the HPLMN. In LTE/SAE roaming, these interoperator IP network connections are handled by IPX, which can be seen as an evolved version of the solution used today in all the commercial 2G/3G PS roaming—that is, GRX as documented in IR.34 [1]. Both GRX and IPX are developed within GSMA and can be used for any kind of IP-based traffic—that is, not just GTP or voice as such. The main benefits of IPX over GRX are the guaranteed end-to-end QoS delivery and access to non-GSM operators.

IPX offers different models as listed below:

1. **Transport**—Layer 3 service, which simply carries packets regardless of the application used.
2. **Bilateral Service Transit**—IPX includes service-level intelligence, such as charging, routing and potentially also conversion and transcoding mechanisms.
3. **Multilateral Service Hub**—As 2. but in multilateral mode allowing one commercial agreement with Hub to open tens or hundreds of partners.

Figure 10.12 illustrates high-level architecture of GRX/IPX where multiple IPXPs (carriers) are connected via peering point(s) to create the whole "GRX/IPX cloud", which is then used to connect various operators.

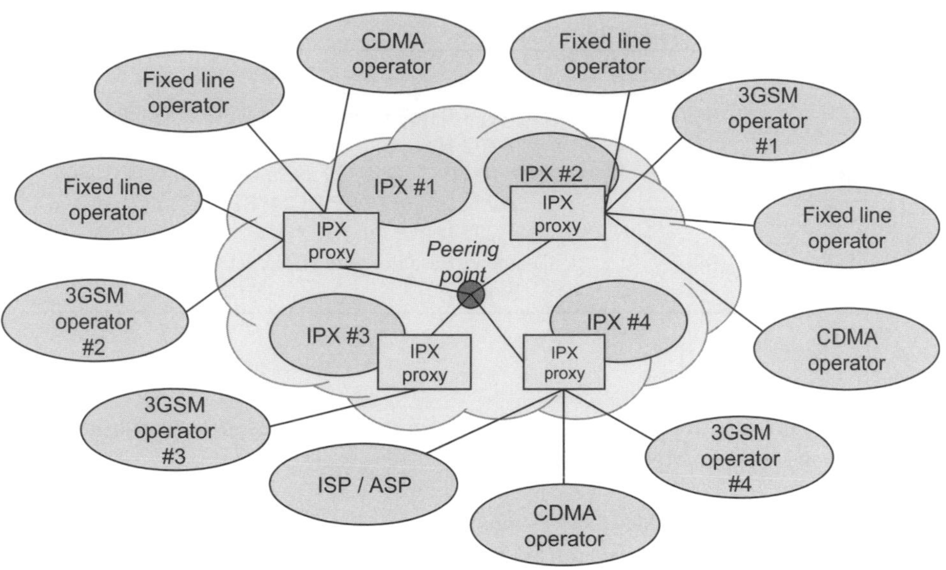

Figure 10.12 The IPX architecture. The internetwork connection can be either direct or it can be formed via HUB/proxy.

It should be noted that IPX is equally applicable to roaming and interconnection purposes—that is, the LTE/SAE roaming towards roaming partners can be handled with the same network infrastructure as via use of the SIP-I [11] or IMS interconnection [12] with interconnection partners. In addition to the international connections, it is also possible to use GRX/IPX for national interconnectivity. As a practical example, the MMS interconnection [13] can be handled by using GRX between the operators, and other services such as SIP/SIP-I based voice interconnection can potentially be deployed between the mobile operators.

Instead of GRX or IPX, it would be possible to use any other IP-based network. One of the most concrete examples of an IP-based network is the Internet, although it does not fulfill all the requirements placed upon a typical interoperator network. For example, demands for secure connections, guaranteed delivery, and end-to-end QoS support are not easy to meet using the Internet. This is the most fundamental reason why no operator has put the Internet forward as a basis for an inter-PLMN network used in LTE/SAE roaming to date.

IPX will handle all the interfaces in LTE/SAE roaming—that is, user media, SIP signaling—and other signaling such as Diameter used between the PCRF nodes via S9 interface will be routed over the IPX network. The end-user traffic of any service, including LTE/SAE roaming, needs to be put inside a tunnel when traversing the IPX network. This eases the routing—for example by masking any private IPv4 addresses used by UE, and it also enhances the general security level of IPX by ensuring that the IPX network nodes are not directly visible for the end users. As the IPX network itself is considered natively secured, the full encryption of traffic offered by IPSec as such is not needed (as it would be in the case of the Internet) so, for example, the more simple and lightweight GRE (Generic Routing Encapsulation) can be used as the mechanism for tunneling the end-user traffic in IPX.

In CS voice roaming, it is currently possible to use Roaming Hubs, which allow the HPLMN to create commercial connections with multiple VPLMNs without a need to go through onerous bilateral agreements and connection setups, which speeds up the introduction of roaming connections. It is very likely that similar mechanisms will also be available in LTE/SAE roaming, handling IP-based GTP and Diameter traffic by some version of the Multilateral Roaming Hub offered by the IPX carrier. At the time of writing the details of LTE/SAE Roaming Hub are not known, however it is clear that there is strong business demand to use the multilateral architecture as the primary model of commercial LTE/SAE roaming.

IPX needs to support potentially huge bandwidth due to customers using LTE/SAE, including interconnection. The need for bandwidth is, naturally, very much dependent on the number of customers but also on other aspects such as what kind of devices they are using and what services are being used. Devices have an impact because, at least in the current 3G market, it is clear that PC users (either via embedded 3G laptops or more commonly via 3G data dongles) are using vastly more bandwidth than customers with a mobile device. This is also likely in the roaming scenario, although the impact of additional roaming charging can obviously have a huge impact on the amount of data used by the typical customer. The type of service has an impact because, as described in Section 10.4.4 "Home Routing," in the case of Home Routing everything is always routed back home via IPX, potentially consuming a lot of bandwidth—for example, in cases of customers watching YouTube videos or downloading files when roaming—unlike the Local Breakout model where potentially only the signaling traffic would end up in the HPLMN and media traffic is routed towards the destination by the VPLMN. The bandwidth offered for the local loop—that is, the IP connection used for all the traffic between the operator and IPX Provider—can be anything from less than 1 mbps to 10 gbps or even more depending on the IPX Provider.

IPX is a managed private IP backbone controlled by the commercial Service Level Agreements (SLAs) that define the level of service such as throughput, jitter, availability, and Mean Time Between Failures offered by the IPX Provider to the operator [14]. This means that deviations from the level of service formally defined in the SLA are subject to possible penalties. Reference [13] defines a set of QoS parameters—for example the availability of IPX to an operator that is connected via a single connection is 99.7%. Upgrading this connection to a dual connection increases the availability to 99.9%. The average monthly packet loss ratio of IPX when using the traffic class AF1 is defined as less than 0.1%. Another example of QoS values defined in [13] is the round-trip time—for example, the delay value for conversational/streaming traffic class between Northern Europe and Southern Europe over the IPX network is listed at 75 ms.

Various Value Added Services such as Border Gateway management, multilateral service connectivity, and application protocol conversion/transcoding, can be offered by the IPX Provider as an additional part of their package. Individual operators have to consider whether these really add value or whether it would be better to keep, for example, the management of Border Gateway elements in-house rather than outsource it to the IPX Provider.

10.4.4 Home Routing

Home Routing is a model where both control plane and user plane are always carried from the VPLMN back to the HPLMN—that is, the VPLMN simply acts as a bit pipe for the HPLMN. All the services come from the HPLMN. This is the model that current commercial 2G/3G PS roaming uses. It means that the traffic is routed from the SGSN (located in the VPLMN) to GGSN (located in the HPLMN) inside the GTP tunnel over the GRX network.

In LTE/SAE roaming, the same model can also be used by accessing SGW and MME in the VPLMN while the PGW is located in the HPLMN. This means that the traffic always ends up

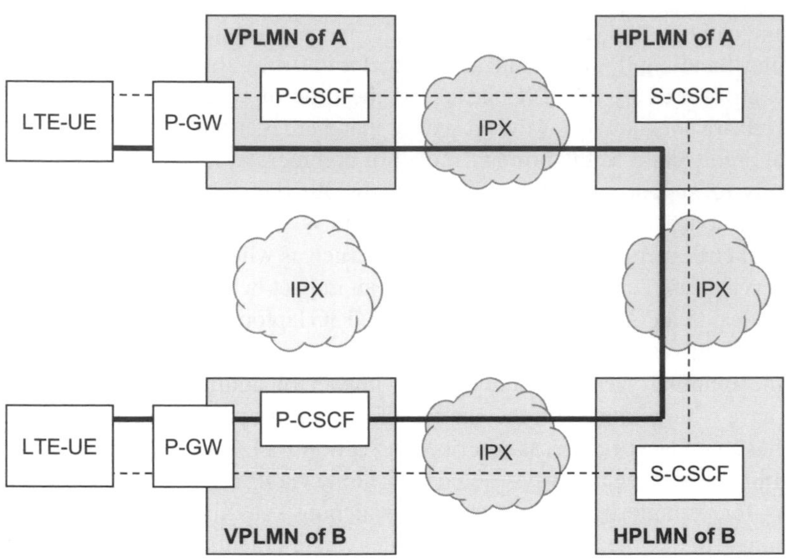

Figure 10.13 Home routing model.

first in the HPLMN regardless the location of the actual recipient which could be roaming in another VPLMN as illustrated in Figure 10.13.

10.4.5 Local Breakout

Local Breakout (LBO) is an architectural model where the control plane gets handled in the same way as in Home Routing, that is, it is always routed back to HPLMN. The user plane, however, breaks out in VPLMN. This allows, for example, the use of various services hosted by VPLMN. Note that, depending on the traffic type, the control plane might or might not exist. This is further detailed in the chapter concerning Service Related Aspects.

In practice, the main advantage of the Local Breakout model is the possibility for routing the user plane in a better fashion. One key service benefiting from the optimal routing of the user plane is VoLTE, because a voice service producing the minimum possible delay forms a very important part of the service. It is obvious that in the case of delay-critical applications, there is a practical need for Local Breakout—for example, if the end-user is roaming in distant location, trying to reach a local number or another roaming user via VoLTE. If Home Routing is used, the voice traffic is always routed from the VPLMN to the HPLMN and back to the VPLMN if a local number is being contacted. Based on common figures listed in IR.34 [1] the typical Round Trip Time, for example, between Northern Europe and East Asia over a GRX/IPX IP connection is 420 ms. Latency figures for voice exceeding 400 ms are normally considered as unacceptable, as ITU-T specification G.114 states. In practice, therefore, adequate optimization efforts are needed in this type of roaming scenario to avoid unhappy customers having suboptimal voice experience when using their fancy LTE/SAE devices abroad.

Local Breakout has been defined in 3GPP already for GPRS roaming as "Visited GGSN roaming". However, it has not been implemented in practice despite its benefits for technical optimization. The most important reason for the fact that commercial 2G/3G PS roaming is implemented currently by Home Routing (i.e., "Home GGSN roaming") is due to the

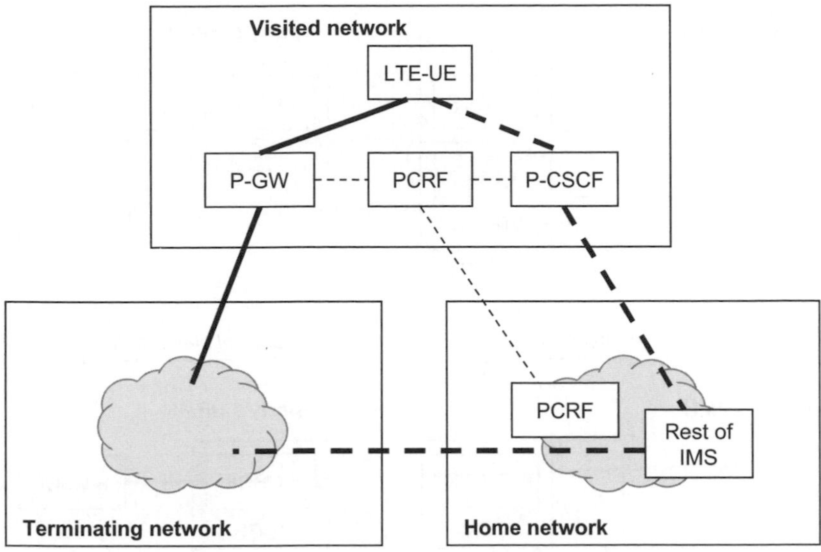

Figure 10.14 Local breakout model.

commercial issues related to the possible misuse and fraud arising from the weaknesses of the model. Another practical reason for this is that the implementation of Home Routing is relatively simple. In cases when no control plane is used (e.g., typical web browsing), the HPLMN is not aware of activities in the VPLMN, apart from those that can be interpreted from the eventual roaming bill.

The lack of visibility by the HPLMN has an effect in LTE/SAE roaming, as shown in Figure 10.14. One possible argument against this statement might be that the IMS services could be run with 3G roaming allowing the operator to have the control plane always available for the HPLMN and still being able to use Home Routing. Unfortunately the amount of IMS traffic in the 3G roaming today is close to zero. However, LTE works in an environment that requires completely new mechanisms to be deployed for the most important service – voice—which means that there was room for new models after the year 2000 when GPRS roaming was introduced. When 3G PS roaming was deployed, GPRS roaming principles did not actually change as the previous principles were still applicable. Now, existing CS voice roaming is not fully applicable in LTE any more, apart from the intermediate CSFB solution. Thus, the introduction of VoLTE drives forward the need for Local Breakout in practice.

It should be noted that the devil is in the detail here: how is it possible to route the traffic from the Originating VPLMN to the Terminating PLMN? During the GSMA work many suggestions were put forward to build on the starting point of the LBO model used in LTE/SAE roaming—that is, P-CSCF, PGW and PCRF nodes all located in VPLMN. At the time of writing [11] presents the "target optimal routing solution" using the best possible route—that is, media flowing directly between Originating VPLMN and Terminating VPLMN using a complete separation of signaling and media as illustrated in Figure 10.15.

This model is technically the most optimized architecture, ensuring the best possible media routing (the shortest path) in all possible roaming scenarios. However, it is not aligned with the existing commercial model used for CS voice roaming. The separation of the control plane and the user plane causes problems for the charging machinery, which today is accustomed to

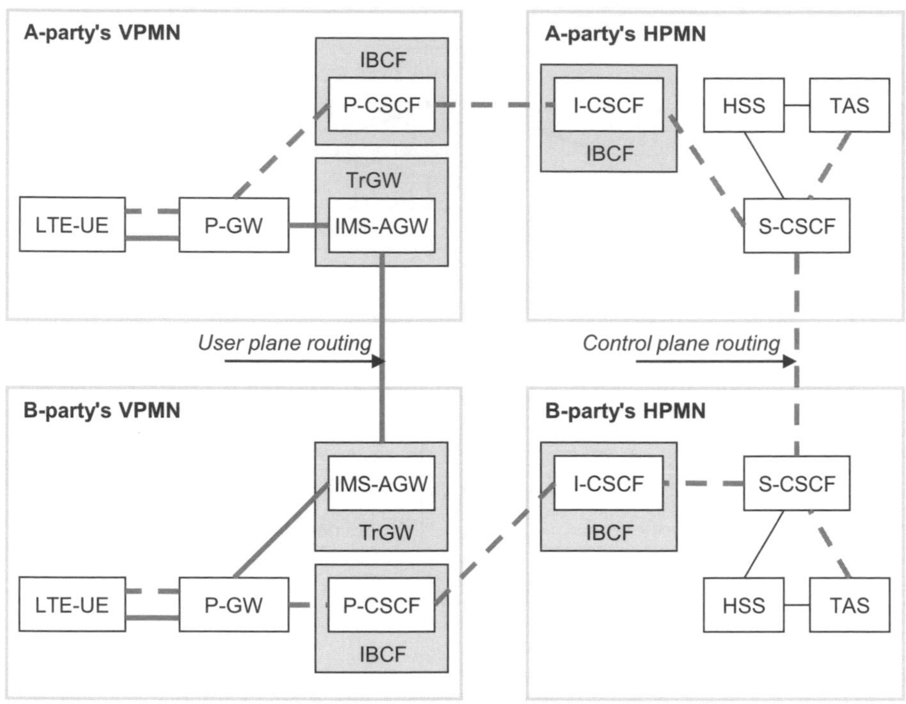

Figure 10.15 Shortest Path Architecture.

signaling and media being coupled. Logically, the next question is whether the current charging model should be evolved and enhanced to accommodate this technically optimized architecture or whether we should select a technically inferior model due to demands of the CS domain-charging model being forced into the IP domain? At the time of writing this billion-dollar question is unanswered.

It is worth noting that current commercial CS roaming uses the so-called partly optimized model, where traffic bypasses the HPLMN of originator but not the HPLMN of receiver if two roamers are calling each other. CS roaming is thus more optimized than the Home Routing but the level of optimization is less than in the shortest path architecture where traffic also bypasses the HPLMN of the receiver.

One possible issue is the fact that when optimizing the routing at the SIP level the routing at the IP level still needs to be taken care of. This means that if the IP level does not work, the SIP level would not work either. IP routing can already offer the shortest possible path between the VPLMN and the terminating operator as a normal inbuilt feature of how the routing works in any IP network in addition to the common public IP address scheme and routing rules deployed across the operators and carriers involved in IPX as documented in IR.34 [1] and IR.77 [15].

10.4.6 Home Routing versus Local Breakout

Tables 10.2 and 10.3 show the expected benefits and drawbacks of Home Routing and Local Breakout.

Table 10.2 Home routing

Advantages	Drawbacks
• Reuses the existing PS roaming model, small impact to the current routing arrangements, agreements, charging models and so on	• Not an optimal model to route IP streams of potentially tens of megabits per LTE user
• Allows handover between 2G/3G and LTE to function without changing the way 2G/3G PS roaming currently works	• In the worst case scenario a service running on top of LTE simply is not usable (too long delay for a service such as voice)
• No need for IMS and PCC architecture to be deployed by VPLMN	• More traffic for the international IP carriers to carry around (=increased costs) • Lawful Intercept more difficult/impossible (depending on IPSec tunneling options used) • Emergency Call more difficult

Local Breakout using optimal routing at IP level does add quite a lot of value allowing, for example, the best possible end-user experience for VoLTE/SAE roaming. Therefore it could be seen as the preferred target solution for LTE/SAE roaming when using services such as VoLTE or RCS.

There is still room for the Home Routing concept in the case of LTE/SAE roaming, though, in order to support those services that require operator-specific arrangements/extensions (such as VPN APN for corporate customers) or that do not offer the necessary visibility for HPLMN due to the lack of a control plane (such as accessing Internet services). Home Routing could also be used by the more advanced/delay critical services where roaming takes place close by—between, for example, neighboring countries. We also have to take into account that, at the moment, many operators are using Home Routing in TDM for some CS roaming scenarios, for

Table 10.3 Local breakout

Advantages	Drawbacks
• Routes media in the most efficient way (potentially even more optimal than in the CS domain today)	• Requires a major change compared to the existing commercial roaming model deployed worldwide
• Allows services such as VoLTE to be usable also when roaming in the worst case scenario	• Requires VPLMN to deploy the full IMS and PCC architecture to support inbound roamers
• Less traffic for the international IP carriers to carry around (=lower costs)	• Requires increased level of trust on behalf of HPLMN to VPLMN on inter-operator charging
• Service based revenue sharing between home and visited operator	• HPLMN might not have visibility on whether the roaming user has actually successfully received the service he/she has paid for (problem for customer care)
• Service based Lawful Intercept in the visited network	
• Better support of Emergency Calls in visited networks (not straightforward when OMR is utilized due to media being forced to "optimal route")	

example supporting prepaid customers. It is expected that in VoLTE/SAE roaming some operators will also be asking for Home Routing to be used in specific call cases to support the same functionality that exists in the CS domain.

The selection of the LTE/SAE roaming model (Home Routing or something else) is performed by the HPLMN, just as in CS voice roaming the HPLMN selects the model using Camel.

One practical aspect of this issue is that, even though a certain operator chooses to use solely the Home Routing in LTE/SAE roaming, it will not be enough if the other operators select the Local Breakout model. In the multi-operator environment, the influence of the large operators has quite a lot of weight in the negotiations about the commercial LTE/SAE roaming cases to be implemented.

Finally it is worth noting that once a model is introduced as a common architecture for LTE/SAE roaming, it is rather difficult to change it afterwards. So trying to come up with a logical way where a simple model is deployed as the first step forward and then, in the second phase, a more advanced architecture takes place, looks good on paper but is quite hard to implement in the real world consisting of tens or even hundreds of individual operators having their own individual product decisions, roadmaps, and cost/benefit analyses.

10.4.7 Other Features

In addition to the well established solution of the GPRS Tunneling Protocol (GTP), 3GPP has defined Proxy Mobile IP (PMIP) as an alternative protocol for LTE. The drawback of the latter is that the established operator that has used GTP would need to deploy GTP → PMIP Inter-Working Functionality and possibly other converters widely in order to support connectivity between operators using GTP and operators using PMIP. In any case, it seems that there is no commercial demand for PMIP-based deployments, so this might not be an issue in practice.

IPv6 is being widely suggested as a recommendable, or even mandated base for the LTE UEs. A number of operators have stated that the launch of LTE/SAE also provides a good opportunity to deploy IPv6 at the same time because operators need to implement a completely new core network anyway. Thus, it is likely that, due to the introduction of LTE, there is a need to take into account the deployment of IPv6 in order to provide correct functionality for the end-user equipment.

Steering of Roaming (SoR) is currently quite widely used in 2G/3G roaming to ensure that outbound roamers register on the preferred visited network when going abroad. This is handled, for example, by that HPLMN updating a list of preference based on commercial or technical grounds such as more advantageous roaming agreement for the HPLMN or a better quality VPLMN network in the SIM card that the device uses to select the VPLMN. The customer is still able to override this manually but a normal customer might not know or care about this.

SoR has caused some dubious or even fraudulent behavior due to some VPLMNs being concerned that they might lose inbound roamers to other operators and thus trying to counter this via different anti-SoR mechanisms, effectively interfering with the steering process by overriding commands from the HPLMN. This kind of activity can also take place in LTE/SAE roaming. As an example, when using Local Breakout, the VPLMN could possibly override the policies sent by hPCRF if this has some benefit for VPLMN. This is obviously activity that is against specifications and the operator trust models but at least in theory could work so perhaps something to keep in mind.

10.4.8 APN Usage

The reference IR.92 [16] states the following: "The IMS application uses an IMS specific APN as defined in PRD IR.88; any other application must not use this APN."

Furthermore, reference IR.88 [2] states that "For Voice over LTE/SAE roaming to work, a well-known Access Point Name (APN) used for IMS services has been defined." According to this reference, the APN name must be "IMS."

Thus the approach of "Single APN" used today by some operators–that is, one common APN for all the services–will not work for VoLTE or any other IMS-based service such as RCS. This decision was made in GSMA after long discussions around the area of APN usage for LTE and VoLTE in order to reduce the number of alternatives to be supported by the operators (for their own customers as well as for inbound roamers potentially using different APN alternatives) and device vendors. Input from interoperator charging, experts also clearly support a dedicated APN for VoLTE in order to allow an easy way to ensure that VoLTE traffic is charged in a proper way, so the decision seems to be feasible.

A dedicated IMS APN provides an easy way to carry out configuration for operators and ensures consistency across networks, helping, for example, charging identification. This means that it:

- Enables a separate billing model from that used for the Internet (it could be configured to be exactly the same, if so wished);
- Enables LBO and thus optimal routing of media, while retaining the option to home route other traffic like the Internet;
- Enables different routing/transit network to be used for the IMS roaming (if Internet APN would be used then inherently the Internet would be used as the interoperator network).

It is possible to utilize other APNs in LTE/SAE roaming for the other services. In practice, it is very likely that any LTE UE will need the capability of supporting simultaneously, for example, "IMS APN" for VoLTE or RCS and "Internet APN" for normal web browsing. "VPN APN" allocated for accessing a corporate intranet will require another APN to be supported. So LTE UEs must have slightly enhanced APN logic, such as an application-specific APN.

A Corporate APN is suitable for its current purpose–that is, accessing pure data services such as e-mail via Home Routing. Using other, nonpure data services such as corporate-hosted IMS (e.g., companies offering their own VoLTE service for their employees including LBO solutions) is highly unclear at the current stage, and it depends, for example, on the UE's capabilities.

"Corporate VPN APN GW" could be set up in each country making it possible for sufficiently large companies to use their own VPN APNs in the Local Breakout model. If this is not feasible, it seems that the Home Routing is the only solution supporting the existing corporate VPN APN solutions.

Web browsing (and any other traffic that lacks a specific control plane) does not provide the HPLMN with any signaling, unlike, for example, VoLTE where the S-CSCF back home is always aware of the situation by checking the SIP/SDP information. From the charging and fraud point of views the Local Breakout poses possible risks because the VPLMN can deliver an incorrect roaming bill to the HPLMN, which does not necessarily have complete information about the real communication activity in the VPLMN.

As a rule of thumb, the IMS APN should never be provided by a UE as the default APN, but instead the default APN in the MME (as downloaded from the HSS) should be used.

The practical reason is that the default APN provided by the network is likely to be more up-to-date regarding the provisioning status of IMS-based services for the subscriber. The UE can, however, provide other APNs such as the Internet. As discussed above, it is possible for the UE to support multiple simultaneous APNs. This means that, for web browsing, the UE could attach to HPLMN PGW via "Internet APN" but for VoLTE it could use VPLMN PGW instead via "IMS APN."

IMS APN is only allowed to be used for IMS-based traffic, whereas non-IMS-based traffic is rejected. One concrete example of this is an LTE customer roaming in Japan. By using the VoLTE service the customer can access the IMS APN offering LBO functionality whereas when using an Internet-based service it needs to access the normal Internet APN, thus forcing Home Routing to take place in a typical scenario.

10.4.9 Service-Specific Aspects

10.4.9.1 General

This section illustrates different service-specific features that have an effect on LTE/SAE roaming. It also analyzes the effects of the solution depending on whether the control plane exists or not.

10.4.9.2 Web Browsing

Depending on the service used, the control plane might or might not exist. This has an important effect on the routing requirements of that service. The main reason for this effect is the need for HPLMN to monitor the service, which the roaming end-user is using in VPLMN.

As Internet services are typically located physically in a particular service provider, and there is thus no interaction with operators via specific control-plane signaling, the use of typical web browsing via the Local Breakout means that the HPLMN never has information about the call activity that is actually happening in the VPLMN. All the traffic flows directly between VPLMN and the Internet. This means that the following, rather important items are not available to HPLMN:

- which service is being used;
- which QoS class is being used;
- how much bandwidth is used;
- how much data is transferred;
- whether the service the user requested is actually being delivered successfully.

For example, if the user later complains to the HPLMN that there are expenses for the use of Service X while the customer was roaming, even though the service never worked, there is basically no way for the HPLMN to prove whether the charging is justified or not. The HPLMN simply has to trust the charging information of the VPLMN. In the real world, this solution has not been proven successful in 2G/3G PS roaming, so its possible problems will apply in LTE/SAE roaming, although the roaming agreement is based on the normal trust relationship of two parties signing an agreement, which is typically augmented by the possibility of a penalty in the case of agreement misuse. In practice this indicates the use of Home Routing for a service lacking the control plane.

Purely technically speaking, it would make sense for web browsing, which could easily mean huge amounts of data being consumed by, for example, watching HD videos, to breakout to the Internet as soon as possible—that is, to use Local Breakout instead of the tunneling of user traffic back to the HPLMN over IPX.

10.4.9.3 Voice

For LTE voice service, it is possible to use CSFB, VoLTE or OTT solutions such as Google Voice. As discussed in Chapter 9 of this book, VoLTE has been agreed as the common long-term target solution of the operators in NGMN and GSMA, whilst CSFB is seen as the intermediate model.

The benefits of using VoLTE instead of CSFB include better quality—for example, related to the transport of Caller ID, which has been identified as a clear problem in current CS roaming from the end-user perspective. Nowadays Caller ID is ripped off completely or some numbers are dropped (or even added) randomly in many cases due to peculiarities of CS roaming related to functions such as least-cost routing over the international TDM lines involving numerous hops. When using end-to-end IP for the voice service, Caller ID should a have better chance to avoid being dropped because it is carried within the SIP message. Using an enhanced "HD Voice" codec (AMR-WB instead of AMR-NB) should also be easier in VoLTE compared to CSFB because VoLTE needs a whole new IP-based infrastructure to be deployed, physically or logically. The main disadvantage of CSFB—that is, additional delay in the call setup time—will also apply to the roaming scenario, so setting up the call should be shorter in "native" VoLTE compared to CSFB.

The main advantage of CSFB is that it keeps the existing commercial CS voice roaming architecture—that is, roaming agreements, charging models and technical arrangements would more-or-less stay intact.

Another way of reusing the existing CS voice roaming architecture for LTE/SAE would be to perform PS/CS conversion in the Originating VPLMN for VoLTE voice, transfer the media over the TDM interoperator network and then in the Terminating PLMN perform CS/PS conversion for the VoLTE recipient. In practice, this highly nonoptimized model would consist of three different interfaces (PS UNI-CS NNI-PS UNI) giving, in theory, the possibility of using VoLTE devices within the operator domain while maintaining the current TDM NNI. However, this is not really considered to be a feasible model due to number of conversions required, and it is not capable of supporting any other services such as RCS. If it is required for whatever reason to stick with the TDM NNI then a better alternative would be to use the standard CFSB solution.

Reuse of the CS roaming model into the VoLTE/SAE roaming model (i.e., routing of media from VPLMN(a) to HPLMN(b)) would be advantageous from the commercial point of view because that would minimize the impact on the existing CS roaming agreements and charging environment. However, it can be questioned whether it makes sense to design a nonoptimal technical architecture for a completely new technology such as LTE/SAE just for the sake of not wishing to touch existing commercial models. This is especially true because the environment that operators have become accustomed to in recent decades is facing a major change due to, for example, increased pressure from various Internet players and regulatory changes. This is a major ongoing debate that will likely continue for quite a while.

One aspect related to the relationship of LTE/SAE roaming with CS roaming is that 3GPP has started a new specific work item called RAVEL, which is studying various issues around copying the CS roaming model into the IMS roaming environment. This work is expected to be finalized in the Release 11 timeframe which means late 2012.

The OTT solutions, such as Google Voice, as an LTE voice service of the operator, requires normal LTE data roaming with a handover to 2G/3G by reusing normal PS roaming. As it is unlikely that HPLMN would have access to any specific control plane in the OTT solution, it should probably be handled in the same way as normal web browsing—that is, by applying the Home Routing model.

If VoLTE is used, it is worth noting that there are some general high-level commercial requirements documented in [11] for VoLTE/SAE roaming:

- Voice call routing for Voice over LTE when the call originator is Roaming should be at least as optimal as that of the current CS Domain. This means that the bearer path for a VoLTE call shall be routed from the Visited network of a Roaming call originator to the terminating network.
- The charging model for the roaming that is used in CS Domain should also be maintained in VoLTE.

In essence, the first requirement above means that Home Routing is not a feasible solution for Voice over LTE because the use of that model means that routing cannot be "at least as optimal as that of the current CS Domain" due to CS voice roaming normally using a model where media is routed from the Originating VPLMN to Terminating HPLMN so it is optimized compared to Home Routing.

The control plane in VoLTE is routed via a dedicated AS (Application Server) in HPLMN, typically called TAS (Telephony Application Server), which is connected to S-CSCF. The TAS takes care of functions such as the support of the necessary Supplementary Services like Call Forward. If VoLTE is used then, in order to enable the transport of XCAP, the PCRF must provide a PCC rule identifying the TAS within the home network. This can be done from the home network over the S9 interface or through a local configuration at the local PCRF.

From the user plane perspective, VoLTE is essentially an IMS-controlled P2P service, meaning that the actual RTP voice media flows directly between the UEs involved without any servers required, in the normal case of two users talking to each other using VoLTE in both ends. In order to support functions such as conferencing, transcoding between different codecs, or PS/CS voice interworking with PSTN, there is a need for network nodes such as MRF, MGW and BGCF to be involved. In essence this means that the support of VoLTE does not necessarily require anything special from the VPLMN side.

In a nutshell, it is tough to find a single solution that would actually fit all the technical and commercial requirements of voice in LTE/SAE roaming. In theory it could be that the mobile industry has to come up with quite advanced/dynamic selection logic in VoLTE/SAE roaming in order to ensure, for example, operator requirements for doing Home Routing in some cases and "Best" Routing for example in a case where VPLMN-A and VPLMN-B are in the same country.

10.4.9.4 RCS

The Rich Communication Suite (RCS) [17] is a combination of various IMS-based services, such as Presence, IM, Video/Image Share and Network Value Added Services (NVAS). Basically the RCS service resembles VoLTE from the LTE/SAE roaming perspective—both use IMS APN and always have the control plane available for HPLMN. It has been specified that when using RCS over LTE access, the voice solution is VoLTE and the APN used is the "IMS APN."

One of the most important difference is that many services that are part of RCS have a specific AS handling. As the control and user plane are located always in HPLMN, Local Breakout applied to these services does not provide much benefit because the user plane has to be routed to HPLMN anyway. For example, Presence/XDM, IM and NVAS require AS in HPLMN, as does Video Share if advanced functions such as one-to-many sharing are used.

It is possible to assign different QoS classes to different services within a single APN by using QCI values, so, for example, the RTP stream related to Video Share session using IMS APN could have a lower priority compared with the RTP stream related to the VoLTE session using the same APN. This would help to ensure that voice always has the top priority.

10.4.9.5 Charging

Interoperator charging experts in the GSMA have noted two key points for VoLTE/SAE roaming:

- The voice roaming charging should be "technology neutral" as far as possible.
- Dedicated APN for VoLTE would allow data bearers supporting voice to be identified and zero rated (subject to billing for calls via other mechanisms).

The first point means that, in principle, VoLTE calls and Short Messages should be included in the existing interoperator charging used in the 2G/3G CS roaming environment—that is, there should not be a need for new dedicated IOTs just for the purpose of using VoLTE-based voice and SMS in the LTE/SAE domain.

The second point means that the easiest (and only proven solution in an interoperator context) would be to use a dedicated APN for VoLTE-related PS traffic—the reason being that technological concepts such as flow-based charging (FBC) are not proven in an interoperator context and will depend on the successful establishment of a number of factors such as local breakout and the S9 interface.

The necessary technical specifications seem to be available for the VPLMN in order to create the Transferred Account Procedure (TAP) file, which is used as a basis for interoperator charging and transfer it to the HPLMN both in the case of Local Breakout and Home Routing used in LTE/SAE roaming. According to the specifications, it is also possible to use PCC architecture for online charging between the VPLMN and the HPLMN for Local Breakout, including the cutoff functionality. Implementing this might require modifications in the operator network.

One item is to ensure the correct combination of charging records from the PS and CS domains, in case of handover to 2G/3G having taken place during a VoLTE session. From the specification point of view, it would seem that the Charging ID can be used to combine records coming from the Visited MSC and PS domain nodes such as SGW, PGW and P-CSCF. In reality this might be somewhat challenging.

10.4.9.6 Control

The reference IR.88 states that "PCC is an integral part of the IMS services in general. If the visited PCRF requires guidance and confirmation from the home network, then Dynamic PCC and the corresponding S9 interface needs to be deployed to exchange policy information between the vPCRF and the hPCRF."

One of the important features of PCC architecture is its ability to support the mandatory roaming cutoff functionality (for example, cutting the traffic if the data roaming bill has exceeded €50). In practice, this could be handled, for example, by PCRF by monitoring the amount of the traffic per user. This is a feature that all the EU-based operators have to support in all PS roaming scenarios, including LTE/SAE roaming, according to European Union regulations, it could support the case for deploying PCRF and indeed the whole PCC architecture.

Moreover, hPCRF can tell vPCRF to follow rules related to QoS classes allocated to different types of traffic, or to block some type of traffic completely.

Whether PCC is actually required or not in LTE/VoLTE/SAE roaming is currently being debated. As an example from the specification field, some of the 3GPP delegates have stated that, when using the normal IMS functionality, it is possible to achieve some level of online charging and gating functionality. There are other opinions that indicate that the delivery of policies from the HPLMN to the VPLMN can be arranged via mechanisms other than hPCRF-to-vPCRF, by using, for example, a fixed policy indicated in the roaming agreement or as a part of the global roaming database IR.21. It should be noted that this topic is still under discussion in the standardization, but the deployment of PCRF nodes by the HPLMN and the VPLMN, including the S9 interface, are seen as mandatory functions for LTE/SAE roaming using LBO by a number of companies to ensure that roaming users would not "overdo their policies." It should be noted that the use of S9 also requires some level of trust from the HPLMN towards the VPLMN because hPCRF obtains information concerning, for example, cutoff from vPCRF.

10.4.9.7 International Forums

As is typical in the field, the requirements, architecture, and the technical details of roaming are handled in 3GPP. Specifications coming out of 3GPP are then reused by other groups to fill in items that are out of the scope of the purely technical standardization work, such as real-world practices, commercial models, common agreements, fraud issues, and so on. In the mobile area these and other items relevant for the commercial deployment of roaming are handled within GSMA.

Work performed in GSMA sometimes also includes profiling of 3GPP specifications—for example when, due to typical political compromise, there are two (or more) options in the specification for handling some item, it is possible for GSMA to have another round of discussions and if possible agree on only one alternative which obviously makes it easier for feasible deployment in the typically rather challenging multioperator environment. That is the reason why it is advisable for operators also to take into account GSMA documentation in the area of LTE/SAE roaming, in addition to the technical specifications originating from 3GPP.

The Next Generation Mobile Networks (NGMN) Alliance has been also involved in discussions related to a common voice solution for LTE. The agreed model is that CSFB will be used as the short-term solution while VoLTE is the long-term solution. As GSMA also endorses this way forward, it can be expected that this is the most probable scenario in the global LTE/SAE roaming environment.

10.5 Charging

Charging procedures are essential in commercial telecommunications networks. The functionality required includes the creation of the charging data based on the principles that the

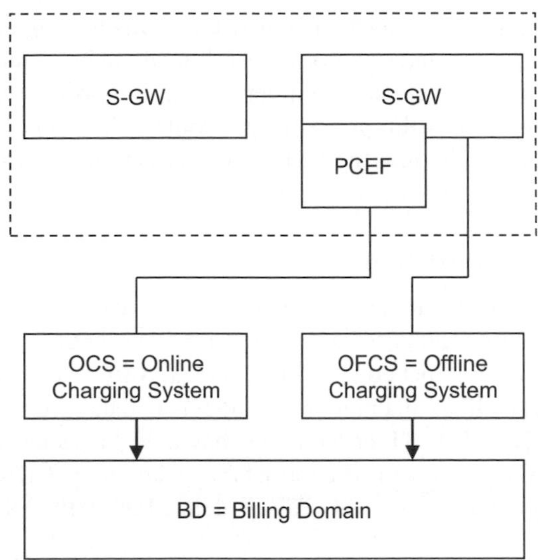

Figure 10.16 The charging of LTE connections can take place in offline and online modes.

operator has adopted. There is baseline legislation for general charging principles in internal network operations as well as for the inter-network connections.

The high-level principle for charging is the collection of the charging data records (CDR) from the connections that the users are creating. In the circuit switched plane, the charging is straightforward as a function of time that the connection is active. In packet switched domain, the charging is typically based on the transferred data.

The network elements are able to collect and store various events that can be charged after the CDR has been transmitted to the typically centralized charging system. The format of the CDR depends on the network.

The charging can be done in real-time (online charging) or later (offline charging). In case of online charging, the Policy and Charging Enforcement Function (PCEF) is used in order to create a real-time interaction with the bearer, session, or service control. Online charging is used for prepaid customers, and it offers the possibility for the real-time event based on session-based charging. In the case of offline charging, the charging information does not affect the service that is charged, and PCEF is thus not needed. Offline charging relates to the post-paid subscriptions, and provides them with event-based or session -based charging in such a way that the cost is charged, for example, monthly. Figure 10.16 clarifies the idea.

10.5.1 Offline Charging

The principle of offline charging is that the CDR arrives to the billing domain after the use of the network resource. The billing domain then post-processes the CDR and creates the charging summary to be utilized in the post-paid customer's billing, for example, on a monthly basis.

The PDN Gateway (P-GW) includes a Charging Trigger Function (CTF) which detects suitable events for charging purposes. The CTF then transforms each of these identified events into separate *charging events*. The charging events are forwarded to the Charging Data Function (CDF), which in turn creates the CDRs with a standard format. The CDFs are then forwarded to the Charging Gateway Function (CGF), which compiles the separate CDR information into the single file and sends it to the billing domain. Figure 10.7 shows the data flow of the procedure with its elements and interfaces. It is worth noting that the implementation of the charging functions of this procedure is flexible in the standardization, and they may thus be integrated, for example, into the S-GW element partially or totally.

10.5.2 Charging Data Record

The Charging Data Record (CDR) is a set of information about the chargeable events. It is formatted in a specific way in order to be understood by the centralized billing domain (BD) where it is sent by the element that collected it. There are various events that can be charged, including bearer utilization like call duration, the amount of data received or sent, and duration of the call setup. The CDRs that LTE generates are based on the packet-switched domain. The CDR is transferred via GTP' protocol. The format for packet-based CDR is standardized in the 3GPP technical standard TS 32.251, and GTP' is defined in 3GPP TS 32.295.

10.5.3 Online Charging

The special characteristics of online charging are that the charging information may affect the service in real time. The online charging information is delivered from the P-GW to the online charging system (OCS), which in turn takes care of the real-time control of the credits of the users.

There is the Charging Trigger Function (CTF) in the P-GW, detecting the events that should be charged—that is, events that include bearer resource usage. The P-GW then converts these events into charging events, which P-GW forwards to the Online Charging Function (OCF).

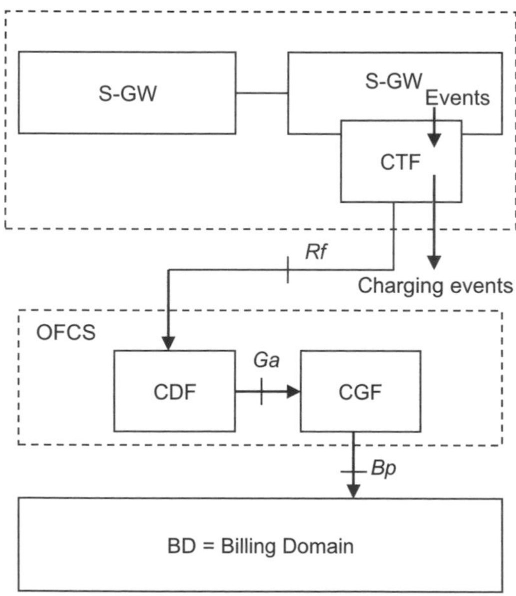

Figure 10.17 Offline charging.

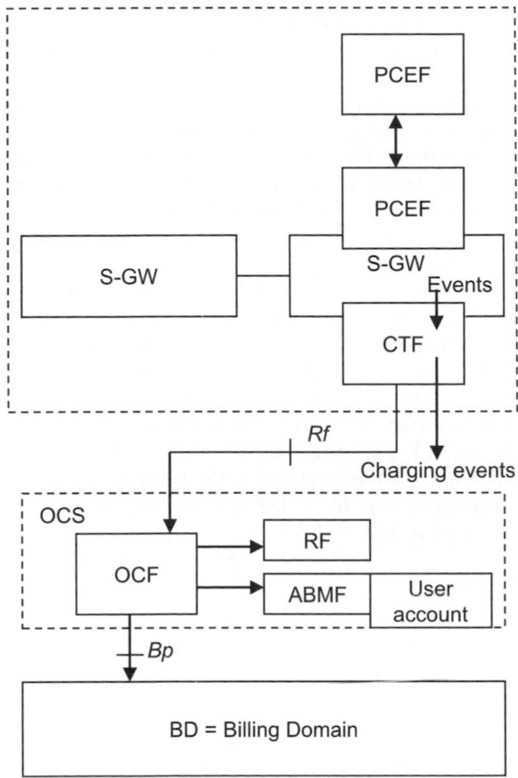

Figure 10.18 Online charging.

This function revises whether the network resource can actually still be utilized by the UE. In order to do this, the OCF interchanges messages with Account Balance Management Function (ABMF). The task of the ABMF is to store the available credits on the user account, and it updates this information along with the utilization of the credits. The OCF also interacts with the Rating Function (RF), which contains the means to decide the cost of the services according to defined tariffs. While there are sufficiently credits, OCS allows the usage of the resources. In contrast, whenever the credits are running out, OSC may interrupt the utilization of the resources, for example, by terminating calls.

The Gy interface between the Policy and Charging Enforcement Function (PCEF) and the online charging system is defined in 3GPP Technical Specification 23.203. The signaling is based on the IETF Diameter Credit Control Application (DCCA) framework as defined in RFC 4006. Figure 10.18 presents the architecture of online charging.

References

[1] 3GPP TS 23.228. (2010) *IP Multimedia Subsystem (IMS); Stage 2*, V. 8.12.0, 3rd Generation Partnership Project, Sophia-Antipolis.

[2] GSMA IR.88. (2010) *LTE/SAE Roaming Guidelines*, V. 3.0.22, GSM, London.

[3] IETF RFC 3588. (2003) *Diameter Base Protocol*, Internet Engineering Task Force, Fremont, CA.

[4] 3GPP TS 29.272. (2011) *Evolved Packet System (EPS); Mobility Management Entity (MME) and Serving GPRS Support Node (SGSN) related interfaces based on Diameter protocol*, V. 8.11.0, 3rd Generation Partnership Project, Sophia-Antipolis.

[5] 3GPP TS 29.274. (2011) *3GPP Evolved Packet System (EPS); Evolved General Packet Radio Service (GPRS) Tunnelling Protocol for Control plane (GTPv2-C); Stage 3*, V. 8.10.0, 3rd Generation Partnership Project, Sophia-Antipolis.
[6] 3GPP TS 29.281. (2010) *General Packet Radio System (GPRS) Tunnelling Protocol User Plane (GTPv1-U)*, V. 8.5.0, 3rd Generation Partnership Project, Sophia-Antipolis.
[7] IETF RFC 5213. (2008) *Proxy Mobile IPv6*, Internet Engineering Task Force, Fremont, CA.
[8] 3GPP TS 29.275. (2010) *Proxy Mobile IPv6 (PMIPv6) based Mobility and Tunnelling protocols; Stage 3*, V. 8.8.0, 3rd Generation Partnership Project, Sophia-Antipolis.
[9] 3GPP TS 29.215. (2011) *Policy and Charging Control (PCC) over S9 reference point*, V. 8.10.0, 3rd Generation Partnership Project, Sophia-Antipolis.
[10] GSMA IR.92. (2010) *IMS Profile for Voice and SMS*, V. 1.0, GSM, London.
[11] GSMA IR.65. (2010) *IMS Roaming and Interworking Guidelines*, V. 5.0, GSM, London.
[12] GSMA IR.83. (2009) *SIP-I Interworking Guidelines*, V. 1.0, GSM, London.
[13] GSMA IR.52. (2003) *MMS Interworking Guidelines*, GSM, London.
[14] GSMA IR.34. (2010) *Inter-Service Provider IP Backbone Guidelines*, V. 4.9, GSM, London.
[15] GSMA IR.77. (2007) *Inter-Operator IP Backbone Security Requirements For Service Providers and Inter-operator IP backbone Providers*, V. 2.0, GSM, London.
[16] GSMA IR.92. (2010) *IMS Profile for Voice and SMS*, GSM, London.
[17] GSMA (n.d.) Rich Communication Suite Technical Realization, http://www.gsmworld.com/our-work/mobile_lifestyle/rcs/index.htm (accessed August 30, 2011).

11

LTE/SAE Security

Jyrki T. J. Penttinen

11.1 Introduction

This chapter presents the security-related functionality of LTE and SAE. It also provides an overview of the aspects that should be taken into account when planning and operating a secure network.

The LTE/SAE network is based on IP, which means that it is vulnerable to the same threats as any other packet network. The main aim of the LTE/SAE operator is to reduce the opportunities for the misuse of the network.

Since the early days of the 3GPP 3G system, security has been identified as an essential part of the whole service. The first Release 99 specifications included 19 new specifications by the SA3 working group, including the main definitions found in TS 33.102 (3G Security—Security Architecture). Ever since, 3GPP has produced advanced specifications for security, taking into account the IP domain, as mobile networks are developing towards IMS and all-IP concepts.

3GPP SA3 has created new specifications for LTE/SAE protection under TS 33.401 (Security Architecture of SAE) [1] and TS 33.402 (Security of SAE with Non-3GPP access) [2]. The LTE system provides confidentiality and integrity protection for signaling between the LTE-UE and the MME. Confidentiality protection refers to the ciphering of the signaling messages. Integrity protection ensures that the signaling message content is not altered during transmission.

All the LTE traffic is secured using the Packet Data Convergence Protocol (PDCP) in the radio interface. In the control plane, the PDCP provides both encryption and integrity protection for the RRC signaling messages that are delivered within the PDCP packet payload. In the user plane, the PDCP performs encryption of the user data without integrity protection. It should be noted that the protection of the internal LTE/SAE interfaces like S1 is left as optional.

The LTE/SAE Deployment Handbook, First Edition. Edited by Jyrki T. J. Penttinen.

11.2 LTE Security Risk Identification

11.2.1 Security Process

The development of security processes has many aspects [3], [4]. The aim of all security measures is to prevent possible attacks by shielding the relevant LTE/SAE interfaces and elements in such a way that outsiders have minimal opportunity to advance fraudulent activities.

The security design of LTE/SAE thus includes features developed according to best knowledge about the current and future methods for attacks together with their technical and business impacts on the network. For instance, security threats like a denial-of-service attack can slow down, or in worst case paralyze a big part of a network and cause limited availability of services, which results in loss of revenue and increases the chances of customer churn. One way to develop up-to-date measures against these security threats is to create a security process.

The first step in security planning is to identify the security threats. Based on the security risk analysis of this phase, the LTE/SAE system is designed and updated accordingly in order to create all possible counter-measures against the identified security risks. This leads to the list of security requirements, and to the specification of the security architecture layout at the system level.

The next step is to take into account the threats at the software level, securing the code as much as possible in the software development processes.

At the end of the security design process, comprehensive security testing is needed by taking into account imaginable attack types in the normal operation of the network, as well as in unstable conditions that are created either intentionally or accidentally.

This example of the security process is logically an iterative activity for the participating parties. This means that, as technologies develop and new methods and ideas for system attacks appear, they should be identified and taken into account at the earliest possible opportunity, to be included in the security planning process so that the network can be updated accordingly to work against the new threat types. As one part of the new security threat identification, a network fraud-monitoring process should be implemented. This provides information about possible new security threats to be taken into account in the security process.

In addition to the security process, it is advisable to execute security audits in the operator networks. This is an important task as there are huge numbers of different combinations of network elements with their different versions and security levels in the end-to-end chain of the mobile networks. Both hardware and software can be audited in cooperation with the network vendor and operator. If vulnerabilities are detected, issues can be corrected by enhancing updated security threat counter-measures.

Figure 11.1 shows high-level items in the LTE/SAE security environment.

11.2.2 Network Attack Types in LTE/SAE

LTE/SAE architecture has special characteristics that should be taken into account in enhanced security planning. LTE/SAE is based on flat architecture, which means that all radio access protocols terminate at eNodeB elements. Furthermore, the IP protocols are also visible in the eNodeB.

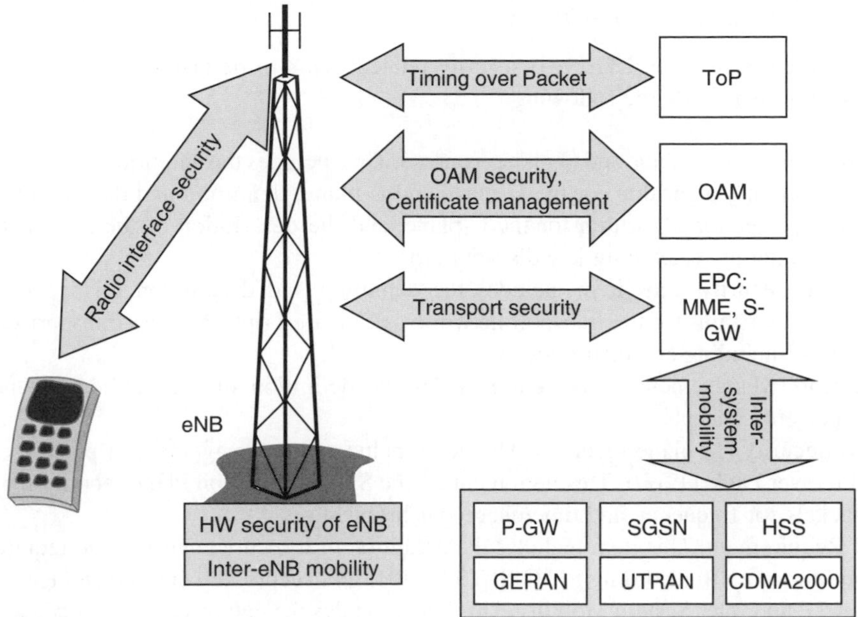

Figure 11.1 The LTE/SAE security chain includes various high-level aspects.

Challenges arise from the architectural realization of LTE/SAE—for example, because it is possible to place eNodeB elements in locations that are more accessible to potential hacker attacks. Furthermore, the LTE/SAE network interworks with legacy and non-3GPP networks that might open unpredictable security holes even if their normal mode of operation would not open these issues. There are also new business environments with networks whose trustfulness is not necessarily known.

Comparing the purely IP-based LTE/SAE architecture with earlier 2G/3G principles, it can be noted that LTE/SAE requires extended authentication and key agreements in order to cope with modern IT attacks. This means that the key hierarchy as well as the interworking security is inevitably more complex than earlier. It also means that the eNodeB element has additional security functionality compared with previous 2G base transceiver stations and 3G node B elements.

The identification of the potential network attack types in the LTE/SAE environment is one of the most essential preventive tasks. As the Home eNB concept basically means that the customer can try to access the hardware and software of the element physically, this has most potential for fraudulent activities [5]. Some possibilities might be the following:

- cloning of the HeNB credentials;
- physical attacks on the HeNB, for example, in the form of tampering;
- configuration attacks on HeNB, for example, fraudulent software updates;
- protocol attacks on HeNB—for example, man-in-the-middle attacks;
- attacks against the core network—for example, denial of service;
- attacks against user data and identity privacy—for example, by eavesdropping;
- attacks against radio resources and management.

11.2.3 Preparation for Attacks

More detailed lists of the LTE/SAE security related items to be taken into account in the security process include the following:

- Air-link security (U-plane and C-plane security). This includes the definition and description of the ciphering algorithm for the U-plane and C-plane, definition and description of the integrity protection algorithm for the C-plane, and the description of the access stratum security signaling (including key distribution).
- Transport security. This item includes the definition and description of ciphering and integrity algorithms for the transport network, and the description of the transport security signaling (including key distribution).
- Certificate Management. This item includes the definition of the public key and key management concepts.
- OAM Security (M-plane security). This item includes the management of plane security.
- Timing over Packet (ToP). This item includes the Synchronization Plane security for IEEE v2 packets for frequency and time/phase synchronization.
- eNB Requirement. This item includes the definition of a secure environment, requirement definition for eNB according to 3GPP TS 33.401, and secure key and file storage.
- Intra LTE and Inter System Mobility. This item includes the definition of security aspects in handover cases (including key distribution).

It should be noted that different planes differentiate the traffic types, and this should be taken into account in security planning. The planes in the LTE/SAE environment are: the U-plane for the delivery of the user data, the C-plane for the delivery of the control data, the M-plane for the delivery of the management data, and the S-plane for the frequency and time/phase synchronization information. Figures 11.2 to 11.5 identify the security-related aspects of these planes.

As will be described further below, IPSec is the 3GPP standardized solution for security on several LTE interfaces. These are the S1-MME and the X2 Control Plane as well as the S1 and X2 User Plane. Security for the Management Plane is not standardized but the use of IPSec or transport security is also suggested. In addition, usage of IPsec in combination with using certificates makes it very difficult for any unauthorized person to gain access to the core network or eavesdrop the traffic between the eNBs and the core network. In this way the integrity and confidentiality of data can be ensured.

11.2.4 Certificates

11.2.4.1 X.509 Certificates and PKI

Digital certificates are used to authenticate communication peers and to encrypt sensitive data. They are essentially for Transport Layer Security and for IPsec support. X.509 certificates contain a public key that is signed by a commonly trusted party. Via this method, the receiving end trusts in the correctness of the public key, as long as a trusted party has confirmed the matching identity by using its digital signature as part of the certificate. The trusting parties and certificates form a trust chain.

The essential problem of trust chains is to get the keys delivered at the stage when the security docs not yet exist due to the absence of keys. The most secure way would be to install the

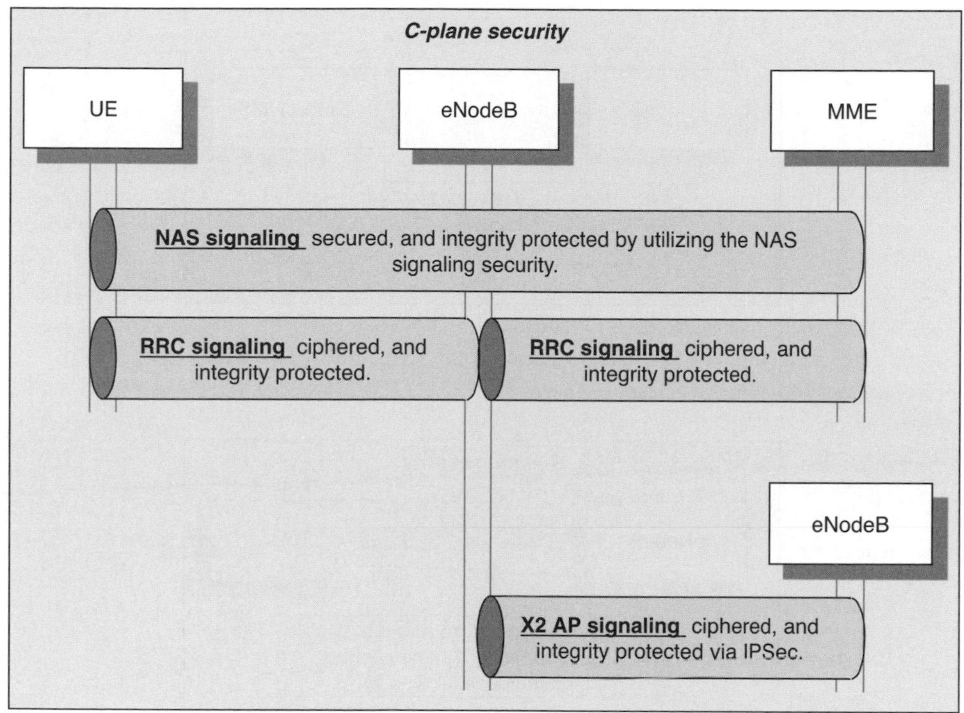

Figure 11.2 The C-plane security principle of LTE/SAE.

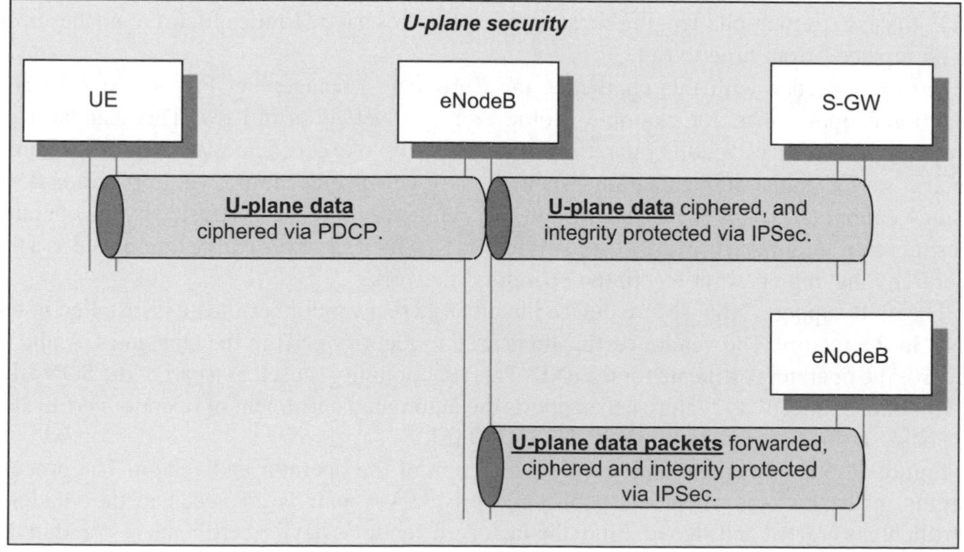

Figure 11.3 The U-plane security principle of LTE/SAE.

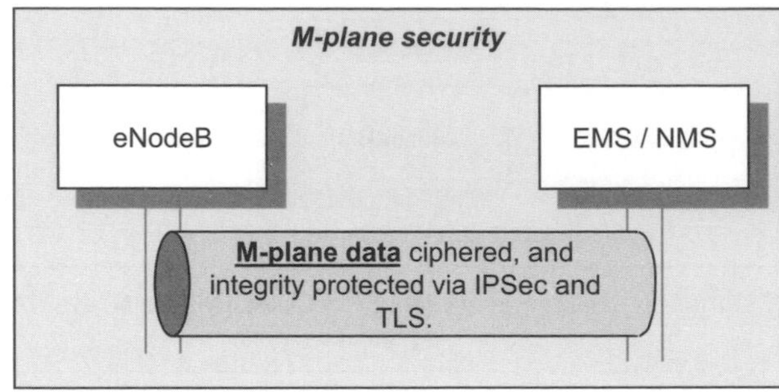

Figure 11.4 The M-plane security principle of LTE/SAE.

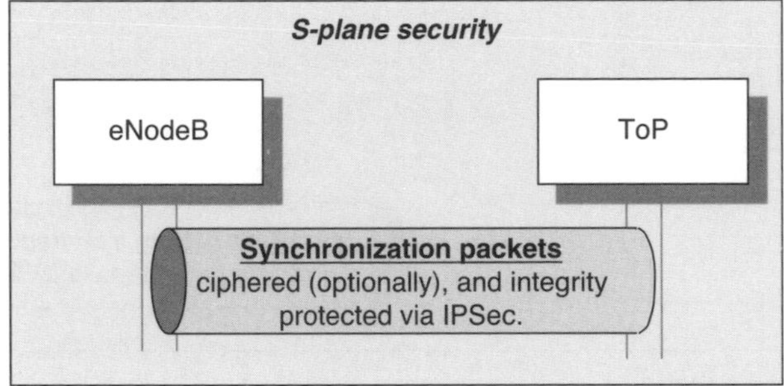

Figure 11.5 The S-plane security principle of LTE/SAE.

required keys locally on site. This solution is feasible for a small number of sites, or for a new network that is going to be commissioned in any case. Nevertheless, for a large network such as LTE, this is extremely challenging because the certificates have a limited lifetime and they need to be replaced from time to time.

In order to cope with this challenge, the Certificate Management Protocol (CMP) is a functional option that, for example, Nokia Siemens Networks utilizes. This standardized protocol provides the capability to retrieve, update and revoke certificates automatically from a central server. Initial authentication (when operator certificates are not yet in place) is done using vendor certificates (which are installed in the factory). These are trusted by the operator Certification authority. As a result, the eNB Public Key Infrastructure can be introduced as a flat hierarchy, having one root CA in the operators' network.

The eNB supports NSN secure device identity where a vendor certificate is installed in the eNB in the factory. The vendor certificate is used to identify eNB in the Operator CA and to receive the operator certificate for the eNB. This functionality is used as a part of the SON LTE BTS Auto Connectivity feature and supports the automated enrolment of operator certificates for base stations according to 3GPP TS 33.310 [6].

Figure 11.6 shows an example of the interaction of the operator and vendor. The process begins in the factory, where the public and private key pair is created, and the vendor's certificate is created and signed. From the factory, a vendor's device certificate is stored in the eNodeB, and it is also delivered to the factory Certification Authority (CA) and factory Registration Authority (RA) as shown in delivery chain (1) of the figure. Next, the product

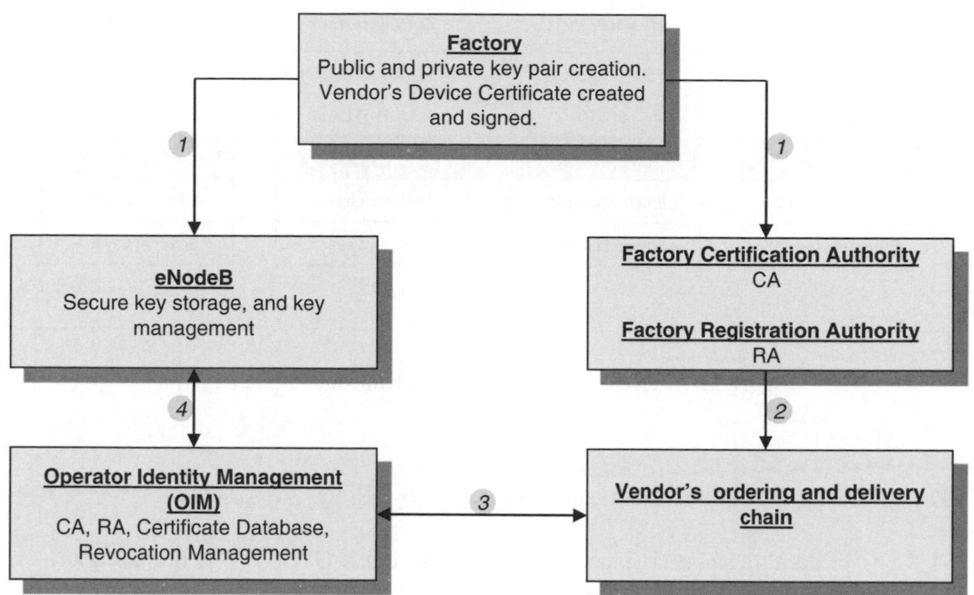

Figure 11.6 The principle of the vendor certificates. This process can be utilized in realistic LTE/SAE network environment.

information consisting of the module serial numbers and vendor's root CA certificate is placed in the vendor's ordering and delivery chain. At this point, the eNodeB is shipped to the operator. Following the process, the vendor's device serial numbers and vendor's root CA certificate is delivered to the operator (3)—that is, to the operator identity management (IDM). Next, the IDM creates an operator node certificate to assure the authenticity of the eNodeB (4). At this point, the operator node certificate substitutes the previous vendor's node certificate. It is now possible to make sure that the equipment is genuine by observing the serial number and vendor's device certification.

11.2.5 LTE Transport Security

11.2.5.1 IPsec and IKE

The eNB follows the rules established by the Network Domain Security/IP Security (NDS/IPsec) architecture defined in 3GPP TS 33.210. The 3GPP TS 33.210 specification introduces Security Gateways (SEG) on the borders of security domains to handle the NDS/IPsec traffic. All NDS/IPsec traffic passes through a SEG before entering or leaving a security domain. The SEGs are responsible for enforcing security policies for the interworking between networks. In this role they also provide the IPsec functions.

It is possible to implement the SEG functionality in dedicated hardware (external SEG) or to integrate it in existing nodes (internal SEG). From the eNB point of view it is not relevant if SEGs are external to the peer-entities or integrated—that is, it is not visible for the eNB. On the eNB side, the IPsec function is integrated into the eNB. The eNB therefore represents a security domain of its own and can act as a SEG according to 3GPP TS 33.210.

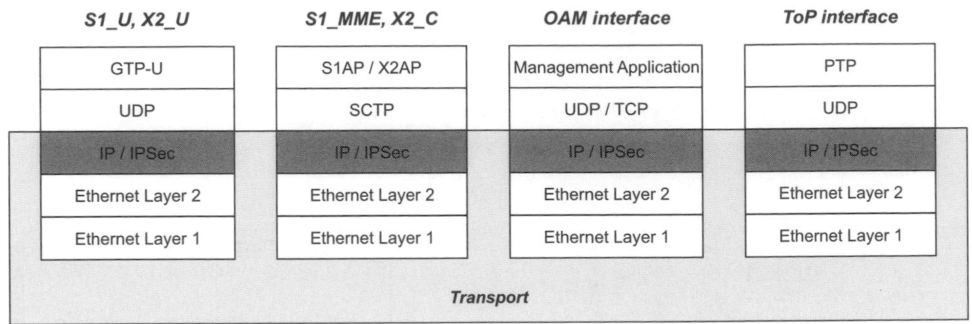

Figure 11.7 eNB Protocol Stacks with embedded IPsec Layer.

The following logical interfaces can be protected by IPsec:

- S1_U: User data transport (U-plane) between eNB and S-GW (GTP-U tunneling);
- S1_MME: Signaling transport (C-plane) between eNB and MME (S1AP protocol);
- X2_U: User data transport (U-plane) between eNB nodes during handover (GTP-U tunneling);
- X2_C: Signaling transport (C-plane) between eNB nodes (X2AP protocol);
- O&M i/f: Transport of O&M data (M-plane) between eNB and O&M System;
- ToP i/f: Transport of ToP synchronization data (S-plane) between eNB and ToP Master.

Figure 11.7 shows the eNB protocol stacks with the embedded IPsec layer.

11.2.6 Traffic Filtering

11.2.6.1 Firewall

It is possible to have the eNB element support a firewall function with the following key capabilities:

- ingress IP packet filtering;
- ingress rate limiting;
- egress rate limiting;
- DoS countermeasures.

11.2.6.2 Filtering for Site Support Equipment

The eNB might provide access to site support equipment (e.g., battery backup units) via additional Ethernet interfaces. Typically, this type of IP-based equipment does not provide its own IP packet filter or firewall. Thus, the site support equipment would be directly accessible if there is no packet filter on the eNB side. Therefore, the eNB provides an IP packet filter service that protects the site support equipment from harmful network traffic, but also protects the network from unintended traffic from this interface.

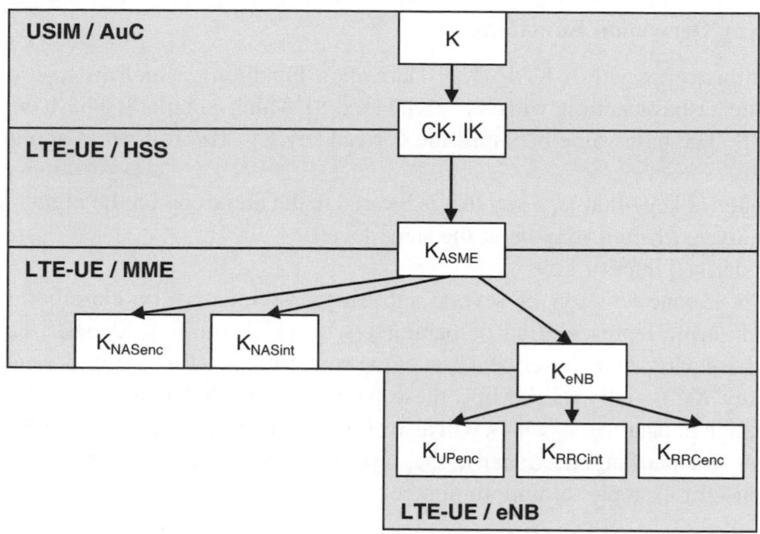

Figure 11.8 LTE Key Hierarchy Concept.

11.2.7 Radio Interface Security

11.2.7.1 Access Stratum Protection (AS Protection)

The following subsection provides information about the key hierarchy and the key derivation function. The key derivation is relevant for normal operation—that is, call establishment—and in case of a handover (i.e., an inter-eNB handover).

11.2.7.2 Key Hierarchy

Figure 11.8 gives an overview about the LTE overall key hierarchy concept. It shows the EPS key hierarchy valid for steady state—that is, if no handover happens. Nodes are represented by frames, and keys are represented by boxes. An arrow represents a key derivation function. If a key will be derived at one node and transmitted to another, then its corresponding box is located on the border of the frames that correspond to the nodes involved.

The part of the key hierarchy that is known to the eNB is called the eNB key hierarchy. It comprises all AS keys—that is, K_{eNB}, K_{UPenc}, K_{RRCint}, and K_{RRCenc}. There is one AS base key K_{eNB} and three AS derived-keys K_{UPenc}, K_{RRCint}, and K_{RRCenc}. The AS derived-keys are used for: UP encryption, RRC integrity protection, and RRC encryption.

The key "K" is the only permanent key. All other keys will be derived on demand using a key derivation function. A key derivation is controlled by a key derivation procedure.

The existence of a key depends on the state in the following way:

- K always exists;
- the NAS keys CK, IK, K_{ASME}, K_{NASenc} and K_{NASint} exists while EMM-REGISTERED;
- the AS keys K_{eNB}, K_{UPenc}, K_{RRCint}, and K_{RRCenc} exists only in RRC-CONNECTED.

11.2.7.3 Key Derivation Functions

Key derivation works with a KDF (Key Derivation Function), which in case of EPS is a cryptographic hashes function, with Ky = KDF(Kx, S), which calculates a hash with a key Kx from string S. The hash value becomes the derived key Ky. The following applies:

- Kx is a superior key—that is, a key that is located in the hierarchy on the higher level (apart from handover, where it may be at the same level).
- Ky is the derived inferior key.
- String S is a concatenation of several substrings, which may be classified as follows: (1) Bound: string representation of parameters to which the key Ky shall be bound to. Usually, these parameters describe a part of the environment—for example, a cell identifier, and the key Ky is only valid while these parameters do not change. (2) Fresh: String representation of parameters which will also ensure different (or "fresh") keys Ky if all other parameters are unchanged. Usually, these parameters are unique for each instance of calculation—for example, a random number.

A cryptographic hashes function provides a result of a fixed size and it is not reversible—that is, it is not feasible (at the current state of the art) to derive unknown parameters if the result and the other parameters are known. In particular, it is not feasible to get Kx, even if Ky and the string S is known.

11.2.7.4 Key Establishment Procedures

There are three basic key-establishment procedures:

- Authentication and Key Agreement (AKA) establishes CK, IK, and KASME on the one hand in USIM and UE, and on the other hand in AuC, HSS, and MME. AKA is a NAS procedure and does not have any prerequisite besides the permanent key K. Please note that MME is the Access Security Management Entity (ASME) for the EPS.
- NAS Security Mode Command (NAS SMC) establishes the NAS keys K_{NASenc} and K_{NASint} which are needed for NAS message encryption and integrity protection. NAS SMC is a NAS procedure and needs a valid KASME as a prerequisite. In addition, the NAS SMC activates the NAS security.
- AS Security Mode Command (AS SMC) establishes the AS keys K_{UPenc}, K_{RRCint}, and K_{RRCenc} which are needed for UP encryption and RRC integrity protection and encryption. AS SMC is an AS procedure and needs a valid K_{eNB} as prerequisite. In addition, the AS SMC activates the AS security.

The establishment procedure for the key K_{eNB} depends on the case:

- At a change to RCC-CONNECTED, K_{eNB} will be derived in MME and transmitted to eNB by the S1AP: INITIAL CONTEXT SETUP REQUEST message.
- On active intra-LTE mobility, K_{eNB} will be derived by a procedure that is shared by the source eNB and the target eNB.

In the case of intra-LTE handover, the key hierarchy differs temporarily, because the K_{eNB} for target eNB will be derived from the K_{eNB} of source eNB.

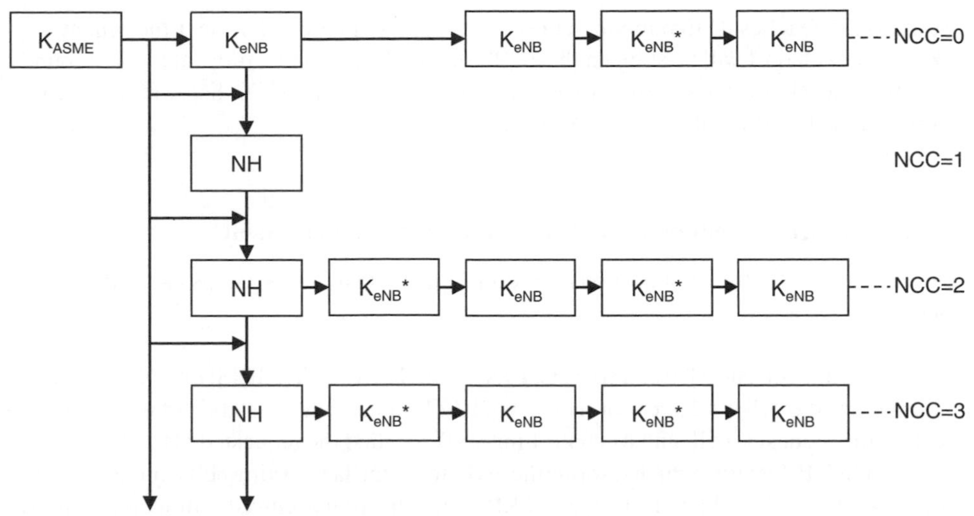

Figure 11.9 Key Handling in Handover.

11.2.7.5 Key Handling in Handover

Figure 11.9 shows the key handling in the handover. The boxes represent keys and the arrows KDF. All the keys of the same row are derived in a single chain of KDFs starting from an initial (yellow colored) K_{eNB} or NH (Next Hop parameter). These chains are called forward chains for short.

At initial context setup, an initial K_{eNB} will be derived from KASME at MME. This starts the first forward chain (NCC = 0). The initial K_{eNB} will be transmitted to the eNB and becomes its K_{eNB} (first blue box on top line).

At handover, a transport key K_{eNB}^* and finally a fresh target K_{eNB} will be derived. Because the derivation uses a cryptographic hashes function, it is not feasible to derive the source K_{eNB} from the fresh target K_{eNB}. Therefore a target eNB cannot expose the security of a source eNB. This is called (perfect) backward security.

However, if derivation happens on one forward chain—that is, if K_{eNB}^* is derived from the source K_{eNB}—then the target eNB keys are no secret for the source eNB, because their derivation is known. This is true in a recursive manner. Therefore, all keys of the same forward chain that are located on the right hand of any K_{eNB} are no secret to this key owner. There is no forward security.

In order to get forward security, the current forward chain has to be terminated and a new one started. This is done by deriving K_{eNB}^* from a parameter NH, which was derived from KASME and not only a source K_{eNB} (a vertical chain of yellow boxes). In an S1 handover, NH is transported by S1AP: HANDOVER REQUEST and applies for this handover. Therefore, this handover attains forward security. This is called forward security after one hop, and it may also be called perfect forward security. In an X2 handover, NH is transported by S1AP: PATH SWITCH ACKNOWLEDGEMENT and cannot apply for this handover (because the new keys are already determined at this point in time) but can at the next one. Therefore, forward security is reached at the next handover. This is called forward security after two hops.

If the NCC reaches its maximum value of 7, it will wrap around at next increment.
Note: forward chain 1 will be skipped (by 3GPP standardization) first time (at the first round of NCC after initial context setup). At each later round (after the NCC has wrapped around) forward chain 1 will usually not be skipped.

11.2.7.6 Security Handling of RRC Connection Re-establishment

If a UE decides to try RRC connection re-establishment, the following steps regarding security happen:

1. The UE transmits, on SRB0, a RRC CONNECTION RE-ESTABLISHMENT REQUEST message to the cell it selected for requesting RRC connection re-establishment. This cell is called the requested cell, and its controlling eNB is called the requested eNB. The message contains a UE-Identity, which informs the eNB about the last serving cell from the UE point of view—the cell in which the trigger for RRC reconnection occurred. This cell is called the serving cell and its controlling eNB is called the serving eNB. Please beware that the serving cell is defined from the UE point of view. Usually the point of view, whether UE or network, does not matter, because the UE and network are synchronized, but a broken RRC connection may cause different assumptions for the UE and the eNB about serving cell. At RRC connection re-establishment, the network will adapt its serving cell assumption to the UE. The UE-Identity is related to the eNB which controls the reported physical (=serving) cell. Please note that this SFS uses the term "requested" for RRC connection re-establishment, and "target" for handover. However, other documents (including some 3GPP specifications) just use "target".
2. The RRC CONNECTION RE-ESTABLISHMENT REQUEST message cannot be authenticated by PDCP MAC-I integrity protection, because it transmits across SRB0. Instead, the requested eNB checks the authentication of the received UE-Identity by comparing a received authentication code, which is contained in the UE-Identity IE and is called shortMAC-I, with an authentication code calculated by the network. Each cell that is enabled to authenticate a RRC re-establishment request applies a dedicated shortMAC-I, because this code is bound to a cell. The following may happen:
   - *If the serving cell is controlled by the same eNB as the requested cell*, the authentication on network side is an internal matter for the requested eNB and may happen on demand.
   - If *the serving cell is controlled by a different eNB from the requested cell*, the authentication on the network side is a matter between two eNBs: A) The calculation of the network side authentication code happens on the serving eNB, because it requires the RRC integrity protection key (KRRCint) of the serving cell. B) The comparison of the authentication codes happens on the requested eNB, because it obtained the shortMAC-I from the UE. The calculation of a set of shortMAC-I and the delivery to another eNB happens in the course of a handover preparation. Background: A RRC connection re-establishment is only possible to an eNB that knows some UE context information—this is a prerequisite motivated by security issues. A requested eNB must therefore either contain the serving cell (see the above bullet) or already be prepared for a handover from the serving cell—that is, contain a handover target cell (this bullet) for the UE. Otherwise, the RRC connection re-establishment request will fail.

The requested cell is controlled by the same eNB as the serving cell, if the UE-Identity is related to the requested eNB—that is, if the physical cell identity reported by the UE-Identity belongs to the requested eNB.

If the RRC connection re-establishment request is accepted, both the UE and the eNB will refresh their AS key hierarchy in the same way as for the handover from the serving cell to the requested cell, but always keeping the security algorithms of the serving cell. The possible cases are the same as for authentication:

- If the serving cell is controlled by the same eNB as the requested cell, the re-establishment procedure is an eNB internal matter on the network side. In particular, key refreshment happens in the same way as for an intra-eNB HO or, if the requested cell is equal to the serving cell, by intracell AS security maintenance.
- If the serving cell is controlled by a different eNB from the requested cell, the re-establishment procedure on the network side is a matter between two eNBs. In particular, the key refresh happens in the same way as for an inter-eNB HO, according to the HO type that prepared the target cell (which exists—see the related description for shortMAC-I). (A) In case of X2 HO, the source eNB needs to calculate a dedicated KeNB* for each cell that will support a re-establishment, and to signal it to target eNB in the course of the handover preparation. This is similar to the shortMAC-I provision described above. (B) In case of S1 HO, the target eNB derives the fresh KeNB from an NH parameter received from MME. Because the NH is independent of the cell, no special measures are needed for re-establishment support. However, the shortMAC-Is that is dedicated to the cell can still be considered as a limiting factor.

In any case, the UE needs to know the NCC parameter for key refreshment. It will be signaled by the RRC CONNECTION RE-ESTABLISHMENT message. This message is unprotected because it transmits across SRB0. Please note: if the X2 interface is not protected by IPsec then the X2AP messages, including the keys, are transferred in plain text. SRB1 (and all later by RRC connection reconfiguration procedure added bearers) will apply the fresh keys immediately.

Please note that in the relationship between the RRC connection re-establishment and handover, from the UE point of view, the RRC connection re-establishment procedure is independent from the handover: The UE will always send its request to a selected cell and, if it is accepted, refreshes its AS keys in exactly the same way according to the NCC parameter received. From the network point of view the RRC connection re-establishment procedure depends on the handover:

- If no handover is prepared, a RRC connection re-establishment can only succeed to the serving (this time from network point of view) eNB, because no other eNB knows the UE context. The UE and network must also share their assumptions about the serving eNB (but the serving cells may differ).
- If a handover is prepared, a RRC connection re-establishment can succeed to both the source and to the target eNB, because they know the UE context.
- The handover source eNB may always expect to be addressed as a serving eNB—there is no difference (apart from the names used) from the "no handover" case above. In contrast, the handover target eNB may either be addressed as the serving eNB or not. If it is addressed as the serving eNB, the UE regards the HO as completed. Otherwise, the UE regards the HO either as uncompleted or does not become aware of it at all, and will report the source cell as the serving cell.

11.3 LTE/SAE Service Security—Case Example

11.3.1 General

LTE/SAE is changing the nature of mobile communications because it is completely based on the IP environment. This also has the potential to increase fraudulent activities. The motivations of such activities might be financial, destructive, or even political. The motivation might simply occur as individuals try to prove their hacking skills.

Modern information technology combined with advanced mobile technologies brings changes and new aspects that might increase vulnerability to intentional attacks by opening new possibilities for fraud.

As an example, base station equipment has traditionally been well protected physically. Radio and transport equipment have been shielded during site construction which is only accessible for authorized personnel. In the future, this type of equipment may be located increasingly in public places and even homes.

On the other hand, the methods for the attacks evolve along advanced tools that are more easily available via Internet distribution. These activities might include increasingly sophisticated attacks performed by IT professionals.

11.3.2 IPSec

For LTE, IPSec with Public Key Infrastructure (PKI) is used as a standardized security solution. PKI is applied to authenticate network elements and authorize network access whereas IPSec provides integrity and confidentiality on the transport route for the control and user plane. The IPSec concept is based on the certificate server, which is the registration authority within the operator's infrastructure. It takes care of the certificates that provide secured IPSec routes between the elements via the migrated SecGW, as shown in Figure 11.10.

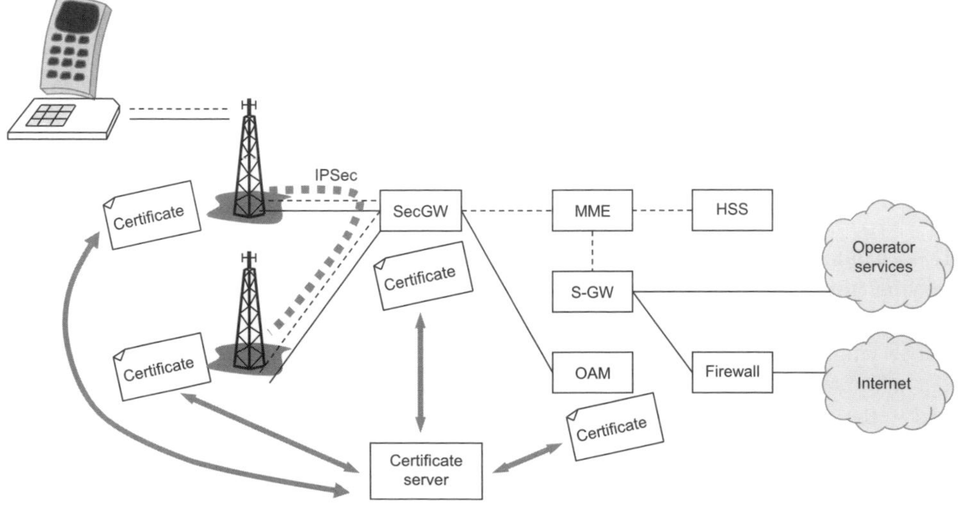

Figure 11.10 The architecture of the combined IPSec and PKI. The light dotted line shows signaling, and solid line shows user plane data flow. The heavier dotted line shows the IPSec tunnel. The communication between SecGW and OAM can be done via TLS/HTTPS.

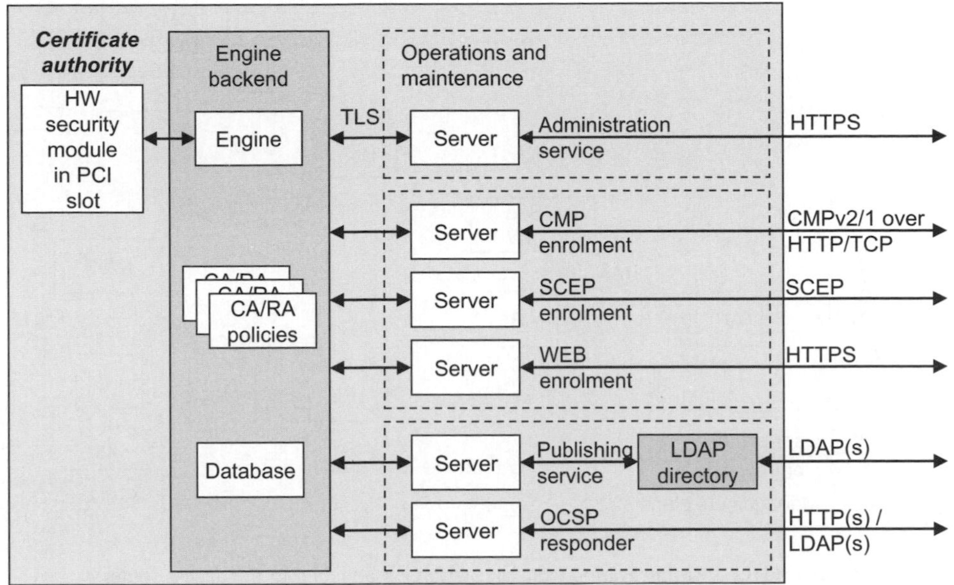

Figure 11.11 The PKI design with the architecture and interfaces.

Security Gateways with high availability, such as Juniper or Cisco, are examples of scalable platforms terminating the IPSec traffic from the eNodeBs and covering fast-growing performance figures for the next years. Insta Certifier is a PKI platform for issuing and managing digital certificates both for users and machines. It provides Certification Authority (CA) and Registration Authority (RA) functionality – that is, manageability for very large PKI environments by introducing centralized management of authentication keys with support for scalable revocation and key updates.

The device identity concept is valid for different entities. LTE/SAE network elements possess identity and are able to authenticate peers and provide their own identities. In addition, there are two kinds of authorities in the security solution. The first one is the Factory Registration Authority, which requests vendor certificates. A centralized vendor-wide CA issues and keeps the certificates in a database. The second one is the Operator's Certificate Authority, which recognizes the vendor's certificates and authorizes requests. This authority issues and manages operator certificates for the network elements.

Figure 11.11 shows the SW architecture and interfaces for the PKI solution.

11.3.3 IPSec Processing and Security Gateway

The LTE standard requires IPSec capability at the eNB level. As an example, Nokia Siemens Networks has integrated IPSec and a firewall at the Flexi LTE eNB. It includes authentication with X.509 certificates. On the other hand, the integrated firewall is more complete than a mere address or port filter, and thus the rules are integrated in the eNB with other management, providing automatic configuration in most of the cases.

In general, the LTE standard requires IPSec capability at the eNodeB elements. Support for the functionality is mandatory but, for the trusted networks (considered reliable by the

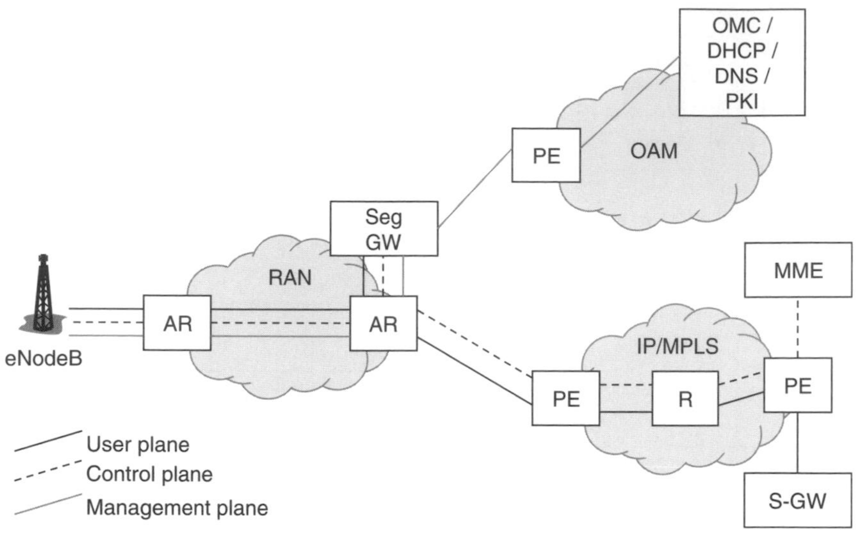

Figure 11.12 An integration example, with GW attached to the access router.

operator), the actual use of IPSec is left optional. In practice, the eNodeB can include both IPSec and firewall, with authentication via the X.509 certificates. The combination of IPSec and firewall provides the possibility to integrate the firewall rules with other network management. This means that there is little or no manual configuration needed.

In the following scenario, presented in Figure 11.12, the SecGW is complementary, placed alongside the Aggregation Router (AR) with both interfaces (incoming and outgoing) directly connected to the AR. The advantage of this scenario is that only minimal changes on the existing network have to be made. Furthermore the scenario allows for aggregation of all interfaces between AR and SecGW into one logical link. Once aggregated, the different kinds of traffic can be separated on Layer 2 by defining corresponding VLAN interfaces on the SecGW. This setup provides a higher level of flexibility and resilience against single link outages.

The SecGW could provide several options to achieve traffic separation:

- *Virtual routers* allow the separation of the routing domains into logical entities with respectively separated routing tables. Each virtual routing entity is handling its directly connected networks and static or dynamic routes.
- *VLANs* are used to separate the traffic within physical links. All security concepts work with physical and logical subinterfaces (tagged traffic on trunk links).
- Definition of dedicated *security zones* as shown in Figure 11.13. These zones are used to logically separate network areas and allow for a more granular traffic filtering and traffic control by defining access control and filter policies.

The VPN design of the segGW can be based either on the single tunnel setup, or a multiple tunnel setup. For the *single tunnel setup*, the traffic of all planes is encrypted with the same encryption set—that is, either a single IKE-SA or a single IP-SEC SA. The setup can be based on the dedicated tunnel interface per eNB, which means that each eNB has its own tunnel

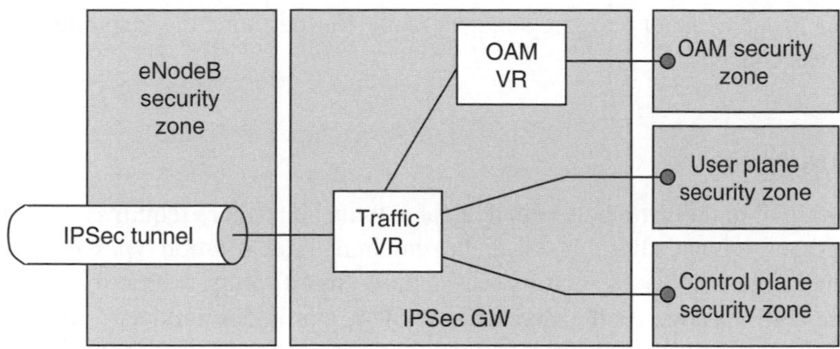

Figure 11.13 The security zone principle.

interface on the SecGW. The setup can also be based on the shared tunnel interface, which means that all the eNBs share one tunnel interface on the SecGW.

For the *multiple tunnel setup*, the traffic for each plane is encrypted with different encryption sets (1 IKE-SA/3 IP-SEC SA). The multiple tunnel setup can be based on the dedicated tunnel interfaces per eNB, or on the shared tunnel interfaces.

11.3.4 Single Tunnel with Dedicated Tunnel Interfaces

The benefit of this solution is that it provides a persistent tunnel interface per eNB. All routes to one eNB point to the same interface, which makes the design easier. The third benefit is the small amount of security associations (SA). As a drawback, the solution requires a large number of tunnel interfaces.

11.3.5 Single Tunnel with Shared Tunnel Interfaces

The benefit of this design is that only one tunnel interface is required per chassis. The eNB inner routes can be aggregated on SecGW. As in the previous case, only a small number of security associations are needed. As a drawback, the scalability of this design is linked to the IP address concept.

11.3.6 Multiple Tunnels with Dedicated Tunnel Interfaces

The benefit of this solution is related to the dedicated tunnel interfaces per plane per eNB. A drawback is that three tunnel interfaces per eNB are required. Another issue is the larger number of security associations. Due to the drawbacks, this is the least feasible solution.

11.3.7 Multiple Tunnels with Shared Tunnel Interfaces

The benefit of this solution is that only one tunnel interface per plane per chassis is needed. In addition, the eNB inner routes can be aggregated on SecGW. As drawbacks, a larger amount of security associations are required. Additional IP network for VPN next-hop table is required if the eNB inner routes cannot be aggregated per plane. Furthermore, the scalability is limited by the IP address concept.

11.3.8 Summary

As the solution of multiple tunnels with dedicated tunnel interfaces requires a large number of tunnel interfaces, it is not advisable. From the remaining three solution types, the single-tunnel design reduces the number of security associations significantly, which requires alignment between the eNB vendors. As the shared tunnel design provides a reduced number of tunnel interfaces per chassis, it is a good solution.

11.4 Authentication and Authorization

In the authentication and key agreement process, the HSS generates authentication data and provides it to the MME element for the processing. There is a challenge-response authentication and key agreement procedure applied between the MME element and the LTE-UE.

The confidentiality and integrity of the signaling is assured via Radio Resource Control (RRC) signaling between LTE-UE and LTE (E-UTRAN). On the other hand, there is Non Access Stratum (NAS) signaling between LTE-UE and MME. It should be noted that in the S1 interface signaling, the protection is not LTE-UE –specific, and that it is optional to apply the protection in S1.

With regard to user plane confidentiality, S1-U protection is not LTE-UE specific. There are enhanced network domain security mechanisms applied that are based on IPSec. Here, protection is optional. Integrity is not protected in S1-U in order to reduce the performance impact.

Let us examine the signaling flow for the authentication procedure, which can be seen in Figure 11.14. In the initial phase of authentication, the MME element evokes the procedure in such a way that it sends the International Mobile Subscriber Identity (IMSI) as well as the serving network's identity (SN ID) to the HSS of the home network of the subscriber in question (2). If the MME does not yet have information about the IMSI code at this stage it is requested first from the LTE-UE (1). It should be noted that IMSI is delivered over the radio interface in a text format, which means that this procedure should be used only when no other options are available.

As a result of the MME user authentication request to HSS, the latter responds with an EPS authentication vector, which contains RAND (random challenge number), AUTN (authentication token), XRES (expected response), and the ASME key (3).

When the MME has received this information, it sends the RAND number and AUTN to the LTE-UE (4). Now, the LTE-UE processes this information in order to authenticate the network (according to the mutual authentication concept). Based on the information received and its own key, the LTE-UE calculates a response (RES) and sends it back to the MME (5).

Both the LTE-UE and the HSS contain the same algorithm, which is used for the calculation of the response with the same inputs. The RES of LTE-UE, and the expected response XRES calculated previously by the HSS are now analyzed by the MME. This means that the MME compares the RES and XRES values. If they are the same, the LTE-UE is thus authenticated correctly, and NAS signaling will be secured. The eNodeB key—that is, K_{eNB}—is now calculated and delivered to eNodeB in order to secure the radio interface for all signaling and data transmission (6).

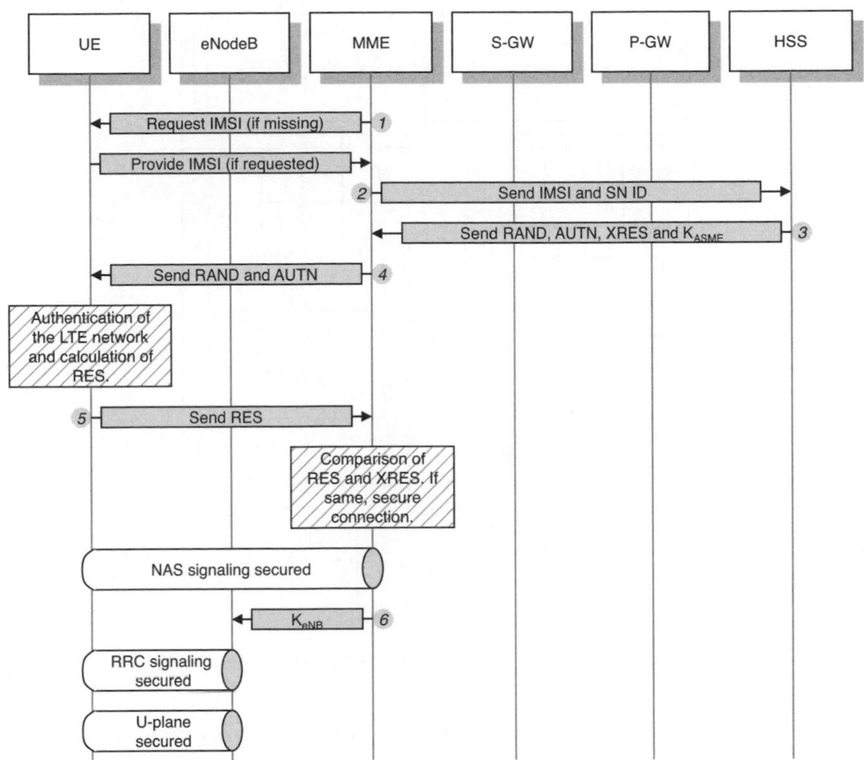

Figure 11.14 The mutual authentication procedure of LTE.

11.5 Customer Data Safety

The normal procedures that are applied in the 2G and 3G environments for the physical protection of the customer subscription data, charging record data, and other confidential information are also used in the LTE/SAE environment. Fraud prevention and monitoring that were applied in previous generations are applicable in LTE/SAE.

11.6 Lawful Interception

Lawful Interception (LI) in general has been designed for authorized and official access to the private communications. The mobile/fixed network operator and service provider can use it to collect delivered traffic that has been delivered and/or identification information for analysis by law enforcement officials. This has been possible for a long time in mobile communications networks. For example, it was designed as a part of the Legal Interception Gateway (LIG) in the first Release 97 of the basic GPRS solution, which can mirror the traffic delivered via the GPRS nodes.

Although it is technically possible to store basically all the details and contents of the data flows, the legal interception process can only be applied in accordance with national and regional laws and technical regulations.

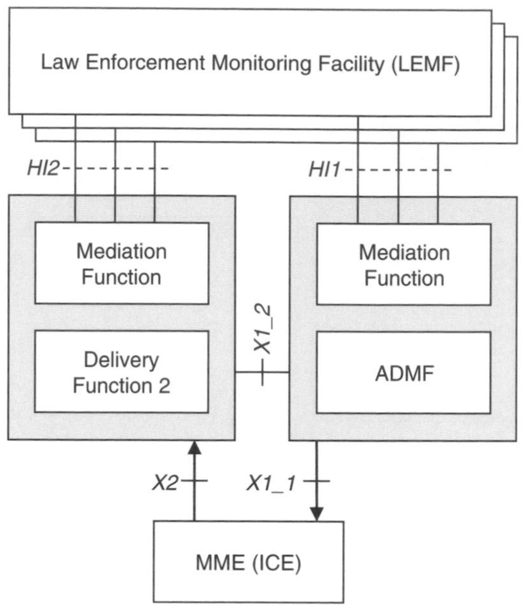

Figure 11.15 The configuration for the MME intercept.

In the case of LTE/SAE, the 3GPP Evolved Packet System provides only IP-based services. This logically means that EPS can be used for the interception of the IP layer's Content of Communication (CC) data flows. The LTE voice connections are considered as IP data flows in this sense via VoIP solutions. If a fallback type of functionality is applied during the LTE voice call in order to continue the voice call as a circuit-switched version, the respective 2G/3G network also contains the LI. In addition to user plane interception, the LI solution of EPS can generate Intercept Related Information (IRI) records in the control-plane messages which identify the called parties, the location of LTE terminal, and other call related information.

The functional architecture of the EPS lawful interception is comparable to the functional architecture of the packet-switched domain of 3G networks of 3GPP. Figures 11.15–17 show the standard defined configurations for the MME, HSS, S-GW and PDN-GW, respectively, for the EPS lawful interception. The key identities for the interception are IMSI (International Mobile Station Identity), MSISDN (Mobile Station ISDN number), and IMEI (International Mobile Equipment Identity) [7]–[9].

As can be seen in Figures 11.15 to 11.17, the MME element only handles the control plane and the HSS only handles signaling. The interception of the content of communication is thus applicable only via the S-GW and PDN-GW elements. In the figures, the Administration Function (ADMF) is functionality that interfaces with the Law Enforcement Monitoring Facilities (LEMF) of the Law Enforcement Agencies (LEA) that request the interception. The ADMF functionality has been designed in such a way that it has a direct interface with the network elements that are intercepted while it keeps the interception-related activities of each LEA separate. This is done in such a way that the ADMF, together with the delivery functions of the intercepted information, is hidden from the Intercepting Control Element (ICE), even in the case of various simultaneous activations on behalf of separate LEAs related to the same subscription.

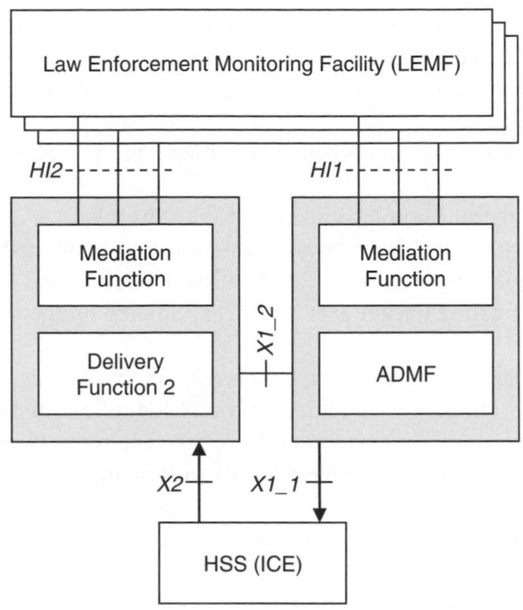

Figure 11.16 The configuration for the HSS intercept.

The physical ICE of the LTE/SAE network is connected to the ADMF via an X1_1 interface. This interface delivers all the interception-related information from each ICE. Each ICE carries out the interception—that is, the activation, deactivation, interrogation and invocation procedures independently. On the other side of the ADMF, an HI1 interface is defined towards the

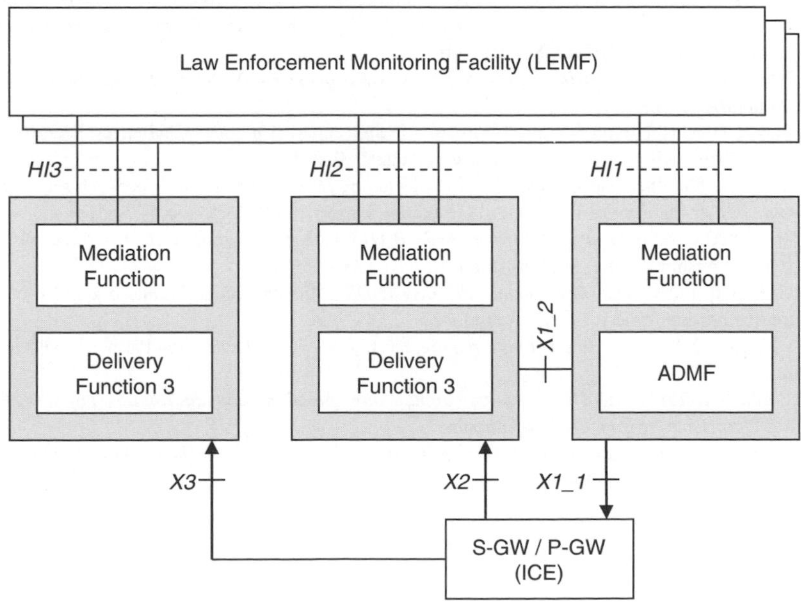

Figure 11.17 The configuration for the S-GW and P-GW intercept.

requester of the lawful interception. For communication between independent delivery functions and the LEA, the HI2 and HI3 interfaces are defined. The task of the delivery functions is to distribute the Intercept Related Information and Content of Communication (CC) to the relevant LEAs.

The following are some examples where Legal Interception is activated:

- there is a change in the location information of the subscriber;
- a terminating or originating short message transfer is being initiated by the target;
- a terminating or originating circuit-switched call is being initiated by the target;
- a terminating or originating packet data service is initiated by the target.

The content of communications (CoC) can be intercepted via the Legal Interception concept from the media plane entities. In addition, various identities related to the intercepted communications can be stored. Some examples of the intercept related information (IRI) that can be intercepted from the subscribers are the following:

- MSISDN (Mobile Sunscriber ISDN Number);
- IMSI (International Mobile Station Identity);
- ME id (Mobile Equipment Identifier);
- event type;
- event time and date;
- NE Id (Network Element Identifier);
- location.

References

[1] 3GPP TS 33.401. (2011) *Security architecture, including IPsec for S1-MME and X2 control plane, S1 and X2 user plane, and management plane, tunnel mode, IKEv2 and authentication by public certificates*, V. 8.8.0, 3rd Generation Partnership Project, Sophia-Antipolis.

[2] 3GPP TS 33.402. (2009) *Security aspects for non-3GPP accesses*, V. 8.6.0, 3rd Generation Partnership Project, Sophia-Antipolis.

[3] Agilent (n.d.) LTE and the Evolution to 4G Wireless—Design and Measurement Challenges. Security in the LTE-SAE Network, www.agilent.com/find/lte (accessed August 30, 2011).

[4] 3GPP TS 33.210. (2009) *Network domain security, including IPsec in tunnel mode between security gateways, IPsec profile and configuration*, V. 8.3.0, 3rd Generation Partnership Project, Sophia-Antipolis.

[5] 3GPP TR 33.820. (2009) *Security of Home Node B (HNB)/Home evolved Node B (HeNB)*, V. 8.3.0, 3rd Generation Partnership Project, Sophia-Antipolis.

[6] 3GPP TS 33.310. (2010) *Network Domain Security (NDS); Authentication Framework (AF)*, V. 8.4.0, 3rd Generation Partnership Project, Sophia-Antipolis.

[7] 3GPP TS 33.106. (2008) *Legal Interception; requirements*, V. 8.1.0, 3rd Generation Partnership Project, Sophia-Antipolis.

[8] 3GPP TS 33.107. (2011) *Legal Interception; architecture, functions and information flows*, V. 8.12.0, 3rd Generation Partnership Project, Sophia-Antipolis.

[9] 3GPP TS 33.108. (2011) *Legal Interception; description of the handover interfaces*, V. 8.13.0, 3rd Generation Partnership Project, Sophia-Antipolis.

12

Planning and Deployment of SAE

Jukka Hongisto and Jyrki T. J. Penttinen

12.1 Introduction

This chapter describes how LTE can be supported by the packet core network of LTE/SAE. The chapter concentrates on 3GPP release 8 LTE/SAE architecture and the Evolved Packet Core (EPC), and it presents potential migration steps from the existing 2G/3G PS core towards fully compatible Rel-8 EPC.

12.2 Network Evolution from 2G/3G PS Core to EPC

12.2.1 3GPP R8 Requirements for LTE Support in Packet Core Network

3GPP Release 8 defines a new EPC for LTE access. The EPC can also be used for other access technologies like GERAN, UTRAN and CDMA2000. Figure 12.1 shows the standard Release 8 architecture with the relevant interfaces between the logical entities.

The Mobility Management Entity (MME) is the equivalent of the SGSN in 2G/3G GPRS networks. In the LTE/SAE network, the MME is a pure control-plane element. It initiates a direct tunnel between the eNodeB and Serving Gateway in order to deliver the user-plane traffic.

According to 3GPP Release 8, the mobile gateway functionality is divided into the Serving Gateway (S-GW) and the Packet Data Network Gateway (P-GW) functionalities. These S-GW and P-GW functionalities can be implemented in the same physical node or in two separate entities. They are connected logically via an open S5 interface, which is called S8 in the case of roaming.

The S-GW and P-GW functionalities are defined as mandatory for LTE radio network deployment. LTE subscribers are always connected to the services via P-GW even if 2G/3G radio access would be used. 2G/3G subscribers may access existing APNs (and possible new APNs) via GGSN or P-GW as the P-GW element also contains the GGSN functionality.

S-GW terminates the LTE core user plane interface towards the evolved UTRAN. An LTE User Equipment (UE) is assigned to a single S-GW at a given point of time. The S-GW acts as a user plane gateway for the LTE radio network in case of the inter-eNodeB handovers, and for inter-3GPP mobility (terminating S4 and relaying the traffic between 2G/3G system and PDN-GW).

The LTE/SAE Deployment Handbook, First Edition. Edited by Jyrki T. J. Penttinen.

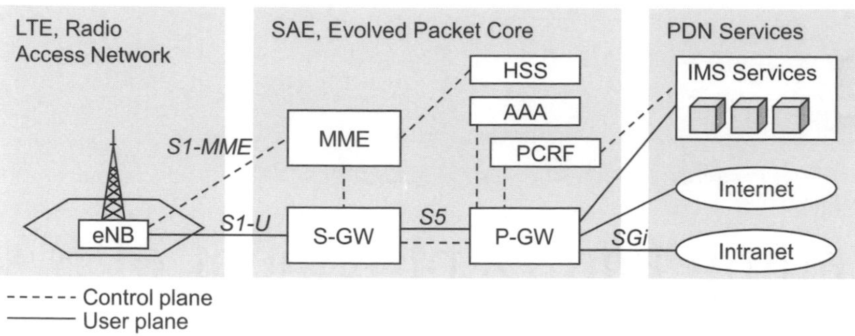

Figure 12.1 3GPP R8 architecture for LTE/SAE.

The PDN-GW acts as a user plane anchor and terminates the SGi interface towards the service networks. It allocates the IP address for the UE. PDN-GW applies policy enforcement to the subscriber traffic and performs packet filtering at the individual user's level (by performing, e.g., a deep-packet inspection). The PDN gateway interfaces with the service provider's online and offline charging systems.

12.2.2 Introducing LTE in Operator Network

Operators who want to start LTE deployment in an early phase will likely begin with a technology trial to test LTE radio and EPC capabilities. The trial network is introduced as an overlay by leaving the existing production machinery intact. When LTE radio is introduced, it is mandatory to introduce the MME and S/P-GW capabilities in the core network, as can be seen in Figure 12.2. In the trial phase, the subscribers are offered the mobility within the LTE radio network. Internet access and selected operator services can be included in the set of services of the trial network.

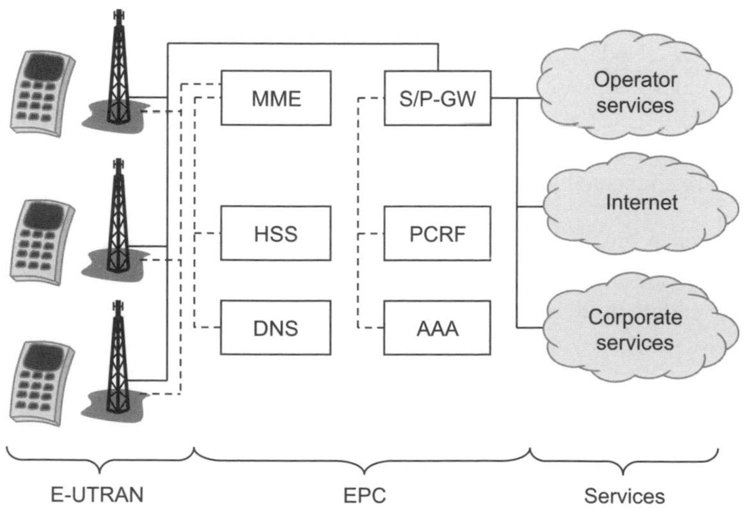

Figure 12.2 EPC for LTE access.

12.3 Entering Commercial Phase: Support for Multi-Mode LTE/3G/2G Terminals with Pre-Release 8 SGSN

When the subscriber moves outside of the LTE network coverage area, service continuity is a necessary requirement for the operators in order to be able to launch commercially feasible LTE networks. This requires multimode terminals to be available that are capable of supporting 2G, 3G, and LTE. The integration of 2G/3G core network with the EPC is needed. Integration is required to allow handovers between the LTE and 2G/3G access networks. The gateway acts as an anchor point for all the subscriber sessions. In other words, the same gateway element serves the subscriber session when the subscriber moves between the LTE and 2G/3G networks. This means that even if the subscriber initiated a session in an area where only 2G/3G coverage is available, the session would be served by the P-GW. In practice, this integration means provisioning the connectivity between the 2G/3G SGSN and EPC.

The simplest option is that the SGSN can be connected to PDN-GW via the Gn interface, which is used for the LTE/3G/2G terminals in order to provide the IP connectivity, and the Gn interface between the MME and the SGSN for the mobility between the LTE and 2G/3G networks. If the Direct Tunnel of the 3GPP Release 7 is used, the user plane traffic goes directly between UTRAN and P-GW. Figure 12.3 clarifies the idea.

In this phase, the operator can use GGSN for the 2G/3G traffic of all the terminals that are not LTE capable. If the operator uses the same APN definitions for both 2G/3G and LTE services, then SGSN has to support the PDN GW selection based on, for example, IMEI or the terminal capability.

12.3.1 Support for Multi-Mode LTE/3G/2G Terminals with Release 8 Network

The deployment of Release 8 SGSN allows the operator to introduce the so-called Common Core where 2G and 3G accesses are also linked to S-GW. The Release 8 QoS model, which means that the network controls the QoS level, is also used for the 2G/3G traffic.

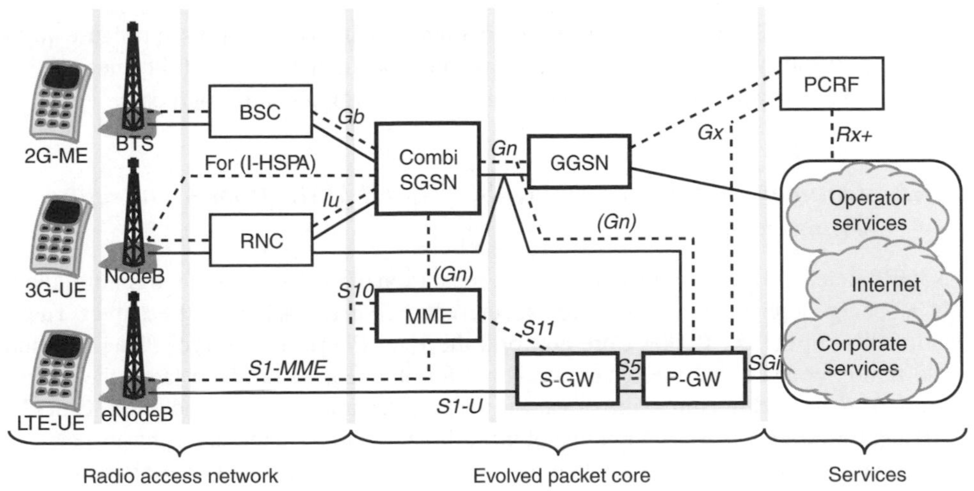

Figure 12.3 LTE/3G/2G interworking with PreRelease 8 SGSN. The control plane interface from NodeB to SGSN is for Internet-HSPA (I-HSPA).

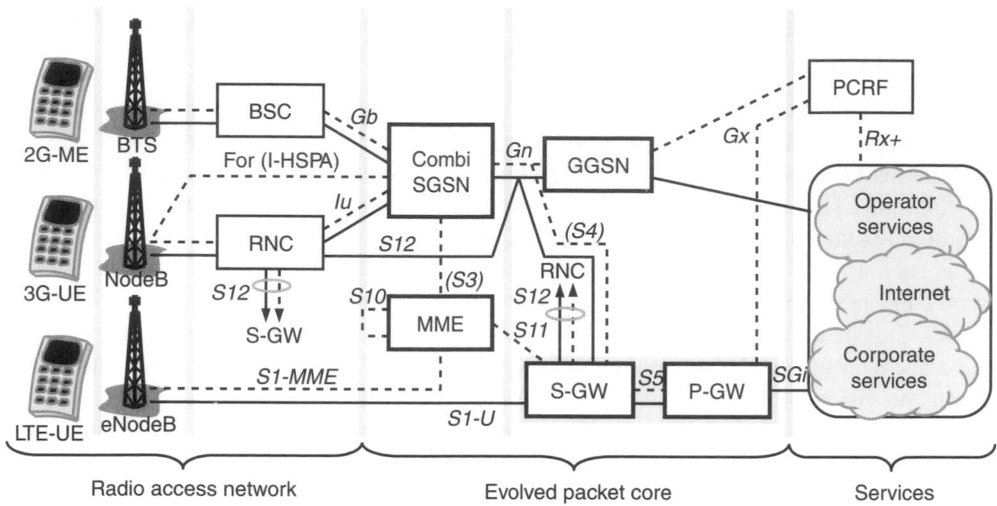

Figure 12.4 LTE/3G/2G interworking with Release 8 SGSN.

When SGSN is upgraded to 3GPP Release 8 level, it will have a new S4 interface towards S-GW, and an S3 interface towards the MME. In this phase, S-GW is used also for the 2G/3G traffic. Furthermore, it is possible to utilize the Release 8 Direct Tunnel (S12 interface) between UTRAN and S-GW. The already existing GGSN of the operator can still be used for the 2G and 3G subscribers. Figure 12.4 presents this phase.

At this point, it is also relevant to have more optimized interworking with integrated SGSN/MME implementation with the following options:

- 3G SGSN and MME for control plane only;
- combined SGSN/MME for all 3GPP accesses.

From the S-GW/P-GW point of view, it does not matter which option is used because all the traffic goes through the S-GW/P-GW anyway, and the intersystem mobility needs to be signaled to the gateway for proper bearer, policy and charging control purposes.

12.3.2 Optimal Solution for 2G/3G SGSN and MME from Architecture Point of View

The 3GPP Release 8 architecture mandates the separation of the User Plane and Control Plane. This is being applied increasingly in the current 3G networks with the Direct Tunnel functionality. In the 2G Packet Core network, the User Plane and Control Plane functions are so tightly coupled that the separation is not feasible. Based on these facts, the optimal solution from the architecture point of view is to apply separate physical nodes for 3G SGSN/MME and for 2G SGSN. The combining of these three functions in the same physical element is technically possible but it means that the same physical element has to support two very different architectures.

In the optimal solution scenario, the assumption is that the 2G traffic growth is moderate compared to the 3G/LTE traffic evolution, and that the operator has 2G SGSN elements that can

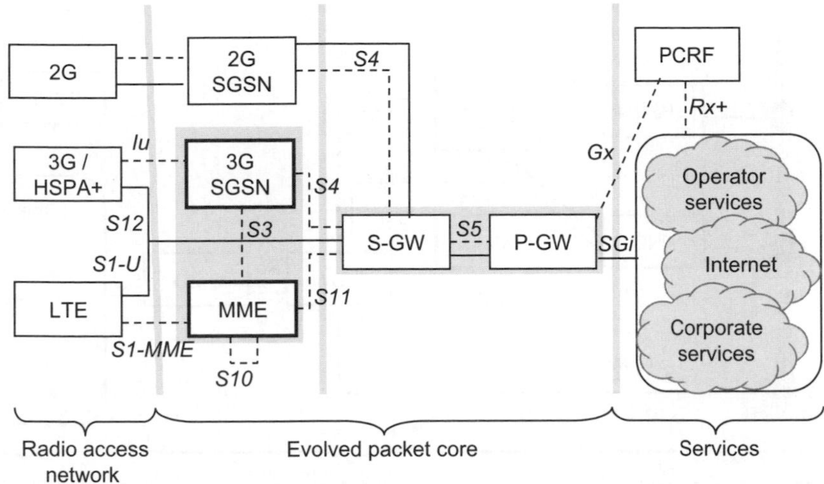

Figure 12.5 Optimized architecture for high-speed 3G/HSPA/LTE traffic.

continue to deliver the 2G traffic. Most of the 2G transport networks may still use the legacy E1/T1 and Frame Relay interfaces but may not be upgraded to support IP-based interfaces.

The flat architecture with a minimum number of user-plane elements has been one of the basic principles of the 3GPP network architecture evolution. As LTE and 3G share common network architecture, this provides an opportunity to align the 3G SGSN and the MME functionalities into a single entity. 3G SGSN, as a pure control-plane element, makes it behave like MME and to offer most benefits as a combined element. On the other hand, S-GW and P-GW handle user-plane traffic in both nonroaming and roaming cases. Figure 12.5 presents this phase.

By separating the control plane and user plane, the operator can build a modernized all-IP network that accomplishes future data traffic and mobility growth with the flat architecture. The benefits of this optimized architecture include the following:

- Most of the user plane traffic of 3G/HSPA/LTE coming from the Internet can be optimally routed based on the operator peering points.
- With combined S-GW and P-GW, the operator can decrease CAPEX and OPEX nearly 50%.
- Combined MME and 3G SGSN allows shared capacity for the LTE/3G subscribers and optimizes the mobility management with the reduction of the signaling traffic when UE moves between LTE and 3G service areas.
- The MME and 3G SGSN can support pooling with geographical redundancy. The pooling location can also be close to MSS in order to simplify the connections to the CS domain (SGs and Sv interfaces).
- Release 8 3G Direct tunnel means that SGSN is meant for the control plane only and the S-GW and P-GW are acting in the user plane.

12.3.2.1 Combined SGSN/MME for all 3GPP Accesses

It is technically possible to combine all the 2G/3G SGSN and LTE/MME functionalities into a single network element. This can, in fact, be a preferred solution if the operator wants to

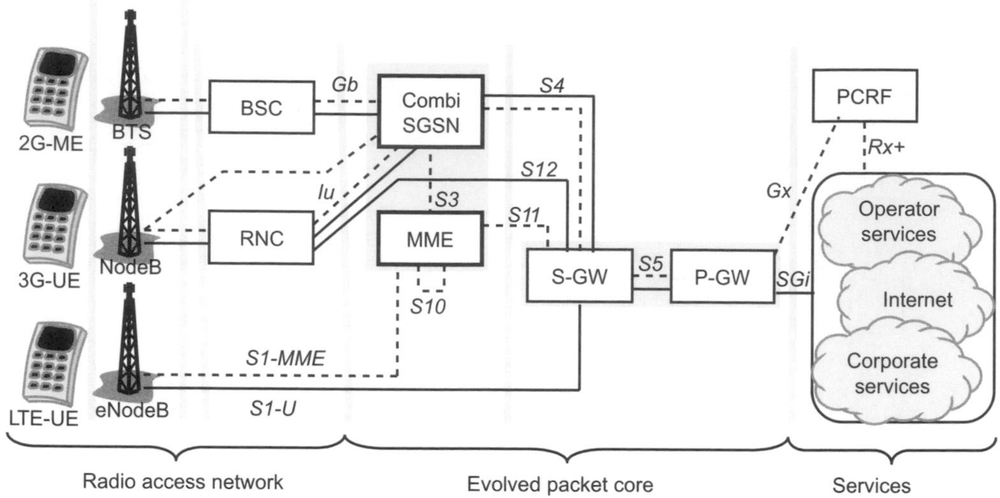

Figure 12.6 2G/3G SGSN and MME functionalities supported in a single element.

upgrade the existing SGSN elements to the new hardware or wants to deploy a minimum number of network elements in the network. Figure 12.6 presents this phase.

There are two alternative paths leading to this combined scenario:

- The MME functionality is introduced on top of the flat-architecture-optimized network element. The 2G and 3G subscribers are migrated to this element as these accesses are supported. This is the preferred option as the 3G and LTE traffics are driving the mobile data evolution and the solutions are to be optimized according to the needs of these accesses.
- The MME functionality is introduced as a SW upgrade to the existing SGSN. This can be considered as an option in case the operator has unused capacity in the existing SGSN elements.

When a single element includes all the SGSN and MME functionalities, it makes subscriber migration from 2G or 3G to LTE easier and also reduces the inter-SGSN-MME signaling. On the other hand, there will only be a few LTE subscribers at the beginning of the operational phase of thc LTE network, which means that not all the combined SGSN elements will be upgraded to support the MME functionality. When the combined SGSN-MME scenario is selected, it is preferred that all the interfaces are upgraded to work in the IP based environment, including the Iu and Gb interfaces.

12.4 SGSN/MME Evolution

12.4.1 Requirements to MME Functionality in LTE Networks

In the 3GPP Release 8 architecture, the MME is a pure control-plane element that takes care of the control-plane traffic handling, session and mobility management, idle mode mobility management and paging.

The flat network architecture, where the eNode-B interfaces directly with the core network elements without an aggregating RNC layer, makes all the mobility management events visible for the core network nodes, as can be seen from Figure 12.7.

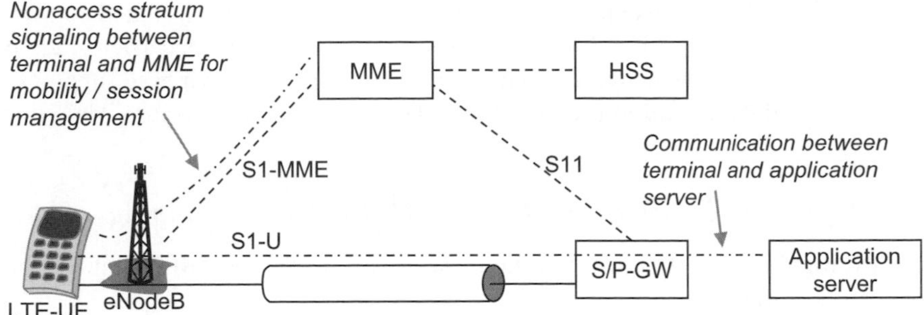

Figure 12.7 Flat network architecture and bad terminal behavior set challenges to MME.

The signaling traffic load is a more severe bottleneck than the user data throughput in networks where a large number of smart phones are used. Always-active services like Virtual Private Network (VPN), e-mail, chat and machine-to-machine applications generate continuous keep-alive signaling. Some "badly behaving" smart phone types try to optimize the battery lifetime by switching the state to idle immediately when there are no more packets being sent or received, which results in a continuous idle-active signaling load to the network.

12.5 Case Example: Commercial SGSN/MME Offering

12.5.1 Nokia Siemens Networks Flexi Network Server

Nokia Siemens Networks Flexi Network Server (Flexi NS) is a control and transaction machine that implements the 3GPP Release 8 compliant MME functionality. The SGSN functionality can be activated via a SW update.

The Flexi NS has been designed for a flat architecture. The ATCA HW platform has been built to cope with the future evolution of LTE/SAE. The SW architecture is optimized to handle signaling and control plane traffic. This allows Flexi NS to implement high signaling capacity in terms of transactions and simultaneous sessions as well as to support a large number of Simultaneously Attached Users (SAU), high bearer and PDP context capacity and 2G throughput capacity (3 shelf configuration).

There are two capacity dimensions where flexible scaling is needed: (1) the user plane versus the control plane, and (2) scaling between the access technologies 2G/3G/LTE.

Flexi NS allows dynamic capacity usage to share the node capacity among 2G, 3G and LTE subscribers. A common database is implemented to handle 2G, 3G and LTE subscriber data. This implementation allows flexible allocation of general purpose HW blades between 2G, 3G and LTE, and makes the element straightforward to dimension. Flexi NS implements session resiliency to allow maintaining the subscriber session despite a unit failure. This means that there will be no interruptions in real-time services in case of failure.

Flexi NS simplifies connectivity towards the radio network and the Gateway (Serving Gateway, PDN Gateway, GGSN). IP addressing virtualization is applied to hide Flexi NS internal architecture and to show one IP address per MME node towards the radio network and the Gateway.

ATCA hardware provides advanced energy-saving options with a low traffic power-saving mode that allows selected CPUs to be shut down when the traffic load is low.

The current DX200 HW of Nokia Siemens Networks SGSN has been designed for a complex interface environment (FR, ATM, E1/T1, IP). The SGSN demonstrates carrier-grade performance and reliability in live networks. The evolution of this equipment base supports Direct Tunnel, and is straightforward to use in flat mobile packet core architecture deployments.

12.5.2 Aspects to Consider in SGSN/MME Evolution Planning

When analyzing the possible network evolution scenarios for SGSN/MME solutions there are three key aspects to be considered:

12.5.2.1 Timing of LTE Introduction

Flexi NS provides an overlay solution for operators who are introducing LTE in an early phase. An overlay solution ideally isolates the LTE introduction from the existing packet core network allowing easy troubleshooting and upgrades in the technology introduction phase.

12.5.2.2 Network Modernization to All-IP

With the transport network being modernized to all-IP and with 3G SGSN functionality becoming available in Flexi NS, the operator can start migrating the 3G subscribers from existing SGSNs to Flexi NS.

12.5.2.3 SGSN Investment Utilization

New software releases continue to be provided for the DX200 based SGSN. DX200 SGSN will evolve into a R8 SGSN that interworks with the MME. SGSN/MME inter-working is critical especially in the LTE introduction phase as handovers to 2G/3G Packet Core need to be supported.

12.6 Mobile Gateway Evolution

12.6.1 Requirements to Mobile Gateway in Mobile Broadband Networks

The mobile broadband gateway needs to scale up to accommodate the growing traffic volumes. This is not enough, however, but also user plane performance is very important. Flat network architectures, where NSN has demonstrated market leadership already in 3G with Direct Tunnel, will change the connectivity architecture of the EPC gateways, making the radio network directly visible to the gateway.

LTE increases the signaling load of base stations as there is no more separate radio network control functionality between the base stations and the gateway. This will increase the amount of signaling per subscriber. When signaling is combined with additional signaling for AAA, online charging and policy control, the overall processing requirements related to signaling will increase substantially compared to those of 2G/3G.

Subscriber density will also be very high due to LTE always-on connectivity, multiplying the signaling by the amount of attached subscribers. All these will impact on the gateway and the MME performance profile. The mobile gateway platform needs to provide high performance,

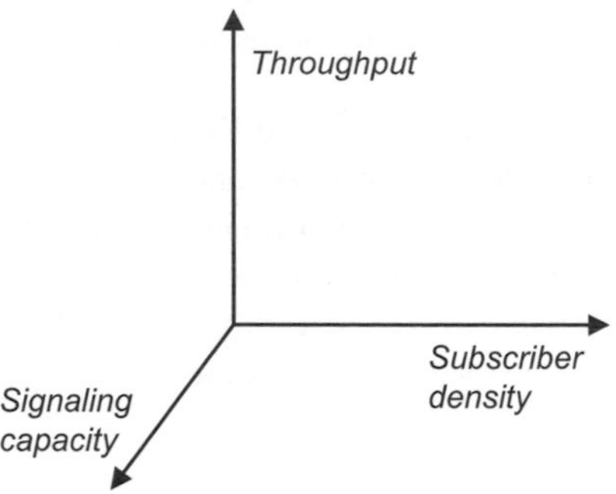

Figure 12.8 Key mobile gateway scalability dimensions.

scalability, and flexibility in all three dimensions, as can be seen from Figure 12.8. Failure to perform in any of them will lead to suboptimal solutions.

12.7 Case Example: Commercial GGSN/S-GW/P-GW Offering

12.7.1 Nokia Siemens Networks Flexi Network Gateway

NSN equipment includes the concept of service awareness, combined with online charging capabilities, which allows differentiated charging for operator-provided services. With increasing demand for open Internet access, there are solutions to track proprietary peer-to-peer applications like file sharing and communication services.

Flexi NG, shown in Figure 12.9, is designed to serve as a high-capacity gateway for mobile broadband networks. It serves 2G/3G, HSPA, I-HSPA and HSPA+ networks as GGSN and LTE networks as S/P-GW.

Flexi NG uses an ATCA platform that fits optimally with current and future packet core and evolved packet core requirements. NSN EPC based on ATCA provides a feasible performance for throughput, signaling and subscriber density. ATCA scales for centralized

Figure 12.9 Nokia Siemens Networks Flexi NG. Printed by courtesy of Nokia Siemens Networks.

and decentralized deployments, providing a flexible platform for potentially evolving topologies.

Flexi Network Gateway implements integrated Deep Packet Inspection (DPI) capabilities. It provides recognition for 300+ protocols and applications including peer-to-peer protocol tracking. The protocol and application signatures are constantly updated in Flexi NG. Product capabilities include service awareness—that is, protocol analysis on L3/L4—and Deep Packet Inspection – analysis on L7/L7+. L7+ analysis includes heuristic analysis that is typically applied to track proprietary protocols like peer-to-peer applications and services.

12.7.2 Aspects to Consider in GGSN/S-GW/P-GW Evolution Planning

Most of the 2G/3G networks still have a very limited number of active PS sessions, typically 10–20% of users have PDP context activated. In LTE there is at least one PS session per subscriber, which means that LTE subscriber growth is directly visible in S-GW and P-GW. Also new services like VoLTE activate more PS sessions, which means that active PS sessions increase by more than 100–200% compared to the number in previous PDP contexts.

12.8 EPC Network Deployment and Topology Considerations

Network topology means a hierarchy of connections between functional units in the network element. 3GPP R8 mandates flat architecture that allows flexible choices in network topology implementation.

It is expected that IPv4 addresses will be exhausted in 2012, which puts pressure on the transformation from IPv4 to IPv6 addressing. The introduction of LTE is a natural point to start IPv6 migration in operator networks. The most important change compared to 2G/3G networks is that every attached subscriber requires at least one IP address due to always-on connectivity. It is a natural choice to deploy new services like operator VoIP and other IMS services based on IPv6 addressing. More addresses are also needed as the LTE network is fully IP based and every eNode-B can have several addresses.

12.8.1 EPC Topology Options

Packet core is typically deployed in a hierarchical topology that includes National sites (GGSN location), Regional sites (SGSN location), Local sites (RNC location) and Base Station sites.

Current 2G/3G deployment consists of a radio network, a packet core network including SGSN and GGSN and a transport network that provides interconnectivity between the sites. Transport networks are being upgraded to IP-based solutions. 3G operators are starting to deploy Direct Tunnel, which allows transmission savings by user plane bypassing the SGSN, which is routed straight between RNC and GGSN.

3GPP R8 architecture allows the network topology to be modernized. Flat architecture becomes mandatory as user-plane traffic is routed directly from eNode-Bs to the S-GW. The S-GW and PDN Gateway can be deployed either at the same site or they can be separated at different sites. The OPEX and CAPEX costs can be minimized with combined deployment and this will likely be the most common scenario. MMEs can also be deployed either in centralized

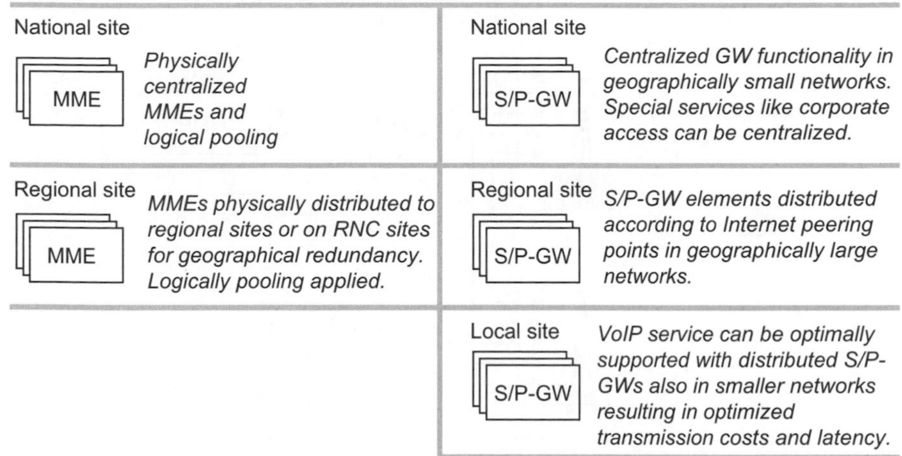

Figure 12.10 Network topology options.

or distributed topology. MME pooling can be implemented irrespectively of the selected topology and it is highly recommended for optimal capacity usage.

Figure 12.10 illustrates different network topology and network element location options in a practical LTE deployment.

12.8.2 EPC Topology Evolution

It is likely that, in the LTE technology introduction phase, the EPC is deployed in a highly centralized manner. The LTE will require new terminals and this will limit the amount of LTE subscribers in the service introduction phase. Initial LTE usage can be compared to existing 2G/3G/HSPA usage with centralized GGSN.

When network usage increases and the operator expands LTE radio coverage, more EPC elements will be needed and the operator can add a number of sites for regional redundancy purposes. With the MME pooling functionality (S1-flex), the MMEs can be deployed in a very centralized manner in a few sites.

Intersystem mobility between LTE and 2G/3G networks may drive SGSN and MME collocation to optimize mobility management and to allow smoother user migration. It needs to be noted that as 2G SGSN continues to handle user plane traffic, highly centralized 2G SGSN is not recommended for latency reasons.

From a connectivity perspective, MME has an Home Subscriber Server (HSS) interface and it is interworking with the Circuit-Switched domain via SGs/Sv connections to MSC, which may drive harmonization of MME and MSC topology.

The P-GW provides a connection to content and service networks including the Internet, operator services, and corporate connectivity. Routing of high-volume traffic needs to be optimized to minimize transmission costs and VoIP traffic needs minimized latency. For these reasons it is likely that in the future gateways will be deployed in a more distributed manner according to peering point locations.

The P-GW has interfaces to PCRF, AAA server and charging system, which need to be considered if P-GW is very decentralized. The P-GW location can also be driven by service/APN usage—for example, corporate VPN or the need to provide connectivity for outbound roamers.

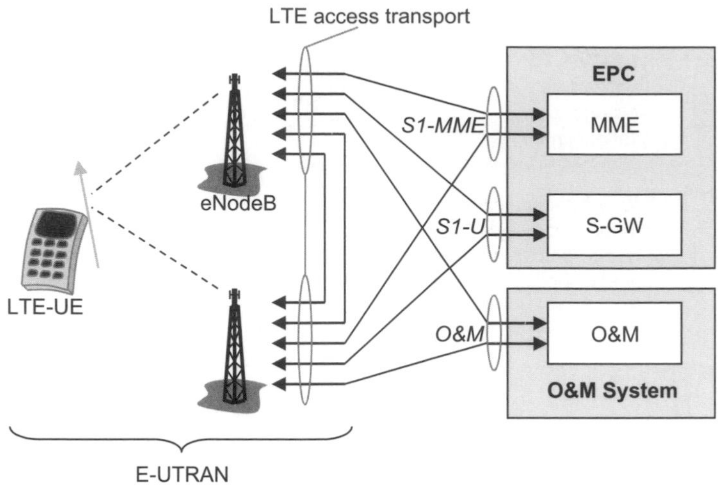

Figure 12.11 LTE access network reference architecture.

12.9 LTE Access Dimensioning

The LTE access network provides interconnection between the LTE—that is, Evolved UTRAN (E-UTRAN) and Evolved Packet Core (EPC) domains. The interfaces to be considered in the access network dimensioning are S1_U and S1_MME between eNodeB, and S-GW and MME, respectively. Furthermore, the X2_U and X2_C between the eNodeB elements for the user and control plane signaling, respectively, belong to the access network. Finally, the interface between the eNodeB elements and the Operations and Management element should be taken into account. In other words, the lines that end at eNodeB elements need to be dimensioned within the access network. Figures 12.11 and 12.12 clarify the access network dimensioning points.

Access network dimensioning requires the relevant inputs, which are the traffic profile, radio network topology and radio interface performance values. The traffic profile provides information about the user-plane traffic per LTE radio cell, handover traffic percentage via the X2 interface, signaling traffic percentage, and transport overhead for the user and control planes. The radio network topology and performance indicates the number of cells of the

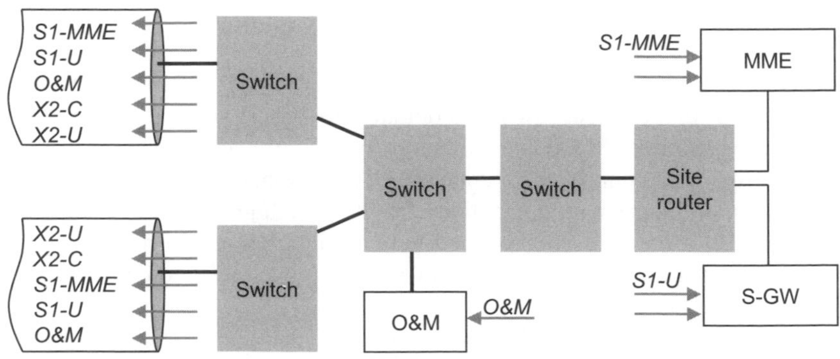

Figure 12.12 The SAE network interfaces for the dimensioning.

eNodeB, the cell throughput (average and peak values are needed), and the radio interface overhead estimate. The result of the dimensioning is the required capacity per logical interface–that is, for the user plane (S1_U and X2_U), control plane (S1_MME and X2_C), and maintenance plane (OAM interface). The final dimensioning indicates the required capacity per whole logical interface (S1 and X2), and the total transport capacity for each eNodeB.

The LTE traffic profile is essential in order to dimension the interfaces correctly. The best estimates can be derived from the operator network statistics of the general data traffic by taking into account the estimated increase of the traffic caused by different LTE applications (VoIP, web browsing, FTP, etc.). LTE-UE penetration as a function of time should be taken into account for a sufficiently long period.

Estimates can be made for the user plane traffic per radio cell in terms of the average data traffic in mbps values generated during the LTE traffic busy period. Note that the busy periods for voice and data traffic profiles typically occur at different times, but in the case of LTE only the general peak period can be considered as the voice service is packet based.

Next, the signaling share for the handovers in the X2_U interface is estimated. The value might oscillate typically between 2–3%, but the final estimate should be made on a case-by-case basis as the network topology does have an effect on the signaling load.

The signaling traffic share in the control plane could be around 1–2% compared to the user plane. This value can be shaped via laboratory and field tests prior to the actual LTE deployment.

The user plane transport overhead in S1_U and X2_U interfaces depends on the utilization of IPsec in such a way that the total transport protocol headers for GTP-U, UDP, IP and Ethernet is 144 bytes with IPsec, and 78 bytes without it, resulting in 25 or 15% overhead with and without IPsec, respectively.

Finally, the control plane transport overhead for the S1_MME and X2_C interfaces can be estimated depending on the use of IPsec. The size of the protocol headers–in this case for SCTP, UDP, IP and Ethernet–is 140 and 74 bytes with IPsec and without IPsec, respectively, causing 179% and 95% control-plane transport overheads, respectively.

As for the radio overhead, the protocol overhead for the PDCP, RLC and MAC is 9 bytes. Depending on the payload packet size, the overhead could be estimated to be around 2% if a mix of packet sizes is assumed.

Based on these assumptions, the user plane, control plane and management plane are dimensioned accordingly. For the user plane, the capacity for the S1_U and X2_U interfaces can be estimated jointly due to the common transport scheduler buffer. It should then be decided if the interface is dimensioned by assuming the aggregated average capacity of all cells, or peak traffics, or a combination of these.

For the control plane dimensioning–for the S1_MME and X2_C, the bandwidth can be estimated, for example based on the level of the user plane traffic. This is a straightforward task when the control plane inputs are estimated as shown previously. For the management plane, a simple rule of thumb is to allocate an additional 1 mbps at the eNodeB. This capacity also includes the transport overhead.

13

Radio Network Planning

Jyrki T. J. Penttinen and Luca Fauro

13.1 Introduction

This chapter presents the network planning aspects of the LTE radio interface. Essential steps and procedures are presented for estimating LTE coverage and capacity as a function of the planned environment type. The LTE link budget is introduced, as well as useful radio propagation models for the estimation of the path loss.

LTE radio network planning is similar to many other, previous mobile communications systems. The goal of this activity is to estimate the coverage and capacity of the services within given parameters and in the planned environment types.

13.2 Radio Network Planning Process

The radio network planning process can be generalized in the following way, as presented in Figure 13.1. Network planning is one item in the deployment project. The final outcome of the radio plan depends on the time frame for the rollout, the target quality, and the capacity offered. Figure 13.2 summarizes the project dependencies. The planning assumptions of LTE can be estimated to be close to the methods and values typically used in HSPA, but there are also differences due to the OFDM and SC-FDMA, and higher data speeds.

The focus of the nominal planning phase is to estimate the required number of sites to provide sufficiently high quality services. Uniform assumptions for the site parameters can be used in each cluster type. The clusters can be divided, for example, into dense urban, urban, suburban, and rural areas. An initial estimate for the required number of sites per cluster type can be made relatively quickly by using a link budget.

Detailed network planning is done on a site-by-site basis, as assumptions regarding antenna directions, down-tilting, power levels, and so on, vary. A tuned radio propagation model and planning software are used at this phase. Digital maps are important as they take into account the topology of the environment, which greatly affects the local predictions regarding the coverage areas. Field measurements can be used to tune the propagation model parameter settings.

Typically, optimization is divided into prelaunch optimization and postlaunch optimization. After actual deployment, optimization can last until the end of the lifecycle of the LTE network.

The LTE/SAE Deployment Handbook, First Edition. Edited by Jyrki T. J. Penttinen.

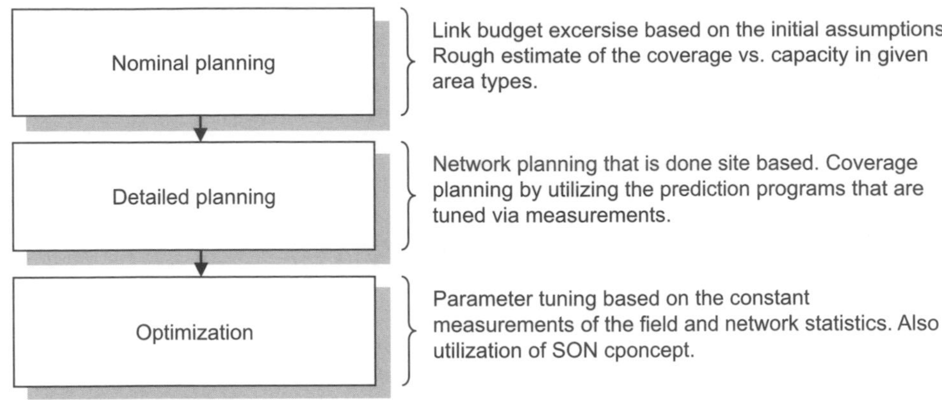
Nominal planning
Link budget excersise based on the initial assumptions
Rough estimate of the coverage vs. capacity in given area types.
Detailed planning
Network planning that is done site based. Coverage planning by utilizing the prediction programs that are tuned via measurements.
Optimization
Parameter tuning based on the constant measurements of the field and network statistics. Also utilization of SON cponcept.

Figure 13.1 The main items of the LTE radio network planning.

The capacity figures change during the operation of the network, which requires the plans to be adjusted. User profiles are investigated constantly at this phase by carrying out regular field tests and by collecting network statistics regarding capacity usage, performance figures, and possible faults.

LTE radio network planning depends heavily on the operator's position in the market. For existing 2G and/or 3G operators, it is essential to reuse the base station sites as much as possible to keep deployment costs at an optimal level. The most logical strategy in this case is to build LTE as an additional layer. At the beginning of LTE operation, the LTE coverage is very limited and it is only possible to provide hot spots of coverage, rather than continuous coverage. This means that the existing 2G and 3G networks take care of the service's continuity. As the LTE network matures, there might be a need to construct or rent additional eNodeB sites. At the same time, existing 2G and 3G capacity could be reduced gradually, providing the possibility for radio frequency refarming—the equipment and radio capacity of previous generations can be reduced while LTE gradually takes more bandwidth from the same frequency according to the bandwidth definitions of LTE, from 1.4 MHz up to 20 MHz. This requires coordination between the radio network planning processes of different systems. Figure 13.3 gives some ideas about the possible system transitions, the 2.1 and 2.6 GHz frequencies being a logical capacity layer solution in the urban environment, and 900 MHz in the rural environment due to its larger coverage areas. Gradual degradation of the earlier systems can occur in such a way

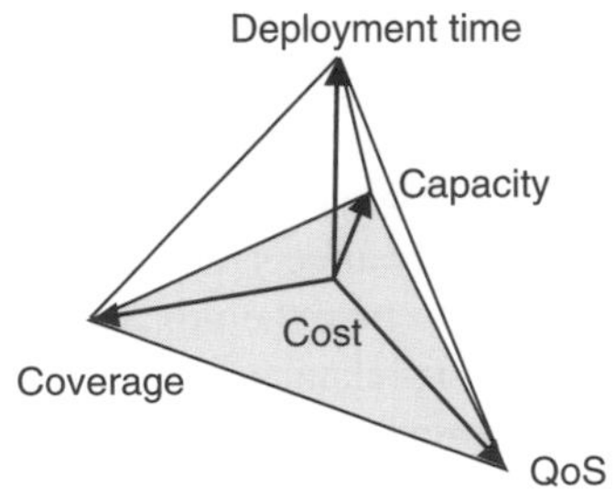
Deployment time
Capacity
Cost
Coverage
QoS

Figure 13.2 The main items that affect on the total cost of the LTE network.

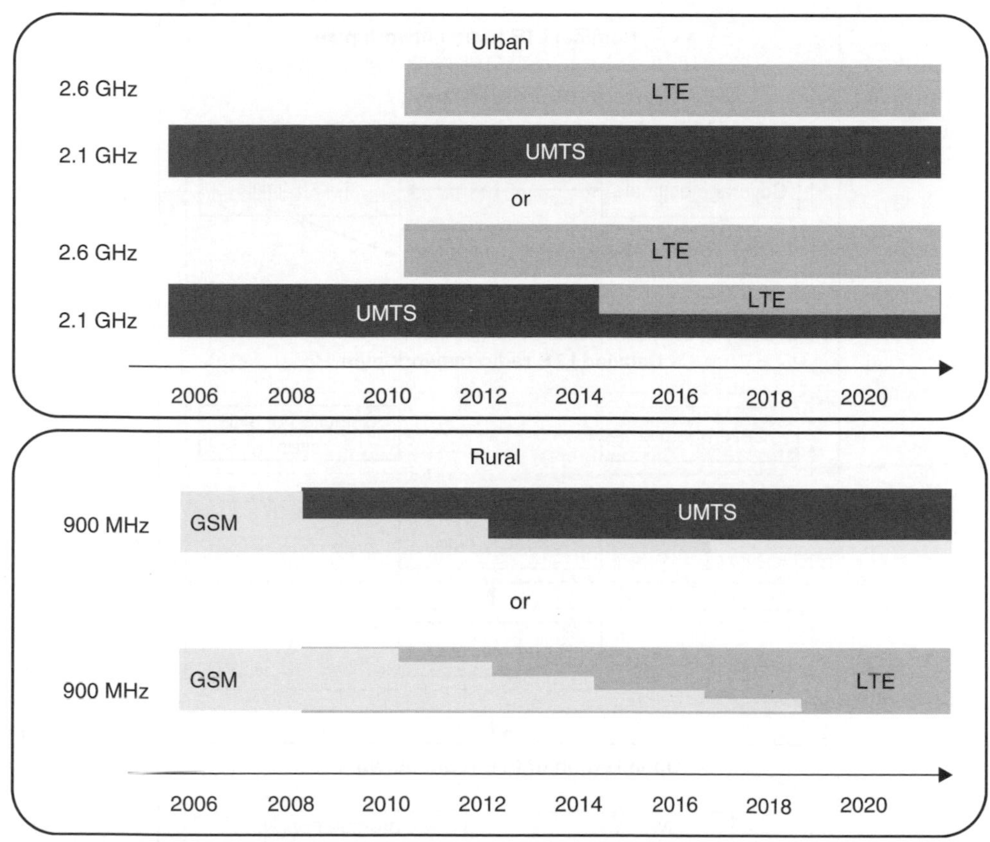

Figure 13.3 Some transition scenarios towards LTE deployment.

that GSM will disappear little by little, and UMTS starts operating at lower capacity, LTE being the parallel system for one of these, until LTE takes over all the traffic. Another realistic scenario could be to remove UMTS gradually while GSM and LTE stay as parallel systems. This latter case is justified due to the vast number of GSM-capable terminals in the market, which provide the basic voice calls and light multimedia whilst LTE/GSM-capable terminals provide the advanced data and part of the voice calls. especially in fallback situations. Advanced spectral efficiency features like GSM's Orthogonal Sub Channel (OSC) and Dynamic Frequency and Channel Allocation (DFCA) support this evolution. More specific examples of transition scenarios are discussed in Chapter 15.

For greenfield operators, the benefit is that the LTE network can be planned optimally way from the start. The drawback is that there is no existing infrastructure for the sites and transmission. Site hunting must thus be executed at an early phase of the planning process by defining the preferred/priority search rings over the planned area. Furthermore, it is essential that there is a constant feedback loop between the planning process and site hunting, as the latter frequently does not produce optimal locations and constant adjustments must be made to locations, antenna heights, and so on. Moreover, the transport and core planning must follow, at least to some extent, progress in radio site hunting before the actual transmission lines or radio links can be constructed physically.

The more detailed radio network task division is presented in Figure 13.4.

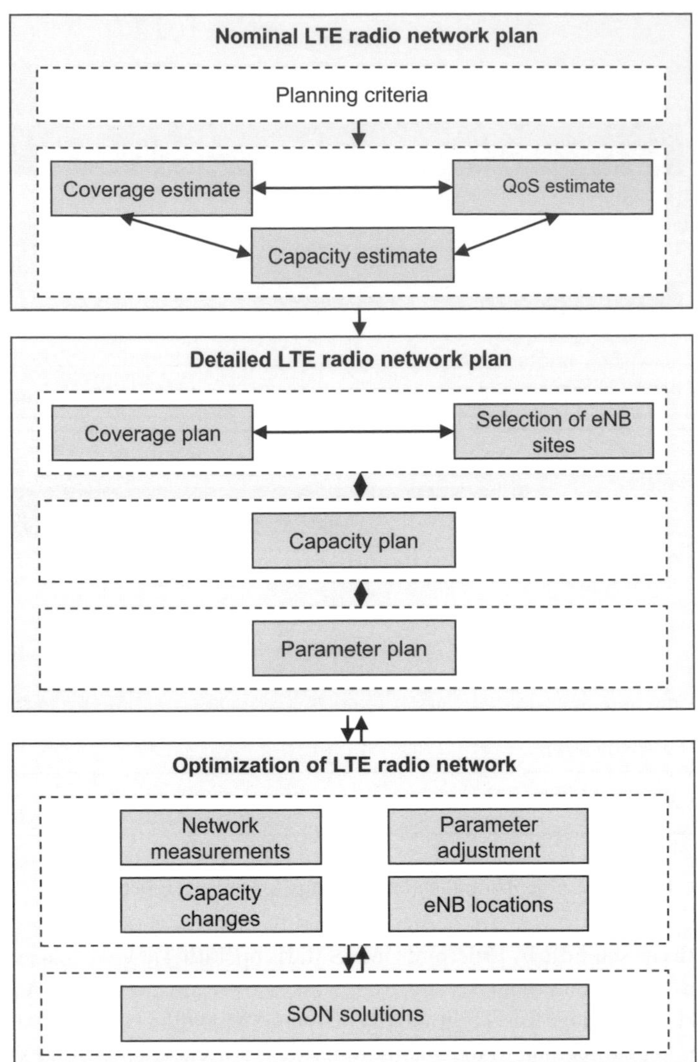

Figure 13.4 The overall LTE radio network planning process.

13.3 Nominal Network Planning

The main aim of the nominal planning phase of LTE is to estimate the required number of eNodeBs as accurately as possible before actual network deployment begins. From this estimate, the total cost of the LTE/SAE network will also be estimated by assuming the required criteria for the coverage, capacity, and QoS. Although this phase does not produce final network architecture plans or more concrete site distributions, it is essential in order to give a realistic understanding about the capital and operational expenditures. The larger the LTE deployment area is, the more important is the accuracy of the estimate of the required number of eNodeBs, and thus the network cost.

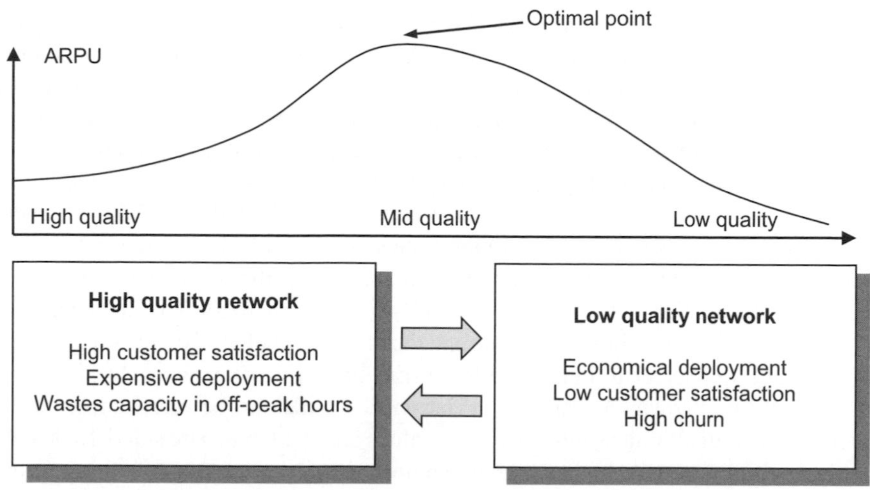

Figure 13.5 The balancing of quality.

13.3.1 Quality of Service

The QoS of the LTE network is better when the blocking of the network is lower. This results in the higher average and peak data rates. Sufficiently high quality balancing is one of the many optimization tasks of the operator: the higher the average throughput is, the more satisfied the customers are—but at the price of higher network deployment costs. It is thus essential to take into account all the relevant technical and nontechnical aspects that affect the average revenue per user (ARPU). Figure 13.5 clarifies the challenges.

The challenge is to find the optimal point that generates the highest ARPU during the lifetime of the LTE network.

Dimensioning of the network is done according to the peak hours and the average load of the network. The load peak inevitably generates lower throughput values per user, or even blocking, when admission control has to limit the entrance of new users. Blocking is designed according to the dimensioning criteria, and taking into account the expected future development of the user load as well as occasional load peaks due to local or special events, and so on. It is important to avoid overdimensioning in order to keep the ARPU close to the ideal level.

The most demanding location for dimensioning is found at the edge of the cells, where the interference level is highest and thus the SINR value is lowest.

The design and deployment of LTE is quite similar to that used for earlier mobile communications systems like GSM and UMTS. The HSPA, in particular, shares many aspects with LTE, both being data solutions for the 3G evolution path.

Similarly to the other general radio network design processes, the radio link budget is the base for the initial planning. It is meant to estimate the maximum allowed functional path loss between the base station and terminal, and it is calculated separately for uplink and downlink. The balancing of the transmission and receiving directions is thus one of the outcomes of the link budget.

The differences between the different transmitting directions are related basically to the different access methods and their minimum power requirements for certain data rates. The useful received power level depends on the presence of the useful and interfering signals compared to the noise level—that is SINR (Signal to Interference and Noise Ratio). The network load does have a direct impact on the SINR value because as the amount of the users grows, the proportion of interference increases. A certain SINR value provides a certain QoS level, giving a location and time-dependent bit rate and bit error rate. As the retransmission rate gets higher, the QoS reduces proportionally, which is experienced directly by the customer. The value also depends on the network load, reducing as capacity utilization increases.

The interpretation of the quality of the coverage area depends on the agreed area location probability level. In general, the location variation is considered to follow a log-normal distribution, which means that the logarithm of the signal level follows a normal or Gaussian distribution. The commonly known mathematical analysis of the respective statistical distribution can thus be applied in the respective quality level estimations. The mean value means that 50% of the samples are above this value and the other half below. For any other percentage for the coverage quality criterion, the relationship between the mean value and standard deviation should be known. A standard deviation of 5.5 dB can be assumed in the typical suburban area by default unless more accurate values are available. The standard deviation is commonly used as a basis for mobile communications coverage predictions, providing information about the confidence levels associated with statistical conclusions.

The relationship between the area location probability and the additional margin that should be taken into account in the link budget can be thus derived from the characteristics of the normal and log-normal distribution, for example by observing the attenuation points (dB) in a cumulative scale that fulfills the required percentage of the area location in the whole area. Furthermore, it can be decided that 90% of area location probability indicates a fair outdoor coverage, whereas 95% is considered as good and 99% provides excellent quality. Table 13.1 summarizes the mapping of the typical values that can be used in the mobile reception, when the standard deviation is 5.5 dB. In addition to the standard deviation, the criteria vary depending on the environment—that is on the propagation slope. A slope of 2 (i.e., 20 dB/decade) represents line of sight in free space. A slope of 3.5 (i.e., 35 dB/decade) is used in Table 13.1 representing a typical urban environment.

The OFDMA of LTE downlink is especially suitable for environments with heavy fast-fading of the radio signals, for example in densely urban areas where the number of multipath propagated radio components is high. The division of the original information into many lower bit rate subcarriers over a wide frequency band ensures that only a small set of subcarriers is lost. The spreading of the data (interleaving) over the band, combined with turbocoded data, protects the data fairly well as it can be recovered efficiently on the reception side.

The drawback of OFDMA is the challenge of the SFN network's Peak-to-Average-Power Rate (PAPR) variations, combined with the nonoptimal power efficiency. For this reason, the transmitter of the LTE-UE, that is in the uplink direction uses more energy-efficient SC-FDMA (Single Carrier Frequency Division Multiplexing).

Table 13.1 The area location probability in the site cell edge and over the whole site cell area for the mobile reception when the standard deviation is 5.5 dB

Area location prob. (minimum coverage target)	Loc probability in site cell edge	Location correction factor	Subjective quality description
90%	70%	7 dB	Fair outdoor
95%	90%	9 dB	Good outdoor, fair indoor
99%	95%	13 dB	Excellent outdoor, good indoor

The combination of OFDMA and SC-FDMA provides coverage that is comparable with previous HSPA networks, the difference being that there are higher data rates in LTE.

One of the typical scenarios is that the operator that already has a 2G and/or 3G license adds LTE service areas gradually as a part of the complete set of network technologies. This means that, especially in the initial phase of LTE deployment and operation, the proportion of LTE coverage is relatively small. The strategy may be to provide LTE coverage only in the most loaded hot spots of the data transmission, providing fast data services and partially load balancing using the Voice via VoIP concept. The fallback solution is thus used largely for handing over already established data and voice calls from LTE, providing that the LTE-UE supports the 2G and/or 3G technologies. It can be assumed that the LTE traffic type is heavily data weighted in the beginning with the data dongles (VoIP typically being possible via Internet and laptop using the headset), and the VoIP service will takes off gradually as there will be more handset models with integrated voice call facilitators (microphone and earphone integrated into the LTE-UE).

An important part of QoS is the continuity of the LTE service. Handover functionality is thus important. The new X2 interface between the eNB elements optimizes the success rate of handovers in LTE compared to previous techniques, where the direct signaling connection between the base stations was missing. In situations where LTE coverage ends but 2G/3G coverage is still available, the connection changes automatically to 2G/3G networks. This does have an impact on the QoS level as the overall data throughput will probably reduce in this case. For the VoIP connection that is changed to 2G/3G CS call, the impact is not dramatic if the fall-back procedure is executed successfully (in fact, the CS voice call quality might be better than LTE VoIP call quality), but the greatest effect will be seen in high-speed data utilization. The applications that are affected most are related to the (close-to) real-time data streaming, when capacity is lower that in LTE networks. In any case, continuous service, even with lower quality, can be considered more important than the breakdown of the call.

13.4 Capacity Planning

The capacity of LTE depends on the services used, the required bit rates per user, as well as the quality of service level. Capacity is related to the bandwidth used. Modulation and coding schemes have a direct relationship with capacity in such a way that, at the cell edge, the capacity offered is lowest due to the highest coding rates and most robust modulation method (QPSK). Nearer the eNodeB, the 16-QAM and 64-QAM can be used with a lower code rate, which facilitates more efficient use of resources for the actual user data. The modulation and coding scheme of LTE is adaptive, which means that the optimal combination is selected. As the interference level of the LTE network increases as a function of the utilization level, the bit error rate also increases, which lowers the throughput. This means that a similar type of "cell-breathing" effect can occur in LTE as occurs in W-CDMA because the cell coverage varies as a function of the number of calls.

As the capacity of LTE depends on various aspects that vary as a function of time and location, one possibility to estimate capacity behavior is to simulate the most probable use cases, assuming a certain distribution of application profiles. Simulation results can show the distribution of the modulation and coding schemes over the simulated area, taking into account interference from neighboring cells. This gives information about expected SINR values and respective throughputs with the given bit error rate, which can be, for example 10%. Performance depends on the distance between the eNodeBs, antenna heights, radiating power levels, and in general on the type of surrounding area and its topology. The most robust

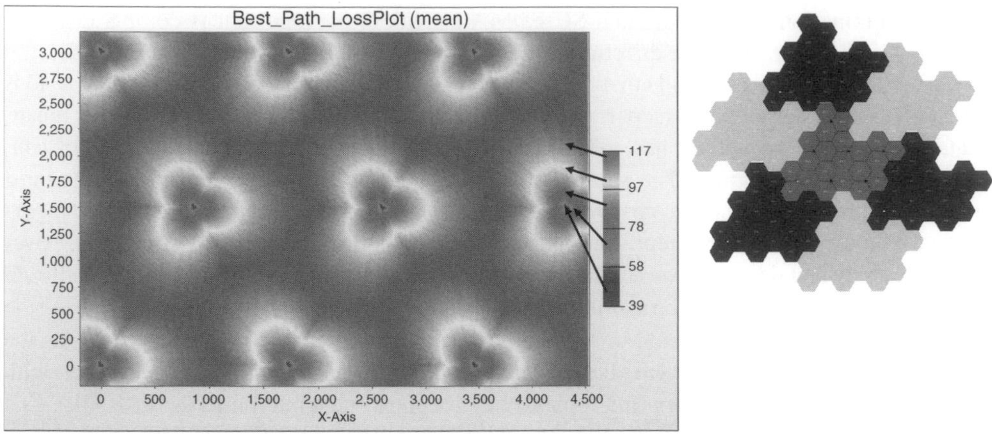

Figure 13.6 An example of a snapshot simulation of OFDM in theoretical (flat) suburban type. The estimate becomes more realistic when the digital map data is taken into account in the path loss prediction.

modulation of LTE, QPSK, typically requires about 0–3 dB SINR, whereas 16-QAM functions with a minimum of about 7–11 dB, and 64-QAM with 12–15 dB, when the coding rate varies between the highest and lowest possible.

The planning of the LTE radio network and the process in general is quite similar to planning used in the case of HSPA. Both use the same frequency planning principles with a reuse pattern size of 1, which means that the users share the same frequency band for the communications. The radio resource management principles are also similar.

The LTE network's capacity and coverage thus varies. In order to estimate the effect, a planning tool is required. There are different ways to model the cases. The basic principle is to assume different data usage cases over the whole investigated area in such a way that the average throughput per user, as well as the variations can be seen via the distribution, for example in the cumulative presentation. The most challenging part is, in fact, to make realistic assumptions regarding traffic types. For this, earlier data from UMTS data services can be used as a basis, in order to guess the possible share of, for example the VoIP, WEB, FTP down/uploading, messaging, and other traffic types.

The performance indicator can be, for example, throughput per user or spectral efficiency as a function of time, SINR, or some other variable. The well known Monte-Carlo method is practical and relatively straightforward, when the static snapshots are repeated sufficiently. The simulation can also be dynamic, although it is more complicated as the mobility of each user has to be modeled. This requires more time and processing power, so the dynamic model is only useful for in-depth investigations of radio research management functionality, whereas Monte-Carlo is useful in general LTE network planning.

Figure 13.6 shows an OFDM snapshot exercise over a theoretical suburban-type area when the path loss has been estimated by taking into account a standard deviation of 5.5 dB. In this specific case, Okumura-Hata has been applied.

13.5 Coverage Planning

The LTE coverage area can be estimated in the simplest way via path loss prediction, although it is much more limited. This method does not reveal the effect of the traffic-type distribution on coverage but it gives an adequate first-hand estimate about the general coverage that can be achieved with a certain offered capacity case (average throughput).

In the very initial phase of planning, even a generalized link budget exercise can be executed in order to estimate the rough, average LTE cell range per area type (dense urban, urban, suburban, rural, or open area). By assuming a certain overlapping percentage for the adjacent cells, this exercise provides the initial estimate of the number of the eNodeB sites required over the planned area.

The coverage planning of the LTE radio network is actually technology dependent. Thus existing radio propagation models can also be used in LTE planning whenever they are designed for the appropriate radio frequencies. The radio path loss prediction per area type gives the estimate for the expected radius of the eNodeB. In order to estimate the maximum allowed path loss, a power link budget can be used by applying the same principles as in the HSPA radio network planning.

13.5.1 Radio Link Budget

The rough estimate for initial coverage planning can be obtained by using the radio link budget. Figure 13.3 shows the general level link budget of LTE. As can be seen, it is quite similar to those of other mobile communications systems.

The link budget gives a relatively good overall and average idea about the maximum achievable path loss between the eNB and LTE-UE. This, in turn, provides an estimation of the average cell radius of the sites. The estimate can be obtained for different environment types like dense urban, urban, suburban, rural, and open area.

The link budget gives the path-loss estimate both for the downlink and the uplink, which makes it possible to balance the transmission and reception of the data streams according to selected criteria. The balance is important in order to provide planned QoS in both directions. In the case of LTE, the importance of balance depends on the application. As an example, for the voice service via the VoIP solution, it is important to provide the minimum needed data flow in both directions in order to have a successful call between subscribers in the cell edge area. For the downloading of, for example, web pages, the data speed in the uplink is not the limiting factor whilst the bit error rate is sufficiently low, so even slow data rates in uplink provides the required acknowledgements for fast data download in the downlink.

The differences between the downlink and the uplink in LTE are related to the different access methods (OFDMA in downlink and SC-FDMA in uplink) and their respective requirements for minimum functional values. The final useful power level depends on the balance of the useful carrier signals and the interfering signals, which is compared to the SINR. A certain SINR value provides a certain quality of service and respective data rate, which depend on the time and location, especially in the case of multi-path propagated radio signals, which cause slow and fast fading as a function of time and location. After all, end-users interpret the quality of the connection based on the useful throughput of the data flow. This takes into account the bit error and respective retransmission, which lowers the useful bitrates even if the erroneous bit rate would be much higher.

The LTE radio link budget is fairly similar to the link budgets used in the other mobile technologies, especially in the case of the HSPA. It is quite useful for nominal planning of the LTE radio network, as it provides a general estimate for the average cell ranges and thus the required number of eNBs within the planned area.

Figure 13.7 shows the principle of the LTE radio link budget. The most important factors in the downlink direction are the output power level of the eNB transmitter (Ptx), the cable and

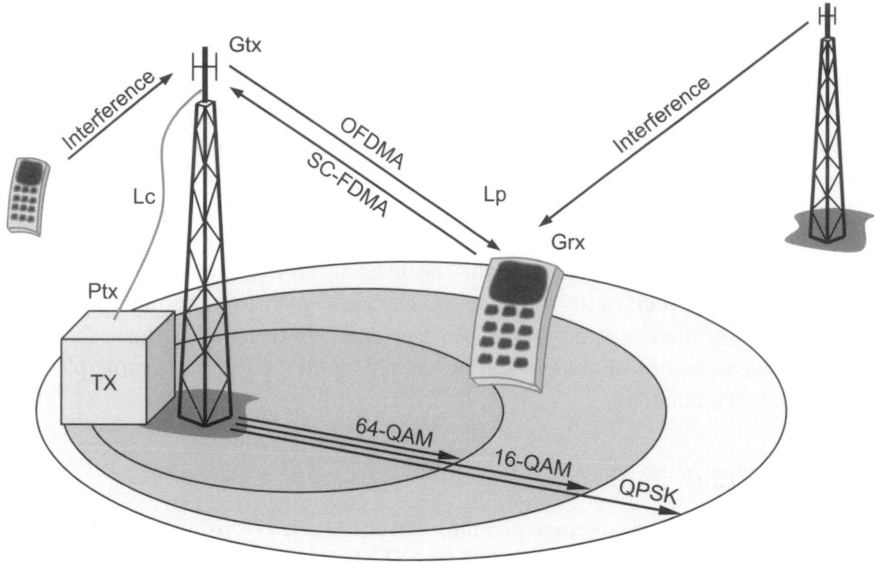

Figure 13.7 The principle of LTE radio link budget.

connector losses (Lc), and the transmitting antenna gain (Gtx). It is worth noting that the active antenna systems (AAS) may become more popular along with LTE deployment. As the AAS contains the front end of the transmitter inside the same antenna element shield, the transmitted content can be delivered from eNB up to the AAS by fiber optics, which eliminates the cable losses.

The radiating power (EIRP, Effective Isotropic Radiating Power) is distributed via the air interface to LTE-UE, which has a certain sensitivity (S), receiver antenna gain (Grx) and noise figure. It should be noted that, in case of the small-sized USB stick –type LTE terminals, the inbuilt antenna causes loss instead of gain. External antennas can be used with LTE terminals, but for the link budget purposes the most likely terminal type should be taken into account, which probably means that inbuilt antennas should be assumed with, for example 0 dB antenna gain or even as slightly negative.

In the uplink direction, the most important factors are the transmitting power level of the LTE-UE, the antenna gain of the terminal and eNB (which are likely the same as in downlink if the same antenna types are used), the cable and connector losses and the sensitivity of the eNB receiver.

The maximum allowed path loss value, L, is calculated for different modes (modulation schemes) by taking into account the minimum received power level requirement for the modes, and by subtracting the emitted and received powers. The estimate is made separately for the outdoors, and indoors by assuming a certain area-type dependent average building penetration loss.

As a rule of thumb, the QPSK modulation provides largest coverage areas, but at the same time it results in the lowest capacity. The 64-QAM modulation gives highest data rates, but only in limited areas compared to the QPSK case. The third possible LTE modulation, 16-QAM, is a compromise between coverage and capacity. The least protected transmission (highest code rates) provide the highest data rates but within smaller coverage areas than the heavier protected codes do. In the accurate coverage and capacity estimates, the respective areas can be calculated separately. LTE includes automatic modulation and coding adaptation (MCS, Modulation and Coding Scheme), which provides the optimal combination of the modulation and code rate at all times. In OFDM, the adaptation could even be done on a timeslot basis but for the practical reason of keeping the signaling load at an optimal level, LTE defines the minimum resolution for the link adaptation to radio block. In practice, this provides a fast enough adaptation for any practical situation.

Well known radio path-loss prediction models can also be applied for LTE coverage estimations when the minimum required power level per investigated mode is known. If the exact practical values are not available, simulation results can be used as a base for the minimum requirements. When the single-frequency network concept is used, it is possible to add a certain SFN gain value for the link budget. It basically means that the eNBs can be constructed slightly further away from each other, or alternatively, the reception level is of higher quality in the cell edge areas than it would be without SFN mode. This concept is valid, for example, for the Multimedia Broadcast concept of LTE (MBSFN).

The link budget calculation is typically based on a minimum throughput requirement at the cell edge. This approach provides the cell-range calculation in a straightforward way. In order to define the throughput requirement prior to the link budget calculation, it is possible to estimate the bandwidth and power allocation values for a single user. This estimate is sufficient for the preliminary radio link budget calculation purposes.

The LTE link budget can be designed by planning sufficiently the essential parameter values shown in Table 13.2 for the downlink, and in Table 13.3 for the uplink. The latter assumes 360 kHz bandwidth utilization whereas the downlink assumes a 10 MHz band for the transmission.

Table 13.2 LTE radio link budget example in the downlink direction

Downlink		
Transmitter, eNodeB	Unit	Value
Transmitter power	W	40.0
Transmitter power (a)	dBm	46.0
Cable and connector loss (b)	dB	2.0
Antenna gain (c)	dBi	11.0
Radiating power (EIRP) (d)	dBm	55.0
Receiver, terminal	*Unit*	*Value*
Temperature (e)	K	290.0
Bandwidth (f)	Hz	10 000 000.0
Thermal boise	dBW	– 134.0
Thermal Boise (g)	dBm	– 104.0
Noise figure (h)	dB	7.0
Receiver Boise floor (i)	dBm	– 97.0
SINR (j)	dB	– 10.0
Receiver sensibility (k)	dBm	– 107.0
Interference margin (l)	dB	3
Control channels share (m)	dB	1.0
Antenna gain (n)	dBi	0.0
Body loss (o)	dB	0,0
Minimum received power (p)	dBm	– 103.0
Maximum allowed path loss, downlink		158.0
Indoor loss		15.0
Maximum path loss for indoors, downlink		143.0

Table 13.3 The LTE radio link budget in the uplink direction

Uplink		
Transmitter, terminal	Unit	Value
Transmitter power	W	0.3
Transmitter power (a)	dBm	24.0
Cable and connector loss (b)	dB	0.0
Antenna gain (c)	dBi	0.0
Radiating power (EIRP) (d)	dBm	24.0
Receiver, eNodeB	*Unit*	*Value*
Temperature (e)	K	290.0
Bandwidth (f)	Hz	360 000.0
Thermal boise	dBW	– 148.4
Thermal Boise (g)	dBm	– 118.4
Noise figure (h)	dB	2.0
Receiver Boise floor (i)	dBm	– 116.4
SINR (j)	dB	– 7.0
Receiver sensibility (k)	dBm	– 123.4
Interference margin (l)	dB	2
Antenna gain (m)	dBi	11.0
Mast head amplifier (n)	dB	2.0
Cable loss (o)	dB	3.0
Minimum received power (p)	dBm	– 131.4
Maximum allowe path loss, uplink		155.4
Smaller of the path losses		155.4
Indoor loss		15
Maximum path loss in indoors, uplink		140.4
Smaller of the path losses in indoors		140.4

The link budget can be planned in the following way, for example in the downlink direction with the 10 MHz assumption as shown in the example above. The radiating isotropic power (d) can be calculated by taking into account the transmitters output power (a), the antenna feeder and connector loss (b), and transmitter antenna gain (c), so the formula is: $d = a - b + c$.

The minimum received power level (p) of the LTE-UE can be calculated as follows: $p = k + l + m - n + o$, using the terminology of the link budgets shown above. The noise figure of the LTE-UE depends on the quality of the model's HW components. The minimum signal-to-noise (or, signal-to-noise and interference, SINR) value j is a result of the simulations presented in [1]. The sensitivity k of the receiver depends on the thermal noise, the own noise figure of the terminal, and SINR in such a way that $k = g + h + j$.

The interference marginal l of the link budget represents the average estimate of the noncoherent interference originating from the neighboring eNodeB elements. The control channel proportion m also degrades the link budget slightly. In the link budget calculations, the effect of the antenna of the LTE terminal can be estimated at 0 dB if no body loss is present near the terminal. In case of the external antenna, the antenna gain increases respectively, but the logical estimate of the average terminal type is a stick model with only its own inbuilt antenna.

The effect of the data speed can be estimated roughly by applying a rule of thumb that assumes a 160 dB path loss in uplink with 64 kbps data rate. Whenever the bit rate grows, the maximum allowed path loss drops respectively. A simple and practical assumption is that the

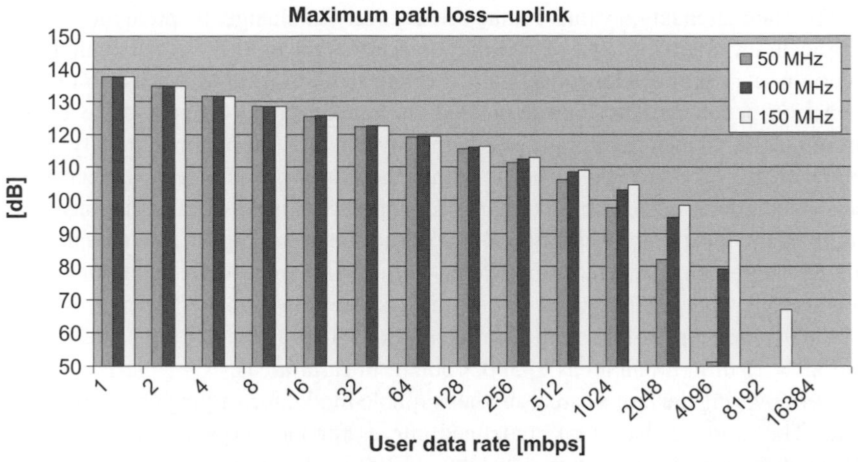

Figure 13.8 A theoretical estimation for the effect of the data rate on the uplink path loss.

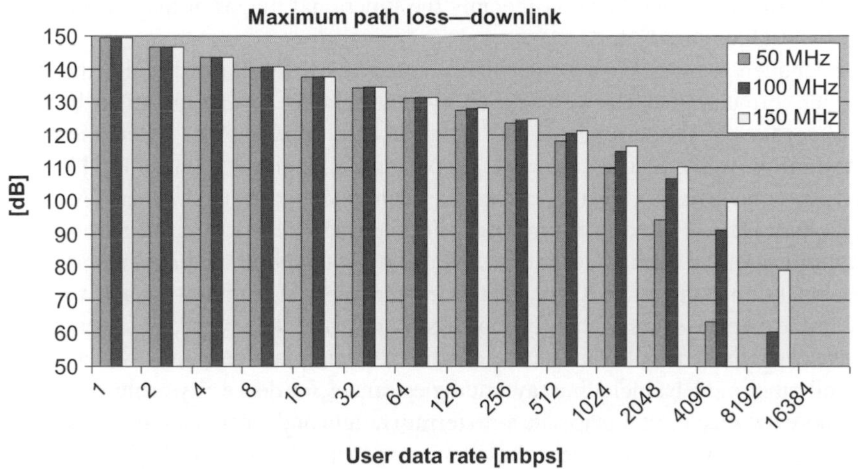

Figure 13.9 The effect of the data rate on downlink.

doubling of the data rate increases the path loss by 3 dB. This can be assumed to work adequately unless the channel coding or modulation scheme changes.

Figures 13.8 and 13.9 show the theoretical calculation for the effect of the data rates.

13.5.2 Radio Propagation Models

Widely known path-loss prediction models include Okumura–Hata and Walfisch–Ikegami, which are also suitable for LTE radio network planning when the functional ranges of the antenna height, operating frequency and maximum predictable cell range are taken into account.

The selection of a suitable radio propagation model in the initial phase of radio network planning depends on the LTE frequency used, the terrain, and the level of accuracy that is wanted. In the early phase of deployment it is essential to validate the functionality and accuracy of the propagation models, in order to do the respective model tunings, for example as for the cluster types and attenuation value. It may be necessary to change the prediction model if the resulting accuracy is still low. One of the limitations is that the radio network planning tools have only a certain set of propagation models, and the digital map data used has also its own effect on the prediction accuracy. The typical method for the coverage estimate is to estimate the maximum functional path loss for the case being investigated (throughput, quality) by defining subregions of, for example 100 × 100 m. Within these areas, a sufficient number of snapshot path-loss calculations are done by taking into account the realistic signal variations of slow and fast fading. If the location probability of this specific subarea is higher than the quality wanted requires, the subregion is selected as functional—otherwise it represents outage of the coverage.

When seeking high accuracy, the size of these subregions should be small, which results logically in the longer path-loss prediction calculations. In practice, the typical raster sizes vary between 25 × 25 m in urban areas to 500 × 500 m in rural areas.

For cases covering the largest area, another suitable model for varying environments is ITU-R P.1546. The model is based on predefined curves for the frequency range of 30 MHz to 3000 MHz and for maximum antenna heights of 3000 m from the surrounding ground level. The model is valid for terminal distances of 1 to 1000 km from the base station over the terrestrial and sea levels, or for the combination of these. This model is especially suitable for large-area estimates, extending considerably the functional ranges of the antenna height and cell radius of Okumura–Hata.

If the frequency or antenna height investigated does not coincide with the predefined curves, the correct values can be obtained by interpolating or extrapolating the predefined values according to the annexes of the model and by applying the calculation principles presented in its Annex 5. The case curves represent field strength values for 1 kW effective radiated power level (ERP), and the curves have been produced for the frequencies of 100 MHz, 600 MHz and 2 GHz. The curves are based on empirical studies of propagation. In addition to the graphical curve format, the values can also be obtained in tabulated numerical format.

It can be assumed that the basic and extended versions of Okumura–Hata as well as ITU-R P.1546-3 models provide a sufficiently good first-hand estimate for the LTE coverage areas and respective capacity and quality levels in the initial network planning phase. Nevertheless, there are several other models, including ray-tracing estimates for dense city centers. These models require more detailed digital map data with terrain height and cluster attenuations. In the most advanced prediction models, a vector-based 3D map is needed. It logically has a cost effect on the planning but it increases the accuracy of the coverage estimate considerably. It can further be enhanced via local reference measurements by adjusting the model's estimate accordingly. As a cost-efficient compromise, 3D models could be used in the advanced phase of the radio network planning in the most important areas.

13.5.3 Frequency Planning

The LTE radio network can be constructed in the same frequency band—that is, the reuse pattern size 1 can be used. As in the case of W-CDMA, it produces a certain level of interference. The benefit of this solution is that the users can take full advantage of the high bandwidth with respective peak data rates.

Another solution is to divide the available LTE frequency band into smaller blocks in order to create higher reuse pattern sizes. As an example, if the operator has been granted a total of 15 MHz of bandwidth for LTE, it can be used completely for all the sites. This option provides with the highest possible data rates, although the average data rate suffers at some extend due to the inter-cell interferences when other users are present. If the 15 MHz block is divided into

5 MHz slices, it makes it possible to use three different frequencies per sector. This means that the reuse size 3 can be applied, and the interference can effectively be reduced. Nevertheless, the peak data rate per user is now only one-third compared to the services offered via the full frequency band of 15 MHz.

The simulations have shown that despite the increased interference levels, the first case, that is reuse size of 1, provides higher capacity. Based on this result, a very deep level neighboring plan is not required in LTE radio network planning. In any case, the overlapping proportions of the network can be optimized by carefully planning antenna tilting and the balance between the site distances and power levels.

13.5.4 Other Planning Aspects

The tracking area planning is one of the items that has an impact on the capacity of the LTE network. If the tracking areas are too small, it results in increased signaling when the LTE-UEs are moving over the tracking area borders. On the other hand, if the tracking area sizes are defined too large, the paging signaling is unnecessarily high and affects on the network capacity. The optimal size of the tracking areas is thus one of the many optimization tasks to be carried out by the operator in order to find a proper balance for the signaling load.

There are some rules of thumb. The tracking areas of LTE could be defined by default as the same as the location areas of 2G, and routing areas of GPRS, if the operator has previous mobile communications infrastructure. It can be assumed that the existing network has had time to mature and a sufficiently good balance of the LA/RA has been found, which can be used for the LTE TA.

Another guideline is that the tracking area border should not be defined in an area that contains heavily moving LTE-UEs, as is the case with motorways.

13.6 Self-Optimizing Network

With the deployment of LTE networks, there will be a clearer need for the Self-Organizing/Optimizing Network concept (SON). The SON concept refers to the complete and automatic optimization of the network via several functions. In fact, it can be claimed that the full spectral efficiency of the LTE network can only be achieved when SON is applied partially or with all possible functionalities. The use of SON saves time and resources in responding effectively to the changing user load profiles, faults, and other dynamically changing phenomena in the network. This, in turn, leads to the cost savings by reducing the expenses both in the deployment (savings of CAPEX, Capital Expenditure) and in the operational phase of the LTE network (savings of OPEX, Operational Expenditure).

SON is a relatively wide concept, and it has been standardized as a joint effort of 3GPP and NGMN Alliance (Next Generation Mobile Networks Alliance). As a result of this common work, there is already a whole set of SON features specified under the 3GPP Release 8 and 9 documents.

The main focus of SON is to minimize the physical work of the technical personnel—to reduce the eNB and element-based tuning that traditionally requires time and effort, including the physical traveling to the site, manual measurements of the faulty elements, software updates, frequency changes, and so on. The list of the enhancements is relatively long, and it logically takes time to design and apply these functionalities to the LTE network. It should also be remembered that the highly skilled technical personnel cannot be replaced completely by the use of automatic devices and software. The SON concept itself also requires management and tuning, and there is always a part of the trickiest problems that requires human intervention

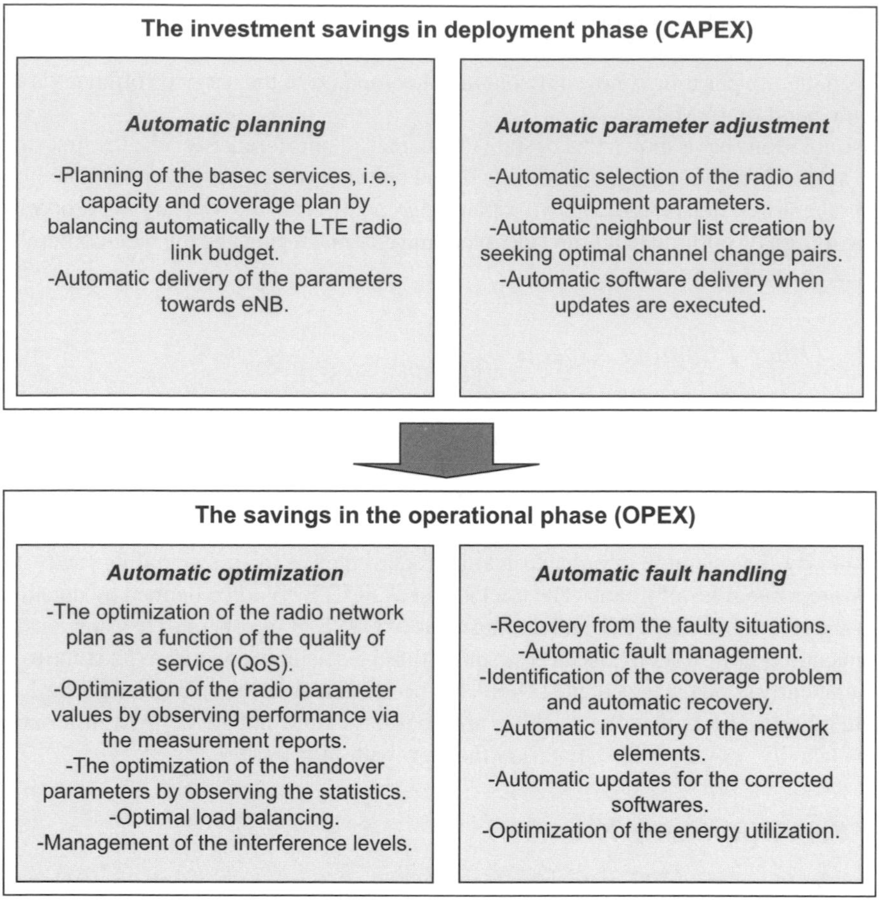

The investment savings in deployment phase (CAPEX)

Automatic planning

-Planning of the basec services, i.e., capacity and coverage plan by balancing automatically the LTE radio link budget.
-Automatic delivery of the parameters towards eNB.

Automatic parameter adjustment

-Automatic selection of the radio and equipment parameters.
-Automatic neighbour list creation by seeking optimal channel change pairs.
-Automatic software delivery when updates are executed.

The savings in the operational phase (OPEX)

Automatic optimization

-The optimization of the radio network plan as a function of the quality of service (QoS).
-Optimization of the radio parameter values by observing performance via the measurement reports.
-The optimization of the handover parameters by observing the statistics.
-Optimal load balancing.
-Management of the interference levels.

Automatic fault handling

-Recovery from the faulty situations.
-Automatic fault management.
-Identification of the coverage problem and automatic recovery.
-Automatic inventory of the network elements.
-Automatic updates for the corrected softwares.
-Optimization of the energy utilization.

Figure 13.10 The high-level idea of the self-optimizing network concept.

to be measured, analyzed, and fixed. In any case, the work has been initiated and it can be expected that the self-optimizing functionalities will reduce the costs of LTE compared to basic solutions. Figure 13.10 shows the basic idea of the SON for the initial and operational phases of the LTE network.

It can be generalized that an aim of SON is to provide a certain level of "plug-and-play" functionalities for LTE operators. A benefit of the SON concept is that it has been standardized and thus the solutions are global for the LTE system. This provides guaranteed support for the concept for the wide deployment. SON would provide much faster deployment and parameter tuning of the LTE network as the proportion of manual work is lower than in previous mobile network environments. This also means that, instead of routine tasks, for example in parameter adjustments, skilful technical personnel can concentrate on more demanding problem solving and on the deeper network optimization that inevitably still remains even with the SON concept.

Reference

[1] Holma, H. and Toskala, A. (2011). *LTE for UMTS. Evolution to LTE-Advanced*, 2nd edn, John Wiley & Sons, Ltd, Chichester.

14

LTE/SAE Measurements

Jonathan Borrill, Jyrki T. J. Penttinen, and Luca Fauro

14.1 Introduction

The same principles that were applied in the 2G and 3G networks can be used in LTE/SAE measurements. Nevertheless, the special characteristics of the evolved network, the more advanced performance, and the more complicated signal processing, require new capabilities from the measurement equipment.

14.2 General

LTE/SAE measurements are needed both in the core (EPC) and the radio (LTE) parts of the evolved packet system as shown in Figure 14.1. LTE terminals also need special attention, particularly in the development phase of the equipment. Field and network measurements are essential but the operational phase of the LTE/SAE network requires regular performance revisions. This can be done with field test equipment as well as by using the new LTE/SAE-specific network statistics measurements directly from the network elements. Network equipment providers tend to add these measurements for basically all of the required types of statistics, including the call-drop rate, blocking rate, and radio resource error rate. Furthermore, the key performance indicators are an essential part of the performance monitoring of the networks, with the possibility to create operator-specific formulas for special measurements [1].

14.2.1 Measurement Points

The possible measurement points of LTE/SAE can be seen in Figure 14.2.

14.3 Principles of Radio Interface Measurements

Radio interface measurements are essential in the deployment and operational phases of the LTE network. The following chapters describe the special characteristics of LTE and the aspects of the measurements.

The LTE/SAE Deployment Handbook, First Edition. Edited by Jyrki T. J. Penttinen.

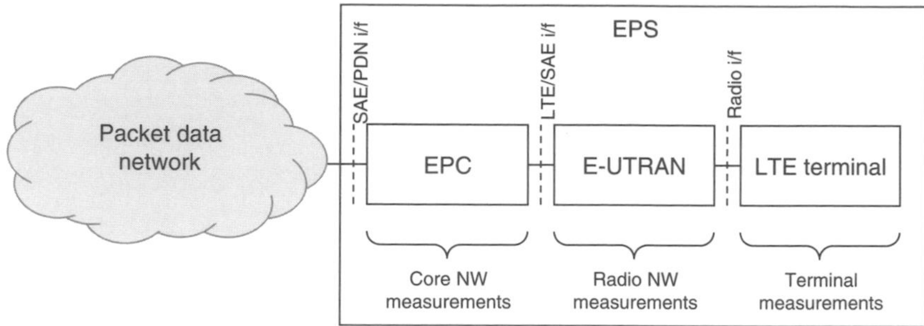

Figure 14.1 The main points for the interface measurements between the core network, the radio network and the air interface.

14.3.1 LTE Specific Issues for the Measurements

The LTE radio interface is based on Orthogonal Frequency Division Multiplexing (OFDM) in the downlink, and on Single-Carrier, Frequency Division Multiple Access (SC-FDMA) in the uplink. This solution provides a scalable bandwidth selection between 1.4–20 MHz. In addition, LTE support both Frequency Division Duplex (FDD) and Time Division Duplex (TDD) modes.

Figure 14.3 shows the principle of the LTE downlink transmission in the time domain. Each resource element corresponds to a single OFDM subcarrier over a period of one OFDM symbol. The channel separation of downlink is defined to 15 kHz in the normal mode of the transmission. In addition, the broadcast mode of LTE can also use a 7.5 kHz raster size. In the uplink, LTE uses SC-FDMA with a scalable band.

LTE functions within a maximum of 20 MHz bandwidth—which is four times larger than was previously the case in the UMTS system—it creates special requirements for the measurement equipment. The HW and SW of the measurement equipment should be able to interpret all the received radio parameters and their values, as well as the performance figures of the measured connection. The latter includes, for example, data transmission speed (mbps),

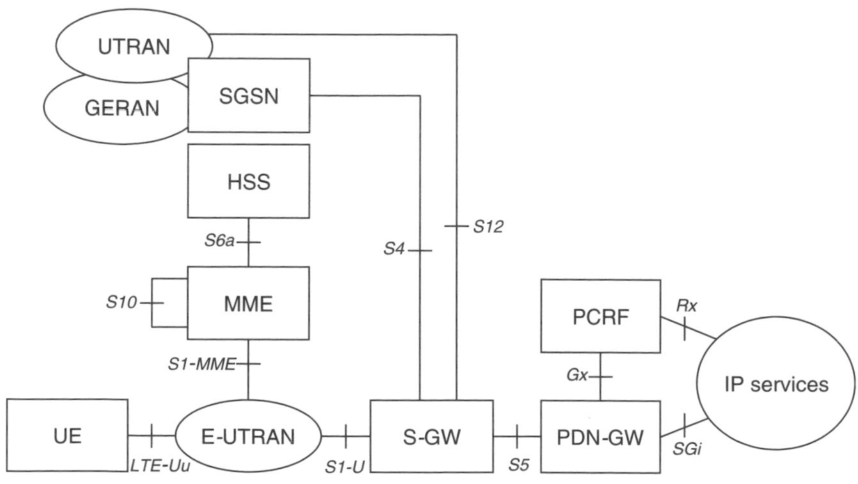

Figure 14.2 The more specific network scheme for the possible LTE/SAE interface measurement points.

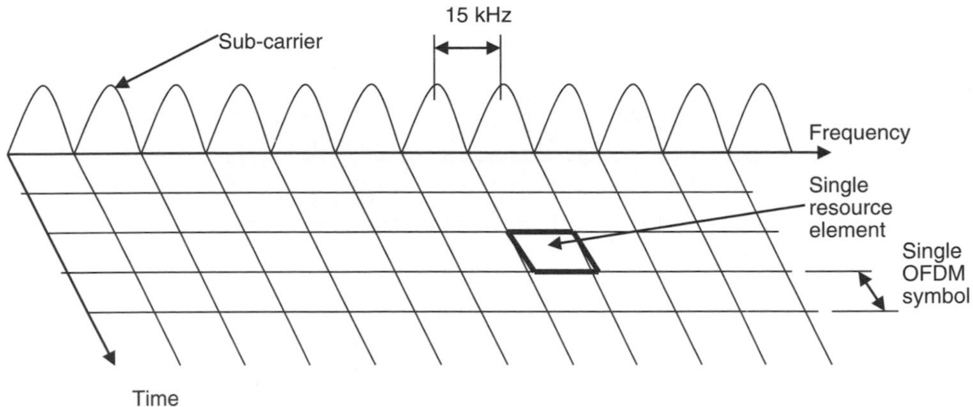

Figure 14.3 The principle of the OFDM downlink.

the latency of the data transmission (s), the jitter, that is the variations of the latency, bit error rate (%), modulation error rate (%), and frame error rate (%).

The measurement needs of the LTE depend naturally on the development phase of the network. In the initial phase of the system, when the equipment is not yet necessarily partially or completely available, LTE simulator is a highly useful tool for estimating the network performance under the planned circumstances. It can be used for the production of performance figures for different traffic scenarios and user profiles, including the combination of different profiles with varying uplink and downlink data transmission by different applications, for example FTP upload and download, web browsing, VoIP traffic with varying codec qualities, and e-mail traffic. In this way, the functionality and performance can be estimated in advance under normal and stressed conditions, to make sure the selected capacity and other technical characteristics of the network comply with the estimated traffic.

The coverage areas can be estimated as a function of the capacity and quality. One of the most useful criteria is the bit error rate as a function of the radio channel type, which can be varied between the line-of-sight types of AWGN (Additional White Gaussian Noise) to highly multipropagated environments in dense city centers, which produce Rayleigh fading as shown in Figure 14.4. The error margin of the estimated performance depends on the accuracy of the simulator's channel models, which can be enhanced via practical measurements.

In the beginning of system introduction, in the early phase of element production, a base station tester is a typical example of the equipment that is needed in the type approval processes. Equally importantly, type approval has to be carried out for the new terminal types. The quality of both hardware and software of the radio equipment needs to be measured in order to avoid problems in the early deployments of the networks. The tester equipment can be integrated into the complete measurement centers, which include a large set of tool libraries for the development and execution of the test cases. The tester package typically includes signal generators and RF test equipment meant for the laboratory conditions.

The measurement equipment can be modular, including a tester platform for LTE and other mobile technologies, in order to ensure common interworking of equipment from different systems in accordance with the standards' definitions. This type of measurement center may include third-party measurement equipment that is integrated with a common control center that takes care of the management of the separate parts of the measurements. This type of solution is especially useful in terminal testing and pre-type approval testing.

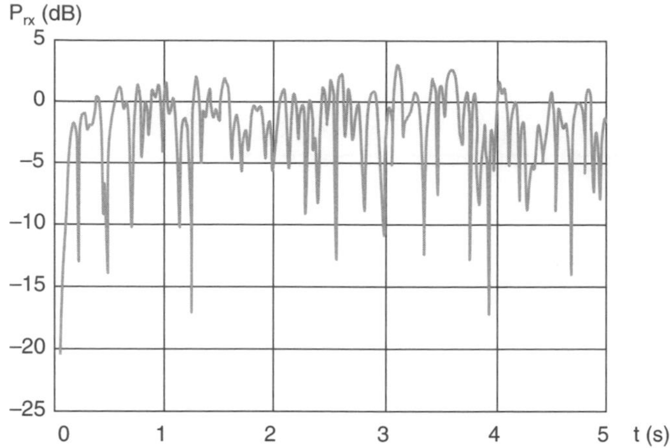

Figure 14.4 An example of the measurement of the received power level when Rayleigh fading is present.

In the implementation phase of LTE, it is still possible to use, for example, LTE signal generators and signal analyzers. Both LTE network and terminal interoperability testing between different equipment vendors can be carried out in laboratory conditions by using signal generators or, in a more realistic environment, in indoor and outdoor test networks by using commercial or precommercial terminals.

In the initial phase of new technology like LTE, it can be assumed that there differences may appear in the functionalities of different solutions of different equipment vendors. This is due to the fact that LTE standards are highly complicated and it is challenging to create such test specifications and requirements to cover all the essential functionalities of the system. In the case of problems in the interworking, one of the most useful tools is a protocol analyzer. It displays and stores the signaling flow at the wanted level between the network elements. The analysis can be done in real-time or by playing the measurement results in the display of the same analyzer. The postprocessing can be performed based on the stored measurements, either with the same equipment or with separate, more specific postprocessing and analyzer tools.

When thc LTE base-station infrastructure is about to integrated and commissioned, the coverage area studies can be carried out with suitable field test equipment. The RF analyzer is one of the most logical selections in the early phase, but as the networks develop, the measurements may be carried out smoothly with vehicle-mounted outdoor receivers, or with portable testers, indoors and in other locations that are difficult to access with a car. One of the examples is Nemo Outdoor field tester, which consists of a common platform for data collection, display and pre-analysis, whereas the actual radio measurements are made with commercial LTE terminals and/ or third-party scanners. Figure 14.5 shows an example of the Rohde & Schwarz frequency scanner, which can be used as one module for typical LTE field tests [2].

14.3.2 LTE Traffic Simulators

For the core network, that is EPC, protocol analyzers are useful in network tuning and optimization, as well as in fault management. Nevertheless, if there is no functional network yet available, or if the planned features require further investigation before introducing them in a

Figure 14.5 An example of the frequency scanner that can be utilized in the LTE radio field tests. Printed by courtesy of Rohde & Schwarz.

live environment due to potential faults, the performance and functionality of complicated scenarios can be investigated via simulators. In the case of LTE, which generates considerably more signaling and traffic load compared to the previous mobile networks, they can include separate high-performance servers for the creation of realistic loads in large areas.

One example of this type of equipment for the LTE performance evaluation is the NetHawk (EXFO), which has, together with Aeroflex, developed an integrated test system, EAST500. The system includes simulator, measurement and analyzer blocks, which can be used for LTE performance on a large scale before commercial traffic is available. The packet contains the radio interface measurement module by Aeroflex and the application tester of NetHawk. The outcome of the investigations gives a realistic estimate of eNodeB performance and capacity behavior in realistic field conditions and with realistic traffic profiles of the users. Because the eNodeB contains considerably more functionalities than UMTS/HSPA NodeB, the measurements and simulation equipment must be updated to support the higher data rates and heavier signaling loads. The data traffic could be estimated to be around ten times more in a typical scenario, which increases the requirements of the analysis accordingly. The equipment thus contains updated IT elements like servers in order to handle the simulated traffic.

Figure 14.6 shows the typical network reference points that can be investigated with the EAST500 equipment. The equipment can be used further for simulating the user data transmission layer as well as the signaling traffic load via realistic terminal emulation. As the

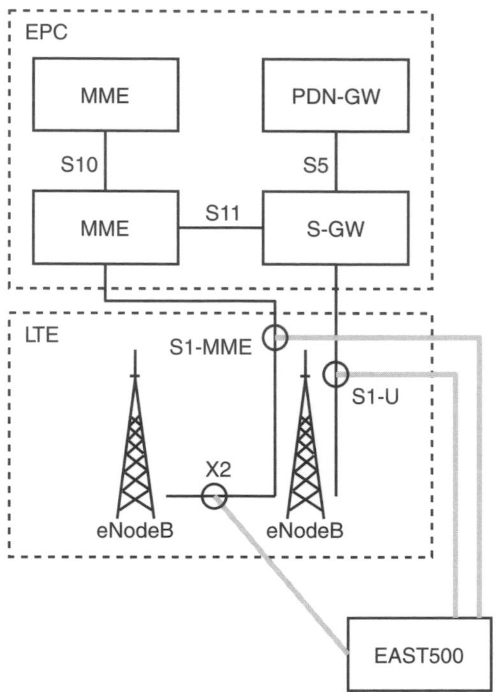

Figure 14.6 The reference measurement points of LTE load simulations.

equipment supports all the LTE protocol layers, it can also be used for the functional testing of eNodeB with varying load, and with different bit error rates that the users sees.

SW upgradeable equipment provides a means to offer LTE functionality without purchasing new hardware. Some examples of such earlier equipment types that will have the required LTE capabilities are the vector analyzer, signal generator, signal emulator, and signal analyzer. The updated SW brings LTE capabilities both in the uplink and the downlink with respective LTE definitions. This equipment can be used for measuring the eNodeB and the terminal, for example in the type approval tests, combined with the baseband signal-generator channel emulator and vector signal emulator. The functionality can also include thc automatic DCI channel coding, channel definitions, power setting and scheduling, as well as the terminal's access to the network.

14.3.3 Typical LTE Measurements

The LTE measurements can be done both in the uplink and the downlink directions. The typical measurement items may be related to the frequency error, transmitted power level, modulation error rate (MER) and error vector magnitude (EVM). The MER indicates the accuracy of the modulation, which can be interpreted in the I/Q diagram as shown in Figure 14.7. The EVM value indicates the performance of the demodulator in noisy environments in a slightly different way. The measurement equipment needs to note the position of the symbol obtained from the carrier wave in the output of the demodulator. This result is then compared with the theoretical location of the symbol in the I/Q axis, and an error in the position is shown as a

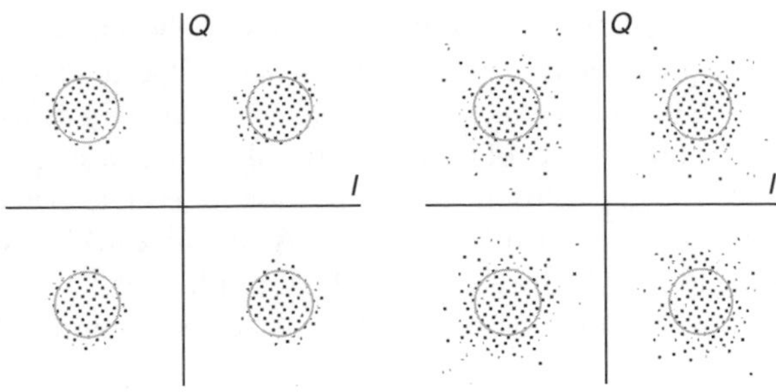

Figure 14.7 An example of the MER levels. The first diagram shows an acceptable MER level, whereas the second diagram might cause already misinterpretations due to the greater variation in the received constellation points.

percentage. Figure 14.8 clarifies the idea of the EVM. The measurements are typically carried out for all of the subcarriers within the selected segment. If the averaging of the results is selected, the maximum and minimum values can also be displayed.

The EVM is closely related to the modulation error rate. In practice, it indicates the signal-to-noise ratio of the digitally modulated signal, and it is measured on a dB scale. Other essential LTE measurements are as follows.

The constancy of the spectrum indicates the amplitude, difference in the amplitudes, phase and group delay for each subcarrier. This measurement is useful for solving OFDM specific issues, like intersymbol errors.

The EVM as a function of the subcarrier graphically shows the vector errors for each subcarrier for a certain symbol or a group of symbols. The measurement can display at the same time the mean values (RMS) and peak values, which provide momentary vector-error measurements.

The EVM, as a function of symbols, shows graphically the vector error ratio for each symbol either for a part or for the whole set of subcarriers.

Channel quality information (CQI) is one of the most important LTE measurement items. It is derived from HSPA definitions, and it indicates channel quality in a single value by combining different quality-related items.

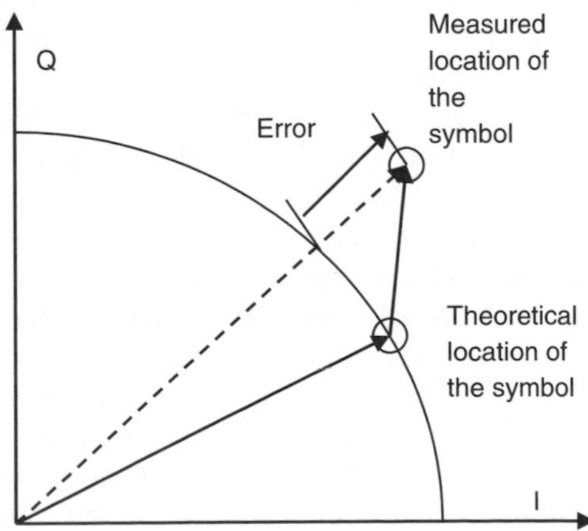

Figure 14.8 The principle of the Error Vector Magnitude (EVM).

Other useful LTE measurements are also the channel's received power level, the utilization level of the band, the leak power of the neighboring channel and the mask of the spectrum source.

Downlink specific LTE measurement set can include the following items: The power versus resource block shows the power level of each resource block over a single subframe or over a larger set of subframes. The impact of the power increase for each resource block can thus be visualized by observing the power distributions. It is also possible to observe simultaneously a set of constellation diagrams, which eases in the fault resolutions.

In the uplink direction, the time-based EVM measurement shows individually for each symbol.

As an example, let's investigate the Rohde & Schwarz R&S FSQ-K100 measurement equipment. It can display the following downlink and/or uplink LTE measurement results:

- power level as a function of time (DL, UL);
- EVM as a function of the carrier (DL);
- EVM as a function of the symbol (DL, UL);
- frequency error as a function of the symbol (DL);
- EVM as a function of the sub-frame;
- the flatness of the spectrum (DL, UL);
- the group delay of the spectrum (DL, UL);
- the difference of the flatness of the spectrum (DL, UL);
- the constellation diagram (DL, UL);
- statistics about the CCDF and allocations;
- EVM per timeslot (DL);
- statistics about CCDF (DL, UL), allocations (DL, UL) and signal flow (DL).

14.3.4 Type Approval Measurements

The 3GPP specification 36.214 [3] describes the measurement definitions of the radio interface of E-UTRAN terminals and network, terminal being either in idle or connected mode. The most important examples are the following.

Reference Signal Received Power (RSRP) is calculated as linear average (in absolute power levels) from those resource elements that are transporting cell-specific information. If reception diversity is activated, the received power level is calculated by averaging the power levels of the separate reception paths.

Reference Signal Received Quality (RSRQ) is defined as a ratio of the power level of the resource blocks (RB) and the signal strength of the whole E-UTRA carrier, which is called the Carrier Received Signal Strength Indicator.

Other measurements are, for example, the received code power level of UTRA FDD mode, received chip-based energy divided by the power density of the band (Ec/No), RSSI and code power of the UTRA TDD mode, the downlink power level of the E-UTRA, the received interference power level, and the thermal noise power level.

The basic measurements of the LTE radio interface are related to the quality of the signals in both the uplink and the downlink. These measurements include MIMO analysis when applicable. In the basic measurements, the measurement equipment should support, in all the cases, the bandwidth up to 20 MHz, which is the maximum value LTE can use. All the modulation types should also be found in the basic set of measurements, that is QPSK, 16-QAM and 64-QAM, as is defined in the 3GPP specification for the 36 series. The OFDM basic

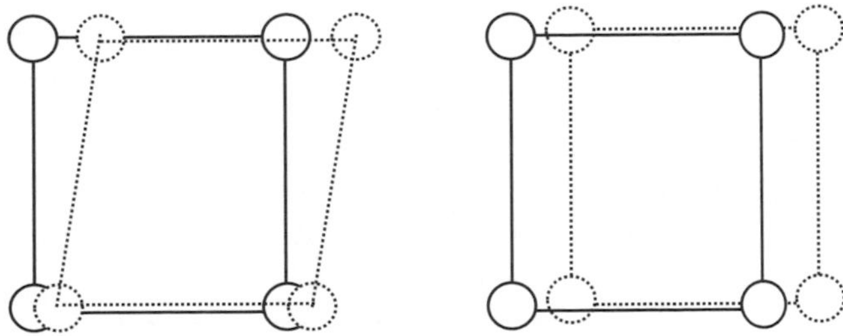

Figure 14.9 Two examples of the possible error behavior in the modulation constellation diagram.

investigations can be related to the quality of the modulation (e.g. Modulation Error Rate, MER) as well as to the other OFDM functionalities like the size of the transport block, the allocation of the resource block, different protection cycle sizes (CRC) and their combinations.

The quality of the modulation of the transmitter can be investigated at a detailed level for example by measuring the EVM value per OFDM carrier, symbol, timeslot and resource block.

14.3.5 Modulation Error Measurements

Interference, nonoptimal radio conditions, and fast-varying signal levels due to, for example, multipath propagation or intersymbol interference, can change the position of the samples in the I/Q constellation diagram. Figure 14.9 shows two examples of erroneous reception. The first diagram shows the twisting of some of the constellation points, and the second one shows the movement of the whole constellation off from its theoretical position. Both can lead to the increased misinterpretation of the values, which leads to an increase in the bit error rate.

Furthermore, nonideal equipment may start to twist the constellation as a function of the signal level—that is, the further away from the 0-point the sample is, the more it may start finding itself off the constellation point as Figure 14.10 shows.

14.3.6 LTE Performance Simulations

Development of an LTE-based communication system sets major challenges for the testing process. Testing of particular protocol implementations is challenging as testing often needs to be carried out in both (simulation-based) host and target environments. There is often also the need to simulate communication towards overlying and underlying protocol layers as these may not be available during early phase testing.

As an example, Iin the CrossNet project, VTT Finland and the University of Oulu (CWC) have developed cross-layer testing and simulation framework together with industrial partners. The framework developed integrates existing industrial tools, the EXFO Nethawk EAST simulator, the M5 network analyzer and the EB Propsim radio channel emulator, to support comprehensive wrap-around testing of LTE-based telecommunication systems [4]. Figure 14.11 shows the principle of the simulator environment. The setup is valid, for example, for the simulation of eNodeB functionality in order to investigate the performance of the system with a multitude of LTE-UEs.

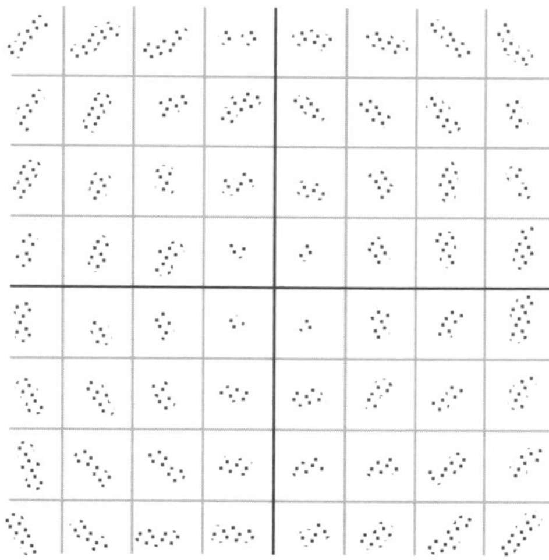

Figure 14.10 An example of the effect of the twisting constellation in the outer part of the diagram, which may be a result of for example component failure of the receiver.

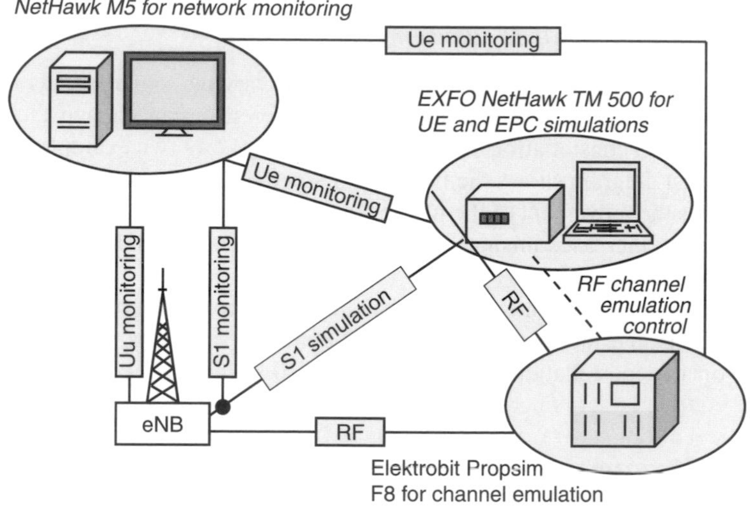

Figure 14.11 CrossNet LTE wrap-around testing environment.

14.4 LTE Field Measurements

The LTE field measurements can indicate a feasible performance of the system in typical radio conditions. The motivation of early technology field trials is to be able to identify inconsistent behavior and ensure that infrastructure and terminal vendors, together with the operators identify, implement and test the necessary corrections and optimize end-to-end performance. This assures that the actual commercial network deployments are executed as smoothly as possible both while technological issues as well as the project management are considered.

14.4.1 *Typical Field Test Environment*

The radio field trial can be located into an area where the EPC, HSS emulator, IMS, and applications are also available. In typical setup, eNBs cover a relatively small area. Nevertheless, the area is normally sufficiently large for the testing of the mobility scenarios. This represents realistic conditions of a live network in terms of interferences. The eNB sites can be colocated with existing 2G/3G sites, which normally requires the antenna systems to be swapped in order to support the LTE-band signal on air.

A suitable system is needed for the operation and maintenance of the eNBs, as it is the basis for parameter, fault, and operations handling, as well as, for example, O&M and SON feature testing. A typical LTE network setup can be the following:

- Evolved Packet Core (EPC)—S-GW; P-GW; MME placed in a same site;
- HSS emulator for user authentication and authorization;
- IMS platform acting as a SIP server for VoIP call support;
- Application Server for the TCP/UDP streaming, FTP transfers, latency tests (e.g. ping-based), and HD video streaming;
- Internet access for internet based scenarios.

In the operator site, the supported elements can logically be located in the equipment room(s) for the purpose of:

- hosting of eNBs and routers; note that the room can also be controlled to manage eNBs of the field remotely by operator's transport network;
- hosting the O&M system equipment that are dedicated to manage LTE eNBs and used for, for example O&M and SON testing.

In addition, the actual eNBs can be installed both indoors and outdoors in such a way that they can be activated individually to form partial or complete radio coverage in the test area. In typical trial cases, part of the eNBs can also be used as a base for the demonstrations.

In detailed field testing, each eNB can be set up in such a way that it either contributes useful signals, or acts as an interference source in the downlink. It should be noted that interfering eNBs can be configured without S1 connections.

In typical field testing, static, pedestrian, and vehicle radio channel types can be measured and analyzed. This requires a plan for the test cases, including the test setup and description of the procedure, the measurement/signal generator equipment setup, and the storing of the performance indicators for the postprocessing and analysis. Multiple sites are required in order to confirm the functionality of the handovers. The coverage area can be checked in the cell edge of the area both in noise-limited and interference-limited situation.

One of the important tasks is to select the LTE terminals. In the early stages of the LTE system, it can be expected that only data-capable USB stick models will be available. In order to make sure that the terminal itself is not the limiting factor, multiple models from various vendors should be considered. This minimizes the effect of possible nonoptimal behavior of some terminals, and the results thus reflect network performance.

As the LTE system is capable of delivering considerably faster data connections over the air interface, the physical end-to-end capability of the complete LTE/SAE network should be designed accordingly. The core side should have sufficiently capacity in order to avoid possible bottlenecks in any of the interfaces. If limiting elements or interfaces occur, it should be known well before the actual LTE performance analysis takes place so that misinterpretations of possible lower performance figures can be avoided.

The easiest solution for the remote control of each LTE/SAE element is the centralized O&M system. If it is not available for trial purposes, it is worth considering a setup of local control systems (PC/laptop with element manager software and direct connection to the element) in each site, which can be used remotely, for example via a VPN/CITRIX type of solution. Such remote connections ease the configuration of the network for each test case. It is also worth noting that the possible interferer sites do not require S1 connections in order to have the interference signal on air. This avoids the additional backhaul load and saves capacity. If the eNBs are connected to the backhaul, for example over a Gbit connection, a single E1 connection of PDH network or other feasible low-capacity connection is enough for the connectivity of the interfering sites to the controlling test network PC/laptop. E1/Ethernet converters can thus be used in the interferer sites.

14.4.2 Test Network Setup

The LTE field test equipment can consist of the drive test tool for generating the data calls and the respective KPIs, the scanner to store the received power level, and the core protocol analyzer. GPS should be included to the radio measurement setups by default in order to correlate the results geographically. Additional troubleshooting tools will probably be needed. These can be more specific signal and protocol analyzers on the core side, and additional analyzer elements for the test-drive tools.

The LTE performance can be measured under varying conditions for the propagation environment (dense urban, urban, suburban, rural, open) and user profiles. The effect of the latter is quite important as the load figures of the network dictate the single user throughput, for example in the average and peak load. In order to create the loading in a controlled way, general models can be applied. These models are available, for example, from 3GPP, which has defined a macro case 1 (with 500 m of inter-site distance) and case 3 (with 1752 m of intersite distance) by assuming ten users per cell as a reference.

For the test cases where interference is present, the interfering signal can possibly be generated via the eNB itself if the equipment provides this type of test signal generation. As an example, NSN eNB can be set up to send artificial signals on the physical DL channels, which can be configured by the so-called PhyTest Interface. For example, in order to generate a 50% interference load, half of the available active PRBs in PDCCH and PDSCH are used to send dummy data. For the full 100% load, all physical channels are transmitting all the time, and all the PRBs are used.

Figure 14.12 shows the idea of the measurement setup when DL interferences are created intentionally.

The uplink interference is generated by the transmitting power of UEs served by the neighboring/other cells to the one considered. The closer the interfering UEs to the cell border the more transmitting power it will use and the lower the path loss to the victim cell, which means increased interference to the UEs in the victim cell.

The uplink interference is challenging to simulate or generate in the field as it varies over time and frequency. A common approach in trials is to generate the UL interference in the field with, for example, a TM500 LTE UE [5], as seen in Figure 14.13, which can be configured in Layer1 measurements to transmit a pseudo-random sequence on the PUSCH with a determined amount of Tx Power.

The total amount of TX power transmitted by the UE is adjusted according to the path loss of the TM500 location and the victim cell in order to generate the desired noise level at the victim

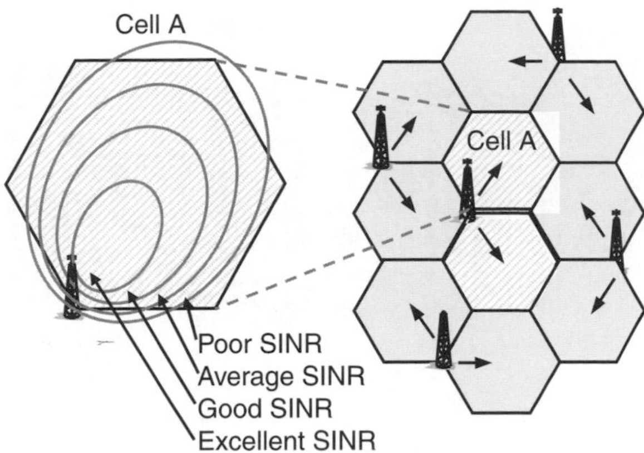

Figure 14.12 A possible field measurement setup for the interference investigations.

cell. As the TM500 is configured in Layer 1 measurements it does not require to be camping on any interferer cell. This allows the TM500 to be strategically placed anywhere in the victim's cell coverage area, increasing the flexibility of the configuration setup and the amount of interference that can be generated. For field test cases, the location in the victim cell and L1 parameters of the TM500 is adjusted in order to generate the noise rise required for the UL-loaded test cases.

Uplink interference can also be generated with a Signal Generator by using the transmission patterns obtained previously. This approach is chosen for early LTE field trials as it is a field-proven concept and as it does not require HW modifications at the eNB site.

Figure 14.14 shows an example of the Vector Signal Generator that has been used in the trial network. Rohde & Schwarz SMBV100A [6] is equipped with an internal baseband generator which allows the generation of LTE digital standards with a sufficiently wide frequency range of 9 kHz to 6 GHz.

It is quite clear that the uplink interference, as presented in Figure 14.15, depends on an "Interfering" UE position—that is the more the interfering set of UEs are located at the cell border, the more transmitting power they will use and also the path loss to the victim cell becomes lower. Consequently, the interference to the victim cell increases.

Figure 14.13 The Aeroflex TM500 Test Mobile for the UL interference creation. Printed by courtesy of Aeroflex.

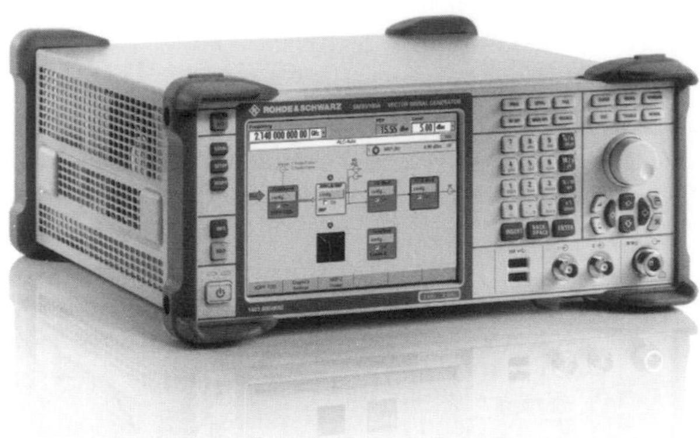

Figure 14.14 An example of the vector signal generator. This specific model is a Rohde & Schwarz SMBV100A. Printed by courtesy of Rohde & Schwarz.

As a basis for the DL performance indicator evaluation, the SINR (Signal to Interference and Noise Ratio) ranges can be fixed, for example in the following way for the at 100% DL load:

- excellent location: SINR $\geq$ 20 dB;
- good SINR (High Location) $12 \leq \text{SINR} < 20$;
- average SINR (Medium Location): $6 \leq \text{SINR} < 12$;
- poor but functional: $0 \leq \text{SINR} < 6$;
- bad SINR (Low Location): ≤ 0 dB.

For the UL, Load Conditions (%) as a function of IoT Range can be fixed, for example as follows: 50% load $\Rightarrow$ 8 dB, 100% load $\Rightarrow$ 11.5 dB.

In the early state of LTE, the terminals have been challenging to obtain, and noncommercial versions have been used. Some of the first versions have been LG branded LD100 and G7, as

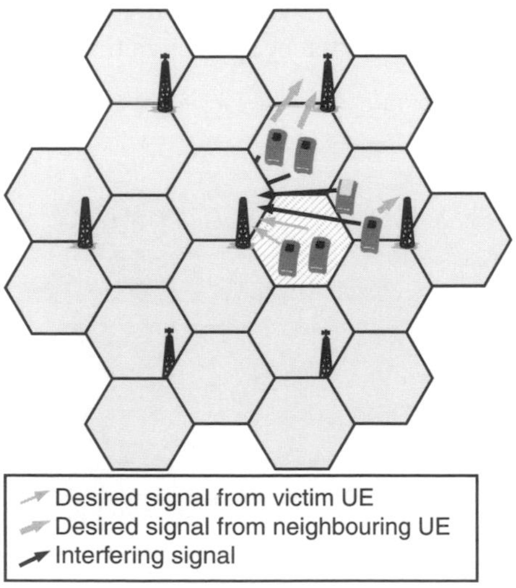

Figure 14.15 The idea of the UL interference creation for the trial test cases.

Figure 14.16 LTE terminal as a form of a USB dongle. Printed by courtesy of Samsung.

well as Samsung branded GT B3710 and GT B3730. Figure 14.16 shows a typical form of the USB dongle-type of LTE terminal.

In the LTE radio test setup, it is important to decide whether a stand-alone terminal as such, or a terminal with an external antenna will be used. The reasoning for the external antenna mounted, for example, on the rooftop of the vehicle, is that it creates receiving conditions that are always similar for each test-drive route regardless of the position of the actual terminal. The coverage area turns out to be larger, though, in this setup compared to the terminal without additional antenna. The benefit of the terminal without external antenna is thus the close to the typical conditions of normal users.

It is also important to take into account the characteristics of the terminal. The most limiting factor is the terminal class. The support of different MIMO antenna ports also has an effect on the data throughput.

The successful execution of a trial requires a project plan and a realistic time and resource planning. The main tasks of the trial are the following:

- initial plan of the scope;
- overall test plan;
- network architecture plan;
- detailed test plan;
- test network deployment;
- test case execution;
- data post-processing and analysis;
- reporting.

As an outcome of the trial, the functionality of the LTE system and the investigated features are evaluated. Also the performance of the network with different test cases is evaluated.

14.4.3 Test Case Selection

The overall LTE test scope may cover the following study items:

Validation Period. This is the first stage test period, with the overall aim of making sure that the network is correctly constructed and that the equipment is functioning as planned. This stage can include the system validation tests, site configuration and design validation tests. The outcome of this stage is information about the general coverage area, with, for example single user throughput in the downlink and the uplink, the functioning and performance of the terminal attach and detach procedures, data streaming performance, possibly in static, pedestrian and vehicular radio channel types. The tests can also include downlink and uplink interference, when its significance is altered from a level without interference up to the maximum level.

Extensive Testing Period. This stage aims to analyze the capacity and throughput. It can include the cases above in a more thorough way, including single user throughput in the downlink and the uplink, deeper data streaming by using the TCP and the UDP, in static, pedestrian, and the vehicular environment by altering the vehicular speed and interference levels. This stage can also contain MIMO cases, and varying capacities by altering the number of the LTE terminals. In addition, the following items are some of the examples that can be included in the test set of this stage:

- Scheduler tests via TCP streaming by varying the number of terminals under different channel types and interference levels.
- End-to-end latency tests for the attach procedure latency. The cases can be varied under different channel types, network load levels, and interference levels. The straightforward way to execute these tests is the use of PING by varying the packet size (small, medium, large). In order to increase the statistical accuracy, several PINGs can be collected by test case in order to present the latency distribution.
- Mobility and handover procedure performance test, including the measurements of the handover success rate and interruption time for different environment and traffic types (TCP and UDP streaming as well as VoIP calls), and by varying the terminal speed. The handover performance can be investigated for the intercell and intracell and site situations. For the voice calls, the Mean Opinion Score (MOS) can also be measured with the appropriate field test equipment.
- Radio features tests by varying the feature parameter settings, for example for the QoS related features, or for the uplink closed and open loop power control variants with a certain number of terminals placed in a certain constellation and moving them between the site and coverage border.
- Application performance tests, including, for example, the success rate and setup speed of the Internet access, IMS VoIP, FTP, video streaming, HTTP browsing, e-mailing and HTTP download
- Robustness and availability tests, including, for example, the attach procedure under different interference levels and the TCP streaming by adjusting the TCP window size.
- Operations and maintenance related tests, including the functionality of the main functionalities of the tool.
- Self-Organizing Network (SON) related tests, including the validation of the correct execution of the commands and the functionality of the SON features.

14.4.4 Items to Assure

In the trial (and commercial) LTE network deployment, the maximum data speed is much higher in the radio interface compared to any other previous mobile communication networks. It is thus important to take into account all the possible bottlenecks of the LTE/SAE network in order to make sure that the end-to-end performance would be at the expected level. It can thus be expected that the data velocity limitations in the transport network can cause problems due to the existence of equipment, interfaces and settings that are not all under the control of the operator.

As an example, there have been experiences of TCP DL single-user streaming that does not exceed a certain, relatively low value compared to the expected theoretical maximum data speed, even under excellent radio conditions. An analysis of the transport network element distribution along the whole communications route is thus advisable. The problem might be derived from the router data speed limitations related to each single interface. As an example, a data speed limitation of 45 mbps can be found in a single gigabit router, which is sufficient to lower the whole end-to-end data speed. Adjusting this limit to the maximum default value that can be, for example 200 mbps, can remove the problem.

The problem can also be related to the sporadic throughput fluctuations for UDP streaming in the uplink, and for TCP streaming in both directions. The fluctuations can be seen even with relatively simple tools like iPerf, which may slow the sudden droppings to some lower data speed level, and slow initiation of the data transmission. This might be caused in nonoptimal performance of the LTE terminal, especially in the initial phase of LTE systems. In fact, there are cases where the LTE terminal occasionally sends a buffer status report "empty" even if it is not so, which can trigger the problem. Also the TCP window size can cause the problem. For testing purposes, a parallel TCP stream with the smallest TCP window size can be used in order to reach throughput stability.

The aim of the trial phase of the LTE is to test the performance of the network, but in some cases this phase also reveals behavior of some of the involved elements that is not directly related to LTE, but needs either parameter adjustments or even complete replacement. Especially, if the core network contains "old-fashioned" solutions, they might have worked well enough for previous data rates, but the higher speed of the LTE might reveal the limitations of those hardware and/or software.

14.5 Evolution Changes the Rules of Testing

The introduction of LTE networks marks the end of the transition of mobile wireless networks from a traditional telephony-style circuit-switched network, which was focused on delivery of voice services, to all-IP network focused on the delivery of data services and applications. This transition will allow operators to deliver diverse broadband services at competitive prices, because of the efficient way in which IP networks use the available bandwidth. This transition has progressed through interim technologies such as High-Speed Packet Access (HSPA), which combine circuit-switched voice with packet-based data. As an entirely packet-based technology, however, LTE will improve the performance of mobile data services at the same time as it allows operators to package mobile services in new ways to make them suitable, from a cost and performance point of view, to a wider range of businesses and consumers than was previously possible.

Although referred to as an evolution, then, LTE represents a significant step forward. This is not only because all traffic, including time-sensitive traffic, is packet-based, but also due to the fact that LTE replaces the established CDMA radio technology with a new radio-access technology based on Orthogonal Frequency Division Multiple Access (OFDMA). Compared to CDMA, OFDMA will allow LTE to support higher data rates up to 300 mbps over the air, as well as lower delay and latency for data packets. In combination with the simplifying of core networks, which is a benefit of the move to entirely packet-based traffic, OFDMA is central to meeting the LTE objective of delivering greater bandwidth at lower cost due to it providing better spectral efficiency. This means that fewer cell sites are required to give the same level of network capacity as previous generation technologies.

Network planners, however, must deal with two major challenges if they are to successfully implement LTE. They must both understand how to optimize the network to meet Quality of Service (QoS) goals, and establish new models for the location and configuration of base stations in order to predict accurately the network coverage they will achieve with the new OFDM radio infrastructure. Effective optimization and network modeling will depend on reliable data obtained from accurate field tests of live LTE networks.

In the following chapter on LTE testing, we will look at the generic LTE test requirements, and then look at specific areas of test methods and requirements. This is broken down into the LTE air interface, the backhaul network (SAE), throughput "end-to-end" testing, Self Organizing Network (SON) technology for reducing test/optimization costs, and live network tests and optimization. In general, testing methods during product development can be separated into "validation tests" and "verification tests." The further category of tests for the production/assembly stage is also considered here.

Validation testing is where a network, subsystem, or element is being tested against a fixed specification (usually a 3GGP specification, or manufacturers own internal performance specification), and normally the outcome is a pass or fail judgment to validate whether the device under test (DUT) is able to meet the required levels being tested and hence is validated as compliant with the test specification. As the test specification is normally written around the operational/performance/design specifications of the DUT, it is essentially a test to validate that the DUT is correctly designed and built and operating according to the specification. Due to this close link between the validation tests and the specifications, the validation tests often have a direct correlation with the system design specifications. This can easily be seen in the 3GPP LTE specifications, comparing the TS36.104 [7] and TS35.141 [8] for base stations, and TS36.101 [9] and TS36.521 [10] for a UE. In 3GPP, the validation tests are called "conformance tests" as they are used to demonstrate conformance of the UE or eNodeB with the required minimum performance specifications.

In the mobile industry, there are two industry bodies that take a lead in providing a forum and method to assist in the validation of handsets. This was set up because of the excessive costs for handset manufacturers and network operators in performing validation of all types of handsets for all different operators. The central principle is to provide a common set of tests, test methods, and test environments, so that a new UE can be extensively tested once by the manufacturer, and the test results would be recognized and accepted by a wide range of network operators. In Europe, the lead body is called the Global Certification Forum (GCF) and today it has a truly global nature. In the USA, the lead body was originally called the PCS Type Certification Review Board (PTCRB), as it was focused on the US 1900 MHz PCS band. Today it has kept the same initials, but has a wider remit to include the needs of North America in the other licensed bands. In addition, there are various national bodies that perform similar tasks at a national level for specific countries that have special needs outside of GCF. Both GCF and PTCRB tests are based on the 3GPP LTE Conformance test specifications TS36.521 [10] and TS36.523 [11]. They then have a set of validated test cases (where the validated tests have been independently confirmed to be correctly implemented on specific test platforms), which are available to the UE manufacturers to perform their conformance testing. It is this independent confirmation on specific test platforms that provides international recognition amongst the network operators of the validity of the testing. In addition, the GCF provides a common accepted framework and set of tests for the performance of live network testing. Again, this reduces the amount of testing required by a UE manufacturer before the network operator needs to make any evaluation or acceptance of a particular device. In recent years, there has also been a trend for some network operators to require additional testing of Inter-Operability Testing (IOT) in the laboratory before live testing on the networks. This has been driven by the increasing complexity of the network architectures and configurations, and those specific network operators have felt it necessary to add an extra level of operational tests specific to their needs over and above the functional testing described in 3GPP.

Verification tests are used during the development phase to measure the actual level of performance. So the actual level is measured, and the tests are performed in a variety of settings and configurations. The verification stage should cover the widest range of conditions, to confirm the performance in the extremes of the design. For example, a UE may be tested for output power at the lowest battery power, or very hottest ambient temperature, to verify the performance in these extremes. But during the validation tests then only a limited number of battery levels of temperature levels would be used. In addition, the verification tests will also include testing under negative conditions, or unexpected/unallowed conditions. Particularly for signaling (protocol) tests, it is important to verify the performance under these negative conditions. Examples of this are the eNodeB scheduler receiving inconsistent measurement reports from the UE such as a very low CQI but very high RSRQ, or the UE protocol stack sending a reject message in response to a mandatory command. In this second example, a reject message is valid, but only in response to an optional command. So although the reject is made correctly, it is an incorrect time to make the reject. So, we must verify how the eNodeB protocol stack responds to this message, to ensure that it has not crashed as it would be expecting a different response from the UE.

Production-line testing is normally a subset of the conformance tests, to verify the correct assembly of the equipment and final check to ensure correct operation. The old phrase "time is money" applies particularly to high-volume production, as time spent testing can slow down the capacity of the factory, or add significant costs due to the cost of the test system and the high value of the "work in progress" of the equipment waiting to be tested. As the testing is normally carried out at the end of the production process, the full value of components and assembly time has already been invested into the equipment, and the manufacturer would like to ship the good to customer as quickly as possible to earn money on the work done. A long or slow production test will reduce revenues for the factory, or increase production costs. The first objective of the production test is to verify correct assembly of the equipment, so no component is missing, mounted incorrectly, or faulty. This can be confirmed using a simple go/no-go test. So the modulation error of the transmitter stage will be either very low if correct, or very high if incorrect. Any small variation in quality would be due to poor design or tolerance on components. So the verification testing is used to confirm quality in all the different conditions, but the production test is a simple check at only one or two levels. During the early production stages (pilot production) more comprehensive testing may be done to establish the normal variations in the production process, but for high-volume testing then only confirmation that the value is within the production tolerances is sufficient. In an extreme example of this, the RF conformance test for a single radio band on a UE may take typically 24 hours of continuous testing to ensure compliance with all required performance specifications, but the production-line test may be reduced to approximately 30 seconds of testing.

The RF test requirements for LTE are given in 3GPP standards TS36.141 Base Station Conformance Test [8], and TS36.521 UE Conformance Specification Radio transmission and reception [10]. These are based on the corresponding performance requirements set out in TS36.101 [9] and 36.104 UE [7] and Base Station transmission and reception. In addition, there are the TS36.571 series of UE Conformance Test specifications [12] that are focused on the positioning technologies used to provide the location information of the UE using information provided by the network. The test specifications TS36.508 [13] and TS36.509 [14] provide the detailed settings and configurations used in UE conformance testing. These are required because the network (and hence the network simulator used in testing) has many possible settings that can all influence the test results. Only by specifying the required network settings can correct and repeatable tests can be made on the UE using the test specifications TS36.521 [10] and TS36.523 [11].

The UE protocol test cases are written by ETSI within Task Force TF160. This industry body is creating an implementation of the test cases using the test description language, TTCN3. This test description is then readily available from the ETSI web site. The test equipment manufacturers then take this TTCN3 code and implement onto their own specific protocol test equipment and systems, to create a test system capable to run these ETSI test cases. Within 3GPP working group of RAN5 the selection and prioritization of the protocol test cases is made, and these requests made to ETSI TF160. The selection of tests is based on the work in RAN1 and RAN2 working groups where the signaling aspects of the air interface are developed.

14.6 General Test Requirements and Methods for the LTE Air Interface

14.6.1 OFDM Radio Testing

OFDM and the use of high-order 64QAM modulation require high linearity, phase and amplitude in both TX and RX modules to prevent intersymbol interference and to enable accurate IQ demodulation. This requires a fast and adaptive EVM measurement capability to track and measure the signals during adaptive frequency channel use. The OFDM signal is composed of up to 1024 individual subcarrier frequencies, each transmitting simultaneously. Testing is made on both the "per subcarrier" performance of each individual subcarrier in the OFDM signal, and then on the "composite" signal where the subcarriers are combined and the overall performance is seen.

The subcarriers require good phase noise performance and accurate phase linearity to prevent leakage across carriers. The frequency mapping and orthogonal properties of OFDM require the "null" in one carrier phase response to be exactly on the peak of the adjacent carrier. Thus, accurate measurement of the phase linearity and amplitude linearity on each subcarrier is important for orthogonal design (resistance to interference) of the system. Any errors in this stage will directly translate into poor link performance, reduced data rates, and poor service.

The OFDM transmissions must also be measured "per resource block" to see how the power levels in each burst are being correctly maintained. Each individual "resource element" (that is, a single OFDM carrier at one time duration/symbol) has a specific power level to be transmitted, and these power levels should be correctly measured across a whole resource block. The power levels per resource element are each specified as relative to a reference power level. Thus the total power of the resource block may be varied, and then the power in individual resource elements may vary relative to a reference power in the resource block. To verify this, it is necessary to measure both the energy per resource block, and then to verify individual resource elements where appropriate.

The EVM measurement for LTE needs careful consideration because of two special features used. Firstly there is the "Cyclic Prefix" (CP), which is a short burst of transmission at the start of each symbol. This is actually a repeat of the end of the symbol, and is used to give a "settling time" to allow for delay spread in the transmission path. If measurement is started too soon in the CP period then there will also be signal from the previous symbol (inter-symbol interference, ISI) included that will corrupt the measurement. The second point is that the symbol transmission has a "ramp" at the start and end to ensure that there is no strong "burst" of power (a strong burst will create high levels of harmonic distortion and interference). So the start and end of the symbol have an up/down ramp on the power. Therefore we must also restrict the period of the symbol that is measured, to ensure that we do not measure in these ramp periods. The technique to address both these issues is to use a "sliding FFT," where the period of the symbol that is measured can be adjusted in time (sliding) to give the best EVM value [15]. This is shown in Figure 14.17.

The effect of the "ramp" can be seen in the measurements. The waveform on the left has no ramp, and so there is a sharp switch on/off between each symbol. This causes a large "spectrum due to switching" emission that is seen as broadening of the output spectrum beyond the desired system bandwidth (in this case 5 MHz). The waveform on the right has ramping enabled, and so there is a much less severe switch on/off between symbols. This has the effect of reducing spectrum emissions. This type of ramping (also called "spectrum shaping") is required to ensure that the transmitter output stays within the allocated frequency band and does not interfere with any adjacent frequencies.

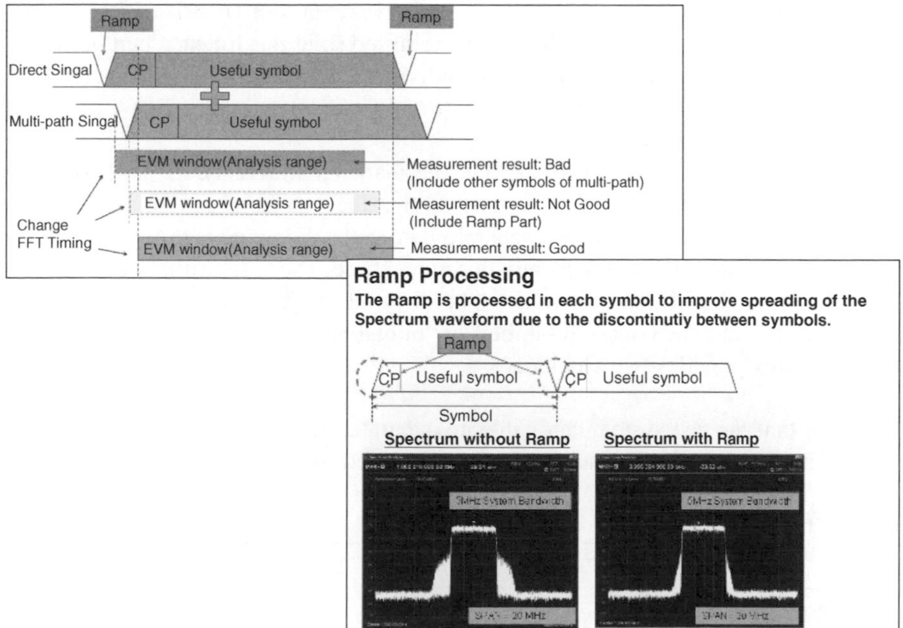

Figure 14.17 The sliding window principle in the EVM measurements.

The remaining LTE RF measurements are all based on the standard measurements made on a wireless cellular transmission, used to guarantee no interference between adjacent networks, and to prevent leakage from the LTE network into an adjacent radio band they may be used for another purpose. These include the spectrum emission mask, spurious emissions, adjacent channel leakage ratio, and power level. These are each individually summarized and shown in the RF test summary at the end of the chapter.

14.6.2 MIMO Testing

In a MIMO system, the coupling from antenna to air characteristics must be fully understood. Data rate and performance of MIMO links depend on how the multiple RF antennas couple to each other. Accurate calibration of antenna paths, factory calibration and then field installation calibration is required to implement a successful MIMO system. During the R&D phase, evaluation of designs is required to confirm the sensitivity calculations required to find critical performance-limiting issues.

The base station transmit antenna array may use specialist phased array techniques (like a Butler Matrix) for accurate control of the phase/timing in each antenna path. This requires accurate characterization of the RF path in terms of electrical path length, coupling and reflections from both ends. This data is then fed into the MIMO adaptation algorithms to enable features such as beam steering. A Vector Network Analyzer would normally be used for complete characterization of the antenna paths.

In testing the MIMO we should consider to test both the baseband processing section and the RF generation/alignment. Both areas need to be tested for both functional test (correct operation) and performance test (optimization of algorithms for best processing and data throughput efficiency). In addition, it is useful to check the "negative test," which is to use deliberately incorrect signals and ensure that they are correctly handled or rejected.

A good approach to developing a MIMO test strategy is to lay out a matrix of each of the areas and stages for MIMO test, and then identify the required solutions for each part of the matrix. The key elements of the matrix are to test the individual subunits (Tx baseband, Tx radio, Rx radio, Rx baseband) separately, and then as integrated Tx and Rx modules. So we can develop a test matrix as shown in Table 14.1.

In a MIMO system it is necessary to calculate the characteristics of the RF path from each TX antenna to each RX antenna. This is required so that the two paths can be separated by the processor and effectively become two separate data paths. To achieve this, the system must accurately measure the RF path characteristics in real time. These algorithms are built into the design of the particular MIMO system used, but they all basically require the accurate phase and amplitude measurement of a "pre-amble" or "pilot tone" that is a known signal. For testing environments this provides two challenges:

1. To ensure that the test system can generate accurate reference signals against which the measurements are made. The accuracy of the received signals must be carefully measured,

Table 14.1 Test matrix

	Tx Baseband	Tx Radio	Tx Module	Rx Radio	Rx Baseband	Rx Module
Functional test						
Performance test						
Negative test						

and the test system calibrated to separate the measurement system uncertainty from the MIMO system accuracy and uncertainty. This way, the exact characteristics of the MIMO system are measured, with minimum influence from the test system. This requires the development of a test method/environment to generate the "reference signals" against which the measurements are made, and then to confirm the measurement method by adjusting the "quality" of the reference and checking that the measured result matches to the change made.

2. The actual RF coupling from transmitter to receiver will affect the measured end to end system performance. So for test environments used in performance measurement, algorithm tuning, Integration and Verification (I&V), and production quality, the RF coupling between antenna must be defined, repeatable and characterized if absolute performance figures (e.g. Mbit/second data rates) are to be measured. This requires the use of suitable fading and multipath test equipment and generation/profiles to create the different coupling between antennas. This can be achieved using static signals (e.g. signal generator based references) for initial checking, and then through baseband fading simulators to verify correct operation at algorithm level, and the finally RF fading simulators for "end-to-end" system-level testing.

The MIMO coding of data blocks (block coding) is based on Space-Time Block coding, where the actual coding of the data is based on both the space (i.e. which antenna) and time (when it is transmitted), and the diversity gains of MIMO are from using diversity of both space and time for each block of data that is sent. So we must measure both the time alignment of each antenna, and then the spatial alignment of the paths between the antennas.

MIMO test requires extensive test and evaluation of the signal processing and MIMO coding algorithms that are used. To do this, we need a "step-by-step" approach in which the individual processing and feedback of the MIMO algorithms can be isolated and tested without the rest of the LTE network. In addition, we need a controlled environment where we can integrate the parts of the MIMO algorithm and run them against "reference conditions" so that we can verify the performance. The verification requires that we set up precise know conditions, both with the RF coupling from transmitter to receiver, but also the measurements and feedback reports being made between transmitter and receiver. To do this we require testing at a pure "baseband" level to check algorithms, as well as testing at RF "air interface." In addition we require precise control of the baseband processes and RF coupling. This is normally achieved by using fading simulators and system simulators. The fading simulator will provide a controlled "air interface" coupling (either real or simulated), and the system simulator will provide a controlled baseband environment (e.g. a controlled UE to test a base station, or a controlled base station to test a UE).

When we are including the "fading" function into the MIMO test, we need to describe fully both the fading of each path (delay paths phase, amplitude and scattering types) and then the correlation between each RF path (the correlation matrix). In the case of "2×2" MIMO, there are four paths, referred to as h11, h12, h21, and h22, and shown above. In an ideal environment for MIMO there is no correlation between the different RF paths, and so the processing algorithms can fully separate the signals from each path and get the full data rate increase. In the "real world" we see correlation between the different paths as they have some similar "shared" routes from transmitter to receiver. In a worst case we have very high correlation between the RF paths as they essentially have the same path. For each of these scenarios we have a "correlation matrix" that describes mathematically how the different RF paths are related, and then we must test, verify and then optimize the algorithms to give the best possible data rate throughputs in each of the different types of RF environment that could be experienced.

An example test environment for an eNodeB is shown in Figure 14.18.

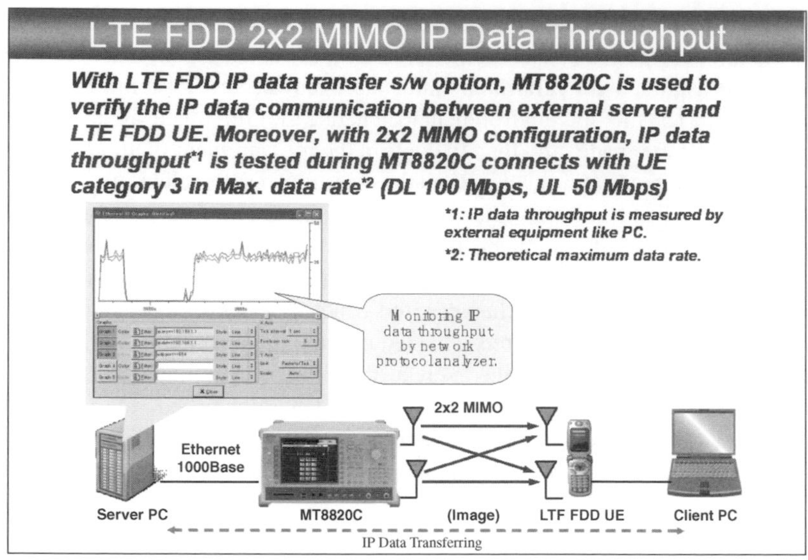

Figure 14.18 Testing in the MIMO environment.

14.6.3 L1 Testing

L1 contains the algorithms and procedures associated with reporting and measurements that then drive the Power Control, Adaptive Modulation and Coding, and MIMO processing capabilities. So from a test point of view we need to verify both the correct measurement are made at the receiver (and then transmitted back to the corresponding element that uses the measurement), and then that the transmitter is correctly reacting to the measurement reports and adjusting parameters accordingly.

Reports made by the UE back to the network include CQI (Channel Quality Indicator), PMI (Precoding Matrix Indicator), RI (Rank Indicator). The CQI is associated with the selection of the AMC (data rate) in the NodeB, and PMI and RI are used by the NodeB to configure the MIMO encoding. In addition, the UE must measure the Reference Signal Received Power (RSRP), and so must correctly identify the antenna specific reference signals and measure power in the individual resource elements containing the reference signals.

The NodeB must adjust the Timing Advance of the UE, so that all UEs are received with the same relative timing (required for an efficient FFT process in NodeB receiver). So the NodeB must detect the timing offset of the UE (time difference due to "time of flight" of the signals that are associated with distance from NodeB, and then timing errors in UE as it tries to lock to NodeB Frame Timing). So it is necessary to test that the UE responds correctly to timing advance commands from the Node B. The NodeB must be tested to ensure correct handling of "out-of-time" UE signals, and correction processes to adjust the UEs reference timing.

Two typical L1 tests are shown below, firstly the Power versus Resource Block (RB), showing the individual power transmitted in each resource block for a single time period (subframe). This will evaluate how the power is distributed across all of the available resource blocks, and hence how the available resources have been set to the correct power level for the receiver based on reports and L1 Power Control algorithms. Secondly, the Power versus RB showing the variation in time of each RB is shown. Each RB is measured over each time period (subframe), and power level is shown by the color of the Resource Block.

14.6.4 L2/L3 Testing in LTE

Layer 2 and Layer 3 testing is concerned with the signaling and message flows between the different layers in the protocol stack. In particular, the correct transfer of incoming messages received at higher levels down to/from the physical link layer (Layer 1) circuits and processes, and the correct handling of messages coming back from Layer 1. In addition, L2/L3 has to perform configuration and state control processes to ensure the UE and network are always in the correct state for communication.

This testing is normally made using a "system simulator" to generate and receive the messages to/from the protocol stack being tested. In addition, the simulator normally has a Layer 1 and physical layer implementation so that it can communicate to the target protocol stack through the Layer1 being used. Optionally, the Layer 1 may be omitted and a "virtual layer 1" may be used to link the Layer 2 and Layer 3 elements of the simulator to the protocol stack.

The system simulator is usually one of the following 3 types, dependant on the object being tested:

- Network Simulator, for UE test;
- UE Simulator, for eNodeB test;
- IP Simulator, for Gateway test.

The different simulators all have a similar architecture, using L1 hardware for PHY layer connection, and then a control environment (usually PC hosted) for L2/L3 and logging/analysis. A typical implementation for UE testing is shown below. For LTE UE test, the script for generating the test messages to/from the UE is written in TTCN-3 programming language.

14.6.5 UE Test Loop Modes

It is normally required to configure special test loop modes, where data being received by a device is automatically retransmitted back by the device. This is to enable the data rates and connectivity to be verified by sending specific data patterns to a device and checking that they are correctly received and retransmitted. This is not the normal operation mode of the devices, and so this special test loop mode is only activated during device test.

A large number of MAC, RLC, PDCP/ROHC and nearly all Data Radio Bearer (DRB) LTE tests need the test loop in the UE. Without the test loop, DRB will not have only limited test coverage and L2 will have insufficient test coverage to ensure correct operation. A single test loop at UE is required for signaling a test in TTCN3, and the closing loop point in UE is set above the PDCP (Packet Data Convergence Protocol) entity. The UE PDCP entities are configured during loop test, ciphering may be on and ROHC possibly configured for DRB test. Ciphering can be set to a dummy algorithm for some tests if necessary.

An example of implementing the functionality into a system simulator is shown below, enabling the simulator to activate the UE into test loop mode and then send test data to/from the UE to verify data throughput and correct operation of the individual signaling layers. Normally, the PHY, MAC, RLC and PDCP functions can all be tested individually and then combined into a throughput and performance test.

14.7 Test Requirements in SAE

SAE is based on an "All IP" network concept, with the ability to use IPv6 to provide the IP protocol. SAE is designed to simplify the core network architecture, provide greater flexibility, and enable simple integration of advanced network technologies such as IMS. The SAE core network architecture is designed to complement existing 3GPP networks, providing interworking and seamless handover to the GPRS and HSPA networks.

The IPv6 protocol was developed from traditional "wired" networks where the common traffic flow problems are due to overload or broken connections. The maximum capacity (bandwidth) of each data link is usually static (the maximum capacity of the fiber or the equipment connected to it), and traffic-flow problems arise from the link being overloaded or effectively zero (broken cables, damaged routers etc.). Capacity issues arises usually from volume of traffic from the users, and the network operator can easily control in a static way how many users are connected to a particular hub and what bandwidth they are offered. In a wireless link, and particularly a fast adaptive link like LTE, the capacity of the "over-the-air" data link is variable according to the propagation environment (RF path loss, distance to base-station, multipath reflections) or according to loading of the cell (the number of users in the cell can vary in real time, without the control of the operator).

So the IP routing, flow control, and QoS mechanisms must be adaptable to varying bandwidth, where the variations are:

1. Fast (RF fading happens in less than 1 second) changes in available bandwidth.
2. High Dynamic range (the level of RF loss can vary 20–30 dB in a few seconds) causing data rates to vary from 100 kbps to 10 mbps in less than a second.
3. Unpredictable (depends on human behaviors and movement of customers within the cells).

This will require the use of IP technologies suited for variable data rates and Quality of Service (QoS) provisioning. To manage the fast variable data rates, LTE uses re-transmission technology in the eNB. This requires buffering and flow control mechanisms into the eNB from the core network to prevent data overflow or loss when there are sudden signal fades that require a high amount of retransmission. So the IP network must be extensively tested for its ability to manage the flow control and re-transmission algorithms. There are buffer status registers within the control elements of the eNodeB, and these must be correctly analyzed by the flow control mechanism to prevent the loss of data during times of peak capacity or overload of the capacity.

The types of services offered in the networks can be classified into four types, the characteristics of which can be seen in Table 14.2.

It can be seen from the above that broadcast services have the most demanding requirements for the network. This is why there is now a strong demand for dedicated new broadcast technologies within a mobile network to meet these requirements. Provisioning of broadcast within the Wireless network has been a key technical challenge in the industry. Attempts have been made to implement MBMS within WCDMA networks, and other proprietary technologies for cellular networks, with very limited success and no widespread adoption of these. LTE offers a new possibility for the integration of broadcast into a wireless network as the OFDMA single-frequency network characteristic is very similar to the DVB-H standard.

Table 14.2 Service classification

Class of Service	Bandwidth	Latency	QoS Requirement	Example
Conversational	low-medium	low	Guaranteed	VOIP/Video calling
Streaming	high	low	Guaranteed	IPTV, multimedia streaming
Browsing	low-medium	normal	Best Effort	Web Browser
Background	medium	normal	minimal	e-mail synchronization
Broadcast	high	low	Guaranteed	Multi-cast

Typical Internet-based applications (such as streaming services and interactive gaming) may have been developed assuming a robust link between client and server, with only limited impairment such as delay and jitter. In an wireless cellular environment such as LTE/SAE, the "handover," fading and mobility can introduce significant delay and variation in data rates (hence in jitter) that are beyond those normally expected from a wireline network. So, the services developed for a wireline network (e.g. VoIP, Instant Messaging) need to be tested in the more demanding cellular environment. We can make such a test of the applications, client and servers using network simulators and traffic impairment simulators, to provide a controlled and repeatable test environment to isolate these effects and evaluate their impact on user experience. When we consider the effects of mobility and cellular environments on IP-based services, we need to verify that the services can work correctly under the demanding conditions of a wireless network.

The network quality assurance testing and monitoring systems must be able to monitor the delivery of each of the types of service, to ensure that customer expectations are met. This involves firstly using test equipment at the individual node/element level to verify the performance of the IP traffic through the network element, such as load test and overload conditions, latency, jitter, and so on. Secondly, the performance of the equipment in the live network should be monitored using live traffic flow logging, protocol decode and analysis tools. This will highlight the interaction between network elements according to parameters such as load, architecture and resource allocations, and allows optimization of the whole network to meet the customer's expectations.

Flow control, buffer status/overload, and bandwidth capacity/overload are normally tested at the individual node level. Testing is performed usually on a single data flow first, to ensure correct implementation of the protocol stacks and connectivity within the node. Then, it is tested with several data streams to verify that the sharing of capacity, multitasking, and flow control mechanisms are working correctly. These first two types are tested within the design specifications, to ensure correct operation according to the design. The third area of testing is then "negative testing," where the node is tested to see how it performs outside of the design specification. This will include providing degraded and bad signals (e.g. high jitter, delays, or out-of-sequence messages) or incorrect protocol messages (which should be correctly rejected), and overloading capacity.

14.7.1 Testing at the Network Service Level

Service-level testing is a critical function embedded into the network operators' operations at a high level. It involves the integration of network probes and monitors into the live network. The probes are able to capture, process and monitor the traffic flows across different links in the network. This data is then stored into a data warehouse and processed to deliver the Key

Performance Indicators (KPI's) to the network operations centre (NOC). These types of systems are often called Service Assurance systems, designed to enable the operator to ensure smooth operation of the network services. They are often integrated into high-level operational systems (for billing, revenue collection, customer marketing and retention programs) to become part of an Operational Support System (OSS) or Business Support System (BSS) solution. Either way, the function is built up from an array of network probes that are collecting, processing, and analyzing the live traffic data to indicate problems and issues in the network. The probes are directly connected to the network links such as S1 (M-GW to eNodeB), S5 (M-GW and S-GW), or SGi (connection to core network and IMS). It may also monitor the S4 interface responsible for management of handovers to an SGSN (for GPRS and HSPA handovers). These links are typically 1GbE or 10GbE fiber-optic links, and so large amounts of data can be flowing and high levels of preprocessing and filtering of data are required in the probe.

The service assurance system is often able to provide an early indication of problems in the network. Typically, it can detect high levels of dropped calls, failed handovers, and other coverage problems that directly affect user experience. The system is used to then trigger the deeper investigation by a drive test team, or "on-site" field investigation of the cell site. The system will detect when certain data links become overloaded, or are degrading traffic flow due to faults or misconfiguration. This is then the trigger to make a further investigation to optimize the links, or again make an on-site visit to the particular cabinet to see if there is some physical damage. At a higher level, the service assurance system can be used to monitor the service given to certain VIP customers, to ensure that any problems are quickly detected and resolved. Also, particular services, network features, or types of handset can be monitored to see if there are any specific problems associated with that handset or service. This again provides an early indication to the engineers and marketing departments in case of problems in the network associated with a certain service or handset.

The service-assurance solutions are usually offered by the network equipment manufacturer as an option on the equipment used in the network. This has the advantage of using access to data within the equipment/node, and not only data flowing across the links. The disadvantage is that they are normally proprietary to the equipment manufacturer. Alternative solutions from third-party providers are also on the market and are used—they have the advantage of being independent of the equipment manufacturer, so they can support networks using mixed vendor equipment, and can often be seen as providing an independent view of the network performance. Both types of solutions are widely used in the industry.

14.8 Throughput Testing

The mobile industry drive for LTE is coming from two key areas of development and innovation—firstly in applications and services driven by innovative new product technologies such as touch screen and user-friendly operating systems, and secondly in providing affordable but attractive subscription packages to encourage use of these innovations. Behind the success of this lies the need to provide users with a high-speed data/browsing experience that enables easy use of the services, but with cost-efficient networks and technologies. To enable this, the new technologies in LTE achieve very high data rates and efficient use of radio spectrum, using OFDMA and Resource Scheduling. But there has been much publicity in the industry on the difference between the marketing departments' advertised data rates, and the actual data rates seen by users and by the technical departments. So it is important to understand and measure the differences and causes between the headline data rates and the real-world data rates. A key testing technology behind this is the measurement of "throughput" on a device to determine the actual data rates achieved in certain conditions in the network. It is widely stated that the success of LTE in commercial networks depends on its ability to provide a high-speed mobile broadband experience to users, which means high data rates, low latency, and reliable data connections. This section will review the technical issues related to throughput on an LTE network, from both Base Station and Handset sides, and then discuss the techniques being introduced in the industry to measure throughput, including new OTA technology. We will also analyze some typical results of throughput testing to explain how this relates to the design of LTE base stations and handsets.

The mobile communications industry has been moving rapidly into the area of mobile data services since the beginning of the twenty-first century, with the introduction of GPRS technology to GSM networks to try to provide a more efficient data service, and with the introduction of HSPA into the 3G networks. The objective of both these technologies is to provide a more efficient use of radio resources (radio capacity) to enable more users to access data services from a single base station, and provide them with higher data rates for download/upload of data/content. With both of these older technologies, the limitation on data rates had been the air interface, as the backhaul connection to the base station provided a much higher data rate (e.g. 2–10 mbps). With the introduction of LTE, where a single base station sector can provide 100–150 mbps download capacity, then there are two key issues to consider: how to provide enough backhaul capacity to the site to support the data capacity of the site; then to have appropriate control mechanisms in the base station to share out this capacity to the users connected in the cell, according to their data-rate demands and the quality of the radio link to each user. As many LTE deployments are being made as an upgrade or extension to existing 3G networks, this end-to-end capacity issue needs to be correctly understood and managed.

14.8.1 End-to-End Network Innovation

The challenge of providing backhaul capacity has been addressed by a change in network architecture and technology in the move to LTE. The LTE network architecture is changed so that the base station is now responsible for all decisions concerning the provision of services to users over the air interface, and it only needs to receive the actual IP data packets for a user and it can manage the delivery locally. This is different from 2G/3G networks where the MSC or RNC was responsible for this data rate decision making, and then provided the base station with data already formatted in network specific protocols for delivery to the user. The backhaul links to the base station have also been changed from ATM fixed capacity links used in 3G to 'All-IP' based signaling for connecting to LTE base-stations. This allows operators to use existing IP network infrastructure to provide convenient and high data-rate links to base stations on the S1 interface. This trend had already started in 3G with the introduction of Ethernet-based Iub links for handling HSPA data.

14.8.2 Base Station Scheduler as Key Controller of Radio Resources

In LTE networks, the air interface technology has been evolved from HSPA technology used in 3G networks. The access technology has changed from WCDMA (a code-based access scheme) to OFDMA (a frequency/time block-based access scheme). This gives greater spectral efficiency (higher data capacity in a given amount of bandwidth), and greater capacity to manage more users and share the radio link capacity between them. The key element in this new scheme is the "scheduler" in the base station. This has the role of deciding how/when each packet of data is sent to each user, and then checking that the delivery is successful or retransmitting if it is not successful. The scheduler is located within the eNodeB, and is responsible for deciding the actual data rates being offered to each and every user, in the uplink and the downlink, on a continuous basis. So, we can clearly see that this critical function must be fully tested and optimized if the LTE network is to give a good user experience.

The scheduler has two different techniques that can be used to prioritize the data flows for certain time critical applications, and to guarantee the data rate for specific applications/users. Firstly, for the downlink, the MAC scheduler in the eNodeB is able to provide Guaranteed Bit Rate bearers (GBR) to the higher layers (i.e. to the RRC layer). This means that, at the low layer MAC, it will guarantee the allocation of sufficient resources to the air interface (i.e. enough resource elements) to ensure that a specified bearer service (i.e. one particular IP flow to one user) will have a guaranteed rate. This could be used, for example, in video streaming or video conferencing, or providing a service to a VIP customer. The S-GW will also impose a maximum downlink aggregated maximum bit rate (AMBR) that limits the data fed into the eNodeB to prevent excessive buffering or queuing of data packets at one particular eNodeB. Secondly, the uplink bearers are given a "priority" level by the RRC, and a Prioritized Bit Rate (PBR). With this process/setting, the capacity is allocated to the uplink bearers according to PBR setting, with the highest priority being served first and lowest priority being allocated resources only when all other higher classes are served. In the next section we will look more closely at MAC and RRC bit rates, testing the throughput, and measuring the differences between them.

14.8.3 L1 Performance vs. L3/PDCP Throughput

The headline performance of LTE networks has been specified (and advertised) as 100 mbps, so it is important to understand what this number means. This is actually the capacity of the base station cell to transmit over the radio link (called Layer 1). This means that the capacity of 100 mbps is shared across all users on the base station according to different parameters. So, in theory, the base station can allocate all of the available capacity to one user so that user has 100 mbps, or allocate different amounts of the 100 mbps to several users on the network. This "sharing" decision is made every 1 ms (the scheduling interval), so the capacity given to each user is changed every 1 ms. However, this is the raw radio link capacity being shared, and this includes not only the new data to be transmitted to each user, but also retransmissions of previous data packets that were not successfully received by users. So, if the base-station scheduler tries to allocate too high a data rate to users (by using higher rate modulation and coding schemes than the datalink can sustain), then more of the capacity can be used up in retransmitting this data again at a lower data rate. Thus there is a balancing act between sending data at highest rate possible (to maximize the use of available radio resources and give highest capacity/data rates), and then sending at too high a rate so more capacity is used in retransmission. In the situation where there are high retransmission rates then, although the air interface is still being used at maximum capacity, the users will experience a much lower data rate (the Layer 3/PDCP data rate). In this case (of choosing first a high data rate and then retransmitting at a lower rate), the users experience of data rate would actually be lower than if a lower data rate were selected to start with. Layer 3 is the higher layer in the protocol stack, and represents the actual data rate achieved by the radio link as seen by an external application. The external applications connect to Layer 3 of the protocol stack via PDCP, which enables IP data to link into LTE protocol stacks.

A key parameter for the base station scheduler when selecting the data rate for each user is the "UE measurement reports" sent back from each user to the base station. These reports are measurements of critical radio-link quality parameters to enable the base station to select the best data rate for each user. LTE also uses a Hybrid Acknowledge Request (HARQ) process to enable the UE to acknowledge the correct reception of each data packet. In this process, each data packet is sent from the base station, and then there is a wait for a positive acknowledgement of correct reception. If the response is negative (incorrect reception) or not acknowledged, then the packet is then rescheduled for a repeat attempt to send, now using a lower data rate that is more likely to succeed.

We can see from the previous section, that the UE's ability to correctly receive each packet of data, together with the ability to provide an accurate measurement of the radio propagation/reception characteristics, is critical to the data throughput performance of an LTE network. Where UE is making inaccurate measurement reports, then the base station will send a larger amount of data at a rate that is too high for the UE, and will then be forced to retransmit this at a lower data rate. Where the UE receiver has a poor implementation, then it will not be able to decode data sent to it by the base station when it has been calculated that the chosen data rate should be suitable. Again, the base station will be forced to retransmit at a lower data rate. Both of these phenomena will have the effect of making the perceived data rate for all users on the network lower than the expected 100 mbps. To ensure that this does not happen, there is now a set of UE throughput test and test equipment available to measure and confirm that a particular UE implementation is performing to the level expected by the base station.

The test environments available are based on the 3GPP Conformance Test Specifications (TS36.521) to ensure quality meets the minimum requirement, and on R&D tools developed for deeper analysis and debugging possible errors in UE implementation. Both systems are built around the use of a Network Emulator to simulate and control/configure the LTE network, and a fading simulator to control/configure the radio link quality between the UE and the network. Using this architecture, it is possible to put the UE into a set of standard reference tests to benchmark any UE against the 3GPP standards and also against other UE implementations. As this is based on simulator technology, it is possible to create precise and repeatable conditions for testing that cannot be guaranteed in live network testing. So this technology forms the basis for comparative testing and benchmarking of UEs and is used by many network operators to evaluate performance of UE suppliers.

Using the controllable nature of simulator testing, the UE developers also use the technology for deeper investigation of UE performance. As they are able to carefully select and control each parameter of the network and radio link, specific issues can be deeply investigated and then performance improvements and fixes can be accurately and quantifiably measured to confirm correct operation. A key aspect to this testing concept is the ability of the test engineer and designer to see both the Layer 1 (radio link layer) throughput and the Layer 3/PDCP (actual user data) throughput. This enables testers and designers better to understand how much throughput capacity is being used for retransmission of incorrectly received data versus actual user data.

Looking at typical results for testing we see that we can provide pass/fail results from the Conformance Test specifications, which provide the baseline for compliance to 3GPP and basic performance. As we go deeper into the testing, we then concentrate on evaluating the throughput at Layer 1 and Layer 3 separately, and the ratio between these. For measuring the throughput we measure the number of Packet Data Units (PDUs) transmitted and the size/configuration of these PDUs. The Layer 1 performance is measured as MAC PDUs, and the Layer 3 throughput as PDCP PDUs, We also need to monitor the UE reports that show the measured signal quality (RSRQ), data reception quality (CQI) and acknowledgement of correct data received (ACK/NACK). These reports are used by the base station to select the optimum format to transmit the next packet of data. In addition, where MIMO is being used, there are two additional reports from the UE to assist the base station in selecting the optimum MIMO pre-coding. These are the PMI and the RI, which report the preferred MIMO matrix to be used for the current multipath environment and the number of separate MIMO data paths that are calculated in the UE.

As the propagation conditions between UE and base station are reduced, we should see the PDU throughput level reduced. At the same time, the UE should report lower RSRQ and CQI to indicate poorer link quality. It is therefore important to monitor the reports and characterize these across a range of propagation conditions. When these conditions are reduced, we should see the base station scheduler selecting lower data rates (modulation type and coding rate) as the response and hence see lower RRC PDU data rates. In an optimum implementation of both base station and UE then the MAC PDU rate should fall at the same rate. As the multipath conditions are reduced, and the cross correlation between different paths is increased, the MIMO based data rate improvement should reduce, also shown as a lower PDCP PDU rate.

Where the PDCP PDU rate is decreased more than the MAC PDU rate, then we should be seeing failed data packet delivery and retransmission. This is monitored via the ACK/NACK reports from the UE, which will turn to more NACK status. As these retransmissions represent a reduction in network capacity and reduction in user perceived data rate, we must aim to minimize this occurring. Fault tracing is made through the above measurements, ensuring that the UE is making correct reports of signal link characteristics and that the base station scheduler is selecting an optimum modulation and coding scheme to suit the channel conditions. The network emulator is used to then change the signal power level and then the CQI and RSRQ calculation can be verified and calibrated to locate any errors.

For testing the eNodeB, the reverse situation is applied, where a UE simulator is connected with a channel emulator. Using the UE simulator, a specific and repeatable set of UE measurement reports can be sent to the eNodeB, together with specific patterns of ACK/NACK, and we can verify that the eNodeB schedules the correct levels of resource to the UE. The UE simulator has the ability to behave as a regular UE and provide normal operation to the eNodeB, but due to the nature of its design as a test tool we can also manually control the procedures and reports from this test UE. This way, the eNodeB is tested in specific conditions, and faults can be more easily repeated and debugged using a systematic approach to isolating each individual report and decision process. The UE simulator is also used to test the eNodeB with "false" measurements or inconsistent data, to ensure that the scheduler algorithm is robust and does not crash in certain unexpected circumstances. As such, the UE simulator is one of the key tools used for design verification and optimization of eNodeB performance, and hence one of the most important tools in ensuring good throughput in an LTE network. The same measurement process as above is used, looking at both the L1 and L3 PDU throughputs. In addition, the UE simulator normally has a "multi-UE" capability to load up to 200 individually controllable UEs onto the eNodeB and then verify the correct sharing of resources between the different UEs.

14.8.4 OTA (Over The Air) Testing

As the performance and data rates for a packet-scheduled data transmission (as used exclusively in LTE, but also introduced in HSPA) are dependent on the air interface coupling between the UE antenna and base-station antenna, there has evolved in recent years a new need for OTA testing. In previous technologies, the influence of the UE antenna was minimal and could be approximated to a fixed antenna gain/loss. So it was always possible (and more reliable and repeatable) to make a direct RF cable connection between the UE and the test equipment. These allowed for very accurate measurement of UE output power and receive sensitivity. However, it excluded the effect of the UE antenna. As modern UE designs incorporate the antenna into the aesthetic design of the device, and at the same time directly influence the possible data throughput rates, it is now necessary to include the effects of the UE antenna in the test. Where multiple antenna schemes are used in a UE (e.g. in receiver diversity, or MIMO) then the coupling between the antennas and the base station via the atmospheric path directly influence the performance and data throughput rate.

The traditional method for measuring this would be through a drive test of a device on a particular route, benchmarking the data rate against a known reference phone (or "golden" device). This would provide a benchmark to see if the performance of the new device was comparable or better in the network than previous devices. With LTE, however, this is not possible as the data rate is changed dynamically by the scheduler according to the performance of handset and network, but also the load on the network. As this can no longer be fixed data rate provided by the network, there is no reliable way to make comparable measurements in the live network. So a new type of test method is required that manages this effect, to evaluate what is the true data rate a user could experience from a particular handset design when used in a real-world environment. The resulting value of the relative throughput performance is called the Figure Of Merit (FOM). This is achieved using a network emulator device that is able to fix exactly the data rates and scheduling offered to the UE in a precise and repeatable way, including all effects such as capacity grants, overload, scheduling, and so on. The network emulator must then be coupled to the UE via its real antennas in a reliable and repeatable way. Several alternative technologies are being developed for this.

The coupling to the UE is made in a shielded chamber, where the UE is isolated from all external RF signals and only experiences the RF signals and paths generated in the chamber. There are then three basic types of chamber: anechoic chamber, reverberation chamber, or isotropic chamber. The anechoic chamber uses a technique where all reflections/multipath are eliminated by design, and then only selected paths are introduced using selective antenna placement and a switch matrix to connect to these selective antennas. The MIMO conditions and paths are then controlled using a channel emulator to provide accurate phase/amplitude variations. The reverberation chamber takes the opposite approach, stimulating many reflect paths within the chamber and then rotating the UE to make an "average" of the many paths that are present. Again, a channel emulator may be used to control the MIMO phase/amplitudes. Finally, the isotropic chamber uses only a single pair of antennas/paths, with all reflections suppressed, and characterizes the antenna pattern of the UE as a stand-alone parameter. In a second stage, the characteristics of the antenna are simulated and the system then uses only the channel emulator to fully configure and control the multipath environment, with a direct cable connection to the UE. At the time of writing, all three methods are being evaluated in trials to determine the comparative results and performance of each method. Each method offers different advantages in terms of cost, size, measurement speed, and accuracy.

14.8.5 Summary

LTE networks are designed for end-to-end IP packet data services, and the air interface is optimized to deliver packet data streams with the most efficient use of radio resources. The mechanisms in the base station and UE create a feedback loop to optimize the selection of most suitable settings for transmitting each individual packet of data. These are based on reporting of channel conditions and adaptation of the OFDMA configuration to match this. Using a laboratory-based test environment based on network/UE simulators and channel emulators, which is an accurate, controllable and repeatable test environment, we can measure the throughput of an LTE link at different points in the air interface protocol stack to see both the actual air interface data rates and also the user perceived data rate. The testing must also monitor the associated UE reports to confirm correct operation, baseline the performance of different implementations, and identify possible areas for further optimization of a design.

14.9 Self-Organizing Network Techniques for Test and Measurement

Self-Organizing Networks (SON) is a new technique being introduced within LTE/SAE as part of the next generation of mobile broadband network technology, and is endorsed by the NGMN alliance as a key requirement for future networks. The objective is to automate the configuration and optimization of base-station parameters to maintain best performance and efficiency. Previously a drive test team would go out into the live network and take a "snapshot" of the performance, then bring this back to the lab and analyze it to improve the settings. Of course more snapshots give more data and hence better optimization, but this drive-test-based data acquisition process is expensive, difficult, and not repeatable. In addition, this is a reactive method to cure problems after they have occurred, and this does not help improve customer experience. Drive test is also heavily used at the start of a network deployment to measure cell coverage and set initial parameters for cell powers, frequencies, and so on, in order to control interference and maximize capacity.

SON should enable network operators to automate these processes using the measurements and data generated in the base-station during normal operation. By reducing the need for specific drive-test data, this technique should reduce operating costs for an operator. By using real-time data generated in the network, and reacting in real time at the network element level, this should enhance customer experience by responding more dynamically to changes and problems in the network much earlier so that users are less affected.

14.9.1 SON Definition and Basic Principles

14.9.1.1 Self Organizing Networks

SON is the top-level description of the concept for more automated (or fully automated) control and management of networks, where the network operator has only to focus on policy control (admission control, subscribed services, billing, etc.) and high-level configuration/planning of the network. All low-level implementation of network design and settings is made automatically by the network elements. The self-organizing philosophy can then be broken down into three generic areas relating to the actual deployment of the network. These are configuration (planning and preparation before the cell goes live), optimization (getting the best performance from the live cell), and healing (detection and repair of fault conditions and equipment failures). Each of these is further explained below.

14.9.1.2 Self-Configuration

This is the first stage of network deployment, and covers the process of going from a "need" (e.g. a need to improve coverage, improve capacity, fill a hole in coverage etc.) to having a cell site "live" on the network and providing service. The stages involved here are roughly:

- planning for location, capacity and coverage of the eNodeB;
- setting of the eNodeB parameters (radio, transport, routing and neighbors);
- installation, commissioning, and testing.

The Self-Configuring network should allow the operator to focus on selecting location, capacity and coverage needs, and then SON should automatically set the eNodeB parameters to enable the site to operate correctly when powered on. This will in turn minimize the installation and commissioning process, and enable a simple "final test" at the site to confirm that the new site is up and running. This will include the optimum setting for power levels, the choice of Cell ID, and correctly identifying the Cell ID of neighbor cells. The neighbor Cell ID must then be used by the eNodeB to negotiate with these neighbors using the ICIC (Inter Cell Interference Control) algorithms on the S2 interface. This is critical to prevent interference where the coverage of 2 cells is overlapping.

14.9.1.3 Self-Optimizing

Once a site is live and running, there are often optimization tasks to be carried out that are more "routine maintenance" activities. As the geography of the area changes (e.g. as buildings are constructed or demolished), the radio spectrum changes (e.g. new cells are added by the operator or by other operators, or other RF transmitters in the same area or at the same tower), then the neighbor cell lists, interference levels, and handover parameters must be adjusted to ensure smooth coverage and handovers. Currently, the impact of such issues can be detected using an OSS monitoring solution, but the solution requires a team to go out in the field and make measurements to characterize the new environment and then go back to the office and determine optimum new settings. SON will automate this process by using the UEs in the network to make the required measurements in the field, and the SON function will report them automatically back to the network. New settings can be determined from these reports. This will remove the need for drive test teams to make such measurements. This concept can also be extended to managing QoS and load balancing by using quality reports to optimize the scheduling algorithms in the eNodeB.

14.9.1.4 Self-Healing (Fault Management and Correction)

When a site is fully operational and active then it is generating revenue and satisfying customers. If there are any problems with the site and it fails to provide a service or coverage then revenue and/or customers are lost, and so the site must be brought back up to full capacity as soon as possible. The third element of SON is to automatically detect when a cell has a fault (e.g. by monitoring both the built in self test, and also the neighbor cell reports made by UEs that are/should be detecting the cell). If the SON reports indicate that a cell has a failure then there are two necessary actions: to indicate the nature of the fault so the appropriately equipped repair team can be sent to the site, and then to reroute users to another cell if possible and to reconfigure neighbor cells to provide coverage in this area whilst the repair is underway. After the repair, SON should also take care of the site restart in a similar process to the site commissioning and testing.

14.9.2 Technical Issues and Impact on Network Planning

To deploy SON in a multi-vendor RAN environment requires standardization of parameters for reporting and decision making. The eNodeB will need to take the measurement reports from UEs and also from other eNodeBs, and report them back into the O&M system, to enable optimization and parameter setting. Where multiple vendor equipment is involved then this must be in a standardized format so that the SON solution does not depend on a particular vendor's implementation.

The equipment vendors who are implementing SON will need to develop new algorithms to set eNodeB parameters such as power levels, interference management (e.g. selection of subcarriers), and handover thresholds. These algorithms will need to take into account the required input data (i.e. what is available from the network) and the required outcomes (including cooperation with neighbor cells).

Furthermore, as SON is also implemented into the core network Evolved Packet Subsystem (EPS), there need to be standards on the type and format of data sent into the core. Inside the core network, new algorithms will be required to measure and optimize the volume/type of traffic flowing taking into account the QoS and service type (e.g. voice, video, streaming, browsing). This is required to enable the operator to optimize the type and capacity of the core network, and adjust parameters such as IP routing (e.g. in an MPLS network), traffic grooming, and so on.

14.9.3 Effects on Network Installation, Commissioning and Optimization Strategies

Equipment vendors have the opportunity now to develop algorithms that link eNodeB configuration to customer experience, allowing fast adaptation to customer needs. Here customer experience is exactly that which is measured by their UE in the network. The challenge is to link RF planning and customer "quality of experience" closer together at a low level technical implementation. The benefit is that the network can adapt to meet the user needs in the cell without additional cost of optimization teams constantly being in the field. The network planners' simulation environment will now need to take into account the SON operation of eNodeB when making simulations of capacity/coverage for the network. As the operator may not directly control/configure the eNodeB, the simulation environment will need to predict the behavior of the network vendors SON function in the network.

The operator's/installer's site test must verify that all parameters are correctly set and working in line with the initial simulation and modeling. This will ensure that the expected coverage and performance is provided by the eNodeB. The SON function will then "self-optimize" the node to ensure that this performance is maintained during different operating conditions (e.g. traffic load, interference). This should reduce the amount of drive testing required for configuration and optimization (in theory reduced to zero), and drive testing is only needed for fault finding (where SON is not able to self-heal the problem). We will see in the later section on live network testing that a comprehensive suite of RF "over-the-air" tests can be made at the time of site installation, so that SON can be correctly configured and verified. It is generally expected that SON may be able to reduce the level of drive test needed to initially configure the network, but will not be able to replace the initial site commissioning/acceptance tests. So the preferred test strategy is to use the initial site tests to further strengthen SON parameter setting.

The potential disadvantage of running SON in the network is that it requires UEs to make the measurements, and relies on enough data being available. The eNodeB is able to command a UE to make measurements and report them, but doing this on a regular basis will have an impact on the battery life of the UE. On current generation Smartphones the battery life is already a limitation, so extra SON measurements can significantly reduce the batter life.

14.9.4 Conclusion

SON can simplify operators' processes for installing new cell sites, reducing the cost/time/complexity to install new sites. It gives an obvious benefit if deploying femto cells, as the operator is not strictly in control of the cell site and needs to rely on automated processes to correctly configure the cell into the network. The running costs of the site are also reduced, as drive test optimization is reduced and site visits for fault investigation and repair can be reduced. All of this leads to OPEX savings for the network by using automated technology to replace manual operations. For the customers on the network, SON will lead to better customer satisfaction as coverage and QoS are driven and optimized by actual customer usage, and there should be reduced downtime or faulty cells. The OSS monitoring systems and SON should work together to automatically detect usage trends and failures and automatically take action in real time to correct errors.

14.10 Field Testing

The two basic types of field testing for an eNodeB are related firstly to the RF power, spectrum and emissions, and secondly to the quality of the modulated waveforms. The RF problems are usually associated with interference issues, and modulation problems are associated with data rates and coverage problems. In this next section we will look firstly at the measurement techniques for modulation quality, and then at the RF power measurements that relate to coverage. Finally we will analyze the test and de-bugging for typical types of hardware failure that lead to RF spectrum problems and interference issues. At the end of the chapter there are two tables to summarize firstly the related field measurements for coverage problems (power and quality issues) and secondly to identify typical hardware failures.

The most convenient and preferred method of validating modulation quality on a live eNodeB is to make "over-the-air" measurements where the actual signal that is being received by UEs is captured and analyzed. Multi-antenna methods such as MIMO (e.g. spatial multiplexing for two-layer transmission) and beam forming are used in LTE to increase the signal-to-noise ratio in a data link, to provide higher data rates or wider area coverage, but they also increase the difficulty of the basic operational and troubleshooting techniques that were much easier with the previous generation of network technology. Spatial multiplexing and beam forming provide the biggest challenges for over-the-air quality measurements, and the dynamic nature of which multi-antenna technique is used at any given moment adds even greater complexity. This is because the feedback reports from the UE (CQI Quality, PMI, and RI) allow the base station to change the multi-antenna processes dynamically to suit the transmission path characteristics. With spatial multiplexing, the different transmitting antennas are seen as co-channel interference sources to a single receiver. This requires very expensive and powerful measurement device with multiple receivers to fully resolve the individual signal qualities. Beam forming can also present problems because it increases or decreases the amount of power received in particular areas on a continually changing basis, making reliable measurements of the signal impossible for a basic measurement device.

These measurements would need to be made in a "closed loop" system such as a UE simulator, where the measurement receiver is able to send reports actively back to the base station, to provide a stable and controlled test environment. While transmit diversity does not present a measurement problem (because multiple antennas' signals can be recovered with a single receive channel), each resource block in the Physical Downlink Shared Channel (PDSCH) used to transmit LTE data can change multi-antenna mode dynamically based on signal conditions per user. When looking at a captured signal with a measuring instrument, it is impossible to know if each resource block uses spatial multiplexing, beam forming, or transmit diversity without fully decoding the PDCCH control channel information. Typically, these measurements are made during development test, in the equipment manufacturers' laboratories, but they cannot be easily replicated in the field on a live network.

These "over-the-air" complications can be avoided by directly connecting the measurement instrument to the eNodeB transmitter using an RF cable link to the instrument. This approach provides the most thorough and accurate measurements of modulation quality, so it is essential in some cases where serious quality issues need to be measured in detail. However, there are a number of limitations associated with direct-connect measurements such as:

- It takes time to open up a shelter or building to get physical access directly to the transmitter.
- If the transmitter has a test port then connecting to the instrument is not a problem. If there is no test port, then you have to disconnect the transmitter from the antenna, which is usually difficult and time consuming. This will also take the cell site "off air" and so will interrupt service in the area.
- If the site uses a Remote Radio Head (RRH) or Remote Radio Unit (RRU), then you need to gain physical access to the RF signal. This may not be difficult if the RRH/RRU is mounted inside a building or on a roof with reasonable access, but if the RRH/RRU is mounted on a tower or inaccessible roof, then you need to climb the tower or otherwise gain access to the transmitter which is usually a difficult and expensive process.

Making field measurements of a live site "over the air" is much easier and faster than making a direct RF connection measurement. Speed is particularly important when troubleshooting a reported customer problem. The test industry has introduced LTE measurement options to handheld base station analyzers and other instruments to perform over-the-air modulation quality measurements. These new measurement options are normally available on handheld spectrum analyzers or as part of a dedicated base station test tool (such as the MT8221B BTS Master base station analyzer from Anritsu). These platforms are specifically designed to support emerging 4G standards such as LTE, including 20 MHz demodulation capability. They are small, lightweight and battery operated, making them easy to use anywhere at a cell site. They also include a suite of measurement capabilities for measuring all key aspects of base station performance, including line sweep, spectrum measurements, interference hunting, and backhaul verification. Another advantage of these handheld base-station analyzers is that they are normally used by technicians and RF engineers testing and verifying the installation and the commissioning of base stations for optimal wireless network performance.

Using an instrument with capability to do the Over-The-Air measurement, it is possible to make measurements on the Physical Broadcast Channel (PBCH), which uses only Transmit Diversity. This allows us to measure the modulation quality without being affected by the influence of MIMO (spatial multiplexing or beam forming) as they are not applied to this transmission channel. This technique allows us to make a quick and simple over-the-air measurement to see if there are any basic problems with the site (e.g. damaged transmitter elements or badly tuned components). Normally, if a failure is found then the related hardware board is swapped for a known good unit. The suspect board is then sent back to a depot or test centre where more detailed test systems are used to find the specific failure cause. Following this, the board is sent back to the factory for repair and then a full test in the normal production test suite.

The field test process is to first set the instrument to the signal of interest, selecting the correct center frequency for the downlink transmission, and attach an appropriate antenna to the instrument through a short cable. Next, find a "sweet spot," a place where the signal strength of the eNodeB to be measured is high and interference, especially from other eNodeBs, is low. Normally a measurement instrument can show the signal strength (using the LTE Sync Signal or SS), as well as interference level (reference channel dominance of one eNodeB over the others). The sweet spot will be a short distance from the transmitter, and near the center of the antenna coverage pattern. If you are too close, the antenna beam will be above you and the signal will be low as you are on the "near side edge" of the coverage pattern; if you are too far the signal strength will also be too low, there will be too much multipath, and too much interference from adjacent transmitters. Being in the center of the sector's transmit beam reduces the co-channel interference from adjacent sectors. The recommended approach for a typical macro site is to start measuring from around one hundred meters away from the antenna, in the beam center, then walk or drive around to find the best available location and record the location with the GPS coordinates.

Omnidirectional antennas are much more convenient for making over-the-air measurements because of their small size and nondirectional nature, but the downside is that they receive less signal and sometimes the signal strength is too weak to make good measurements. A good tradeoff is to make initial measurements with an omni antenna, and then if a problem is detected connect a larger Yagi directional antenna, which will provide higher signal strength. Rotate the directional antenna around 360 degrees to find the best measurements; while usually this is when pointed directly at the transmit antenna this is not always the case due to multipath effects. If the directional antenna can be rotated so that EVM levels are reduced below the measurement specification, then this is a good indication that the transmitter is working correctly and any signal problems are coming from external interference. The 3GPP modulation quality specifications for LTE transmitters are 8% EVM or less at 64QAM modulation scheme. This level can be applied when making a direct connect measurement as the signal path and interference levels are well understood and controlled by the RF connection cable. In the case of an over-the-air measurement, an additional margin should be added to take account of the uncertain signal path. As a general rule, readings under 10% are good, as the contribution from the atmospheric path would not be more than 1–2% distortion of the signal.

As an additional note, it is good practice in the industry to find a sweet spot and take an over-the-air measurement when the base station is first commissioned. This will provide a benchmark value to maintain going forward, and to use as a repeatable reference measurement in case of future problems. By recording the GPS location of the measurements, it is quickly replicated in the future for comparison purposes. A typical example would be to evaluate the site interference levels after additional buildings or radio sites have been installed by other radio network operators in an area of rapid urban development.

14.10.1 LTE Coverage and Power Quality Measurements

There are two basic techniques for evaluating the coverage of an LTE cell site. Firstly, there is the "normal operational"' method where the Reference Signal received power (RSRP) is measured by a UE. This takes place in the normal operation of a UE in the network, and is reported back to the network by the UE when commanded to do so. This is a normal operation procedure and is used in the eNodeB and network operations to manage the settings and power levels in the cell. The second is to use a specific measurement instrument to make a more detailed and accurate measurement, allowing the recording and analysis of additional coverage parameters.

RSRP is measured on the specific eNodeB transmission called the Reference Signal. This signal has the property of being specific to each antenna on the cell, allowing a UE to determine from which antenna the signal came in the case of a multi-antenna site. Multi-antenna sites are normal in LTE as they are required for the support of MIMO (Spatial Multiplexing or beamforming), transmit diversity, or MBMS. As the Reference Signal is antenna specific, the receiver is required to decode and analyze the signals to determine from which antenna the Reference Signal is coming and hence how to correctly decode.

Once the UE has correctly interpreted the Reference Signals, the power per resource element is measured (as each Reference Signal occupies a single resource element per resource block) and reported as RSRP. The UE is also required to measure the total received signal power in the radio band (RSSI), and should report this parameter. Finally, the RSRP and RSSI ratio is calculated in order to show the signal quality in terms of the ratio between Reference Signal and total received signal. This is reported as the Reference Signal Received Quality (RSRQ).

This normal RSRP and RSSI reporting is used for measurement on all cells, and is the technique used for "drive test." In this type of testing, a live UE is connected to the network and then driven around specific areas/routes to record the signal level (with GPS location) so this can be mapped into planning tools. This then gives a map of the coverage in each cell, and allows planners to evaluate power levels, antenna direction/tilt, and hence to optimize the coverage and interference levels to give the best possible performance. The advantage of this method is that it accurately evaluates the exact signals seen by the UE and reports made by a UE. The problem is that make a measurement requires the UE to be connected to the network (not always possible if the site has a problem) and the measurements are restricted to the level accuracy of the UE. As an alternative, it is possible to use a "scanner" to perform the RSRP/RSSI/RSRQ measurements. These have a more accurate and sophisticated receiver that can correctly identify the different antenna reference signals and measure the power according to the 3GPP method. The key advantage for this instrument is it has much better accuracy in measurement than a UE due to it being built from purpose-built components, not limited by size/cost/power restrictions of a real UE. Such a scanner is usually relatively highly priced, and used in complete drive-test systems that aim to capture all possible network information when debugging intermittent problems in a live network.

The cell's radio coverage can also be measured by observing the Synchronizing Signal (SS) of the LTE base station (eNodeB). This is an additional method for understanding the downlink signal coverage quality for LTE networks. Demodulation and analysis of this signal is available in handheld instruments such as the MT8221B BTSMaster [16] as presented in Figure 14.19.

The approach outlined in this section is more convenient and less expensive than drive test systems, as most network operators have only a few such systems. This analysis will explain in more detail how a simple handheld instrument can use the eNodeB synchronization signal (SS) power to estimate LTE coverage, as well as how to interpret these measurements.

14.10.1.1 Using SS Power to Estimate Total Power and Coverage

The SS can provide an accurate alternative for the power measurements because SS power is a static value that is related directly to the eNodeB maximum output power. When all the Resource Blocks are occupied and no power control is used (such as in the 3GPP LTE Test Model 1.1 signal1), the P-SS and S-SS subcarriers are at the same power level as the other subcarriers. Then SS power per subcarrier and symbol (also called Energy per Resource Element or EPRE in the standard), is 0*log10(subcarriers) lower than the total power. At 10 MHz, this formula yields a value of [27.78 dB] which means that the SS power is [27.78 dB]

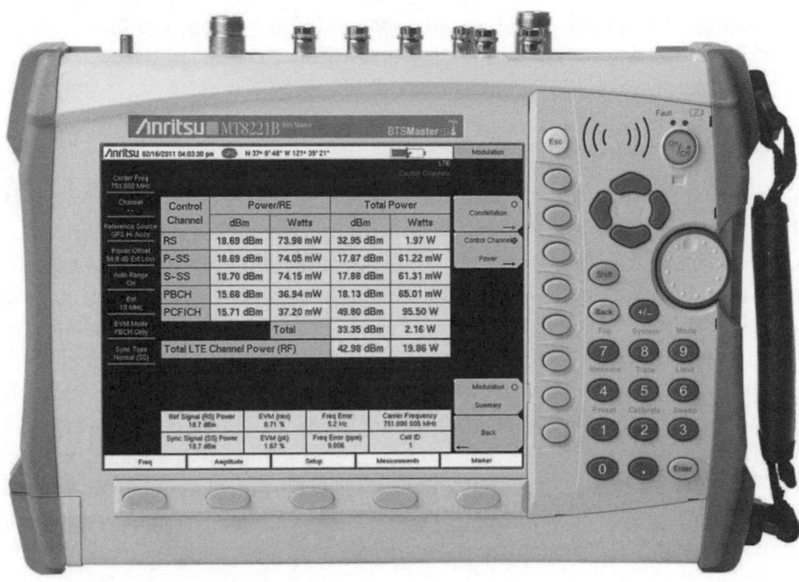

Figure 14.19 An example of portable LTE measurement equipment (MT8221B BTSMaster). Printed by courtesy of Anritsu.

below the maximum channel power. Table 14.3 shows the relationship between SS power and maximum output power at all applicable LTE bandwidths, for the nominal SS power setting where all REs have the same power level. Table 14.3 presents the principle.

However, the SS power can be adjusted in the eNodeB. The values in Table 14.3 will vary based on this adjustment—if the SS power is 1 dB lower than nominal, then the ratio between SS power and total power will be about 1 dB higher. For the most accurate measurements, use the Test Model 1.1 signal from the eNodeB. To measure the power on live traffic in an active cell, you need to know what the SS Power setting is on the eNodeB, and make this correction to the above table. The final ratio between the rated output power and the SS Power is then required to scale the data in the coverage map correctly.

For example, with a 10 MHz channel bandwidth and a 40 W (46 dBm) output power (total for all transmitters), the expected SS Power would be [46dBm]—[27.78dB] equal to [18.2 dBm] or (33.3 mW). If the eNodeB is set to provide SS Power of [21.2 dBm], this requires an additional offset of 3dB on the coverage map. Given a known relationship between the SS Power and the

Table 14.3 Total output power as a function of SS power (for nominal SS power setting)

Bandwidth	Number of Resource Blocks	Maximum Total Output Power/RS EPRE (dB)
1.4 MHz	6	18.57
3 MHz	15	22.55
5 MHz	25	24.77
10 MHz	50	27.78
15 MHz	75	29.54
20 MHz	100	30.79

maximum output power, and a UE sensitivity specification (which is based on the maximum output power), we can determine an equivalent sensitivity based on the SS power.

14.10.1.2 Interpreting an SS Power Map

The quality of downlink coverage at a particular location can be determined by comparing the measured SS values with the UE receiver sensitivity specifications as defined in the LTE standards document or with some other sensitivity specification set by a network operator. We need to make an adjustment for the relationship between the SS Power (which we are measuring), and the maximum output power (which is what the UE sensitivity specification uses). The required sensitivity in Table 14.4 varies with the bandwidth and frequency band. For example, the table provides a reference sensitivity of [−94 dBm] in the 10 MHz bandwidth for Band 3, so the corresponding SS power would be [−94 dBm] −[27.8dB] equal to [−121.8 dBm].

The captured data can be loaded onto coverage maps and color-coded to show how the coverage varies across the cell site, using known thresholds to change the color between good and bad coverage areas. The power level used on the coverage map is based on the relationship

Table 14.4 Reference sensitivity for QPSK mode, in dBm (from 3GPP TS 36.101 V8.10.0 (2010–06) [9])

Band
Downlink
Frequency
(MHz)
Channel bandwidth (MHz)
20 15 10 5 3 1.4
1 (FDD) 2110–2170 −94 −95.2 −97 −100
2 (FDD) 1930–1990 −92 −93.2 −95 −98 −100.2 −103.2
3 (FDD) 1805–1880 −91 −92.2 −94 −97 −99.2 −102.2
4 (FDD) 2110–2155 −94 −95.2 −97 −100 −101.7 −105.2
5 (FDD) 869–894 −95 −98 −100.2 −103.2
6 (FDD) 875–885 −97 −100
7 (FDD) 2620–2690 −92 −93.2 −95 −98
8 (FDD) 925–960 −94 −97 −99.2 −102.2
9 (FDD) 1844.9–1879.9 −93 −94.2 −96 −99
10 (FDD) 2110–2170 −94 −95.2 −97 −100
11 (FDD) 1475.9–1495.9 −97 −100
12 (FDD) 728–746 −94 −97 −99.2 −102.2
13 (FDD) 746–756 −94 −97
14 (FDD) 758–768 −94 −97 −99.2
17 (FDD) 734–746 −94 −97
33 (TDD) 1900–1920 −94 −95.2 −97 −100
34 (TDD) 2010–2025 −94 −95.2 −97 −100
35 (TDD) 1850–1910 −94 −95.2 −97 −100 −102.2 −106.2
36 (TDD) 1930–1990 −94 −95.2 −97 −100 −102.2 −106.2
37 (TDD) 1910–1930 −94 −95.2 −97 −100
38 (TDD) 2570–2620 −94 −95.2 −97 −100
39 (TDD) 1880–1920 −94 −95.2 −97 −100
40 (TDD) 2300–2400 −94 −95.2 −97 −100

between the output power implied by the SS measurement and the UE sensitivity specifications. A [− 124 dBm] threshold equates to a sensitivity of [− 96.2 dBm], which is roughly in the middle range of the sensitivity specifications. This sensitivity level is for the QPSK operation, which will be used at the cell edge. However we also want an indication of where faster data rates are possible. The power levels measured can allocated other color steps based on estimates of the additional power required for using higher level modulation formats such as 16QAM, 64QAM, and even spatial multiplexing (MIMO). LTE also has a variable amount of error protection coding, so there are many more combinations that can be shown in a simple graph. In general, however, higher power allows the more complex modulation format and less error protection, which allows faster data transmission to that user. While these thresholds are not a precise determination of what modes should be used by the UE, it is a useful indication of performance levels that may be possible (e.g. QPSK, Transmit Diversity, high coding rate required; or 64 QAM, Spatial Multiplexing, low coding-rate possible). Technicians and engineers can easily drill down to the underlying data when they need the actual measurement values.

While the discussion above focuses on thresholds for 10 MHz channel bandwidth, note that as the bandwidth changes, the relationship between SS power and maximum power changes the same as the sensitivity. When the channel bandwidth is doubled, the sensitivity is 3 dB higher, and the relationship between SS power and maximum power also increases by 3 dB. This means that the same thresholds are usable for all bandwidths. A list of computed SS power measurements that indicate the same sensitivity limits as Table 14.4 are shown in Table 14.5.

It's important to note that the eNodeB scheduler usually implements power control in which higher power is used to transmit data to UEs on the edge of the cell and lower power to UEs near the middle of a cell. This is done to minimize possible interference problems within the cell, and to minimize interference to adjacent cells (used in conjunction with the ICIC algorithms). Combined with the fact that scheduler algorithms are eNodeB vendor specific, it becomes impossible to predict exactly what transmission scheme (modulation, coding, and MIMO mode) will actually be used for any specific power level. The transmission scheme will even dynamically change during a data session, even from a fixed location, based on the CQI and PMI/RI reports from the UE. These factors indicate that the different thresholds should be used as a general indication of downlink signal quality, rather than a precise measurement.

Key Performance Indicators vs. LTE Field Measurement

Table 14.6 provides a guide for RF field measurements related to power and quality issues in the network. The different RF measurements are related to usual problems that are reported by users. Such a guide is normally used by field engineers to locate and confirm problems in the cell. This is used when investigating poor coverage reports, aiming to relate the RF measurements to KPIs that are measured in the network monitoring system. A field engineer who has prior knowledge of the failed KPIs that are reported from the NOC will then aim to use RF measurements to confirm the cause, and identify the required actions to fix the problem.

LTE Field Measurement vs. BTS Field Replaceable Units

Table 14.7 provides a guide of the RF field measurements that are possible and relates them to usual causes when a hardware failure in the eNodeB is suspected. If a cell site has been checked for correct settings and parameters but still has failure reports or low KPIs, then usually a hardware failure is suspected and the faulty module should be replaced. Using the guide, it is possible to quickly identify the most likely hardware failure related to a failure in testing against the specifications.

Table 14.5 SS Power required for QPSK sensitivity, for nominal SS Power setting. Bold values indicate levels different from the10 MHz channel bandwidth

Band
Downlink
Frequency
(MHz)
Channel bandwidth (MHz)
20 15 10 5 3 1.4
1 (FDD) 2110–2170 −124.8 −124.8 −124.8 −124.8
2 (FDD) 1930–1990 −122.8 −122.8 −122.8 −122.8 −122.8 −123.1
3 (FDD) 1805–1880 −121.8 −121.8 −121.8 −121.8 −121.8 −122.1
4 (FDD) 2110–2155 −124.8 −124.8 −124.8 −124.8 −125.3 −125.1
5 (FDD) 869–894 −122.8 −122.8 −122.8 −123.1
6 (FDD) 875–885 −124.8 −124.8
7 (FDD) 2620–2690 −122.8 −122.8 −122.8 −122.8
8 (FDD) 925–960 −121.8 −121.8 −121.8 −122.1
9 (FDD) 1844.9–1879.9 −123.8 −123.8 −123.8 −123.8
10 (FDD) 2110–2170 −124.8 −124.8 −124.8 −124.8
11 (FDD) 1475.9–1495.9 −124.8 −124.8
12 (FDD) 728–746 −121.8 −121.8 −121.8 −122.1
13 (FDD) 746–756 −121.8 −121.8
14 (FDD) 758–768 −121.8 −121.8 −121.8
17 (FDD) 734–746 −121.8 −121.8
33 (TDD) 1900–1920 −124.8 −124.8 −124.8 −124.8
34 (TDD) 2010–2025 −124.8 −124.8 −124.8 −124.8
35 (TDD) 1850–1910 −124.8 −124.8 −124.8 −124.8 −124.8 −124.1
36 (TDD) 1930–1990 −124.8 −124.8 −124.8 −124.8 −124.8 −124.1
37 (TDD) 1910–1930 −124.8 −124.8 −124.8 −124.8
38 (TDD) 2570–2620 −124.8 −124.8 −124.8 −124.8
39 (TDD) 1880–1920 −124.8 −124.8 −124.8 −124.8
40 (TDD) 2300–2400 −124.8 −124.8 −124.8 −124.8

There are several minor exceptions, pointed out in Table 14.3: 0.5 dB high for band 4 with 3 MHz BW, 0.3 dB high for 1.4 MHz bandwidth (most bands) and 0.7 dB low bands.

Table 14.6 Guide for LTE RF field measurements. ‘x’ means probable, and ‘xx’ means most probable case

Key performance indicators vs. test	Sync power	RS power	BW, ACLR	EVM (peak)	EVM (average)	Freq error	Rx noise floor	OTA EVM
Call/session blocking								
Power shortage	X	X		X				
Resource block shortage			X	XX	XX			
UL interference			X				XX	
Call/session drop								
Radio link timeout	X	X		X	X	X	X	X
UL interference			X					
DL interference	X	X		X	X	X		X

Table 14.7 Guide for LTE RF field measurements

Test vs. BTS field replaceable units	Freq. ref.	Signal gen.	MCPA	Filters	Antenna	Antenna down tilt
Sync power		X	XX		X	
RS power		X	XX		X	
Occupied BW		X	XX	XX		
Adjacent channel leakage ratio (ACLR)		X	X	XX	X	
Spectral Emission Masque (SEM)		X	X	XX	X	
Error Vector Magnitude Peak (EVM pk)		X	XX			
Error Vector Magnitude (EVM)		X	X	X	X	
Frequency Error	XX					
OTA EVM		X	X	X	X	X

14.10.2 Guidelines for LTE Measurements

14.10.2.1 LTE Occupied Bandwidth

Figure 14.20 shows an example of the "LTE-occupied bandwidth" screen. This is a measurement of the total spectrum used by all the sub-carriers. The occupied bandwidth contains 99% of the signal's RF power.

Guideline–direct connect:

- less than defined LTE bandwidth (1.4, 3.0, 5.0, 10, 15 or 20 MHz).

Consequences:

- leads to interference with neighboring carriers;
- dropped calls;
- low capacity.

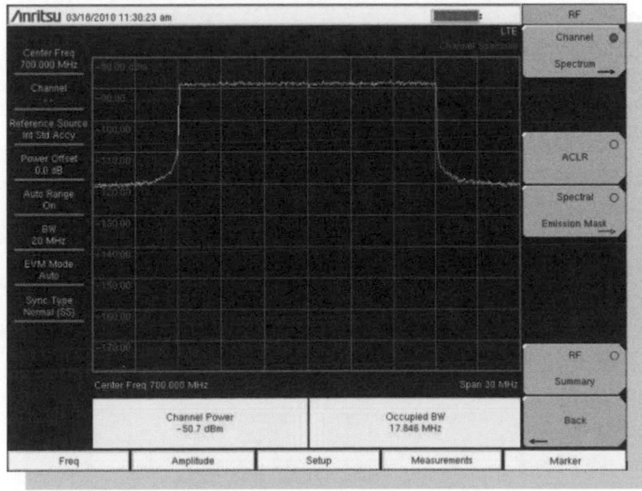

Figure 14.20 An example of the Occupied Bandwidth measurement.

Common faults:

- Tx filter;
- MCPA;
- signal processing;
- antennas.

14.10.2.2 LTE Adjacent Channel Leakage Ratio (ACLR)

Figure 14.21 shows an example of the ACLR measurement with a single-carrier displayed. This measures how much of the RF carrier leaks into neighboring RF channels. The measurement checks the leakage into the closest (alternative) LTE channel.

Guideline—direct connect:

- – 45 dBc adjacent channels;
- – 45 dBc alternate channels.

Consequences:

- interference;
- low capacity;
- blocked calls.

Common faults:

- Tx filter;
- MCPA;
- baseband processing cards;
- cable connectors.

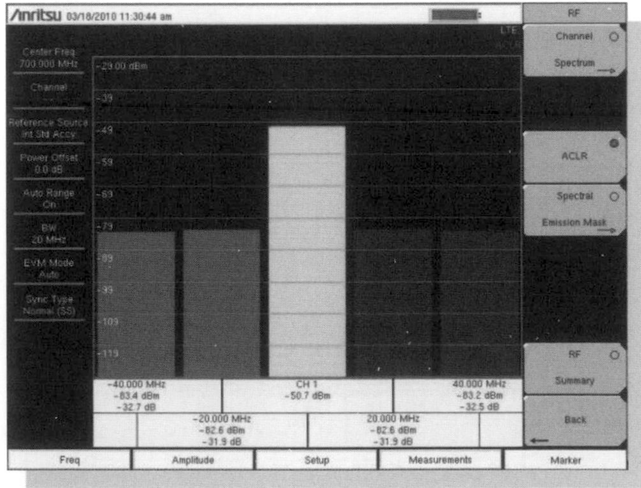

Figure 14.21 An example of ACLR measurement.

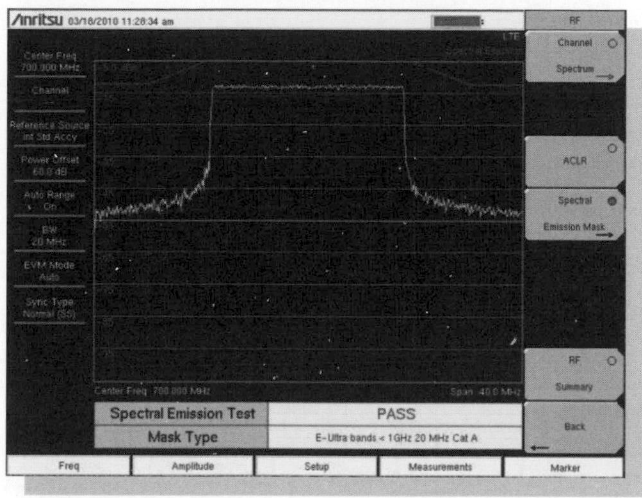

Figure 14.22 An example of the SEM measurement.

14.10.2.3 LTE Spectral Emission Masque (SEM)

Figure 14.22 shows a Spectral Emission Masque (SEM) measurement, which checks closer frequencies to the signal than ACLR does, looking at leakage at the edges of the allocated band into the neighbor channel. It also is sensitive to absolute power levels. The related compliance is usually a requirement of regulators.

Guideline—direct connect:

- Must be below masque.

Consequences:

- interference with neighboring carriers;
- legal liability;
- low data rate.

Common faults:

- check amplifier output filtering;
- look for intermodulation distortion;
- look for spectral regrowth.

LTE Error Vector Magnitude (EVM)

Figure 14.23 shows an example of the LTE EVM measurement. This measurement shows the ratio of distortion in the actual signal compared to a perfect signal. EVM measures the PBCH if there is no data traffic and the PDSCH if there is traffic.

Guidelines—direct connect:

- QPSK: −17.5%;
- 16 QAM: −12.5%;
- 64 QAM: −8%.

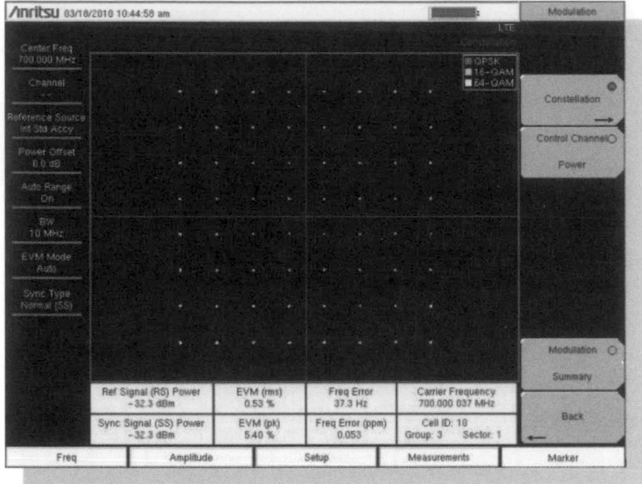

Figure 14.23 An example of the Error Vector Magnitude measurement.

Consequences:

- dropped calls;
- low data rate;
- low sector capacity;
- blocked calls.

Common faults:

- distortion in the baseband processing cards;
- power amplifier;
- filter;
- antenna system.

LTE Frequency Error

Figure 14.24 shows an example of the LTE Frequency Error measurement. This measurement checks that the carrier frequency is precisely correct one. This is a regulatory requirement in many countries.

Guideline:

- OTA with GPS;
- +/− 0.05 ppm wide area BS;
- +/− 0.10 ppm local area BS;
- +/− 0.25 ppm home BS.

Consequences:

- calls will drop when mobiles travel at higher speed;
- in some cases, UE cannot handover into or out of the cell.

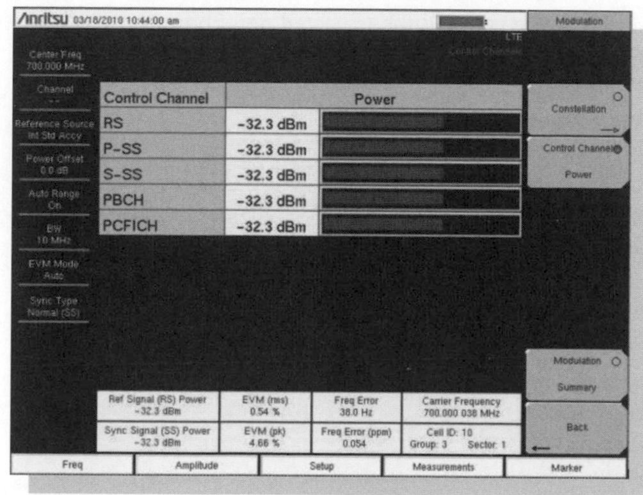

Figure 14.24 An example of the LTE Frequency Error measurement.

Common faults:

- reference frequency in eNodeB;
- frequency distribution system through backhaul (failure to synchronize);
- GPS failure in case GPS is used.

LTE Synchronization Signal Scanner

Figure 14.25 shows an example of the Synchronization Signal Power measurement. This measurement indicates which sectors are present at the current location. The assumption

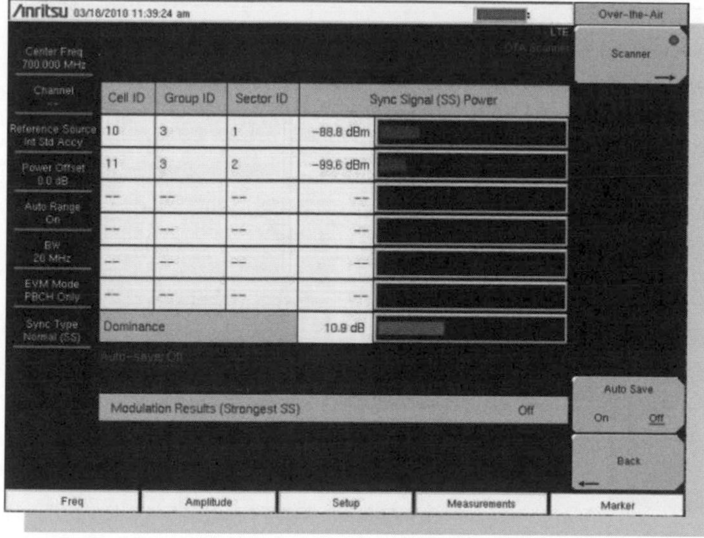

Figure 14.25 An example of the LTE Synchronization Signal Scanner measurement.

is that there should be active cell and the expected neighboring cells. If there are too many strong cells present, it means that the ICIC requires frequency avoidance, which reduces the cell capacity.

Guideline—direct connect

- or fewer codes;
- within 10 dB of dominant code;
- over 95% of the coverage area.

Consequences:

- low data rate;
- low capacity.

Common faults:

- antenna down tilt on neighboring cells;
- damaged antennas on neighboring cells;
- control channel power setting too high in neighboring cells;
- illegal repeaters within the cell coverage.

LTE OTA Modulation Quality Testing

Figure 14.26 shows an example of the LTE OTA Modulation Quality testing—that is valid EVM OTA measurement. In general, MIMO presents a challenge to measure EVM OTA. In practice, there is a need to measure EVM of PBCH which has transmit diversity but no MIMO. The use of PBCH measures should be done for both paths of Tx1 and Tx2.

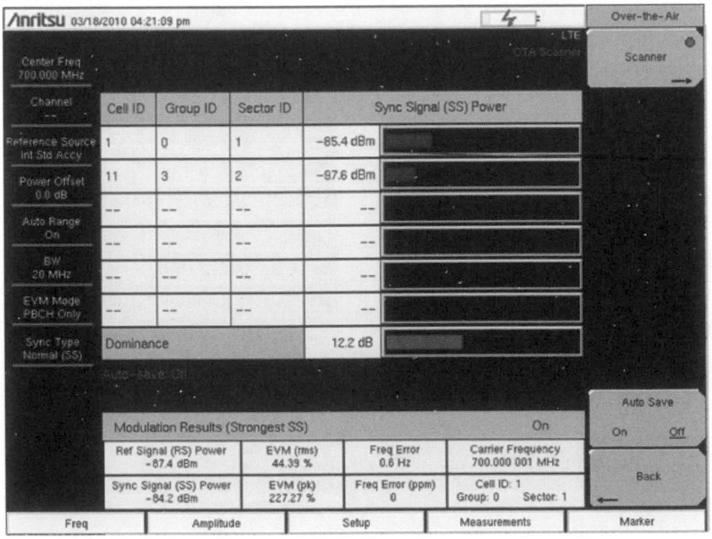

Figure 14.26 An example of the LTE OTA Modulation Quality testing.

OTA Modulation Quality testing:

- valid signal quality measurements can be made OTA;
- guidelines are established from a known good base station;
- must be measured in a valid location.

Valid OTA location:

- OTA Scanner validates location where dominance is more than 10 dB.

If pass, becomes desirable spot:

- note GPS location;
- becomes default location for future OTA Modulation testing;
- record and create OTA Pass/Fail limits.

References

[1] 3GPP TS 36.101. (2010) *User Equipment (UE) radio transmission and reception*, V 8.12.0, 3rd Generation Partnership Project, Sophia-Antipolis.

[2] Rohde & Schwartz (n.d.) Technical data sheets of Rohde & Schwartz, www2.rohde-schwarz.com/en/products/test_and_measurement/mobile_radio (accessed August 30, 2011).

[3] 3GPP TS 36.214. (2009) *Evolved Universal Terrestrial Radio Access (E-UTRA); Physical layer; Measurements*, V. 8.7.0, 3rd Generation Partnership Project, Sophia-Antipolis.

[4] Nethawk (n.d.) *EAST simulator*, https://www.nethawk.fi/products/nethawk_simulators (accessed August 30, 2011).

[5] Aeroflex (n.d.) Infrastructure Test System TM500 TD-LTE Single UE, 3GPP TD-LTE Release 8 Test brochure of Aeroflex, www.aeroflex.com/ats/products/prodfiles/datasheets/TM500%20TD%20LTE%20Single%20UE.Iss2.pdf (accessed August 30, 2011).

[6] Rohde & Schwartz (n.d.) SMBV100A Vector Signal Generator Specifications, www2.rohde-schwarz.com/file_15878/SMBV100A_dat-sw_en.pdf (accessed August 30, 2011).

[7] 3GPP TS 36.104. (2011) *Evolved Universal Terrestrial Radio Access (E-UTRA); Base Station (BS) radio transmission and reception*, V. 8.12.0.

[8] 3GPP TS 36.141. (2011) *Evolved Universal Terrestrial Radio Access (E-UTRA); Base Station (BS) conformance testing*, V. 8.11.0, 3rd Generation Partnership Project, Sophia-Antipolis.

[9] 3GPP TS 36.101 (2010) *Evolved Universal Terrestrial Radio Access (E-UTRA); User Equipment (UE) radio transmission and reception*, V8.10.0, 3rd Generation Partnership Project, Sophia-Antipolis.

[10] 3GPP TS 36.521. (2010) *Evolved Universal Terrestrial Radio Access (E-UTRA); User Equipment (UE) conformance specification; Radio transmission and reception; Part 1: Conformance testing*, V. 8.6.0, 3rd Generation Partnership Project, Sophia-Antipolis.

[11] 3GPP TS 36.523. (2010) *Evolved Universal Terrestrial Radio Access (E-UTRA) and Evolved Packet Core (EPC); User Equipment (UE) conformance specification; Part 1: Protocol conformance specification*, V. 8.6.0, 3rd Generation Partnership Project, Sophia-Antipolis.

[12] 3GPP TS 36.571. (2010) *Evolved Universal Terrestrial Radio Access (E-UTRA); User Equipment (UE) conformance specification; UE positioning in E-UTRA; Part 1: Minimum Performance conformance*, V. 1.0.0, 3rd Generation Partnership Project, Sophia-Antipolis.

[13] 3GPP TS 36.508. (2010) *Evolved Universal Terrestrial Radio Access (E-UTRA) and Evolved Packet Core (EPC); Common test environments for User Equipment (UE) conformance testing*, V. 8.6.0, 3rd Generation Partnership Project, Sophia-Antipolis.

[14] 3GPP TS 36.509. (2011) *Evolved Universal Terrestrial Radio Access (E-UTRA) and Evolved Packet Core (EPC); Special conformance testing functions for User Equipment (UE)*, V. 8.7.0, 3rd Generation Partnership Project, Sophia-Antipolis.

[15] Anritsu (n.d) Technical data sheets of Anritsu, www.anritsu.com/en-GB/Products-Solutions/Test-Measurement/index.aspx (accessed August 30, 2011).

[16] Anritsu (n.d.) MT8221B BTSMaster technical data brochure. http://downloadfile.anritsu.com/RefFiles/en-US/Services-Support/Downloads/Brochures-Datasheets-Catalogs/Brochure/11410-00441.pdf (accessed August 30, 2011).

15

Recommendations

Sebastian Lasek, Dariusz Tomeczko, Krystian Krysmalski, Maciej Pakulski, Grzegorz Lehmann, Krystian Majchrowicz, and Marcin Grygiel

15.1 Introduction

The history of mobile network rollouts seems to reflect the advances of mobile communication systems, the milestones of which are marked by appearance of new generations of wireless technologies. Taking Europe as an example, notice that the early analogue wireless networks started to appear in the mid-1980s (NMT, 450 MHz band), followed by GSM, with a major expansion around 1995 (900 MHz band, then also 1800 MHz). Next came UMTS (early 2000s, 2100 MHz band), enhanced with HSxPA technologies in the middle of the 2000s. It was common practice for the mobile operators to try to cover traffic hotspots first, and then gradually extend the coverage to the whole population (which was also quite often mandated by the regulators). With the advent of LTE things look more complex. This chapter addresses the multitude of innovative approaches to rolling out this technology in practice. As discussed in Section 15.2, it is not obvious that the operators intend to replace existing legacy wireless networks with new one, built only around LTE technology. It is more likely that the LTE component will be used to complete and enhance the overall wireless service portfolio of the operator but will not replace the networks that are already in existence.

The flexibility of LTE technology makes for a break in the rigid association between radio access technology, operational frequency band, and system bandwidth. These aspects are discussed in Section 15.3. Additionally, the advances in electronic component miniaturization make it possible for manufacturers to offer multiband multistandard terminals, capable of taking advantage of simultaneous coverage of multiple types of wireless networks. Another important factor that has to be considered is the status of definitive solution for voice calls over LTE, which is still unclear, causing the mobile operators to be reluctant to switch off their GSM or WCDMA networks in areas of planned LTE coverage. Therefore, GSM functionalities that allow smooth spectrum refarming and minimizing the performance gap between 2G and LTE become crucial in the multi-Radio Access Technology (RAT) environment. Some selected solutions are covered in Section 15.4. Another aspect of network evolution is the network sharing idea, involving several mobile network operators. Regardless of actual technical solution put into place, the basic principle behind this is that certain network elements provide

The LTE/SAE Deployment Handbook, First Edition. Edited by Jyrki T. J. Penttinen.

service to more than one operator. These functionalities are presented in Section 15.5. When migrating towards new technology, the hardware-related aspects have to be addressed as well, so that expenditure is minimized. Hardware migration path is the subject of Section 15.6. Mobile backhaul evolution towards an all-IP environment, presented in Section 15.7, is an inseparable element of LTE deployment due to its inherent datacentric characteristics.

Deployment of the next generation communication system on top of existing ones always raises issues related to interworking. Coexistence of different technologies brings challenges concerning how to ensure efficient cooperation. There is no exception for LTE being setup in the ecosystem of for example GSM/EDGE and WCDMA/HSPA. Therefore, along with introduction of LTE, there is a high demand for techniques that enable optimal interworking between systems. Aiming to meet this challenge, a number of features designed to either allow or further enhance cooperation between systems in question have been standardized by 3GPP. Selected mechanisms are discussed in Section 15.8.

15.2 Transition to LTE—Use Cases

15.2.1 Total Swap

When considering new generation system deployment, seemingly the most straightforward approach is to replace all of the legacy wireless technologies with LTE. This approach has several positive aspects. Having to deal with only one type of RAT makes operation and maintenance much simpler for the operator. The logistics would be simplified (i.e. the number of different components, pieces of software or configurations to keep track of), the training costs of personnel will be much lower (i.e. a single team, no need to maintain GSM/WCDMA-related skills). The flat LTE architecture with fewer nodes to purchase and take care of is an additional argument in favor of going all-LTE. It has also to be noted that many manufacturers of mobile equipment might consider phasing out some or a majority of products belonging to GSM or WCDMA portfolio in the predictable future. Another advantage of a single technology network lies in the fact that the potential coverage area of pure LTE operator will not be restricted by the availability of multimode multiband terminals, which may be especially important in the early phases of LTE technology development. If an operator is going to roll out LTE in a frequency band not used so far by GSM or WCDMA, it will be spared from the burden of interference coordination between the legacy systems and LTE. The fact of having a purely LTE network may also be a key differentiator for the operator, facilitating selection of marketing strategies. Going all-LTE is a natural and straightforward solution for a greenfield operator. It is therefore reasonable to assume that a certain number of the operators will be using LTE-only network from the beginning, or will migrate towards a pure LTE network over time.

15.2.2 Hot Spots

A viable alternative to the total swap of the radio access technology is a stepwise approach, where the new technology will be deployed in a noncontiguous manner, aiming to cover only certain areas. Typically the areas of interest are traffic hotspots that will benefit the most from the capabilities offered by LTE. Covering the whole country with LTE, especially rural and sparsely inhabited areas, where the existing GSM technology in 900 MHz band may be still justifiable, could be economically unacceptable unless lower frequency bands like Digital

Dividend are available for LTE (cf. Section 15.3.1.1). Most of the operators will start deploying their LTE networks in the areas of high traffic demand, where the expenses caused by new network equipment will be quickly compensated. A natural candidate for early or isolated LTE deployment is Femto technology, where small LTE-capable Home eNodeBs (HeNBs) will be deployed indoor in high traffic areas like offices or public spaces (malls, stadiums). It is quite reasonable to think about deployment of HeNBs in residential areas (much like today's WLAN Access Points), to take care of stationary indoor users wishing to have a wireless alternative to broadband Internet access. The presence of LTE in isolated spots needs special care, as the service should be offered only to stationary users or practical inter-RAT mobility management mechanisms have to be put in place to enable the users to enjoy continuous coverage. The hotspots covered by the LTE network may continue to be covered by GSM and WCDMA, or the operator may intend to actually rely on LTE as the sole radio access technology. Whatever the case, partial coverage by the LTE network needs some attention from the radio planning perspective, because inevitably the inter-RAT interference needs to be managed inside and on the boundaries of hotspots. While selecting locations for LTE eNodeBs, the mobile operator may wish to select new sites, or reuse existing ones. In this context one may distinguish two different scenarios with specific implications for mutual interference between systems, namely uncoordinated and coordinated. They are addressed in Section 15.3.2.

15.3 Spectrum Aspects

15.3.1 General View on Spectrum Allocation

In Release 8 of LTE standard, 23 frequency bands were specified for LTE use—that is 15 paired bands have been reserved for FDD and eight unpaired bands for TDD. The list of LTE frequency bands is still growing as new ones are added during the standardization process. Currently, over 30 frequency bands are supported. The need for new LTE band allocations comes, among other things, from growing demand for mobile broadband and predictions that without new spectrum mobile broadband networks will hit capacity shortages soon.

Table 15.1 shows LTE operating bands defined in Release 8 [1]. As can be seen, LTE can be deployed in bands that are currently used by 2G or 3G technologies as well as in new bands like for example 2600 MHz (IMT Extension band) or 700/800 MHz (DD - Digital Dividend spectrum).

LTE supports a wide range of channel bandwidths including 1.4, 3, 5, 10, 15 and 20 MHz. Thanks to flexible channel bandwidth, operators will be able to introduce LTE in existing GSM/WCDMA bands—that is 850, 900, AWS 1700/2100 (Advanced Wireless Services), 1800, 1900 and 2100 MHz gradually, hence with minimized impact of the squeezed bandwidth on the performance of legacy systems. For example with the refarming of 1.4 or 3 MHz only, a baseline LTE system can be deployed. Obviously, a system operating with such a limited spectrum would not deliver the peak throughputs of LTE deployed in 10 or 20 MHz channels. However, it presents an attractive migration path to the full-speed LTE system. Once the traffic served by the legacy technology goes down or the spectral efficiency of such a system is improved then more spectrums can be freed up for LTE.

In case of the new spectrums like 2600 MHz or 700/800 MHz it will be much easier to take a full advantage of LTE as the new spectrum will allow operators to use larger channel bandwidths. Furthermore, 700/800 MHz will enable LTE to be deployed more efficiently over large geographical areas and improve in-building coverage. 2600 MHz and Digital Dividend spectrums can be considered as a good combination of spectrum bands because

Table 15.1 Paired and unpaired frequency bands for LTE (status Release 8)

Frequency Range		Mode of Operation	Bandwidth (MHz)	Band
Uplink [MHz]	Downlink [MHz]			
699–716	729–746	FDD	2 × 17	Band 12
704–716	734–746	FDD	2 × 12	Band 17
777–787	746–756	FDD	2 × 10	Band 13
788–798	758–768	FDD	2 × 10	Band 14
824–849	869–894	FDD	2 × 25	Band 5
830–840	875–885	FDD	2 × 10	Band 6
880–915	925–960	FDD	2 × 35	Band 8
1427.9–1447.9	1475.9–1495.9	FDD	2 × 20	Band 11
1710–1755	2110–2155	FDD	2 × 45	Band 4
1710–1770	2110–2170	FDD	2 × 60	Band 10
1710–1785	1805–1880	FDD	2 × 75	Band 3
1749.9–1784.9	1844.9–1879.9	FDD	2 × 35	Band 9
1850–1910		TDD	1 × 60	Band 35
1850–1910	1930–1990	FDD	2 × 60	Band 2
1880–1920		TDD	1 × 40	Band 39
1900–1920		TDD	1 × 20	Band 33
1910–1930		TDD	1 × 20	Band 37
1920–1980	2110–2170	FDD	2 × 60	Band 1
1930–1990		TDD	1 × 60	Band 36
2010–2025		TDD	1 × 15	Band 34
2300–2400		TDD	1 × 100	Band 40
2500–2570	2620–2690	FDD	2 × 70	Band 7
2570–2620		TDD	1 × 50	Band 38

network capacity demands can be well addressed by the 2600 MHz layer whereas a good LTE coverage footprint will be delivered by means of DD-based layer.

LTE spectrum deployment scenarios will vary over the regions of the world (see, for example, www.gsacom.com and http://gsmworld.com). For instance, in Europe, APAC (Asia Pacific), MEA (Middle East Africa) and Latin America, 2600 MHz will be an important band for LTE. Many operators from these regions already acquired 2600 MHz spectrum and in many countries auctions of 2600 MHz band are planned. DD or smooth migration of 900/850 MHz to LTE are other options operators are looking at to offer broadband services, especially in rural areas. For example, in Germany LTE has been launched in the 800 MHz band to provide broadband services to the areas of the country where there is no high-speed Internet available. In fact, mobile operator was obliged by its 800 MHz license conditions to deploy LTE first in areas without DSL (Digital Subscriber Line) access before making countrywide rollout.

There is also growing interest and momentum towards deploying LTE in refarmed 1800 MHz spectrum due to many advantages this band is offering (see below for more details). The 1800 MHz band is widely available throughout Europe, APAC, MEA, and some regions of South America.

In 2008, US Federal Communications Commission (FCC) auctioned 62 MHz of spectrum in the 700 MHz band for LTE services and the commercial LTE networks working in this band are already on the air. Other spectrum variants for LTE deployments in the US are AWS, 850 and

1900 MHz bands where currently systems like WCDMA/HSPA and GSM are in operation. In Japan, the 2100 and 1500 MHz spectrums are available for initial LTE deployments. Later on the expansion of LTE to other frequency bands—for example to 800 and 1700 MHz—is planned.

Regarding LTE TDD, significant spectrum resources suitable for this mode are at 2300 MHz and in the 2500/2600 MHz band. In the first one, depending on the region, up to 100 MHz of contiguous band is available. In the latter, which is located in the IMT Extension center gap, 50 MHz of spectrum is provided. New 2300 MHz spectrum has been allocated in China, India, Korea and in other countries, especially in Asia. Chinese operators, in addition to introducing LTE in the above-mentioned bands, may also introduce it in the spectrum that is now used by the TD-SCDMA system that is 1880–1900 and 2010–2025 MHz. LTE TDD is expected to be commercially launched in 2011–12, initially in China and India. China and India will be the key markets for LTE TDD but when TDD mode becomes a globally accepted technology one can expect that LTE TDD networks will be deployed in other parts of the world as well. Numerous TDD LTE trials were conducted or are planned in other regions of the globe, for example in Europe (Ireland, Poland, Germany, Russia), Australia, and in the US. The Global TD-LTE Initiative, launched at MWC '11 (Mobile World Conference), is another indication of increased interest in LTE TDD outside China and India. This initiative, aiming to accelerate LTE TDD ecosystem development, is driven not only by Asian operators but, for example, by European and US ones too.

15.3.1.1 Digital Dividend (DD)

Digital Dividend refers to the potential reduction of the radio spectrum required to deliver terrestrial TV services that results from the transition to digital TV. The reason for this reduction is the ability of digital TV to provide a greater number of TV channels in a given amount of spectrum compared to analog TV. The switch from analog to digital TV allows significant amount of spectrum to be carved out for wireless broadband in the UHF band. Three blocks of digital dividend spectrum have been identified by the World Radio Communications Conference 2007 (WRC 07) to be reallocated to mobile communication (www.gsacom.com, http://gsmworld.com).

In Europe and MEA (region 1) the spectrum range identified for mobile broadband services is from 790 to 862 MHz while in both Americas (region 2) the frequency range from 698 to 806 MHz has been identified for this purpose. In Asia (region 3), in China, India and Japan, the 698–862 MHz spectrum has been defined whereas other Asian countries will follow region 1's spectrum band—that is 790 to 862 MHz. The availability of the DD band is country specific and depends on national time schedules of switchover to digital TV. For instance, in the US or Germany, the DD spectrum has been auctioned and large-scale commercial LTE systems have been launched there already. In other countries, the process of vacating the 700/800 MHz radio frequency spectrum will take more time and will extend to 2013–14 and beyond.

Because of its good signal propagation characteristics, Digital Dividend band is an excellent band for extending coverage in, for example, rural areas and providing good indoor coverage in urban areas in contrast to higher frequencies such as 1800/1900 or 2600 MHz. Therefore, to offer wider mobile coverage, fewer base stations must be deployed, meaning that, for example, LTE services can be delivered at much lower costs. Operators will be able to reuse their existing 2G/WCDMA 900 MHz sites more efficiently when deploying LTE in the 700/800 MHz band. For the same or even better coverage no new sites will be needed, which will speed up the rollout of the LTE networks. Apart from good coverage properties, the DD band provides good transmission capacity for broadband services as channels can be assigned to operators easily in 10 MHz or bigger chunks.

Spectrum at 700/800 MHz will be the key band for mobile broadband networks enabling them to reach as many subscribers as GSM networks managed to do today. The majority of commercially launched LTE networks nowadays operate in 700/800 and 2600 MHz. Many technical trials are ongoing or are scheduled in the near future in all regions of the world using the 700/800 MHz spectrum. Governments of many countries are considering auctioning the DD band soon to stimulate the expansion of mobile broadband communication.

Since the DD band is nearly aligned within and between the regions of the world, it will provide equipment makers with global economies of scale, enabling them to lower the cost of devices and network infrastructure. Harmonization of the DD spectrum will help to facilitate international roaming.

15.3.1.2 900 MHz

The 900 MHz band is the most harmonized and ubiquitous band available today worldwide (www.gsacom.com and http://gsmworld.com). In fact, most countries across the world work on 900 MHz, which is currently used predominantly for providing GSM based mobile services. However, due to the data traffic explosion and ongoing subscriber migration from GSM to WCDMA there is a very strong business case and momentum for rolling out 3G services on top of the 2G spectrum.

Like Digital Dividend, 900 MHz provides the benefit of increased coverage and subsequent reduction in network deployment costs compared to higher frequency bands. Furthermore, 900 MHz offers improved building penetration and is particularly well suited for delivering mobile services to rural areas. This makes 900 MHz a highly strategic spectrum band that may help operators to extend mobile broadband service coverage by leveraging the advantages of lower frequency bands.

In total 2×35 MHz is used by GSM 900 MHz (Standard GSM and Extended GSM). The operators, however, typically have less than 10 MHz of consecutive bands allocated, which implies that, in most the cases, only a 5 MHz LTE carrier could be deployed. Figure 15.1 presents the 900 MHz spectrum distribution over EU operators and reveals that 80% of the whole GSM 900 MHz spectrum has been assigned to operators in blocks smaller than 10 MHz.

In contrast to a single WCDMA carrier that requires 5 MHz[1] of spectrum, one will be able to deploy LTE in smaller for example 1.4 or 3 MHz channel bandwidths. Therefore, thanks to the inherent bandwidth flexibility, LTE can also be introduced in networks where the size of available spectrum blocks would be not sufficient for WCDMA deployment or current GSM traffic would not allow for enough frequencies to be carved out for WCDMA. Later on, once more spectrum will be released as a result of declining GSM traffic, these squeezed LTE deployments could be smoothly transformed to more broadband variants offering better peak throughputs and cell capacity. To overcome the problem of the defragmented spectrum, operators may also agree to rearrange band allocation in such a way that, for example, 10 or 15 MHz carrier allocations will be more feasible. This will increase the attractiveness of that band for LTE implementation as, apart from the coverage advantage, capacity requirements will also be better addressed when LTE is configured with higher channel bandwidths.

[1] The actual spectrum needed for WCDMA carrier may be lower in certain deployment types, for example 4.2 MHz in a coordinated scenario.

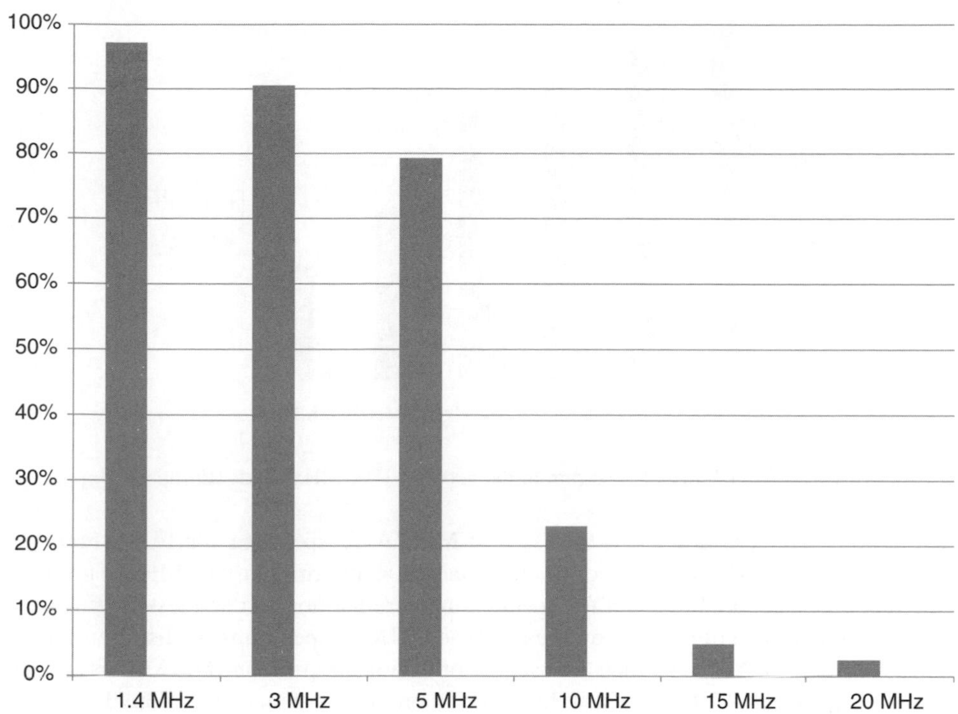

Figure 15.1 Availability of spectrum in the 900 MHz band (European Union).

Finally, reusing existing GSM sites and other assets will allow LTE deployment costs to be significantly reduced.

It can be expected that mobile networks operators will migrate their 900 MHz band also to LTE even though nowadays 900 MHz band is used for WCDMA refarming only. In particular, operators that do not have Digital Dividend licenses might consider the 900 MHz band as an enabler for the ubiquitous LTE coverage.

The process of spectrum liberalization that has been started by regulators in many countries will also pave the way for the potential deployment of LTE at 900 MHz. The liberalization means that the frequency licenses are no longer related to certain technologies. This allows the deployment of the new mobile systems (e.g. WCDMA or LTE) in the same bands that have been used for existing and legacy systems (e.g. GSM 900 MHz and 1800 MHz).

15.3.1.3 1800 MHz

A number of operators consider the deployment of the LTE system in the 1800 MHz spectrum. Those who are most keen to refarm to 1800 MHz are European mobile network operators and 1800 MHz may become main band for LTE deployment in Europe. Apart from Europe, 1800 MHz has a potential to be refarmed to LTE in, for example, APAC or Latin American countries (www.gsacom.com and http://gsmworld.com).

At the time of writing this book there is one commercial LTE network in Poland operating in this frequency band and LTE 1800 MHz trials are ongoing in many other countries. One can therefore expect that, by the end of 2011, and at the beginning of 2012 there will be more commercially available LTE networks running in 1800 MHz.

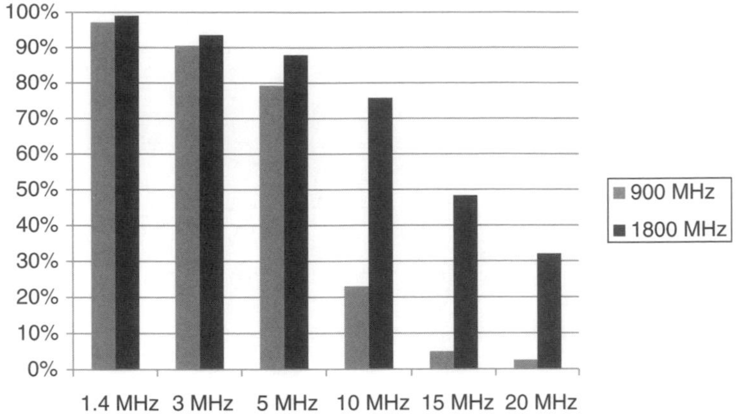

Figure 15.2 Availability of spectrum in the 900 and 1800 MHz bands (European Union).

The 1800 MHz spectrum provides a 2×75 MHz wide spectrum for FDD (Frequency Division Duplex) mode, which is often much easier to refarm than 900 MHz. This is due to the fact that spectrum allocated for operators is more harmonized and less fragmented in 1800 MHz. Figure 15.2 illustrates how the 1800 and 900 MHz spectrums are distributed among European operators. As can be seen almost 75% of overall spectrum in 1800 MHz is allocated in chunks bigger than 10.4 MHz,[2] and 20.4 MHz in one slot is available in more than 30% of total spectrum meaning that maximum LTE throughput can be provided. As far as 900 MHz frequency band is concerned only in 23% of overall allocations the size of assigned spectrum chunks exceeds 10.4 MHz whereas 20.4 MHz wide allocation exists very seldom (only in 3% in analyzed spectrum allocation schemes).[3]

In many countries, the 1800 MHz band is still partially not utilized and new auctions are scheduled (or have been scheduled recently) so that the mobile operators may acquire additional bandwidth for mobile broadband use.

The radio propagation characteristics of the 1800 MHz frequency carrier are other motivators for LTE in 1800 MHz. LTE 1800 MHz offers coverage advantage over LTE 2600 MHz, please refer to Figure 15.3 which shows example cell ranges for 900, 1800, 2100 and 2600 MHz frequency variants.

By reusing existing GSM 1800 MHz sites good urban LTE coverage footprints, including indoor ones, can be provided without the deployment of additional sites. Indeed, as GSM 1800 MHz site-to-site distance in urban clusters was reduced to meet capacity requirements, the coverage offered by the 1800 MHz network layer will be sufficient to serve indoor mobile broadband subscribers as well.

In the rural areas, too, good and seamless coverage can be provided by the 1800 MHz band. This might be valid for those operators that were initially granted a GSM license in 1800 MHz and who started to build their GSM network in layouts dense enough to compensate the higher propagation losses compared to 900 MHz. In countries where, in the first stage, LTE was launched in the 2600 MHz band, the 1800 MHz layer can be used to extend the geographical accessibility of LTE across the country.

[2] An uncoordinated case is assumed here, which means that for, e.g. 10 MHz LTE allocation 10.4 MHz of spectrum has to be dedicated. For more information please refer to Section 15.3.2.1.

[3] Based on spectrum assignment in all EU countries.

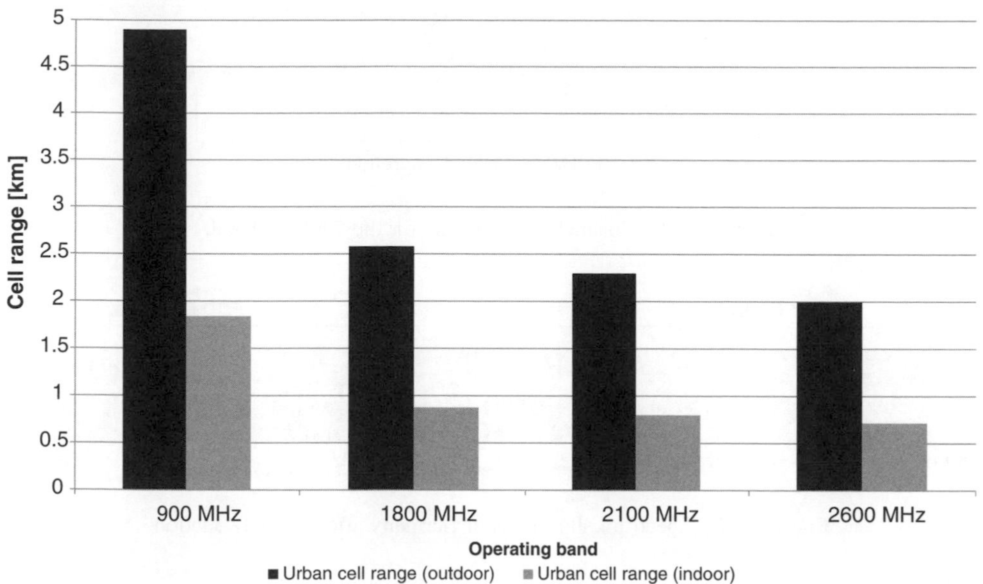

Figure 15.3 Example cell ranges for various frequency bands.

LTE 1800 MHz can be considered as a capacity layer as soon as any other system, such as LTE 800 MHz or LTE/WCDMA 900 MHz system, is built in the same area to offer a coverage layer for broadband services.

Reuse of the existing antennas that cover the frequency range of the 1800 MHz band through a variety of antenna-sharing techniques can also bring additional benefits in terms of reduced CAPEX/OPEX. Apart from the costs of adding new antennas and feeders, leases must often be renegotiated and changes in the site design must be approved before new antenna can be installed on an existing site. Of course, one should not forget about drawbacks resulting from antenna sharing like, for example, the cost of combining equipment or additional insertion losses (for more details see Section 15.6).

Availability of the multiradio BTS supporting efficient GSM-LTE RF sharing in 1800 MHz and the fact that the LTE 1800 MHz device ecosystem is ready should encourage operators to select the 1800 MHz band for mobile broadband usage.

Summarizing, the 1800 MHz band that is today widely used for GSM might become a key band for LTE due to the many advantages that it offers. It will be very important for LTE deployments, especially in countries where no additional spectrum is available or where early access to 800 and 2600 MHz will be not possible. Deployment of LTE in 1800 MHz can be viewed there as a part of transition strategy between HSPA and LTE operating in a new (e.g. DD, 2600 MHz) spectrum once it is made available.

The 1800 MHz band may be considered as a complementary band to 700 and 2600 MHz as it will provide additional capacity but with better coverage than the one provided by the 2600 MHz layer.

15.3.1.4 2600 MHz (IMT Extension Band)

The 2600 MHz band has been identified by ITU as an additional band needed for 3G/HSPA systems as a result of the accelerating increase in the traffic volume being carried by mobile

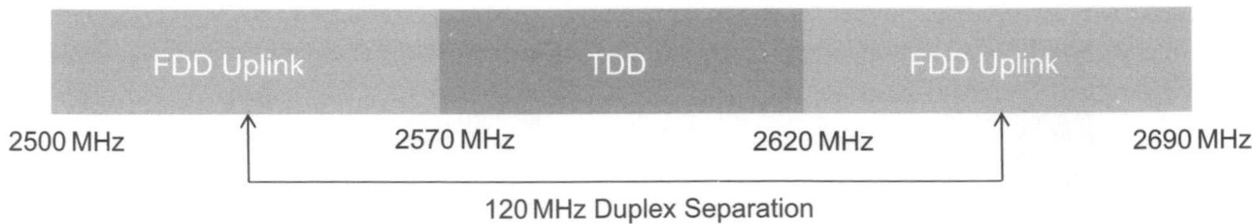

Figure 15.4 Channel arrangements for the 2.6 GHz band.

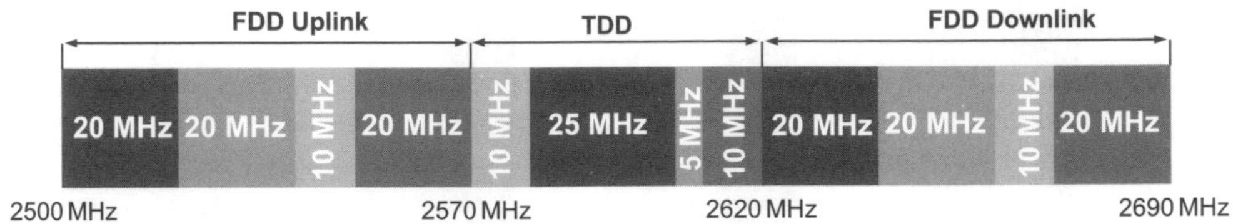

Figure 15.5 Spectrum allocations in Germany after 2.6 GHz auction.

broadband networks. As a result, 2×70 MHz of paired FDD spectrum (2500–2570 MHz for uplink and 2620–2690 MHz for downlink) and additionally up to 50 MHz (2570–2620 MHz) for an unpaired TDD band were made available as can be observed in Figure 15.4 (www.gsacom.com, http://gsmworld.com).

The IMT extension band gives operators access to the full 20 MHz channel, which is needed to achieve highest LTE data rates. To date, operators in Sweden and Norway have been granted licenses in 2.6 GHz and launched LTE networks in that band. The 2.6 GHz spectrum has also been awarded for example in the Netherlands, Germany, Austria and Finland. Figure 15.5 shows an example of the LTE spectrum allocation in 2600 MHz band.

There are more auctions of 2600 MHz spectrum planned in the near future, not only in Europe but also in Latin America. In Colombia, one operator has acquired 2600 MHz band and plans to deploy LTE in this band. In APAC, LTE trials at 2600 MHz are ongoing in many markets including, for example, Indonesia, Australia, and Singapore. In Hong Kong 2×15 MHz blocks of 2600 MHz spectrum have been allocated to mobile operators already. In MEA, for example in South Africa and Saudi Arabia, trials are under way. It can therefore be concluded that 2.6 GHz is becoming a key LTE band for many regions of the world and as a globally common band will facilitate LTE international roaming and will bring economies of scale in equipment and handset production.

Of course, due to high propagation losses, building full LTE coverage using 2600 MHz would be too costly and impractical. This means that, to achieve good country-wide coverage, a lower frequency band is needed. For that reason a combination of 2.6 GHz band with, for example, a DD band or refarmed 900 MHz will provide excellent balance for the operators with 2.6 GHz spectrum offering potential for higher capacity and with a lower frequency spectrum enabling higher cell ranges and improved indoor coverage.

15.3.1.5 Other Spectrum Variants

The UMTS core band at 2.1 GHz, used across the world, will also be a potential band for LTE deployment. Most of the operators were awarded more than one 5 MHz carrier in this frequency band but many do not use all the carriers yet. Unused carriers can be either switched on to increase the capacity of WCDMA system or they can be intended for LTE introduction, meaning that WCDMA and LTE would coexist in the same band (www.gsacom.com and http://gsmworld.com).

The Advanced Wireless Services (AWS), 1700/2100 MHz band, used in America predominantly for 3G deployments, has also been identified as suitable band for LTE. The combination of AWS-1 (part of AWS spectrum), which is located at 1710–1755 MHz for the uplink and 2110–2155 MHz for the downlink, and 700 MHz spectrum can be exploited to provide sufficient capacity and wide coverage for LTE services. Extensions to the 90 MHz of AWS-1 band (AWS-2, AWS-3) are also planned to be used for mobile broadband.

The 850 and 1900 MHz bands that are currently used by GSM and WCDMA may also be refarmed to LTE in future, for instance after capacity on 700 MHz and AWS spectrum exhausts.

The wireless industry has seen explosive growth in the demand for data services over the past several years. To keep pace with this growth, more spectrum will definitely be needed for mobile broadband services. Regulators will therefore be under increasing pressure to facilitate access to new frequency bands in order to foster the proliferation of broadband communication. To meet future capacity requirements and to extend the coverage of broadband systems, there are also other bands identified for mobile broadband services. For instance, the 450–470 MHz band, which is suitable for providing excellent coverage, and 3.4–3.6 GHz band, which would be used to create additional capacity, have been allocated to mobile services as well. It might be expected that spectrums will remain a hot topic in the years to come.

15.3.2 Coexistence with GSM

If the LTE technology operates in the same frequencies as the GSM network, or frequencies that are adjacent to it, in a given region two major deployment scenarios could be distinguished:

- an uncoordinated case, where there is no rule with respect to sites' location of GSM and LTE RAT;
- a coordinated case, where all GSM and LTE sites are collocated and the level of inter-system interference is minimized thanks to the inherent characteristics of such deployment and additional hardware as well as software features.

Since the level of inter-RAT interference is entirely different in these two scenarios, they have to be analyzed separately. The former option is described in Section 15.3.2.1 while the latter one is described in Section 15.3.2.2.

15.3.2.1 Uncoordinated Scenario

The principal problem of the uncoordinated deployment is caused by the near-far effect as presented in Figure 15.6. This means that the UE needs to increase its transmit power to reach a faraway base station receiver, which causes inadvertently high level of received uplink interference in the area nearby the base stations belonging to other radio technology. The same phenomenon also occurs also in the downlink direction due to distance-induced propagation losses.

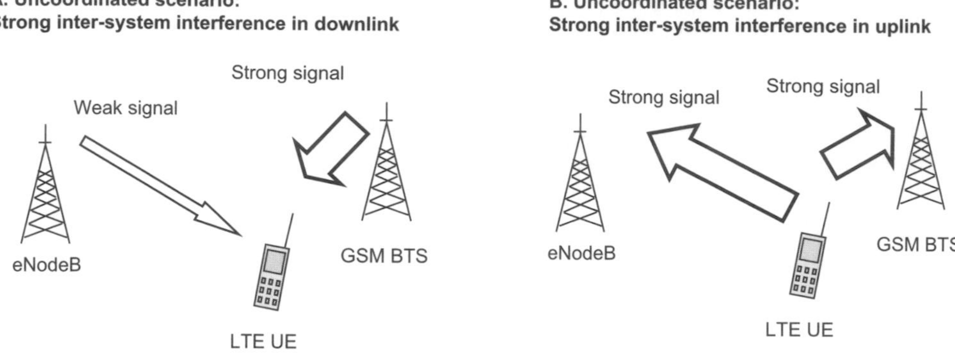

Figure 15.6 LTE-GSM coexistence: uncoordinated scenario.

Table 15.2 LTE-GSM carrier-to-carrier separation in the uncoordinated scenario

LTE channel bandwidth (MHz)	Carrier-to-carrier separation in uncoordinated scenario (MHz)
20	10.3
15	7.8
10	5.3
5	2.8
3	1.8
1.4	1

Due to unmatched signal-strength values coming from different systems the carrier-to-carrier separation between center of edge GSM frequency and LTE center frequency has to ensure that the mutual interference dependency between two involved systems is minimized. The separation values proposed in [2] are presented in Table 15.2.

15.3.2.2 Coordinated Scenario

The optimal way of using the spectrum in dual technology deployment in the same band is the so called coordinated case where all GSM and LTE sites are collocated with matched azimuths of GSM and LTE sectors. In the coordinated scenario the transmitted signals belonging to both radio technologies originate from the same point (roughly, as both systems may use different physical antennas), so are subject to the same magnitude of propagation losses (slight differences in the frequency band notwithstanding). Consequently, they might be received at similar levels. Moreover, mobiles closed to the site location usually operate at lower power levels so they are sources of less severe interference. Therefore, in such a scenario, certain control over intersystem interference could be achieved. Hence, it is possible to reduce the necessary carrier-to-carrier spacing between GSM and LTE in comparison with the uncoordinated scenario where there is no strict rule with respect to GSM versus the LTE site location. The actual ratio of signal to interference in both directions will, of course, depend on the

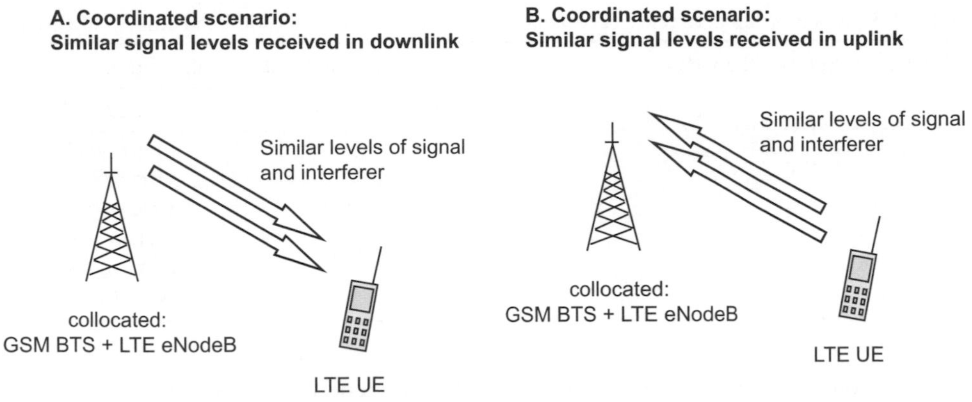

Figure 15.7 LTE-GSM coexistence: coordinated scenario.

particular configuration of each system (e.g. initial transmit power and presence of power control). Figure 15.7 shows the interference situation in coordinated deployment cases. LTE UE is shown, but obviously the same conclusions also apply for a GSM UE.

In order to optimally deploy the coordinated scenario, contiguous set of frequencies refarmed to LTE have to be taken from the GSM spectrum in such a way that the outer part of the spectrum belonging to the given operator will be still used by GSM while the inner part will be devoted to the LTE. Hence, legacy arrangements with respect to assignment of frequencies at the spectrum borders between different operators will be still valid.

In order to define the proper carrier-to-carrier spacing in the coordinated deployment the levels of interference coming from

- LTE eNodeB to GSM MS;
- LTE UE to GSM BTS;
- GSM BTS to LTE UE and;
- GSM MS to LTE eNodeB.

have to be jointly analyzed. The influence of interference coming from LTE to GSM is described in Chapter 15 (*See* Interference from LTE to GSM) while the effect of a GSM signal interfering with LTE is presented in Chapter 15 (*See* Interference from GSM to LTE). For the purpose of the analysis presented in the sections below the following nominal transmit power levels were assumed:

- GSM BTS 43 dBm;
- LTE eNodeB 43 dBm;
- GSM MS 33 dBm;
- LTE UE 23 dBm.

Interference from LTE to GSM

The interference dependency between interferer and victim operating in adjacent frequencies could be described with the help of the Adjacent Channel Interference Rejection (ACIR), which is affected by both the Adjacent Channel Leakage Ratio (ACLR) and Adjacent Channel Selectivity (ACS). They reflect the imperfection of the transmitter that is transmitting out of the

Table 15.3 Relationship between LTE channel bandwidth and transmission bandwidth configuration

LTE Channel Bandwidth (MHz)	Transmission Bandwidth Configuration	
	Number of Resource Blocks	Spectrum (UL/DL) (MHz)
1.4	6	1.08/1.095
3	15	2.7/2.715
5	25	4.5/4.515
10	50	9/9.015
15	75	13.5/13.515
20	100	18/18.015

wanted band and the imperfection of the receiver that is not perfectly filtering out the received out-of-band signal. The following equation describes this relationship:

$$ACIR = \frac{1}{\frac{1}{ACLR} + \frac{1}{ACS}} \tag{15.1}$$

The calculation of ACIR for GSM-LTE interference is provided, for example, in [2].

The transmission bandwidth configuration is defined as the highest possible transmission bandwidth for the given nominal LTE channel bandwidth. It could be expressed by the maximum number of available Resource Blocks [3]—as presented in Table 15.3.

The maximum signal strength of the LTE carrier outside of the channel bandwidth is bounded by Spectrum Emission Masque (SEM)—defined separately for downlink (eNodeB) [3] and uplink (UE) [4]. However, the exact type of transmit filter to be used either in eNodeB or in the mobile station is implementation specific. The way that the signal is filtered in the transmitter affects the value of signal strength measured in the guard region inside the LTE channel bandwidth. Consequently, it has direct influence on the level of intersystem interference coming from LTE eNodeB/UE to GSM MS/BTS if these systems operate on adjacent frequencies—as depicted in Figure 15.8. Realistic filter implementation usually is an embodiment of a tradeoff between overall LTE signal quality and out-of-band emissions.

It may be assumed that GSM receiver could attenuate the 200 kHz signal centered at 200 kHz frequency from the serving one by 18 dB and in case of interferer centered at 400 kHz

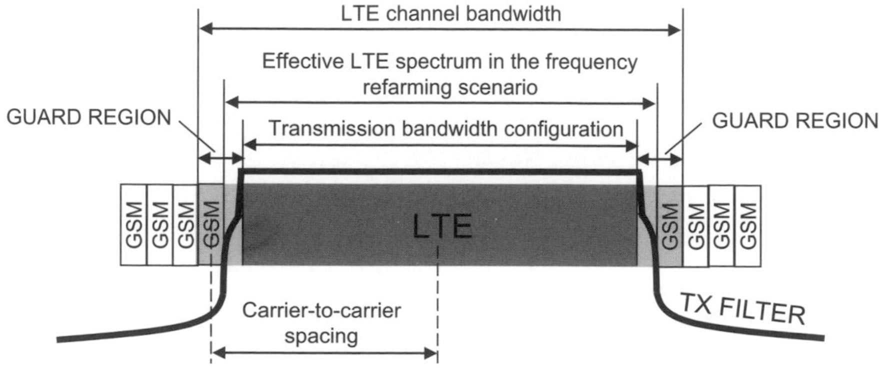

Figure 15.8 LTE-GSM coexistence: carrier-to-carrier spacing.

Table 15.4 Example ACIR values for selected carrier-to-carrier separation values

Carrier-to-Carrier Separation between GSM and LTE 5 MHz	ACIR [dB]		ACIR [dB]	
	LTE signal power matched with LTE SEM		LTE signal power with realistic Tx filter assumptions	
	LTE eNodeB interfering GSM MS	LTE UE interfering GSM BTS	LTE eNodeB interfering GSM MS	LTE UE interfering GSM BTS
2.7 MHz	35	38	48	40
2.6 MHz	32	27	43	39
2.5 MHz	17	24	38	37

frequency by 50 dB with respect to the received interference power measured at the antenna connector that is exactly as it was standardized for intra-GSM operation [5]. Hence, these values represent the Adjacent Channel Selectivity figures for the GSM receiver.

If the level of example LTE 5 MHz signal in the most part of guard band is close to the maximum level allowed by SEM, the ACLR of LTE transmitter will be significant and consequently the ACIR in GSM MS would be as low as approximately 17 dB for 2.5 MHz GSM-LTE carrier spacing and above mentioned ACS figures as presented in Table 15.4.

Such a low level of ACIR could effectively degrade the instantaneous C/I in the GSM, for example from 20 dB to 15 dB—as presented in Figure 15.9—and consequently seriously affect the quality in terms of for example FER and MOS for voice transmission in GSM.

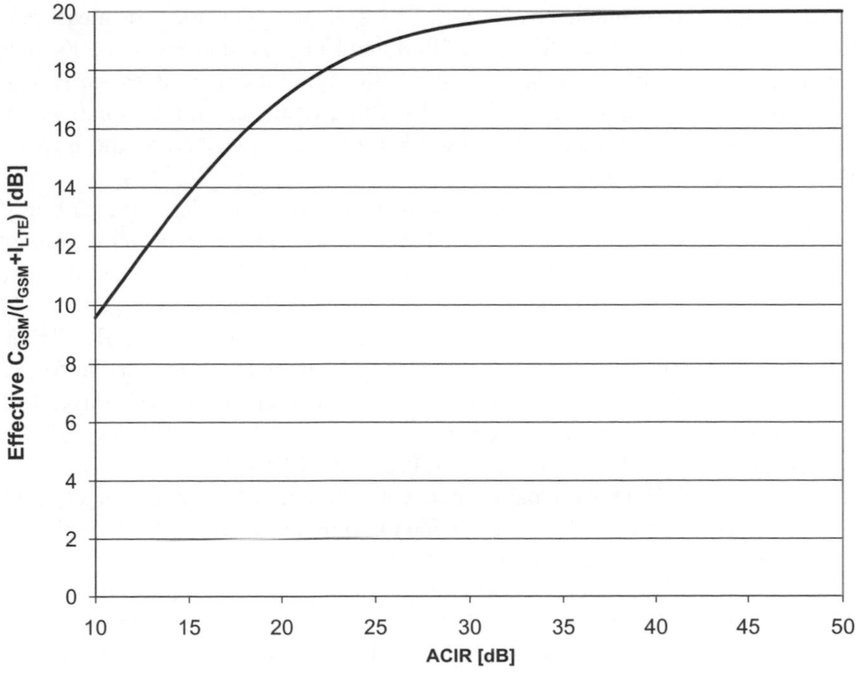

Figure 15.9 Effective instantaneous C/I for $C_{GSM}/I_{GSM} = 20$ dB.

Taking into consideration, however, some realistic representation of the LTE transmit filter, the typical ACLR will be lower than the one for LTE signal power matched with SEM. Due to this fact, ACIR higher than 30 dB may be achieved for 2.5 MHz GSM-LTE carrier-to-carrier spacing (please refer to example in Table 15.4). The way in which this value will transform into quality degradation of GSM depends on the C/I region where the GSM network is operating and the GSM frequency-hopping scheme. In typical cases, however, this degradation should not be visible from GSM performance point of view—in both DL and UL—especially if advanced GSM features are implemented where resources are assigned according to the real interference situation in the network. Particularly useful in the context of interference control in the frequency refarmed GSM network could be, for example, Dynamic Frequency and Channel Allocation (DFCA) functionality where radio channels in terms of frequency and timeslot are assigned in an interference-aware manner. Moreover, the effect of temporal interference peaks coming from edge LTE PRBs could be effectively diminished with the help of regular quality based handovers—even if the network is operating at moderate C/I. For further details on DFCA please refer to Section 15.4.7.

Interference from GSM to LTE

The level of acceptable interference coming from GSM depends on the acceptable quality degradation in the edge RBs, which, in turn, is affected by, amongst other things, the overall SINR statistics in the LTE network. Regarding UL direction, the performance of the PUCCH channels situated closest to the interfering GSM signals has to be given particular consideration. In the so-called sandwich allocation where LTE band is inside the GSM spectrum, the GSM transmission will affect PUCCH performance on both edges of the LTE spectrum and for this reason PUCCH hopping does not provide major countermeasure against GSM interference. In many cases, however, the observed GSM signal strength values—as measured by the GSM BTS receiver—are very low. Since the GSM and LTE base stations are collocated in the coordinated scenario, it is likely the case that also GSM signal strength values—as perceived by the LTE eNodeB—will already be quite low before filtering the incoming signal. An example uplink signal strength distribution in 900/1800 MHz band as received by the BTS in a live network is presented in Figure 15.10

For 2.5 MHz GSM-LTE carrier spacing, the instantaneous noise level increase in the UL PRB experiencing GSM interference integrated over 180 kHz PRB may be of the order of a few dBs for −80 dBm level—as presented in Figure 15.11 for LTE eNodeB Noise Figure equal to 5 dB [2]. However, this increase could reach 13 dB if the GSM signal was equal to −70 dBm.

In the context of downlink-related interference, it is beneficial to locate BCCH channels farther from the LTE part of spectrum than frequency-hopping channels. The performance of the LTE receiver benefits from power control used in the non-BCCH GSM layer that leads directly to interference reduction as perceived by the receiver. An example of GSM DL signal strength distribution at the MS is presented in Figure 15.12.

For a median level of DL GSM signal strength equal to −70 dBm, approximately 7 dB noise increase is expected in the DL PRB experiencing interference assuming a 12 dB Noise Figure for LTE mobile [2]—as presented in Figure 15.11.

The reduction in the interference level could be also achieved with the help of properly optimized Power Control functionality and DTX feature with the help of which during silent period voice frames are not transmitted over the air interface. Note also that interference-aware scheduler in eNodeB should—at least up to the certain extent—compensate negative effect of inter-system interference coming from GSM network by avoiding allocation of interfered

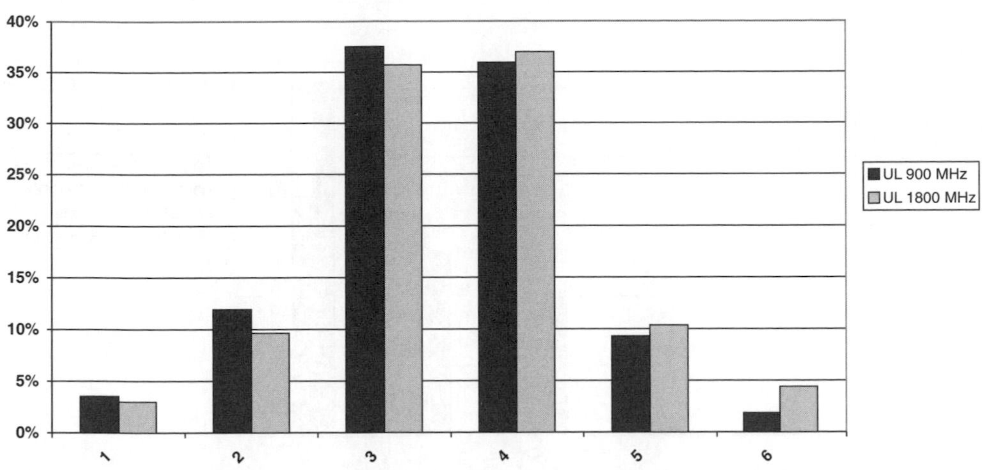

Figure 15.10 Distribution of GSM signal strength samples (uplink).

frequencies. Such a need may especially arise due to for example instantaneous peak in the GSM voice traffic in the considered cluster.

Conclusions on the Spectrum Requirements

Examples of carrier-to-carrier separation figures in coordinated scenario for various LTE channels are presented in Table 15.5 [6].

It is expected that a satisfactory performance may be achieved from both LTE and GSM networks once, for example, 2.5 MHz carrier-to-carrier separation will be applied in the

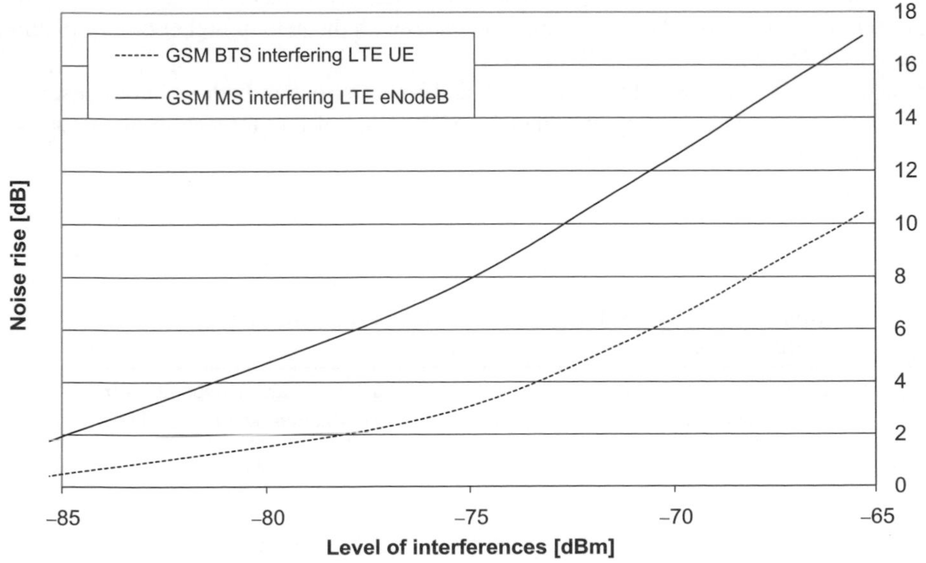

Figure 15.11 Noise increase due to GSM interference.

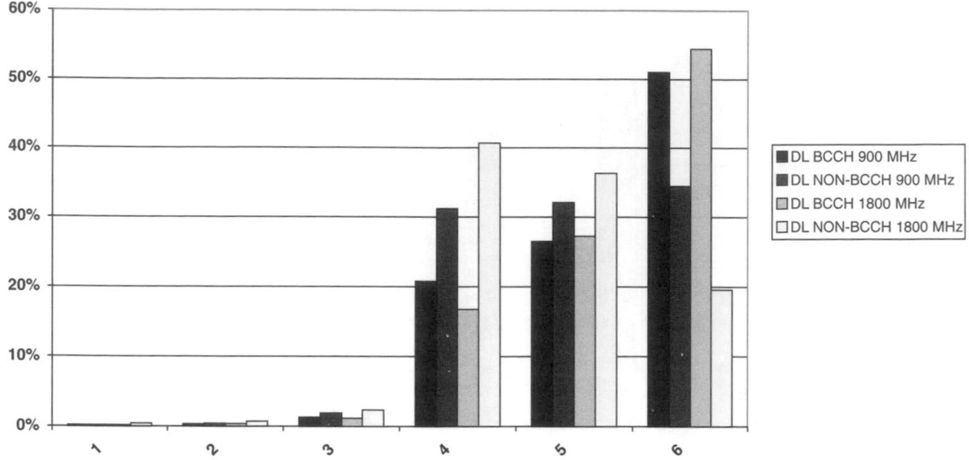

Figure 15.12 Distribution of GSM signal-strength samples (downlink).

coordinated scenario in case of the LTE 5 MHz channel bandwidth leading to the effective LTE spectrum equal to 4.8 MHz—as described in Chapter 15 (*See* Interference from LTE to GSM) and (*See* Interference from GSM to LTE). The final performance figures will however strongly depend on many factors, such as:

- SINR conditions;
- clutter type;
- frequency band;
- load of the network;
- traffic profile time correlation between both interfering systems;
- performance of Tx/Rx filters (e.g. Adjacent Channel Protection values in GSM for interference coming from LTE that are modulated with the help of higher order modulation like e.g. 32QAM);
- transmit power levels used by the mobiles and base stations in both technologies;
- availability of features aiming at reduction of the level of interference in the network;
- performance of the LTE scheduler in eNodeB;
- PUCCH configuration.

Table 15.5 LTE-GSM carrier-to-carrier separation requirements in the coordinated scenario

LTE Channel Bandwidth (MHz)	Carrier-to-Carrier Separation in Coordinated Scenario (MHz)
20	9.3
15	7
10	4.8
5	2.5
3	1.6
1.4	0.8

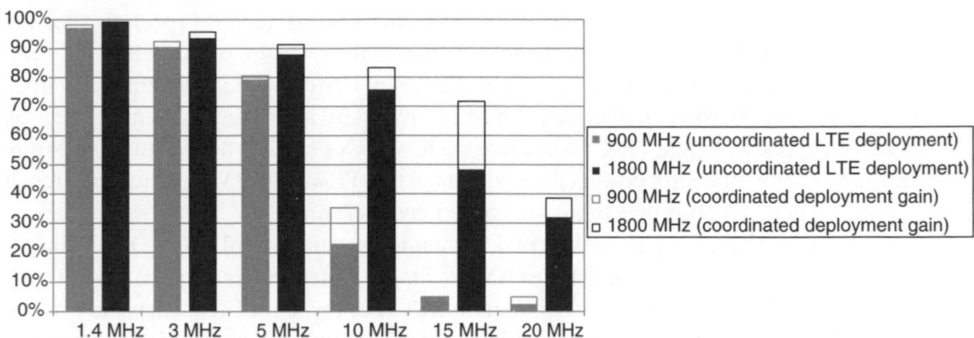

Figure 15.13 Spectrum availability gain due to the coordinated deployment of LTE.

The reduced LTE spectrum requirements for the coordinated scenario, unlike the uncoordinated case, directly address the problem of scarce spectrum resources available for LTE refarming. Please see Section 15.3.1 for more details. In Figure 15.13 the availability of contiguous 900/1800 MHz spectrum chunks for a given LTE channel bandwidth is presented for the uncoordinated refarming scenario together with spectrum availability gain coming from coordinated deployment. The analysis was based on the spectrum assignment in 27 member states of the European Union.

It should be also noted that, in many cases, LTE will be deployed in a coordinated way only in a part of the network—for example LTE will be introduced in the band occupied previously by GSM only in a single cluster of cells. In such a situation certain geographical buffer zone has to be introduced between LTE and GSM sites using the same frequency resources. The exact size of such buffer zone depends on many factors—most of them match with the list provided above in the fully coordinated case. Hence, the resulting buffer distance will vary a lot depending on the actual network scenario. It may be even longer than 10 km in some cases, in order to achieve optimum performance of both the GSM and the LTE networks.

Another aspect that has to be taken in account in the frequency planning is the channel raster. It equals 200 kHz in GSM and 100 kHz in LTE. The carrier frequency must be a multiple of the channel raster, which might result in the unavailable frequency allocations, even if carrier-to-carrier separation rules were satisfied.

15.4 Effect of the Advanced GSM Features on the Fluent LTE Deployment

Deployment of HSPA at 900 MHz is ongoing in many countries. The main driver for the introduction of HSPA at 900 MHz is the better coverage provided by this band in comparison to the coverage that can be achieved at 2100 MHz. As far as 1800 MHz spectrum is concerned it is expected that many operators will reuse that spectrum for LTE. LTE at 1800 MHz would benefit, for instance, from better coverage than could be achieved at 2600 MHz, and would also allow the operator to reuse its existing antenna systems. Regardless of the band at which refarming will be done there are some challenges the operators may face on the way towards the mobile broadband. Given that today still a lot of voice traffic is carried over GSM it must be assured that refarming of the spectrum does not degrade the QoS offered by the GSM network to below the acceptable limit. To address the negative consequences of the reduced spectrum left for GSM after refarming, the implementation of features boosting GSM spectral efficiency needs to be considered. The hardware efficiency which is the measure of how efficiently HW resources (e.g. TRX) are used also becomes important Key Performance Indicator (KPI) in the context of refarming. Hardware efficiency can be improved by increasing the number of users that can be carried on by single TRX. Once there are less TRXs/carriers needed to serve certain GSM traffic then more transmit power can be allocated to LTE or WCDMA, which may result, for example, in better indoor penetration for broadband services.

Features such as Frequency Hopping (FH), dynamic Power Control (PC), Discontinuous Transmission (DTX), Adaptive Multi-Rate (AMR), Single Antenna Interference Cancellation (SAIC), Orthogonal Sub-Channel (OSC), common BCCH or Dynamic Frequency and Channel Allocation (DFCA) are potential means to achieve very high spectral and hardware efficiency for GSM voice traffic.

This section focuses on the common BCCH, AMR, SAIC, OSC and DFCA features, as they are the powerful tools facilitating efficient GSM spectrum refarming. Solutions addressing poor link performance of the FACCH and SACCH signaling channels are also briefly described. Their activation may significantly lower the Call Drop Rate (CDR) in the networks exploiting AMR with tight frequency reuse.

15.4.1 Common BCCH

The multiband operation offers an attractive solution for mobile network operators who are allowed to deploy their GSM networks in more than one frequency band, for example in 900/850 MHz and in 1800/1900 MHz.

By applying multiband operation, good coverage footprint in a lower frequency band can be built whereas a higher frequency band can be used to provide extra capacity.

There are two options for operating a multiband network. The first one requires a separate BCCH layer in each frequency band. In the second option, which is facilitated by common BCCH functionality, only a single BCCH layer for both frequency bands is needed. This is because the common BCCH feature allows the integration of resources—that is TRXs—from different frequency bands into one cell while having the BCCH carrier configured in only one of them (typically in the coverage layer, that is 850/900 MHz) as depicted in Figure 15.14. The basic requirement for a common BCCH feature is that resources across all bands are colocated and synchronized.

Due to the nature of the broadcast control channel concept in the GSM standard, the spectral efficiency of the GSM networks is limited by the frequency reuse of the BCCH layer. Indeed, the lack of a downlink Power Control and Discontinuous Transmission on the BCCH carrier

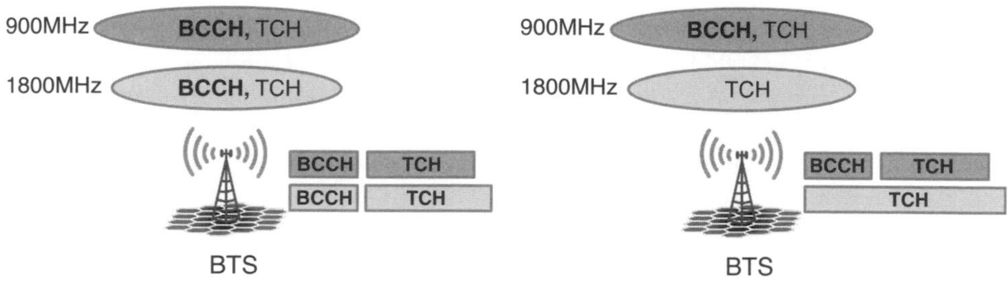

Figure 15.14 Common BCCH deployment.

Table 15.6 Common BCCH vs. non-common BCCH deployment—TRX configuration

	Non-common BCCH		Common BCCH	
	900 MHz	1800 MHz	900 MHz	1800 MHz
Spectrum	7.6 MHz	9.8 MHz	7.6 MHz	9.8 MHz
BCCH layer	15 channels	15 channels	15 channels	0 channels
TCH layer (guard channel between BCCH and TCH layer)	22	33	22	49
max TRX configuration assuming EFL = 18%	5/5/5	7/7/7	5/5/5	9/9/9

implies that BCCH frequency reuse has to be relatively large to ensure reliable reception of the broadcast (including system information), common control, and traffic channels.

Consequently, by eliminating the second BCCH layer in multiband networks, the spectral efficiency of the network can be significantly increased. The common BCCH feature might be considered as an attractive solution for those operators who are planning to introduce other systems such as WCDMA or LTE into bands that are currently used by GSM technology. When rolling out WCDMA or LTE in the 900/850 MHz and/or the 1800/1900 MHz bands, the available spectrum for GSM reduces, which means that the low spectral efficiency of the BCCH layer will become a more relevant issue. Let us consider the dual band network scenario with spectrums of 7.6 and 9.8 MHz in 900 and 1800 MHz correspondingly. Assuming that 15 frequencies are needed on BCCH layers to provide acceptable network performance remaining 22 and 33 frequencies (with one guard frequency between BCCH and TCH layer) are available for TCH pool—as presented in Table 15.6. We also assume a random frequency hopping scenario with tight frequency reuse pattern equal to 1/1 in that scenario.

If allowed Erlang Fractional Load [7] for network operating with AMR FR/HR mix 40%/60% is equal to approximately 18% then the maximum site configuration that can be supported in 900 MHz is 5/5/5 (including BCCH TRX). If we take the same assumptions for 1800 MHz, then the maximum TRX configuration in that band will be 7/7/7. Let us consider here, implementation of common BCCH functionality with the BCCH carrier at 900 MHz. In such a case all 49 frequencies in the 1800 MHz band can be allocated to hopping layer allowing for cell capacity extension that is for assumed EFL up to nine TRX can be deployed in 1800 MHz layer of the cell. Fifteen channels for BCCH should provide satisfactory performance on the BCCH layer. However, in some network scenarios Antenna Hopping may also be required for BCCH TRX—Section 15.4.5 provides further details on Antenna Hopping functionality.

If LTE 5 MHz[4] is going to be introduced in the 1800 MHz band, spectrum for GSM services must be squeezed to 4.8 MHz, so the total number of frequencies is reduced to 24. As 15 channels are required for the BCCH layer, BCCH and TCH frequencies are separated by one guard frequency, eight channels only can be used to create mobile allocation list. This means that, according to the assumptions made, only up to three TRX (with BCCH TRX) can be installed in the cell. Once the common BCCH feature is activated, the capacity offered within that layer is significantly enhanced. In fact, the maximum site configuration could now be 5/5/5. This is possible because there will be no BCCH carriers in the 1800 MHz layer. It is worth noting that if as a consequence of, for example, refarming, the number of available GSM

[4] Actually, even 4.8 MHz would be enough for LTE deployment in a coordinated scenario. For more details please refer to Section 15.3.2.2.

frequencies drops below the minimum required for BCCH reuse, then activation of common BCCH functionality will be indispensable.

As one might notice based on the example given, all frequencies that do not belong to BCCH frequency band can be used more efficiently—that is they may be planned with a tighter reuse pattern. Moreover as only one time slot needs to be spent on broadcast channels (FCCH, SCH and BCCH) and common control channels (PCH, AGCH, RACH)[5] additional capacity is freed for voice or data traffic. For instance, with the Orthogonal Subchannel feature activated, up to four voice users can be served on that saved timeslot.

Dual band cells with BCCH layers in each frequency band have independent pools of the SDCCH channels. Therefore, if in one of the frequency band the congestion on SDCCH layer occurs then there is no possibility to use the SDCCH resources that might be available in the second one. With common BCCH feature only one common pool of SDCCH channels have to be created for both layers and due to the trunking gain there might be less time slots needed for SDCCH traffic than if it would have been needed in the network operating with separate BCCH layers. Consequently, the saved timeslot(s) can be used to carry the user plane traffic. Table 15.7 illustrates the outcome of SDCCH/TCH dimensioning for selected cell configuration(s) for both: an non-common and a common BCCH case.

The main conclusion from this SDCCH/TCH dimensioning exercise is that by merging sets of the channels into the trunk pool, the capacity of the system can be increased as the result of the improvement in the trunking gain. It is worthwhile mentioning that the calculation of TCH capacity for common BCCH scenario was made with the assumption that resources on 900 MHz TRXs and on 1800 MHz TRXs are accessible to all users in the cell. If this is not the case, that is

Table 15.7 Common BCCH vs. non-common BCCH deployment—SDCCH/TCH capacity

<table>
<tr><th rowspan="2"></th><th colspan="2">Non-common BCCH</th><th>Common BCCH</th></tr>
<tr><th>900 MHz</th><th>1800 MHz</th><th>900 MHz/1800 MHz</th></tr>
<tr><td>Configuration (only FR considered)</td><td>2/2/2</td><td>2/2/2</td><td>4/4/4</td></tr>
<tr><td>Traffic per subscriber on TCH</td><td>25 mErl</td><td>25 mErl</td><td>25 mErl</td></tr>
<tr><td>Traffic per subscriber on SDCCH</td><td>4 mErl</td><td>4 mErl</td><td>4 mErl</td></tr>
<tr><td>Blocking probability of TCH</td><td>2%</td><td>2%</td><td>2%</td></tr>
<tr><td>Blocking probability of SDCCH</td><td>0.5%</td><td>0.5%</td><td>0.5%</td></tr>
<tr><td>Number of channels</td><td>16 (14 TCH + 2 SDCCH)</td><td>16 (14 TCH + 2 SDCCH)</td><td>32 (29 TCH + 3 SDCCH)</td></tr>
<tr><td rowspan="2">Total carried traffic/no. of subscribers that can be supported</td><td>8.2 Erl/328 subscribers</td><td>8.2 Erl/328 subscribers</td><td rowspan="2">21.04 Erl/841 subscribers</td></tr>
<tr><td colspan="2">16.4 Erl/656</td></tr>
<tr><td rowspan="2">Total carried signaling traffic/ no. of subscribers that can be supported</td><td>2.73 Erl/682</td><td>2.73 Erl/682</td><td rowspan="2">8.1 Erl/2024</td></tr>
<tr><td colspan="2">5.46 Erl/1364</td></tr>
</table>

[5] Provided that extra capacity for CCCH is not needed.

the coverage provided by 1800 MHz layer is limited and not available in the whole cell area then, of course, trunking gain will be decreased and thus total carried traffic will be lower.

Another benefit of the common BCCH feature is the reduced number of neighboring relations, which contributes to the better accuracy of the measurement reports sent from the MS in CS dedicated mode. Firstly, fewer neighbors means that there is more time available for measurements of each particular neighbor and therefore measurements performed and reported by MS are more accurate. Secondly, the measurement report limitation (up to six cells with the highest RXLEV amongst those with known and allowed BSIC can be reported)[6] is better addressed because the limited space in the measurement report is not spent for reporting 1800 MHz neighbors. Consequently, better quality due to a decreased handover rate can be expected. Moreover as the number of cells and neighbor relations will be reduced after conversion to the common BCCH-based architecture, the network operation and maintenance will be simplified.

15.4.2 AMR Full and Half Rate

AMR introduces the set of codec modes, which have a different level of protection against errors and which are automatically selected according to their robustness to suit the conditions of the radio channel. Owing to better error-correction capability against channel errors for the lowest AMR FR codec modes, such as 5.9 or 4.75 kbps, good speech quality can be provided even in poor radio conditions. For example, with AMR FR 4.75 codec mode, FER remains under 1% down to C/I = 3 dB, whereas approximately 8.5 dB is needed for the same performance with Enhanced Full Rate (EFR) codec.

This makes AMR FR an attractive solution that provides increased spectral efficiency for the networks operating with the spectrum that has been narrowed as a consequence of refarming, for example, to LTE. Moreover, as lower C/I targets can be set for AMR terminals, tighter PC thresholds can be used and therefore overall level of interference, also intersystem one, can be minimized. Figure 15.15 depicts BTS Tx power profile for EFR and AMR with two different set of PC parameter settings. In general, one can observe that AMR requires less power to be transmitted by BTS compared to EFR and that with the aggressive[7] PC parameter settings even more downlink bursts are transmitted with lower power.

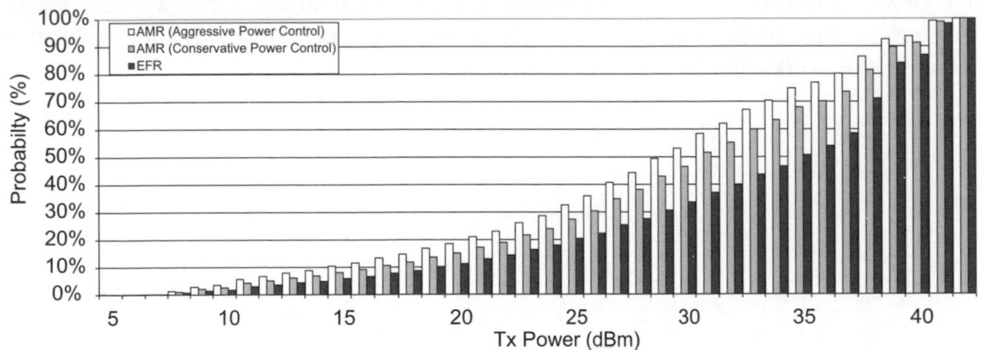

Figure 15.15 Cumulative distribution functions of BTS Tx power for AMR with different PC settings policy referenced to EFR.

[6] Unless enhanced measurement reporting (EMR) is implemented.

[7] Aggressive means that tighter PC thresholds are used.

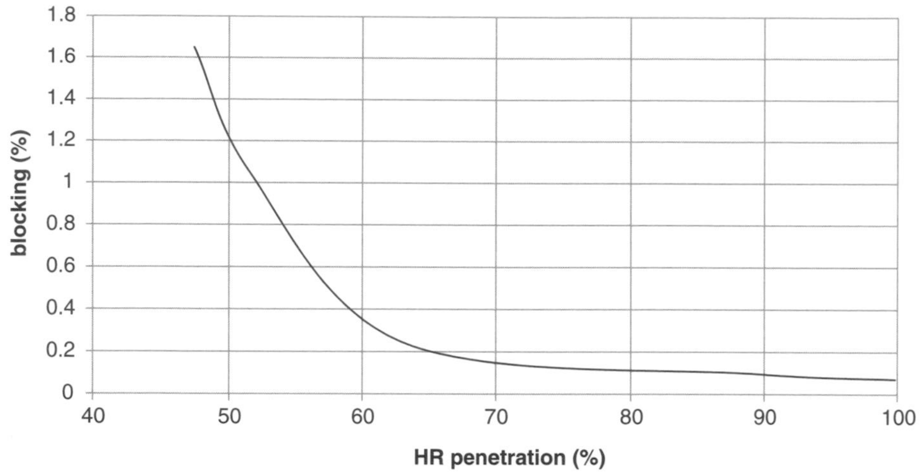

Figure 15.16 Blocking on radio interface vs. AMR HR penetration.

The robustness of the associated control channels is another important issue that needs to be addressed when improvement of the spectral efficiency in the network is desirable. Section 15.4.8 provides further information on selected signaling enhancement features.

With AMR, another channel mode has been introduced: AMR HR. The gross bit rate of its codecs is only 11.4 kbps whereas for FR channel mode codecs gross bit rate of 22.8 kbps is used. This entails that for the same codec mode (highest codec mode for AMR HR is 7.95 kbps) HR channel mode has considerably fewer bits for channel coding and therefore it has higher C/I requirements. The effect of AMR HR utilization on network performance indicators like voice blocking rate and average Frame Erasure Rate (FER) at constant load conditions is illustrated in Figures 15.16 and 15.17. Voice call was regarded as blocked when there were no resources to serve it. Average Frame Erasure Rate (FER) was calculated as the ratio of the number of erroneous speech frames to the number of all transmitted speech frames. It can be seen that by increasing AMR HR penetration, blocking on radio interface was significantly lowered and no new TRX was needed to serve the traffic that was so far not admitted to the network due to lack of resources. Therefore the use of AMR HR channel mode can be considered as an efficient way to increase network capacity and TRX utilization.

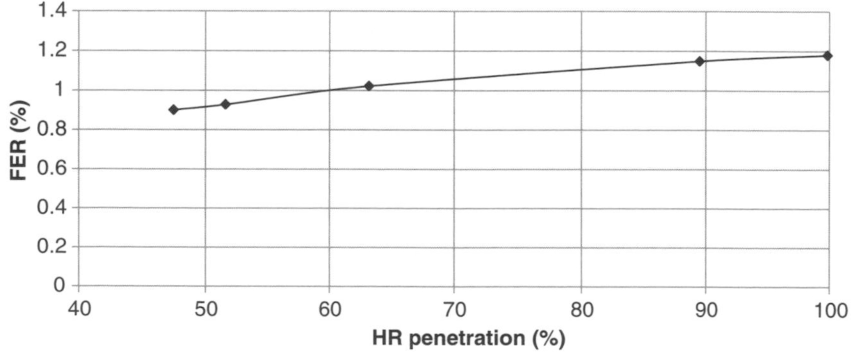

Figure 15.17 Average FER vs. AMR HR penetration.

It should, however, be kept in mind that AMR HR requires higher C/I conditions to achieve comparable FER performance as AMR FR. To keep the quality at the desired level, only a fraction of traffic can be served in AMR HR mode. Assuming that our quality target is average FER below 1%, it can be observed from Figure 15.17 that at AMR utilization up to 60% the average FER performance target is met. But, when more users are in AMR HR mode, the average FER starts to increase and goes beyond 1%. AMR HR utilization can be higher at lower load operation points, as the better C/I distribution enables the AMR HR usage.

The number of timeslots occupied in the network is reduced when HR channel mode is used, meaning that less interference is generated on radio interface inside GSM system but also towards other systems that may coexist with GSM in the same band.

15.4.3 Single Antenna Interference Cancellation

Single Antenna Interference Cancellation is a generic term for single antenna receiver techniques that by means of signal processing attempt to cancel or to suppress interference from co-channel cells. Mobile terminals with SAIC capability can tolerate higher interference levels than non-SAIC terminals. This allows network operators to deploy tighter frequency reuse factors, which is inevitable once the GSM spectrum is going to be refarmed to LTE or WCDMA. Mobiles that indicate SAIC capability to the network are called DARP (Downlink Advanced Receiver Performance) phase 1 in the 3GPP nomenclature.

The SAIC receiver has better Raw BER performance than non SAIC receiver in the same radio conditions and thus better RXQUAL values are reported by SAIC mobile stations. Since the power control is triggered by, amongst other things, RXQUAL values, better quality samples reported by SAIC capable mobiles drive down the transmission power in the downlink direction. The reduction in the power transmitted by the BTSs reduces the overall interference in the network, which in turn improves the performance of the conventional non-SAIC mobiles as well.

As SAIC terminal penetration increases, the downlink power control algorithm is able to reduce the BTS power more efficiently. This can be seen in Figure 15.18, which shows the cumulative distribution functions of BTS transmit power as the SAIC terminal penetration rate

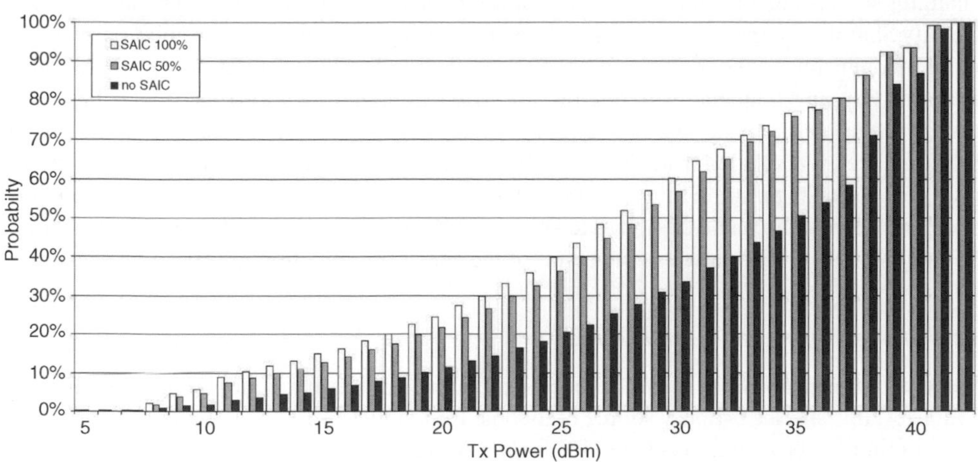

Figure 15.18 Cumulative distribution functions of BTS Tx power for different SAIC terminal penetration.

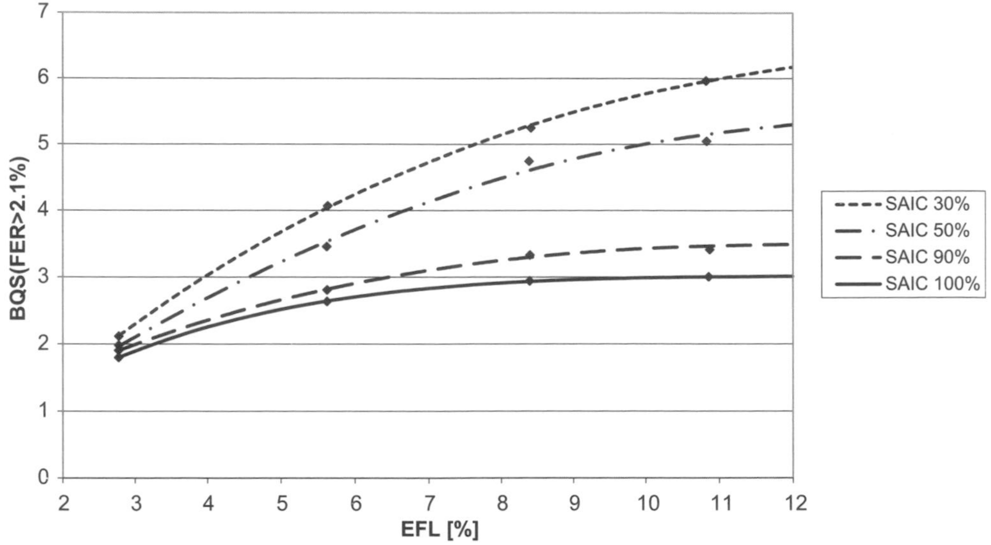

Figure 15.19 System capacity versus SAIC terminal penetration rate.

increases from 0 to 100%. One can note that the percentage of bursts transmitted with lower power increases once SAIC-capable mobiles are present in the network. At the same time, the share of bursts transmitted with higher power decreases compared to the reference case with non-SAIC terminals only. The reduced transmission power in GSM will also contribute to the better performance of services provided by other systems like, for example, LTE, which are operating in the same frequency bands as GSM (for more details about GSM and LTE coexistence please refer to Section 15.3.2).

In addition to investigating the effect of SAIC mobile penetration on the TX power distribution, system-level simulations were performed to examine impact of SAIC mobiles on system capacity. Figure 15.19 shows the percentage of unsatisfied users measured for changing load conditions in terms of EFL. Speech quality has been evaluated by means of Bad Quality Samples (the percentage of samples with FER higher than 2.1%). Simulations were run in inhomogeneous network cluster with average site $2 \times 2 \times 2$ configuration, 1/1 reuse pattern and nine hopping frequencies defined in the MA (Mobile Allocation) list. Only the performance of speech service on the hopping layer was taken into account–that is, BCCH performance was excluded from this study. It can be noted that as the SAIC terminal penetration rate goes from 0 to 100%, the percentage of the BQS samples for given load condition decreases. This gain can be exploited to keep speech quality at a satisfactory level even though EFL in the network increases as the result of the spectrum reduction in refarming scenarios.[8]

15.4.4 Orthogonal Subchannel

Voice, apart from SMS, is the key revenue-generating service in the mobile industry. In many countries, it is still the primary source of income for mobile operators and it is expected that it will continue to be an important service for many years to come. However, when taking into

[8] EFL is directly proportional to the traffic and inversely proportional to the available bandwidth.

account that revenues from voice are declining as well, it is evident that operators need to find a way to cut the costs of providing voice services over the GSM network.

In order to improve the cost efficiency of GSM voice services, an innovative feature called the Orthogonal Sub-Channel (OSC) has been developed. The OSC uses improvements in link performance provided by the AMR and SAIC features. It enables up to four users to be allocated on the same radio time slot, which significantly increases the BTS hardware efficiency. With AMR FR, a maximum of eight users can be allocated on the same TRX—once AMR HR is activated, the capacity of TRX can be doubled and up to 16 calls can be handled on one TRX. OSC again doubles capacity of TRX and this technique allows up to 32 half-rate connections to be supported per TRX.

This means that, in contrast to a non-OSC scenario, either less radio timeslots or even TRXs are needed to serve the same CS traffic, or a higher traffic volume can be carried on without new TRX being deployed. As far as the technical realization of OSC is concerned, in downlink it is based on QPSK-like modulation where the users are mapped onto QPSK constellations in such a way that first bit in a QPSK symbol is assigned to one user (subchannel 0) and the second bit to another user (subchannel 1). The subchannels are mutually orthogonal and can be received as a legacy GMSK signal (legacy GMSK SAIC handsets can receive them separately). In uplink direction, however, the traditional GMSK modulation is applied in combination with 2×2 multi-user MIMO technique.[9] Simultaneous reception of both users at the BTS requires antenna diversity and interference cancellation algorithms for separating of two orthogonal sub-channels. Different training sequence codes have to be assigned to terminals simultaneously sharing the same channel in order to optimize the link performance, in both uplink and downlink directions. Deployment of OSC requires mobile handsets that support AMR and SAIC. Gains from OSC strongly depend on the penetration of SAIC-capable mobiles, traffic distribution in the cell and the quality of the network which is the function of for example network load and frequency reuse pattern.

In case of AMR HR mode, although the same speech codec modes as in AMR FR mode have been defined, higher C/I values are required to achieve similar speech performance. Therefore transition to AMR FR is needed when voice connection in AMR HR mode suffers from bad radio conditions. On the other hand when the quality of AMR FR voice call is good, handover to AMR HR is strongly recommended as the interference generated in the network by AMR HR connection would be reduced (less timeslots are on air for the same number of connections with AMR HR than with AMR FR). Moreover, there is no gain in MOS speech quality when the Link Adaptation operates already on the highest AMR FR codec mode above certain C/I value. Equivalent MOS performance may be provided at that time by AMR HR. Similarly, OSC has higher C/I requirements compared to the AMR HR. This means that a dynamic switch between OSC and non-OSC mode according to current radio conditions is required. In addition to the quality-based criterion, due to the fact that the powers received from multiplexed OSC users in the BTS should not differ too much,[10] the dedicated mechanisms ensuring balance between the RXLEV of the paired OSC connections are needed. It is the task of Radio Resource Management (RRM) algorithm to select, based on quality and path loss criterion, appropriate users for OSC pairing, and then, by means of the power control, to keep their uplink RXLEVs close to each other. To pair users on the same channel the multiplexing HO, which is counterpart of packing HO in case of AMR FR to AMR HR transition, has been introduced.

[9] MIMO—Multiple In Multiple Out.

[10] The RXLEV balance between two connections occupying paired OSC-0 and OSC-1 subchannels must be maintained within a certain dB window (e.g. 10–15 dB) for sufficient quality.

When either the quality of the connection served on any OSC subchannel falls below a defined threshold or uplink RXLEV differences start to be too high, demultiplexing HO is triggered to AMR FR or AMR HR. Demultiplexing HO is executed to preserve adequate speech quality when radio conditions get worse.

The OSC concept is applicable to both AMR half-rate mode and AMR full-rate mode offering the capability of multiplexing of up to four AMR half-rate connections on the same radio time slot in Double Half Rate (DHR) mode and up to two AMR full-rate connections on the same radio time slot in Double Full Rate (DFR) mode.

System-level performance evaluations for OSC (with only DHR mode enabled) are presented below. The results are obtained from dynamic network level simulations. The simulations have been run over a macro cluster, and a typical urban 3 km/h (TU3) propagation model has been assumed. The network cluster consists of 123 cells from urban and suburban areas with the average configuration of 2 TRX per cell. The BCCH layer was not in the scope of this simulation campaign. The simulated scenario represents an interference-limited scenario with a tight 1/1 frequency reuse pattern and 9 TCH hopping frequencies. Penetration of SAIC compliant terminals was set to 50%. Mean call duration was fixed at 90 s. Power control and discontinuous transmission (DTX factor 50%) in uplink and downlink were enabled.

The blocking rate has been monitored to evaluate capacity gains resulting from OSC activation. The call was considered to be blocked when all available traffic channels in a cell were in use and the new connection could not be established. In a simulated scenario, approximately 15% of capacity gain has been achieved as can be noticed from Figure 15.20. This improved HW efficiency provided by OSC has been achieved at the cost of only slight speech quality degradation, which has been evaluated in terms of average Frame Erasure Rate (FER) (see Figures 15.21 and 15.22).

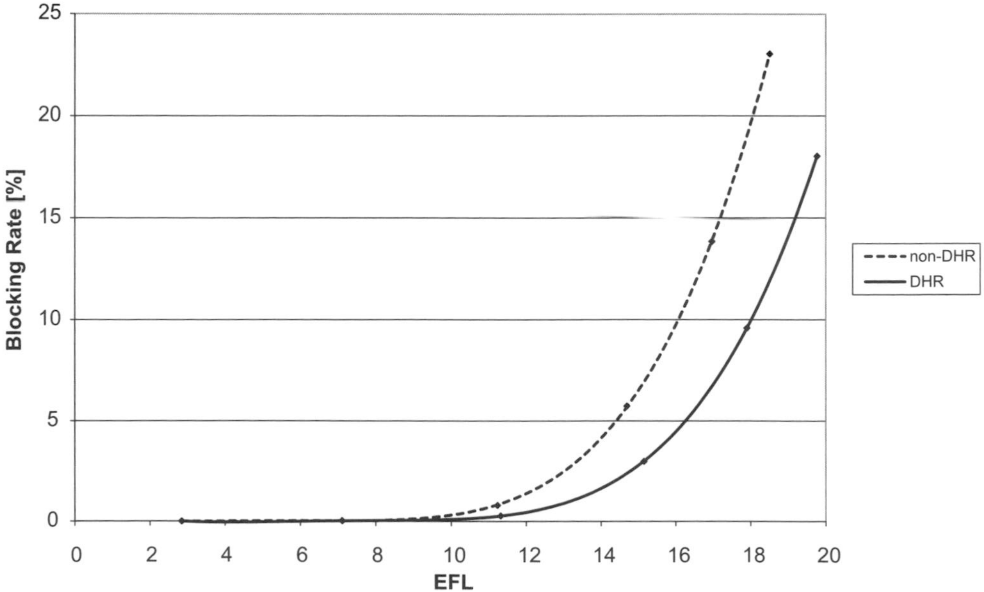

Figure 15.20 Blocking rate in non-OSC and OSC scenario.

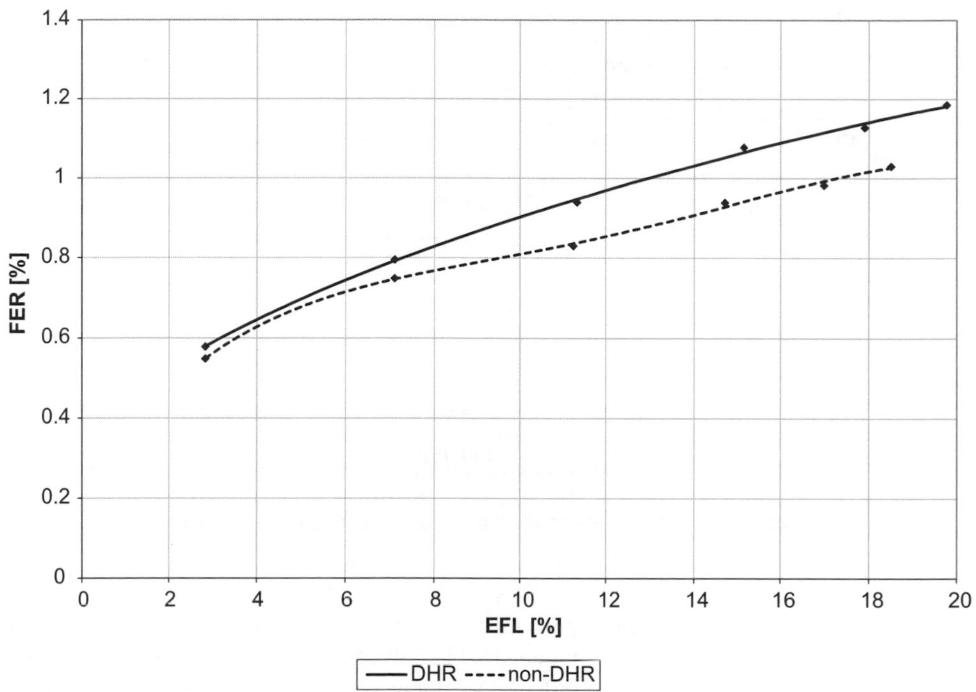

Figure 15.21 Average downlink FER in non-OSC and OSC scenarios.

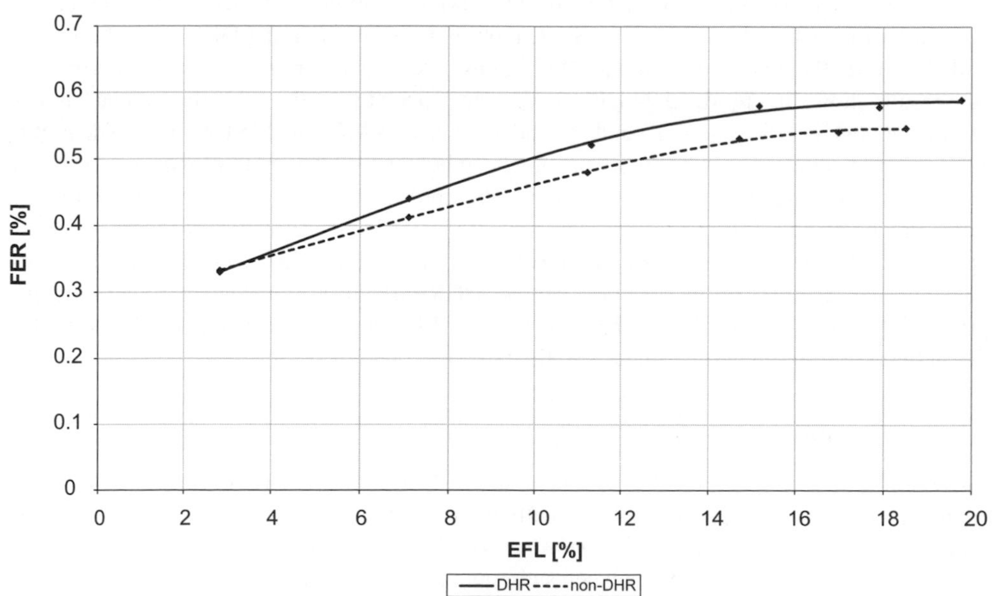

Figure 15.22 Average uplink FER in non-OSC and OSC scenarios.

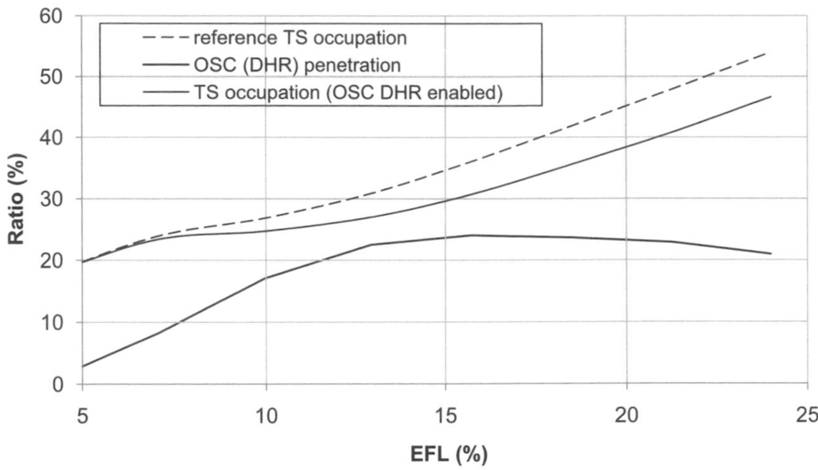

Figure 15.23 Timeslot occupation in non-OSC and OSC scenarios.

Owing to increased HW efficiency with OSC, fewer radio time slots are needed to serve the same traffic as in the non-OSC case. Figure 15.23 shows the time slot occupation ratio as a function of network load in terms of Erlang Fractional Load. The reference TS occupation denotes scenario with AMR only and variable AMR HR penetration, reaching almost 100% at very high network load conditions. The configuration of two hopping TRXs per cell and 1/1 frequency hopping with nine frequencies in the MA list was studied here. It can be observed that the activation of OSC results in a reduction of the TCH time-slot occupation by around 15% when a typical OSC penetration rate is assumed. Consequently, it contributes to the reduction of interference caused by the GSM system to LTE in both the uplink and the downlink directions.

In order to verify the impact of the SAIC mobile penetration on the DHR mode utilization and therefore the magnitude of the OSC gains, additional simulations were performed. The results shown in Figure 15.24 indicate that the proportion of the users that can be handled by the network in OSC mode depends on the number of SAIC capable terminals. The higher SAIC mobile penetration is, the more potential candidates for DHR multiplexing can be found in the cell and hence, for the same multiplexing/demultiplexing criteria, higher DHR utilizations may be achieved. It can be seen that overall cluster DHR utilization has increased from approximately 20 to 43% as the penetration of SAIC terminals reached 100% (i.e. there were no conventional receivers in the network). Higher DHR utilization translates to lower blocking probability (see Figure 15.25) for the same network load or more traffic that can be offered to the network without negative impact on the blocking rate.

15.4.5 Antenna Hopping

The well known frequency hopping functionality in the GSM network introduces certain frequency diversity and consequently directly improves the link-level resilience against deep fading dips. Different fast-fading profiles of various frequencies in the hopping scheme help to achieve decorrelation of fading between subsequent bursts, and hence improve the link-level performance. Deep fading could be also minimized by alternating the transmit antenna on a burst-by-burst basis, which introduces certain space diversity and artificial Doppler spread,

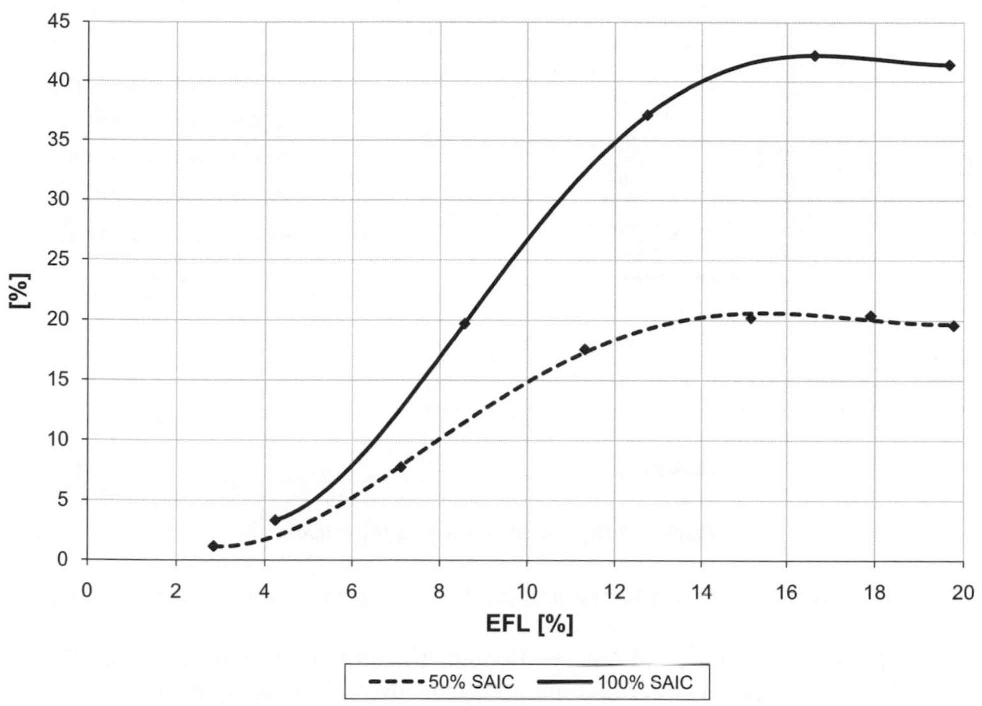

Figure 15.24 DHR utilization for 50 and 100% penetration of SAIC mobiles.

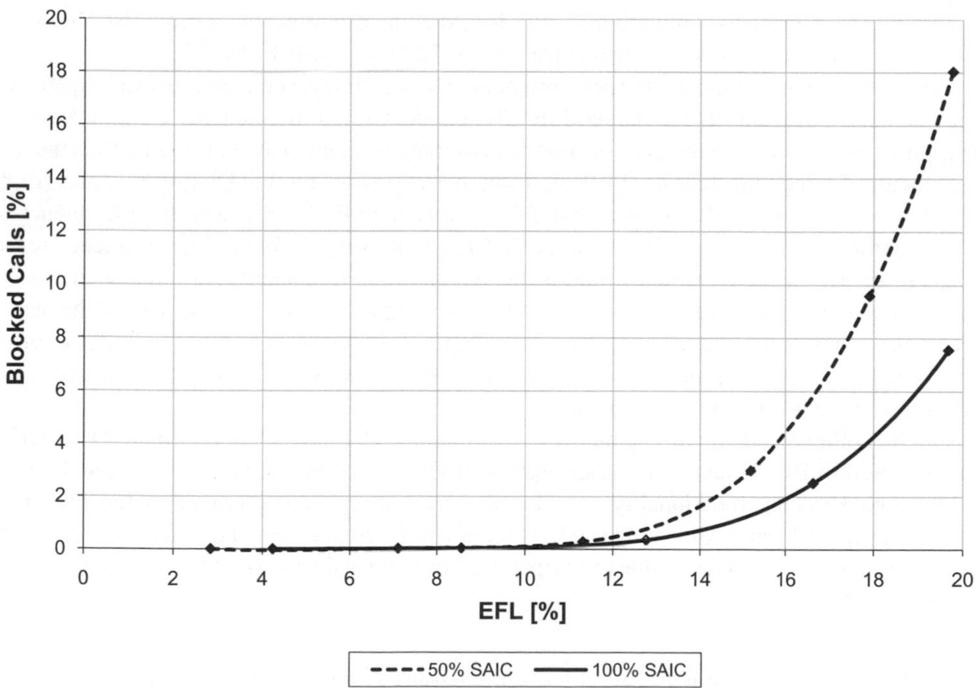

Figure 15.25 Blocking rate for 50% and 100% penetration of SAIC mobiles.

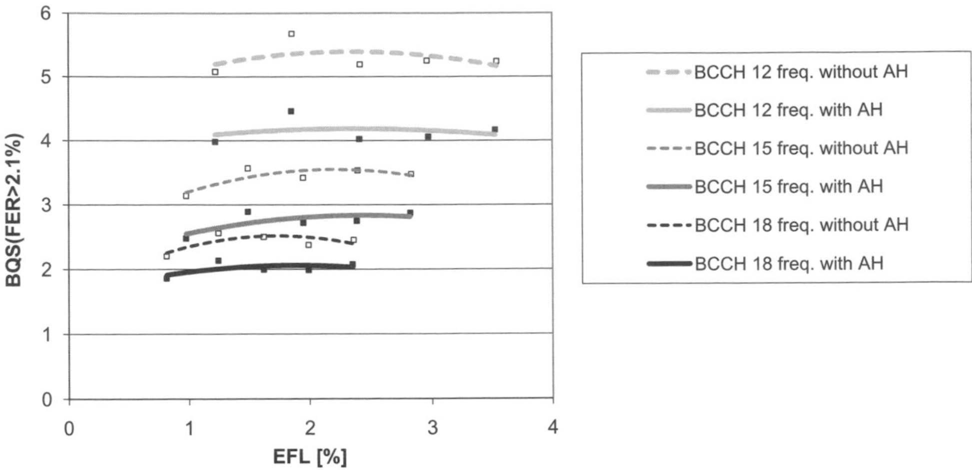

Figure 15.26 Antenna Hopping performance.

leading to decorrelation of fast fading profiles. Such a scheme is used in Antenna-Hopping functionality.

Since Frequency Hopping and Antenna Hopping benefit from the same radio propagation effect, the gains of these two features are not cumulative. In fact if the number of hopping sequences is high, the gain of Antenna Hopping on top of Frequency Hopping is very limited. Therefore, Antenna Hopping is especially applicable in BCCH layer where standard RF hopping is not used.

Figure 15.26 presents simulation results for Antenna Hopping for various BCCH plans deployed in a real network scenario consisting of 123 cells with 100% FR.

Thanks to Antenna Hopping, the performance of the BCCH layer consisting of 15 frequencies may be nearly the same as for standard BCCH scheme with 18 frequencies without Antenna Hopping. Moreover, with the help of Antenna Hopping it might be even possible to achieve a satisfactory Frame Erasure Rate (FER) performance expressed by Bad Quality Samples (BQS (FER > 2.1 %))[11] that is BQS lower than 5% for very tight BCCH reuse with 12 frequencies.

A significant improvement in downlink BCCH quality may enable tighter frequency reuse plans in the BCCH layer, which would be beneficial in the frequency refarming scenarios as more frequencies could be refarmed to other technologies. In some cases, part of the saved frequency carriers may be added in the list of hopping frequencies in GSM in order to offer additional capacity. Consequently, overall GSM quality would not suffer from spectrum reduction in network migration scenario.

On the other hand, quality gains due to Antenna Hopping also facilitate OSC DHR introduction in BCCH layer (for a description of OSC DHR please refer to Section 15.4.4).

Antenna Hopping functionality may be also treated as a complement to the Dynamic Frequency and Channel Allocation (DFCA) feature as it provides direct quality improvements in BCCH layer where DFCA does not operate. For further details on DFCA please refer to Section 15.4.7.

[11] BQS(FER > 2.1%) represents the ratio of 1.92 s speech samples with frame erasure rate (FER) higher than 2.1% (more than two erroneous frames out of 96); BQS(FER > 2.1%) = 5% means that there were more than two bad frames in 5% of all observed 1.92-second intervals.

15.4.6 EGPRS2 and Downlink Dual Carrier

In order to reduce the PS performance gap between GSM and LTE there is a strong need to introduce performance boosters in GPRS/EDGE network that will help to increase the throughput as perceived by the user attached to the 2G network. This is especially needed if contiguous LTE coverage is not available (which would be the most common case in the first LTE deployments) and simultaneously significant amount of LTE UEs have also GSM capability.

Such PS performance boosters namely: Downlink Dual Carrier (DLDC) and EGPRS-2 were introduced in 3GPP Release 7. With the help of DLDC functionality DL TBF for PS transmission may be allocated on two independent carriers, not on one—as in previous releases. Hence, achievable peak downlink throughput could be doubled.

In EGPRS-2, the physical layer of EGPRS was changed. New modulations (up to 32QAM) and coding schemes were defined and grouped into two levels: EGPRS-2A and EGPRS-2B. Moreover, turbo codes and higher symbol rate were also incorporated in the EGPRS-2 functionality. In the uplink linearized GMSK pulse shape wider than the standard was also defined. Work on pulse-shape improvements over legacy LGMSK in DL direction has not been concluded yet in 3GPP [8]. For further details related to these functionalities, refer to [7].

The EGPRS-2 theoretical peak throughput per timeslot could be doubled in comparison with legacy EGPRS transmission and moreover it could work on top of Downlink Dual Carrier functionality. EGPRS-2 improves also delays as perceived by the user. In Figure 15.27 simulation results for EGPRS-2B with wider pulse shape (EGPRS 2B-W) together with DLDC are presented. Simulations were performed in BCCH layer consisting of 17 frequencies in real network scenario with 112 cells in the SMART simulator (for a further description of the SMART simulator please refer to [7]).

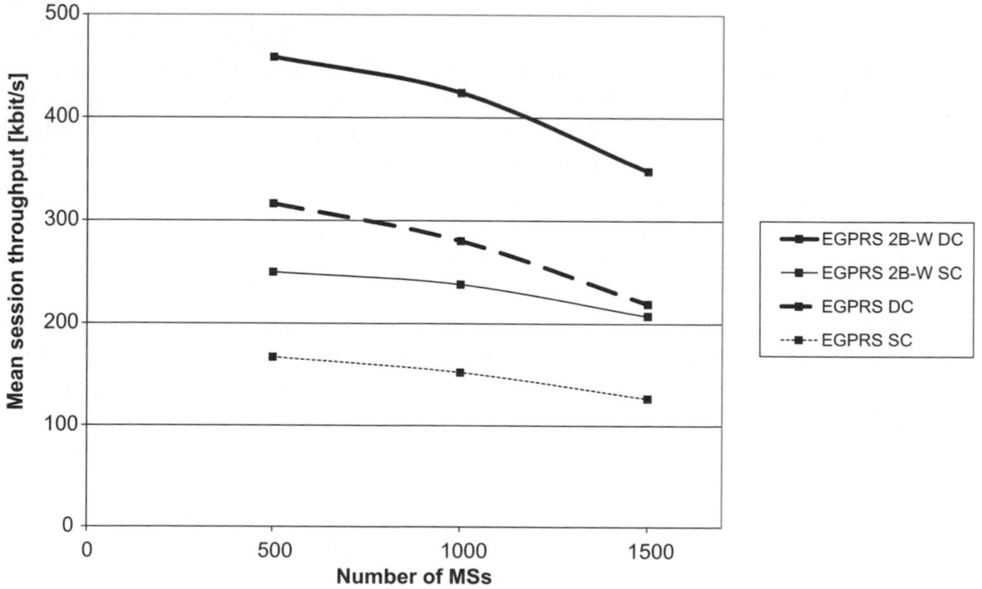

Figure 15.27 Performance of EGPRS 2 together with DLDC.

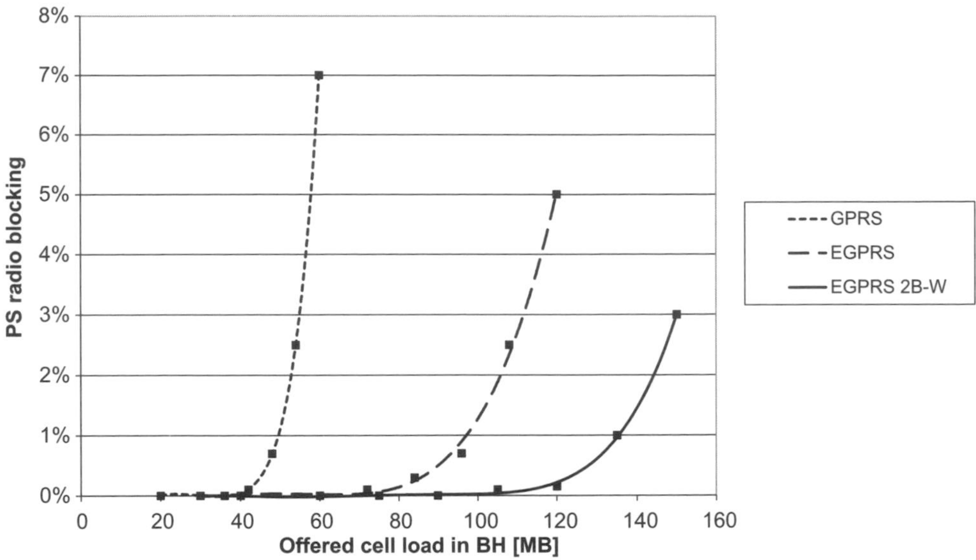

Figure 15.28 PS radio blocking—EGPRS 2 vs. legacy systems.

In the simulated scenario, the mean session throughput in heterogeneous network deployment could be nearly tripled in comparison with legacy EGPRS if both EGPRS-2 and DLDC are implemented together, which makes the enabling of these two functionalities attractive in the multi-RAT environment.

As a result of DLDC and EGPRS-2, PS resources and consequently system throughput may be distributed more efficiently between network users leading to either significant improvement in mean user throughput (as presented in Figure 15.27) or higher system load without compromising the user perceived quality. In Figure 15.28 simulated PS radio blocking in a single cell for HTTP service is presented. Any given user performs 5 HTTP transactions each of which leads to 1 MB data download following by a short time break of 12 s, which aims to model the real user behavior while browsing the WWW pages.

Thanks to EGPRS-2B offered cell load could be increased by approximately 35% while maintaining the PS radio blocking at the same level as in pure EGPRS scenario (assuming the same amount of resources to be used by PS services).

Apart from hardware efficiency improvements presented above, EGPRS-2B may be also regarded as a spectral efficiency feature. In Figure 15.29, the mean session throughput for EGPRS 2B-W (with wider pulse shape) as well as EGPRS service is depicted for BCCH layer with 12 frequencies and for the TCH layer (1TRX/cell) with tight 1/1 reuse and only six hopping frequencies.

According to the simulation results, use of EGPRS 2B-W compensates the limited number of frequencies available for PS transmission in TCH layer and helps to achieve better or similar performance in terms of throughput in comparison with EGPRS used in BCCH layer. This advantage of EGPRS-2 functionality will be especially beneficial in network migration scenarios where part of the GSM spectrum is refarmed to other technologies like LTE, and will help to assure proper performance of the GSM network operating in reduced spectrum allocation.

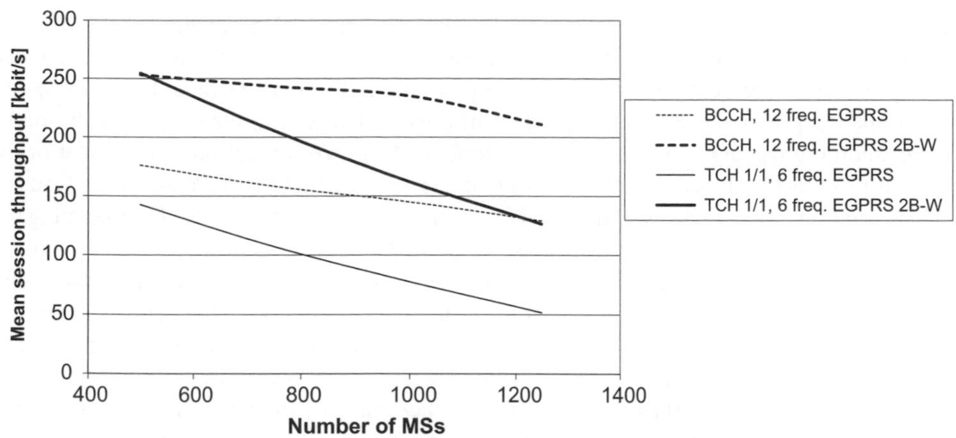

Figure 15.29 Spectral efficiency of EGPRS 2.

15.4.7 Dynamic Frequency and Channel Allocation

The radio frequency spectrum is a precious, limited natural resource for which demand fairly exceeds supply. The notion of spectrum scarcity causes that spectral efficiency (which quantifies the traffic volume that network can serve per MHz) is a critical metric and the techniques that allow reuse of the existing spectrum in the most efficient way have always been a top priority for mobile network operators. Dynamic Frequency and Channel Allocation as an advanced spectral efficiency solution is essential, especially in the case of spectrum refarming for WCDMA/HSPA or LTE needs.

Unlike standard random frequency hopping (which averages the interference throughout the network) Dynamic Frequency and Channel Allocation (DFCA) does not randomize interference but controls it and hence significantly increases spectral efficiency. The control of interference is achieved by means of cyclic frequency hopping and intersite synchronization.

With DFCA the actual frequency of each TRX comes from MA list and assigned MAIO and is changed to another one, every TDMA frame. The synchronization between the frames of the TRXs ensures that the two calls using the same MA list will always use different frequencies: they use the same timeslot number and the same MA list but they do not interfere because they have different MAIOs allocated. If these TRXs belong to the same site, they are synchronized as usual. If the TRXs belong to different sites, they need to be synchronized in order to control the interference between them.

DFCA allocates radio channel as a combination of timeslot, MA, MAIO, and TSC. It also calculates an initial DL and UL power reduction to ensure that the only required power level is used from the very beginning of the connection (before standard power control action). When a channel is to be allocated for the incoming call (call setup or handover), DFCA evaluates all the possible candidates and then selects the most suitable channel. For this reason, the C/I is estimated for each available radio channel. The aim is to choose the best channel, in terms of both incoming interference (the disturbance that the new call would be suffering) and outgoing interference (the disturbance that the new connection would cause to calls in the other cells). C/I for potential target channel is estimated based on the mobile measurement reports. For the cells not included in the latest measurement report, C/I estimation is obtained by collecting the long-term neighboring cell Rx Level measurement statistics from all mobile measurement reports in a cell. These statistics are stored in a BSC as the Interference Matrix (IM). In this way the DFCA is able to estimate C/I for all the potentially interfering cells even if some of them were not included in the latest measurement report. For the neighboring cells that are included in the latest measurement report, C/I is estimated based on the provided by the mobile Rx Level value of the serving and neighboring cells as well as the applied DL power reduction. For the rest of the surrounding cells the statistical C/I estimation (data from the interference matrix) can be used.

There are four C/I estimations calculated for each available radio channel. The incoming C/I describes the interference coming from existing connections that would affect the new call for which the channel assignment is being performed. The new connection may also cause interference to existing calls using the same or an adjacent radio channel. This is examined by determining the outgoing C/I for every potentially affected existing connection. The incoming and outgoing interference are estimated separately for downlink and uplink directions. The C/I estimation relies on the fact that the interference relations in a DFCA network are stable, predictable, and controlled due to the use of site synchronization and cyclic frequency hopping. The C/I is determined by combining information from measurement reports, interference matrix and BSC radio resource table where the current status of radio resources occupation is stored. The difference between estimated C/I and the target C/I value (required C/I) is calculated. This difference is computed for incoming and outgoing C/I for both UL and DL directions. Amongst these four C/I differences (calculated for the strongest interferer) the lowest one is determined. This most restrictive C/I difference is then used in the channel selection procedure.

Having determined the most restrictive C/I difference for all the candidates, the channel with the highest C/I difference is checked against soft-blocking limit. The soft-blocking limit is the configurable threshold in dB to assure that the new connection would not generate or would not suffer from too strong interference. A channel request that breaks soft-blocking limit is served on non-DFCA TRXs of the cell (typically BCCH TRX). This soft-blocking validation is an admission control mechanism in DFCA networks.

DFCA also takes care of training sequence code assignment. After a suitable channel has been found, the BSC determines the most appropriate TSC by checking all the interfering connections. The BSC searches for the training sequence code (TSC) that has been used by an interfering connection characterized by the highest C/I relation. This means that for the selected MA, MAIO and time slot combination, the C/I differences are examined TSC by TSC. Training sequence code selection aims at avoiding the worst-case situation where a significant interference source uses the same training sequence code as the new connection. A conflict with a significant interferer would cause the receiver to obtain incorrect channel estimation and consequently lead to link-level performance degradation. For that reason TSC used by the connection with the highest C/I difference is selected by DFCA algorithm, to guarantee that the new connection will not suffer from undesired effects coming from TSC cross-correlation.

Presented above channel selection process is used for any service type (e.g. EFR, AMR HR, SDCCH, (E)GPRS) with its service specific C/I target and C/I soft-blocking thresholds.

DFCA allows different QoS requirements to be set (by means of different C/I target thresholds) for different type of services. This is very important because the robustness of these services to interference varies significantly due to different channel coding. For example, an AMR connection can tolerate a much higher interference level than an EFR one while still maintaining comparable performance in terms of speech quality. For (E)GPRS services, higher C/I requirements are usually set to benefit from higher modulation and coding schemes (MCS). For SAIC-capable mobiles able to provide sufficient quality at higher interference levels, a special configurable offset is applied to connection-type-specific C/I target and C/I soft-blocking thresholds.

In high network-load conditions, most resources are occupied and DFCA does not have much freedom in channel selection. Therefore it is beneficial to increase half-rate usage in the high load scenario even if there are still enough hardware resources available in the cell. For that reason the forced half-rate mode has been introduced with DFCA. The forced HR mode is a method for channel-rate selection between full rate and half rate, based on the average of the estimated incoming DL C/I values. The average DL C/I ratio provides a good benchmark of the load of DFCA frequencies, and hence it can be used to force HR allocation. A forced HR mode is triggered if the averaged C/I drops below configurable threshold.

The main advantage of DFCA, compared to other channel allocation methods, is that DFCA dynamically adapts to changing interference conditions. By controlling interference the DFCA functionality provides the best possible way of radio channel allocation, with the highest possible quality.

The performance of the DFCA has been evaluated by network-level simulations comparing DFCA performance with a reference case that used random frequency hopping with 1×1 reuse. The real network scenario was created by importing network layout of a real network in a city center and its surroundings. An uneven traffic distribution was used to achieve the cell loads from the real network. In both the reference and the DFCA cases, nine hopping frequencies were used. Considering 12 frequencies used for BCCH layer, 4.2 MHz in total was used in that network.

As presented in Figure 15.30 activation of the DFCA feature allowed for increase of HR usage by 35 percentage points. This gives a huge benefit in terms of hardware efficiency as the same traffic can be served by 2/3 number of TRXs (comparing to the nominal number of hardware resources).

With RF hopping, HR penetration becomes saturated at the level of 63%. Above this level, radio quality becomes so poor that it does not allow for further HR allocation even though load conditions are fulfilled. With DFCA much better radio conditions allows for 95% of HR usage (only very small portion of calls cannot be allocated as HR due to bad radio quality).

Higher HR usage is possible thanks to better overall C/I conditions due to interference control ensured by DFCA. On top of this the other Key Performance Indicators (KPIs) like Dropped Call Rate (DCR) and Bad Quality Samples (BQS) decreased.

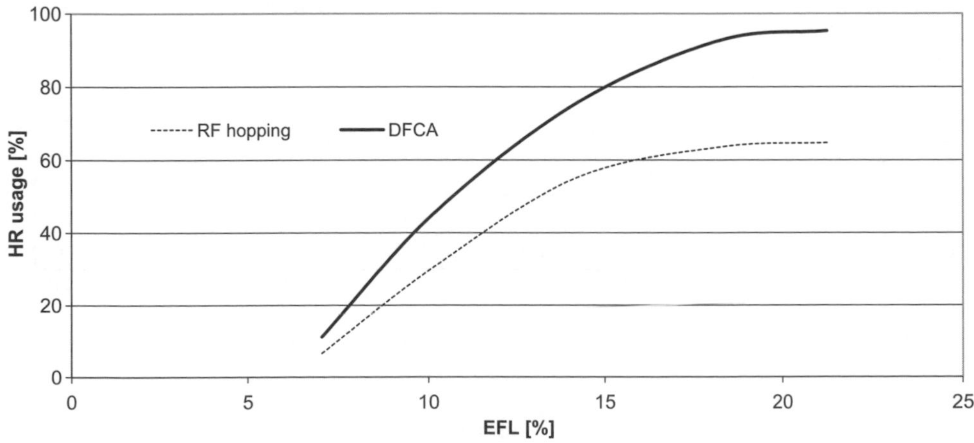

Figure 15.30 HR usage in case of DFCA and RF hopping.

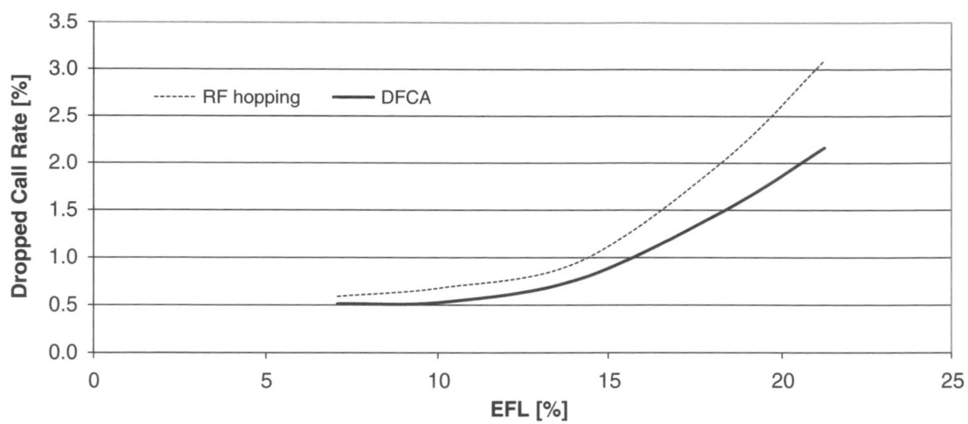

Figure 15.31 Dropped Call Rate in case of DFCA and RF hopping.

As illustrated in Figure 15.31, simulation results show that DFCA reduces DCR (there is an approximately 20% quality gain at 14% of EFL). In addition to the improved DCR, the simulations clearly indicate an improvement in terms of BQS. Bad Quality Samples were calculated as the percentage of SACCH frames with the Frame Erasure Rate above 2.1% to all samples. This definition means that the BQS reflects speech quality that is perceived by the end user.

A BQS of 5% is considered as the outage threshold above which the end user starts to experience unacceptable speech quality. At 13% of EFL, a simulated network with RF hopping hits BQS thresholds of 5%. At the same traffic load but fairly higher HR penetration DFCA provides 20% better DL speech quality that is BQS of 4% (Figure 15.32). The number of measurement samples with bad quality is correlated with the level of interference that is reduced by DFCA, which allocates a channel impacting ongoing calls in the least harmful way.

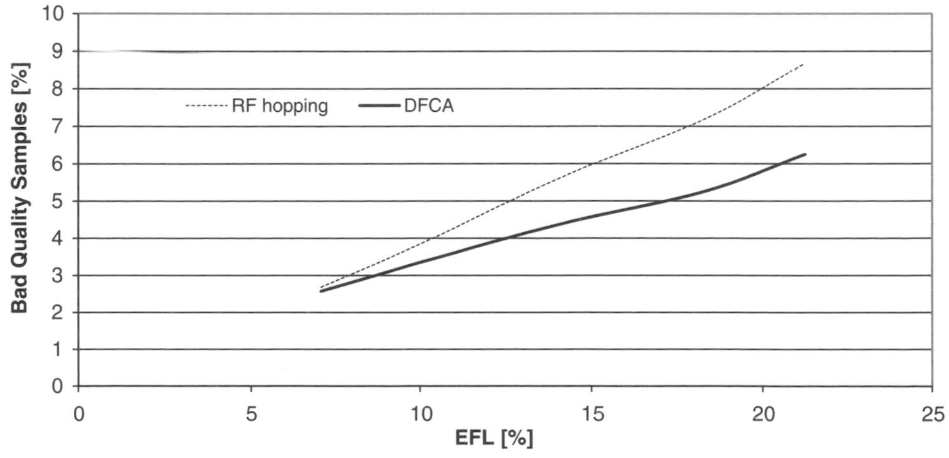

Figure 15.32 DL Bad Quality Samples in case of DFCA and RF hopping.

The performance figures from simulations show that the DFCA functionality provides significant spectral efficiency gains. This means that the DFCA allows operators to serve the same traffic with the squeezed frequency spectrum while keeping quality at satisfactory level. In this way DFCA is the best spectrum efficiency booster that allows spectrum refarming for the other technologies like WCDMA/HSPA or LTE.

Dynamic Frequency and Channel Allocation and Orthogonal Sub-Channels

DFCA is a spectral efficiency feature that improves the quality of the network throughout the intelligent, interference-aware allocation of radio resources. Improved quality of the network results from the fact that, firstly, DFCA assigns a channel that provides an adequate performance for given service. Secondly, DFCA ensures that the impact of the connection to be allocated on the selected channel towards the ongoing ones is minimized. OSC has higher quality requirements compared to, for example, AMR HR, and to trigger the pairing of users into OSC mode, both of them have to be in good radio conditions (among other criteria).

Figures 15.33 and 15.34 show downlink and uplink RXQUAL distribution collected from the network cluster without and with DFCA activated. It can be noticed the percentage of good RXQUAL samples significantly increased after DFCA activation.

The number of measurement samples with good RXQUAL results from the level of interference that is reduced by DFCA.

If we assume that OSC de-multiplexing threshold is set to RXQUAL = 4, then the percentage of RXQUAL samples at which OSC may operate changes from 85% to 97% for the uplink direction and from 75% to 88% for the downlink, which is the limiting link here. Consequently, the boost in the good quality samples may be converted to higher OSC penetration in the network. The combination of DFCA and the OSC features may therefore be considered as a very attractive way to increase both HW and spectral efficiency, which are the primary concerns that need to be addressed when refarming to LTE/WCDMA.

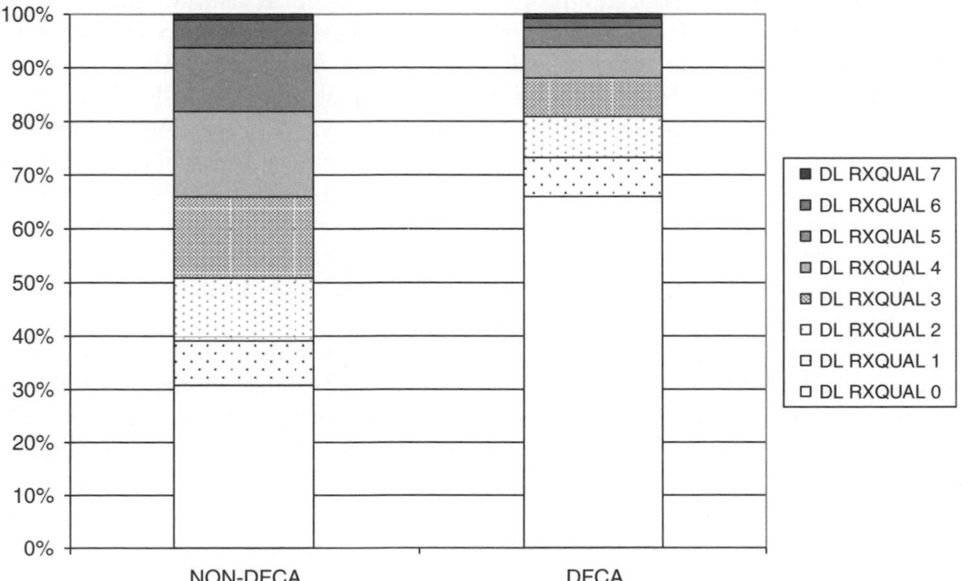

Figure 15.33 DL RXQUAL improvements due to DFCA.

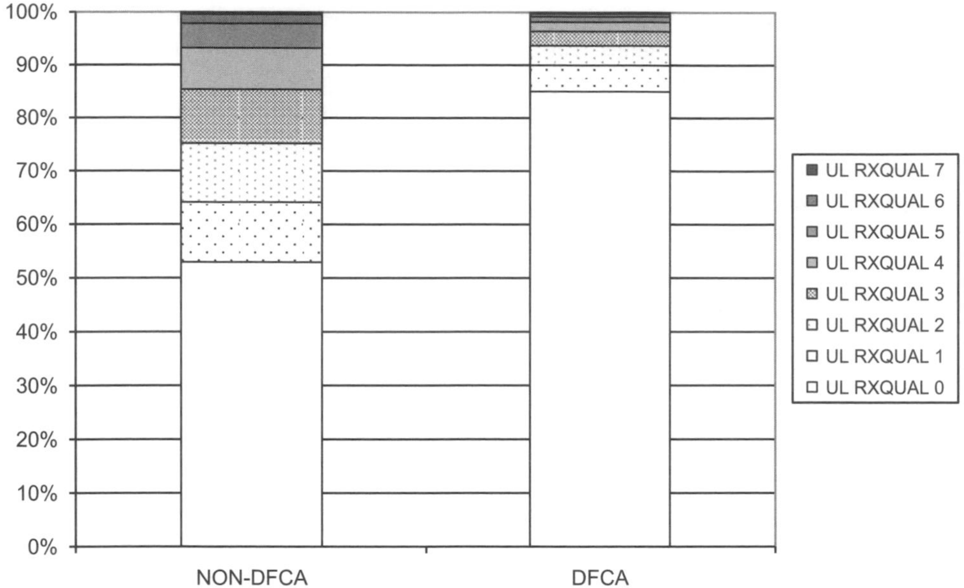

Figure 15.34 UL RXQUAL improvements due to DFCA.

15.4.8 Signaling Improvements

Advanced techniques that allow for the operation of GSM traffic and signaling channels in squeezed (narrowband) spectrum are essential to ensure the proper operation of a GSM-installed base for GSM spectrum refarming for LTE needs.

Originally, in the 3GPP standard, the robustness of the signaling FACCH and SACCH channels was designed to match the error-protection schemes of the standard codecs (full rate, half rate, and enhanced full rate). With the introduction of AMR speech codecs the robustness of the traffic channels has been considerably improved. The most robust AMR codec modes, AFS 4.75, AFS 5.9, provide sufficient voice quality even at very low C/I levels. However, signaling has not been improved by AMR and therefore SACCH and FACCH protection schemes cannot cope with the same C/I conditions as AMR codecs. Thus, with usual call drop indication mechanisms, applied in AMR networks, calls may be indicated as "dropped" from a signaling point of view, although acceptable quality speech communication is still possible. This is especially important for the spectrum refarming that causes/forces GSM network to operate in narrowband scenario with typically lower C/I conditions and where it is crucial to adjust signaling robustness to the AMR capabilities.

There are several proprietary solutions to adjust signaling performance for AMR capabilities. Changes to 3GPP GERAN standard have also been introduced to improve the error protection schemes for FACCH and SACCH. "Repeated Downlink FACCH" and "Repeated SACCH" have been defined in the 3GPP GERAN specification TS 44.006 [9] for a Rel-6 compliant MS that is capable of performing joint decoding of two successively repeated FACCH/SACCH blocks. The MS indicates support for these features via the MS Classmark 3 Information Element called Repeated ACCH as defined in 3GPP TS 24.008 [10]. The repeated DL FACCH/SACCH procedure foresees a serial repetition of the FACCH/SACCH blocks. The MS can then enhance its link performance by attempting soft decision combining of the repeated FACCH/SACCH blocks. According to link-level simulation results, the expected gain in decoding

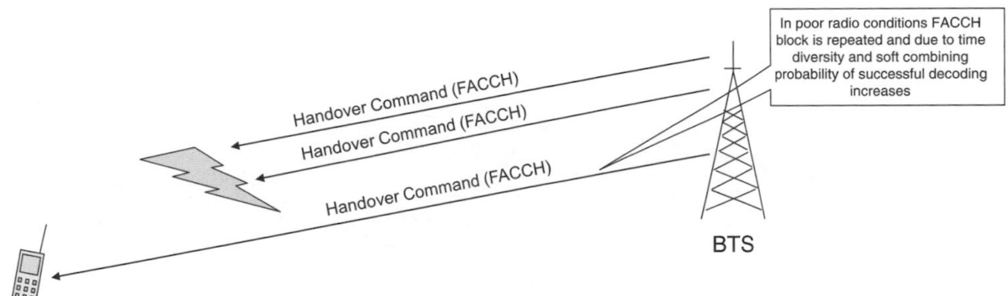

Figure 15.35 Repeated FACCH procedure.

performance is about 3 dB (for legacy MS) to 5 dB (for Rel-6 MS, which performs soft combining). This link-level improvement reduces the number of lost FACCH/SACCH blocks and hence decreases the number of dropped calls due to radio link failure or handover failure.

Repeated FACCH is based on the principle that, when repeated FACCH blocks are combined in the receiver, a decoding gain is achieved thanks to the time diversity between the transmissions.

With the Repeated Downlink FACCH (Figure 15.35), the duplicate FACCH frame is sent using the speech codec's tolerance for the stolen speech frames (for signaling purposes). These stolen speech frames can be used by the receiving end to provide a second chance for successful decoding. The network may retransmit FACCH block eight TDMA frames after sending the original FACCH block (or nine TDMA frames, if an idle frame or a SACCH occurs in between). If the mobile receives both FACCH blocks but is unable to decode either of them independently, then it attempts joint decoding of two blocks (if MS is Release 6 compliant). With a legacy MS, soft combining is not possible but diversity gain is achieved when the legacy mobile attempts to decode both FACCH frames independently.

The Repeated SACCH, like the Repeated Downlink FACCH, makes use of time diversity between repeated transmissions to increase the likelihood of successful decoding. For SACCH reception on the downlink, the MS attempts to decode a single SACCH block first (without attempting to combine it with the previously received block). If decoding fails, the mobile requests SACCH repetition. In the next SACCH frame mobile attempt decoding, after combining it with the previously received SACCH frame. For SACCH retransmission in the uplink, MS first checks if Repeated SACCH is ordered by the network (by checking the flag in the last DL SACCH block) and if that previous SACCH block was not already retransmitted. The mobile then transmits a duplicate frame instead of the originally scheduled frame (Figure 15.36).

To adjust the signaling performance to AMR capabilities, wireless network equipment providers have developed several proprietary solutions. Amongst these solutions there is an option to reflect different robustness of AMR against interference in separated settings for Radio Link Timeout for AMR and non-AMR calls. The purpose of Radio Link Timeout (RLT) is to force release of the call when speech quality is so poor that the majority of subscribers would have manually released the call anyway. Considering that AMR calls are more robust than EFR calls, the common RLT value for AMR and non-AMR calls may cause the AMR calls to be unnecessarily dropped from the signaling perspective (while the speech quality is still at acceptable level). For this reason, separate RLT thresholds for AMR and non-AMR calls have been implemented. The RLT threshold for AMR should be at a much higher value than for EFR. Trials in a real network proved that the RLT increase (e.g. from 20 for EFR to 32 for AMR) significantly reduces call drops, maintaining speech quality at the acceptable level.

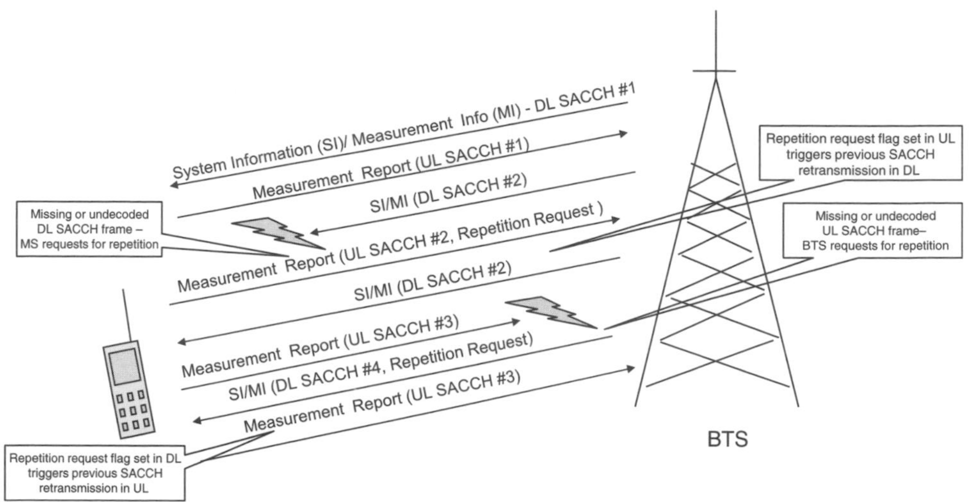

Figure 15.36 Repeated SACCH procedure.

The next proprietary techniques for fine-tuning signaling performance to AMR capabilities improve the C/I for the signaling channels by applying temporarily higher DL transmit power on FACCH and SACCH bursts. The 3 dB improvement in C/I offered by Repeated FACCH for handsets that complied with Release 5 and earlier is not enough to give reliable operation with the most robust AMR FR codec so, for legacy mobiles, the FACCH/SACCH Power Increment feature is used and temporary increments transmit power on signaling channels above the level ordered by the power control procedure. Power Increment is a practical and fully backwards compatible option to improve the probability of successful decoding of the FACCH/SACCH frames. Under normal radio conditions both FACCH and SACCH are transmitted using the power commanded by the power control algorithm for the traffic channel. In poor radio conditions, however, based on a proprietary trigger mechanism (e.g. RXQUAL based, FACCH repetition based or MS codec request based), FACCH and SACCH are transmitted independently of each other with power increased by, for example, 2 dB in relation to the currently used TCH transmit power commanded by the power control. To avoid the undesired effect of permanent overpower on the interference level experienced on TCH channels, the increased power to FACCH and SACCH only occurs if radio conditions degrade below a certain quality threshold, or FACCH repetition is required, or MS requests the lowest downlink codec (which indicates poor radio conditions).

Another proprietary solution used to adjust the signaling performance to AMR robustness is the reduction of handover (HO) signaling message length. Handover Command (which is used in the intercell HO) and Assignment Command (sent in intracell HO) are the longest messages in handover procedures and therefore affect handover performance the most. The longer these messages are, the more FACCH blocks are required to convey them over the radio interface. The greater number of FACCH blocks required for Handover/Assignment Command transmission the greater chance that one of these FACCH blocks will be lost and finally MS will not be able to decode this message. By minimizing the message length, the performance of signaling transactions is more reliable, which improves HO success rate and overall call drop rate.

Handover/Assignment Command message length reduction is achieved by the BSC manufacturers by removing the Information Elements (IEs), which are redundant and optional in the given scenario. For example, if the Cell Channel Description Information Element is not present, the MS uses the same hopping frequency allocation as in the old cell. This allows the BSC to remove 17 octets from the Handover/Assignment Command reducing number of FACCH blocks for Handover/Assignment Command transmission by 1 FACCH block effectively improving performance in terms of the HO success rate and the call drop rate.

15.5 Alternative Network Migration Path (Multi-Operator Case)

Limited multi-operator network sharing has taken place since the onset of cellular technology, but it involved sharing of passive elements and was limited to isolated cases—for example site sharing at the highest building in town. More advanced sharing was not popular, as the operators considered the network hardware as their major asset and the regulators tended to block network sharing. The situation has, however, changed in recent years, and this change in attitude has been caused by several factors:

- economic downturn, making operators reluctant to spend money, paired with lowering ARPU;
- no increase in ARPU likely, as the new technologies will principally offer data transfer at flat rates;
- scarcity of the spectrum to be allocated;
- the need for a wide contiguous frequency band to achieve the high data throughputs that the LTE technology is capable of providing (due to signaling overhead and lower flexibility of scheduler the LTE deployments at 1.4 or 3 MHz bandwidth may not achieve satisfactory throughput requirements—cf. also Section 15.3);
- technological advances, both in LTE and in legacy radio access technologies, allowing greater flexibility (e.g. carrier aggregation) and high efficiency (e.g. MIMO, EGPRS-2) to be achieved;
- an expected explosion of traffic (of several orders of magnitude), calling for investments to be able to handle the traffic in the years to come;
- cost-cutting measures, calling for outsourcing some of the part of MNOs (Mobile Network Operator) activities;
- new business models coming into market (managed services offered by the mobile equipment manufacturers, third-party companies to own and manage networks);
- extremely quick development of the subscribers' base in the developing countries, with specific constraints of extremely low ARPU and high cost to reach subscribers (lack of power grid, costly backhaul etc.);
- "green" pressure on operators to limit their environmental footprints.

The site CAPEX and OPEX contribute significantly to mobile operator expenses, so quite naturally the operators are eager to seek some savings there. Moreover, LTE is designed in a way that facilitates network sharing (all-IP, flexible spectrum, flexible frequency band, carrier aggregation), and network-sharing mechanisms were built into standards since the beginning. Due to scarcity of the continuous frequency resources it may happen that the common-frequency network sharing methods will be the actual enablers of LTE technology, allowing the operators not only to achieve satisfactory levels of LTE services (i.e. having deployed the LTE over at least 5 MHz band, ideally 10 or 20 MHz), but also to be able to use LTE and GSM technologies concurrently. Figure 15.37 depicts a situation in which two operators can benefit from a network-sharing scenario. The case described relates to the 900 MHz band, and assumes an uncoordinated

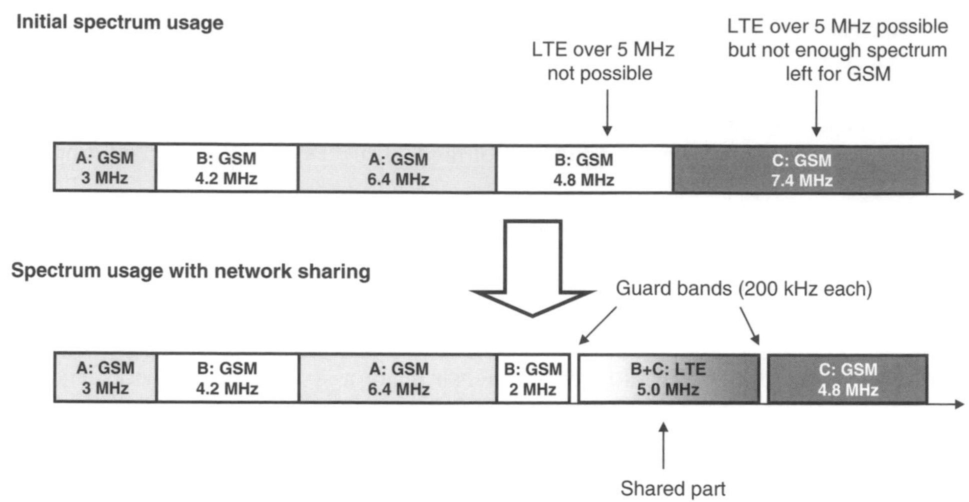

Figure 15.37 Frequency refarming—enabling 5 MHz LTE and continuation of GSM operation.

scenario (i.e. the existing sites will not be reused by LTE on a one-to-one basis). Considered is a situation in which spectrum reshuffling is not possible, so the operators have no way to rearrange the allocation of sections of the spectrum. Operator "B" has two separate parts of the spectrum (9 MHz in total), and operator "C" has a contiguous piece 7.4 MHz wide (the third operator, "A," is shown for the sake of completeness). All operators use the spectrum to offer GSM services, but wish to introduce LTE over as wide band as possible, while continuing to offer GSM. It can be seen that both "B" and "C" are facing the following challenges:

- Operator "B" cannot deploy LTE over a 5 MHz bandwidth as it lacks sufficient contiguous frequency resources. Network sharing is, in fact, the enabler for 5 MHz LTE.
- Operator "C" can deploy 5 MHz LTE, but is left with insufficient resources to continue to offer GSM services (at least in the 900 MHz band), as the quality will be compromised.

Note that it was assumed that, in order to offer satisfactory GSM services, it is required to have at least 15 channels in the BCCH layer (3 MHz wide frequency band) (please refer to Section 15.4.5). It is also necessary to consider the guard bands required for uncoordinated cases (as presented in Table 15.5). If both "B" and "C" wish to offer LTE over 5 MHz, they need to have 5.4 MHz of bandwidth, as two 200 kHz wide slots are needed as guard bands. When a shared band is used for LTE, both operators still have enough bandwidth to offer GSM services. Note, however, that in special cases, certain GSM features (e.g. common BCCH) may be needed to ensure satisfactory operation of the GSM network depending on the spectrum arrangement in other bands, such as 1800 MHz—for further details please refer to Section 15.4.

15.5.1 Introduction to Network Sharing Variants

Network sharing may use one of these main principles:

- Participants use individual frequency carriers:
  a. Multi-Operator Base Station Subsystem (MOBSS), solution applicable for GERAN—-shared is the BTS equipment and BSC resources.

Table 15.8 Network-sharing compatibility matrix

UE 3GPP release	Network Sharing Method per Technology					
	MOBSS (GERAN)	MORAN (WCDMA)	MORAN (LTE)	MOCN (GERAN)	MOCN (WCDMA)	MOCN (LTE)
Below Rel. 99	+	n/a[b]	n/a[b]	+/ − [a]	n/a[b]	n/a[b]
Rel. 99, 4 and 5	+	+	n/a[b]	+/ − [a]	+/ − [a]	n/a[b]
Rel. 6 and 7	+	+	n/a[b]	+/ − [a]	+	n/a[b]
Rel. 8, 9 and 10	+	+	+	+/ − [a]	+	+
Rel. 11	+	+	+	+	+	+

[a] Supports only single PLMN. Support of NITZ and Equivalent PLMN may be required.
[b] WCDMA UMTS starts with Rel. 99, LTE starts with Rel. 8.

  b. Multi-Operator Radio Access Network (MORAN), specific for WCDMA and LTE systems—shared is the (e)NodeB equipment and RNC.
- Common frequencies are used by all participants:
  a. RAN-based sharing:
    i. Multi-Operator Core Network (MOCN) for GERAN (partial support from 3GPP Rel. 10, full support from Rel. 11).
    ii. MOCN for WCDMA (supported from 3GPP Rel. 6).
    iii. MOCN for LTE (supported from 3GPP Rel. 8, i.e. as long as LTE exists).
  a. Roaming-based sharing:
    i. National roaming.
    ii. Geographical roaming.

The virtual operator case (MVNO—Mobile Virtual Network Operator) is excluded from this analysis. There are also other network-sharing scenarios like, for example, Gateway Core Network, where certain core network nodes are shared while the remaining ones are operator specific [11], regional roaming cases, or various other hybrid solutions.

It has to be kept in mind that network sharing is a complicated and multidimensional issue, and many operators might be reluctant to participate in it. Operators may face problems in finding the right partner for various reasons—besides not all such joint ventures will be possible due to anti-trust laws and regulatory issues. One should not neglect the likely problems in making two different organizations to cooperate, but one should also expect many side technological issues: separate tools, different KPIs monitored, different processes (e.g. dealing with site building). An additional issue is the fact that not all UEs may be capable of supporting certain sharing-related features. The detailed compatibility matrix is presented in Table 15.8 (" + " signifies full support, " − " lack of support, and "+ / − " denotes support with some reservations that are explained below the table; "n/a" stands for 'not applicable').

15.5.2 MORAN and MOBSS

MORAN/MOBSS functionality allows several operators to share physical radio access nodes, while each participating party owns and maintains its own elements of the mobile network. In particular, Core Network components are not shared, as presented in Figure 15.38.

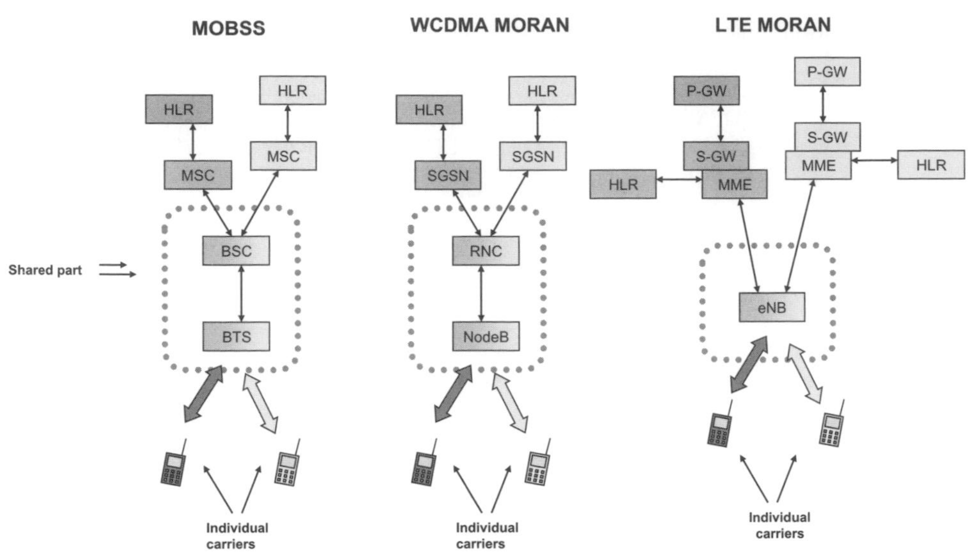

Figure 15.38 MOBSS/MORAN sharing scenario.

Operators sharing physical equipment continue to use their own licensed frequencies. Each operator has its dedicated radio resources, so they can have their own PLMN identities, carrier frequencies, and Cell ID. The other network parameters can be individual, or common. Each operator's specific end-user services are available in the area covered by the shared network. The sharing is fully transparent to the mobile stations, as they see their operators' PLMNs being present at the normally used frequency bands. This is further enhanced by the fact that each operator uses its own specific PLMN ID and the UEs continue to display own operator logos. Use of MORAN/MOBSS functionality does not prevent any of the participating operators from simultaneously operating their own dedicated RAN networks. The presence of the MORAN/MOBSS feature requires some modifications in the user-plane or control-plane message handling. The shared radio access node needs to correctly identify the message destination for each user. The differentiation takes place on the basis of the UE's registered PLMN. Note that the cell identities have to be expressed in CGI (Cell Global Identity) format, to assure them to be unique amongst operators (a Cell ID is required to be unique only within a single registration area in the given network).

Effectively, in MORAN/MOBSS solution, the following elements of the network can be shared:

- (e)NodeB (MORAN) or BTS (MOBSS) equipment (baseband resources, battery backup, air conditioning etc.);
- backhaul equipment (not necessarily in all cases);
- the site-related elements (buildings, towers, cabinets, antennas, feeders);
- processing capacity in control nodes (RNC, resp. BSC).

Some actions related to the network operation also need to be coordinated between participants: network planning and dimensioning, network rollout and maintenance, equipment upgrades (software actualization, capacity upgrades). Each of the participants, however, is individually responsible for providing licensed frequencies, and providing and operating their own core networks. They should also define and implement their own cell parameters (e.g. cell selection/handover thresholds, neighbor lists), and provide their own specific services to the end users. Each of the operators carries out its own performance and fault monitoring.

The principal benefits of applying MORAN or MOBSS sharing solutions are immediate reduction in equipment cost and OPEX savings. For a greenfield operator it significantly reduces the costs of entering the market, as both the volume of the equipment and the commissioning efforts are greatly reduced, as is the time needed to find and acquire the needed sites. An additional advantage is the possibility of conveniently and effectively covering the areas of low traffic. The MORAN/MOBSS allows participating operators to maintain control of their own network traffic in terms of quality and capacity, and does not restrict the operators in signing individual roaming agreements with third parties (thanks to having an individual PLMN ID). This type of network sharing leaves a secure exit path to each operator's own dedicated radio access equipment, in case of traffic volume increase or any unfavorable change of sharing policies by other participants. The additional benefit of MORAN/MOBSS is full support of this solution by all terminals. The implementation of a MORAN and MOBSS solution may, however, be hindered by a potential lack of coordination amongst sharing operators.

15.5.3 MOCN

The MOCN functionality allows several operators to jointly operate some elements of the radio access network, but the principal difference is that they will be using shared carriers, as opposed to usage of individual separate carriers in MORAN/MOBSS. Figure 15.39 depicts Multi-Operator Core Network sharing scheme.

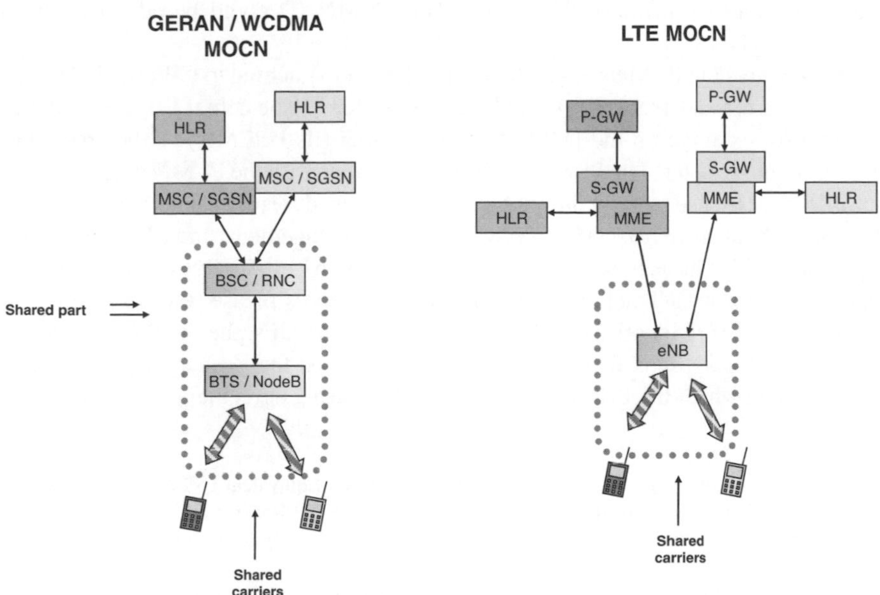

Figure 15.39 MOCN sharing scenario.

This has immediate implications both for the mobile stations and for the shared components of the network. The used carrier frequency may be a regular frequency used by one of the operators, but also may be a separate one, dedicated strictly for the shared part of the network. Note that the participating operators may decide upfront to refarm their GSM/WCDMA/LTE networks to facilitate MOCN-based network sharing.

There is also a challenge to identify the network in the system broadcast, as mobile stations of each participating operator shall properly obtain the shared network identity. The shared nodes need to have mechanisms for policing the traffic of each operator according to agreed levels of service, and also have to be able to route the user and control plane data correctly so that they reach the appropriate destination Core Network. The MOCN feature was added to the WCDMA in the Rel.6 and was built into LTE standard since the beginning (Rel. 8). The GERAN standards offer partial support for the MOCN starting from Rel.10, with the intention of making full support available in Rel.11. The Core Network components are not shared, although the name of the functionality might suggest otherwise.

MOCN functionality has been made possible thanks to the inclusion of multiple PLMN IDs in the cell broadcast message. In LTE this has the form of a "PLMN identity list" present in SIB 1 and containing up to six entries. Each LTE UE can read all the information contained there, so it immediately becomes aware of the existence of MOCN. By default, the first-listed PLMN is the primary PLMN in that cell. The situation is slightly different in WCDMA, where the support of multiple network identities is a later add on. The usual PLMN identity present in the MIB has become a "Primary PLMN identity," and a new optional information element has been added to the MIB, carrying a "Multiple PLMN ID list," which contains up to five additional PLMN-IDs. Additional related modifications have been added to SIB 3, 11 and 18 (PLMN-specific access restrictions, multiple PLMN identities of neighbor cells). All WCDMA UEs compatible with Rel. 6 and above can detect and select the correct PLMN-ID, and consequently can connect to their home Core Networks, albeit using the licensed frequency of another operator. For pre-Rel.6 WCDMA UEs, the multiple PLMN ID list is not visible, so these UEs can only read the primary PLMN ID. This PLMN ID could be added to the list of equivalent PLMNs.

The list of equivalent PLMNs is sent to the mobile after attaching to GPRS or EPS, Location Area, Routing Area and Tracking Area Update procedures. The actual list sent to the mobile may depend for example on the IMSI of the subscriber (its HPLMN) or the area where the mobile is going to register. The PLMNs in the list, together with the PLMN of the network that sends this list, are treated equally during the mobility procedures [12]. The same list is used for MM (Mobility Management), GMM (GPRS Mobility Management) and EMM (EPS Mobility Management) [13]. The provision of roaming services in certain areas is facilitated by the forbidden Location Area/Tracking Area concept—for details please refer to [12].[12]

If the shared network identity is not forbidden[13] for these UEs, they will be able to connect to the shared eNodeB. It is therefore advisable to use, as the primary PLMN, a different identity from that which the PLMNs used before the sharing agreement as the mobile station

[12] Apart from the procedures described above, a new option of an Equivalent HPLMN (EHPLMN) list that replaces the HPLMN derived from the IMSI was introduced in 3GPP Release 6 [23.122]. The network the PLMN of which is the highest one in the EHPLMN list is selected by the mobile during PLMN selection process. The EHPLMN list is stored in the (U)SIM card.

[13] If the Location Area/Routing Area/Tracking Area Update is rejected by the network with the cause "PLMN not allowed," such PLMN is included in the forbidden PLMN list permanently stored in the SIM card. This entry could be deleted only if the LAU/RAU/TAU triggered manually by the user is accepted.

would remove from the list of Equivalent PLMNs those identities that are on the list of forbidden PLMNs. A new functionality is needed to allow these users, who are unaware of the existence of multiple Core Networks, to reach the one that is appropriate for them. This is called a Core Network Node redirection. At UE attach, the RNC assigns a random CN (other decision logic can also be applied, e.g. weighted round robin) to which the initial NAS message will be sent. If the selected CN is not appropriate for that UE, the relevant MSC sends a Redirection Request message back to the RNC. The RNC probes another CN, and does it until a CN that can serve the attaching UE is found. Similar equivalent procedures are relevant for registration to the PS domain. The resolved association UE-CN is remembered by the RNC. Therefore the pre-Rel.6 WCDMA UEs can also access their home Core Network in MOCN environment, but the attach procedure may take somewhat longer time [11]. Moreover, control over PS and CS registration is needed for pre-Rel. 6 WCDMA UEs. Coordination of these registrations aims at assuring that the same Core Network operator serves the user from the PS and CS perspective. This could be achieved either with the help of an optional Gs interface that connects MSC and SGSN, or via dedicated functionalities in the RNC if the Gs interface is not implemented [11].

The MOCN feature for GERAN network, planned to be made fully available in 3GPP Rel. 11, will be done along the principles used for WCDMA case. In Rel. 10, where a partial support of MOCN has been introduced, the air interface is left unmodified, and therefore no multiple PLMN list will be introduced, thus a single PLMN ID will continue to be broadcast. It is naturally recommended that this single shared PLMN identity is a new one, seen by the mobile stations of each operator as an equivalent PLMN. The association of the appropriate Core Network will be performed by the BSC using the rerouting mechanism similar to that of WCDMA. This procedure is practically invisible to the user, as the access request is not rejected while BSC probes the available Core Networks. In order for this procedure to work, the support for the A-flex and Gb-flex is necessary. As the radio interface remains untouched, all GSM mobile stations are compatible with MOCN solution as standardized in Rel.10. It is planned to extend the GERAN standard to support full MOCN in Rel. 11 [14]. The multiple PLMNs (up to 5 additional ones) will be introduced, using Extended BCCH feature. Additionally, each PLMN may optionally be associated with the list of appropriate Access Classes to be used by mobile terminals while trying to access particular PLMNs.

In order to hide from the end user the fact that their mobile terminal is currently served by the shared component of the network, the Network Identification and Time Zone (NITZ) feature may be used [15]. This unctionality is often used to set the clock to the local time zone for travelers, and also to update the clock to the daylight-saving time changes. It also allows the serving network to send the network name within mobility management procedures. The mobile terminal may use this name instead of the one stored in its memory. However, the network names stored on the SIM card take precedence over those sent via NITZ or stored in the mobile station memory. In a shared network, NITZ can be used to send different network names for subscribers of participating operators. It is possible to associate at least 10 PLMN identifications with the same operator name. While NITZ is not a mandatory component of a network sharing scenario, is quite a useful functionality from the point of view of the end user.

As all operators share a single physical piece of hardware, only one of the operators (or a selected third-party operator) can actually install and manage the shared radio access nodes—that is, do network planning, install and maintain the node, collect alarms, keep track of performance counters, and so on. Other operators have to negotiate specific agreements but, in fact, the shared elements will not be fully integrated in their network. The specific areas where the operators have to reach a satisfactory degree of consensus are related to network rollout (planning, dimensioning, topology, provisioning of transport/backhaul) and operation (OAM, upgrades). It is also necessary for the participants to agree on the specific details of financial contribution scheme of CAPEX and OPEX (e.g. equal share, or in function of traffic volume). The specific details of the operation of the shared network depend on solutions available in each case—for example, the mechanisms for load steering between the operators sharing the RAN will depend on whatever exists in the installed equipment (e.g. equal versus prioritized capacity sharing) and dedicated features designed for MOCN. This may take form of solutions that will allow a certain capacity to be dedicated for each operator—for example in form of guaranteed bit rate per operator plus some floating capacity allocated on per-need basis. This policing functionality may also be linked to admission control mechanisms, aiming at proactively avoiding the congestion.

Note that also the backhaul link can be either common or separated, the former calling for appropriate packet network load-balancing functionality.

One of the principal benefits of MOCN is that the participating operators can jointly use part of their bandwidth to create a much wider overall bandwidth, making it possible to offer much higher data rates for individual users. Additionally, trunking gain comes into play, as the usage of a single larger band is more effective throughput-wise than having two halves of that band used separately. Moreover, the signaling overhead is more favorable for wider bandwidth allocations. If the 1.4 MHz band is used for LTE deployment, only slightly more than half of the band is available for traffic channels scheduling (51%), whereas in 5 MHz deployment 63% of the resources can be allocated for actual traffic. This difference originates mostly from different guard band overhead (cf. Table 15.3) and the specific layout of PBCH and synchronization signals.

15.5.4 National Roaming, Geographical Roaming and IMSI Based Handover

The national roaming feature offers another possibility for operators to provide coverage to their subscribers[14] without having physical equipment deployed at a certain area, as depicted in Figure 15.40.

The underlying idea is the same as in case of roaming in international network, and requires the participating operators to sign a roaming agreement. As a result, the PLMN ID of the hosting party is added to the allowed PLMN list of the subscribers of the operator in question. Such an agreement can be reciprocal, or not, depending on the decision of the participants. In the national roaming, the roaming agreement covers the entire country. In case of geographical roaming the roaming area covers only selected parts of the country—these could be for example rural areas, isolated islands, and so on, but the participating operators will also own and operate their own independent networks. The UEs operating in such an environment need to apply different mobility procedures when inside the visited area and when at the border between home and the visited network. Normally the area where the user is allowed to roam is delimited by forbidden zones (e.g. routing or tracking areas), causing the roaming user to reselect to another PLMN (the home PLMN in this case).

[14] National roaming does not require any particular support from mobile stations, but there may be some limitations with respect to support for national roaming in very old GSM phase 1 mobiles.

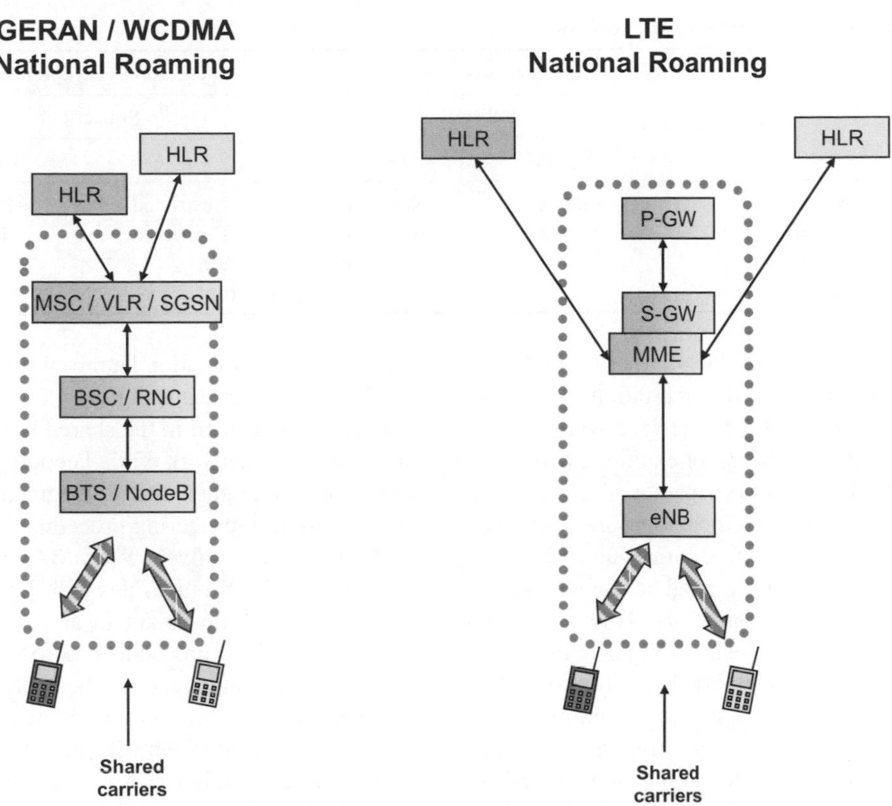

Figure 15.40 National roaming scenario.

In idle mode, access control of a particular group of users to certain cells may be accomplished with the help of Core Network procedures related with handling of Equivalent PLMN functionality. In order to benefit fully from the PLMN-aware traffic steering, the dedicated mode has to be integrated in this process as well. In this case, certain support from Radio Access Network is needed. In order to retrieve the home PLMN of the end user, IMSI identity—where MCC (Mobile Country Code) and MNC (Mobile Network Code) are embedded—could be used. Having the user's IMSI available, BSC[15] or RNC could route the traffic—that is, perform the handover only towards predefined cells (either 2G or 3G) that were designated as valid cells for the group of users that the given subscriber belongs to. In order to optimize this procedure, certain restrictions may be applied to measurement of the neighbor's cell. In LTE the support for such a scenario is embedded in the "Index to RAT/Frequency Selection Priority" (RFSP Index) concept where a UE-specific RFSP index could be retrieved from HSS and sent to eNodeB via the MME and it could subsequently be used to apply UE-specific RRM strategies including the evaluation of potential handover target cells. Note too that the applicability of IMSI-based traffic routing may be extended beyond the MCC and the

[15] Due to the sensitivity of IMSI it is generally not sent via air interface. Instead, BSC may receive IMSI information from the MSC in, for example, a Common ID or Handover Request message. Note, however, that providing such information by MSC is not mandatory unless a Dual Transfer Mode (DTM) is implemented in the given network [32].

Table 15.9 Regional roaming scenario

LAC	Resources of operator A		Resources of operator B	
	Subscriber		Subscriber	
	Op. A	Op. B	Op. A	Op. B
2G Urban LACs	Allowed	Restricted	Restricted	Allowed
3G Urban LACs (2100)	Allowed	Restricted	Restricted	Allowed
2G Rural LACs	Shared spectrum			
3G Rural LACs (900)	Shared spectrum			

MNC part of the IMSI—that is, to the MSIN (Mobile Station Identification Number)—if only reasonable users' differentiation may be performed due to such an extension.

Typically, IMSI-based handover functionality may be implemented in the shared network environment where, for example, two operators share the GSM network while keeping their WCDMA networks separate. To avoid making national or geographical roaming agreements on the 3G side, the operators may introduce a PLMN-aware traffic steering procedure in the GSM. This policy means that subscribers of operator A will measure only the WCDMA cells of operator A and they could be handed over only towards these cells. Similar rules will apply for subscribers of operator B. This functionality will be especially important in areas where WCDMA coverage is ending, leading usually to frequent inter-RAT procedures performed at the edge of 3G network. In the intra-GSM scenario, IMSI-based handover could be applicable for example in geographical roaming case where only certain part of the network is shared by the operators (e.g. in the rural areas) and there is a strong need to route the traffic reasonably in the areas where shared and non-shared parts of the networks conjunct. An example of a geographical-based roaming agreement between two different operators partially sharing the GSM and WCDMA networks is presented in Table 15.9.

15.6 Hardware Migration Path

Migration towards LTE requires extra expenditure from operators on hardware modernization. In order to support LTE technology, new base station hardware needs to be implemented as legacy 2G/3G technology orientated equipment already installed on sites, is not LTE capable. New hardware will complement the existing network infrastructure, which in a majority of cases is going to be maintained to support 2G/3G systems in parallel with LTE deployment. Thus, LTE introduction will imply additional investments and increase of operational expenditures (OPEX) because of more hardware used in the network. At the same time the operators are forced to reduce costs and improve operational efficiency of the network. Amongst others, this is due to continuously increasing energy costs, which are already the major component of the operator's OPEX and will increase in future.

Different LTE deployment scenarios are currently analyzed (e.g. hot spot areas only, continuous coverage in city areas, etc.), however, in most scenarios, reuse of existing 2G/3G sites is considered. This is the solution that allows the operators to save money, which is usually spent on the construction of new sites; on activities such as site acquisition, site design, civil works, and so on. In case of 2G/3G sites reuse, the base station hardware dedicated to LTE, as well as the antenna system, will be colocated with other technology equipment. In other cases (e.g. for greenfield operators), LTE sites will be built from scratch unless other operators' infrastructure is shared. Hereafter, only the reuse of existing sites will be considered, as it is likely to be applicable to the majority of the operators and it is directly related to hardware migration path.

15.6.1 Colocated Antenna Systems

There are several alternative methods available to colocate different technology antenna systems. Thus, the operators again have some room to optimize LTE deployment costs. First of all, the selection of a preferable method depends on whether the colocated systems will operate in the same frequency band or in different frequency bands.

15.6.1.1 Colocated Systems–Different Frequency Band

The colocated antenna systems can be composed of either separate single-band antennas or can be built using multi-band (dual-band or triple-band) antenna types (Figure 15.41). Hence, LTE-related single-band antenna must be installed in addition to the existing one or it needs to be replaced by multi-band antenna type. In both cases, some site reconstructions must be done to allow the system extension.

With multi-band antennas the necessary amount of antennas per site is minimized, however, the reduction of the visual impact is still a big challenge in a majority of the extension scenarios.

When multi-band antennas are used, the same azimuth applies for all related RATs. The network plan for the existing system should not be changed, so the planning process for the LTE system has to take care about such limitations.

Each of the frequency bands used on site requires dedicated feeder cables to be installed on site, which is another aspect to be considered when modernizing the existing antenna system. By using diplexers or triplexers, the requested amount of feeder cables per sites can be reduced, as shown in Figure 15.42.

Diplexers (triplexers) are filter units that combine and separate the outputs of two (three) transmitters of different frequency bands into a common feeder cable. They also perform the reverse function of accepting a received signal of different frequencies and splitting them to their respective receivers. In some multi-band antenna types, there is a diplexer (triplexer)

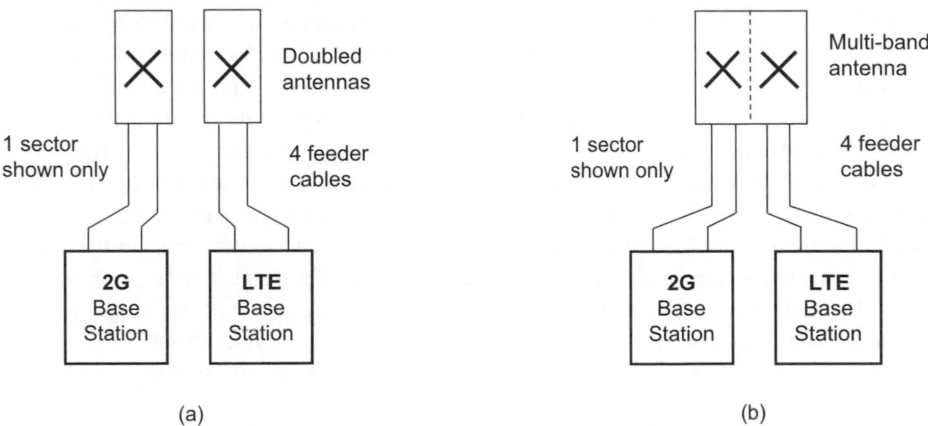

Figure 15.41 Colocated antenna systems (base stations work in different frequency bands); (a) separate single-band antennas, (b) multi-band antenna.

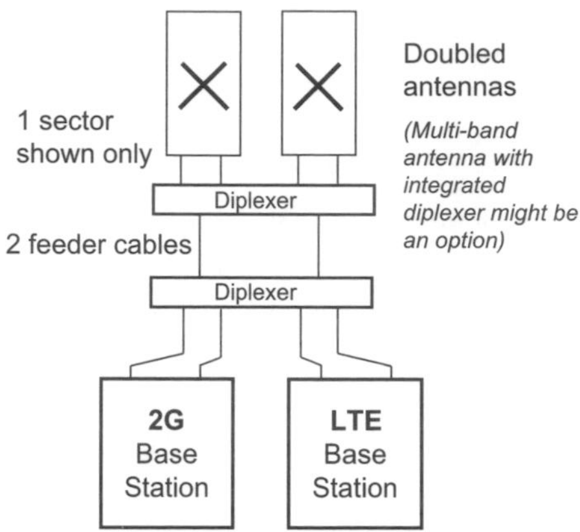

Figure 15.42 Colocated antenna systems with common feeder cables (base stations work in different frequency bands).

already embedded in the antenna panel. Usage of such antenna types makes the external diplexer (triplexer) mounted close to the antenna unnecessary. It should be remembered that additional losses in downlink and uplink direction are introduced when those new elements are applied in the antenna system. Thus, the impact of such solution on LTE planning as well as on colocated technology performance needs to be considered. With separate single-band antennas as well as with multi-band antenna types it is possible to support multiple streams transmission in the downlink direction for the LTE system (e.g. 2×2 MIMO). This is because LTE signal can be transmitted via both antenna lines available in typical antenna system.

15.6.1.2 Colocated Systems—The Same Frequency Band

The existing antenna system can remain untouched if LTE is deployed in a frequency band that is used by RAT already operating on site—for example in a so-called frequency refarming solution. In that case, colocated RATs may use a common antenna system. As no changes are introduced to the antenna system, the operator can rollout LTE much faster than with dedicated or multi-band antennas. Moreover, the operator can save money that would be spent on the modernization if the existing antenna system had to be changed. With common antenna system both colocated technologies must use the same antenna azimuths and antenna tilts.

Typically, two-branch reception diversity is applied in a base station to improve receiving system sensitivity. For that purpose, in each of the configured on-site sectors two antenna panels (or single cross-polarized antenna) need to be connected via two antenna lines to the base station equipment. Downlink transmission can be realized via a single antenna panel or via both antenna panels. In case of a 2G system, there are usually several transmitters (TRXs) being used in each sector. Since the signals coming from all TRXs need to be combined to the antenna line, additional attenuation is introduced into the transmit path by combining equipment. The attenuation increases with the number of combined TRX outputs, hence the option to minimize combining losses is to split transmit paths into two separate branches connected to two antenna panels.

In case where colocated RATs are going to use a common antenna system, available transmit (TX) and receive (RX) paths need to be shared between the systems. The RX signals from both antenna panels need to be distributed between two pairs of receiver branches: to primary

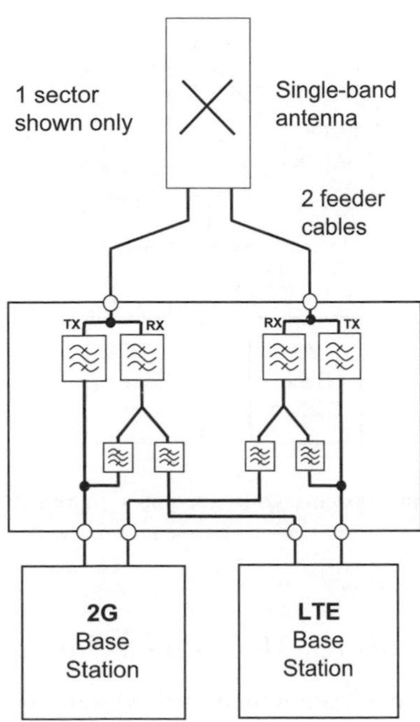

Figure 15.43 Colocated antenna systems with external RX/TX path combiner (base stations work in the same frequency band).

receivers (RX) and to diversity receivers (RX-D) of colocated base stations. As two antenna lines are available, it is possible to connect colocated systems to the separate transmit branches. Hence, downlink losses are not increased for LTE system. However, for 2G all the TRXs have to be combined into the single TX path, which can imply, in some cases, increased combining losses comparing to the double-TX antenna solution. An important aspect is that, due to the single transmit path assigned for each of the systems, LTE cannot support multiple-stream transmission. Anyway, an external TX/RX duplexer must be installed to separate the TX and RX paths before they are connected to the base station equipment. Figure 15.43 shows an example of an external hardware solution that allows both the colocated systems to use common feeders and antennas.

In some cases, it might be possible to use common antenna system without extra hardware being installed. Modern base station types offer in their frontend modules RX output signal, which can be used for external purposes, for instance it can be delivered to another base station as RX diversity signal. Using such RX output signals, the diversity paths can be exchanged between both colocated base stations. The RX path to primary RX receivers as well as the transmit path remain exclusively within its "mother" base station hardware (Figure 15.44).

It should be remembered that the RX performance of both base stations may change in comparison to standalone scenario. This is due to the fact that the RX diversity path will suffer certain degradation as it goes through additional active and passive HW components of colocated base station. The TX paths are not affected because each base station transmit via its own TX path; however, for 2G base stations there is again the need to combine all the TRXs into a single transmit path.

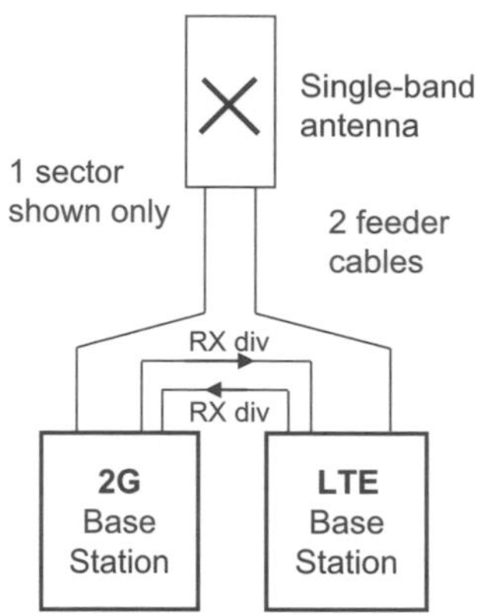

Figure 15.44 Colocated antenna systems with RX paths shared (base stations work in the same frequency band).

15.6.2 Colocation with Shared Multi-Radio Base Station

The constant development of mobile communication systems has accelerated upgrades in the area of the base station technology. LTE could be deployed using software definable radio, where instead of dedicated hardware there are software components applied to support base-station related functions (e.g. channel coding, modulation, filtering, etc.). In parallel wideband transmission and reception technology is introduced, using multi-carrier power amplifiers and wideband receivers. In a traditional base station, single-carrier technology is used with dedicated transceiver and power amplifier units. In a multi-carrier unit, the number of carriers can be freely configured (including various multiple transmission schemes), enabling flexible capacity expansion and reduced power consumption. Multiple signals are transmitted through a broadband power amplifier, which has its power shared amongst defined carriers. Multi-carrier technology was first applied in 3G base stations, but nowadays it is reused for LTE as well as for the 2G system. Since base station functions are realized by dedicated software components it was possible to develop common hardware platform for 2G, 3G, and LTE technology. It means that all the technologies can be handled using the same set of hardware elements. Moreover, different RATs can operate in parallel with carrier signals being transmitted and received via the same radio frequency unit. As a result, no additional hardware needs to be installed to allow a common antenna system to be used by different technologies operating on site. However, since radio frequency unit is designed to support a single-frequency band, LTE must be deployed in already used on site frequency band to take advantage of minimum hardware configuration (Figure 15.45).

LTE deployment gives the operators a unique opportunity for low-cost 2G hardware modernization, especially in case of LTE being deployed in 2G frequencies. Owing to multi-radio hardware, both technologies can work on the same hardware platform, which brings OPEX savings on spare parts, energy consumption, maintenance activities, and so on. Since the existing antenna system can remain untouched, LTE deployment can be carried out with minimum investments and within a short time frame. That is why it should be considered by the operators as the optimum solution, especially when frequency refarming is one of the options.

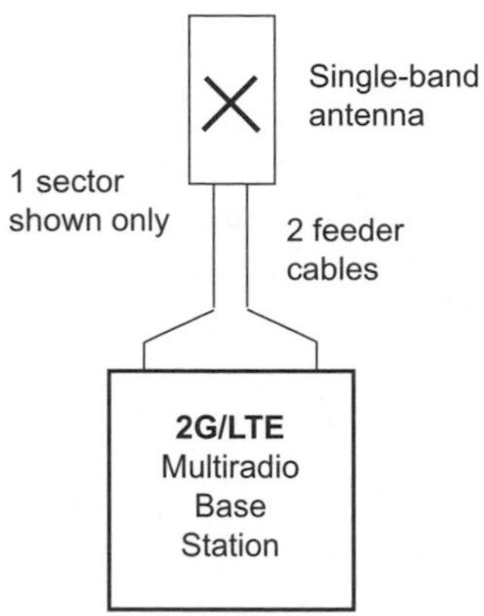

Figure 15.45 Colocation with shared multiradio base station (base stations work in the same frequency band).

15.7 Mobile Backhaul—Towards "All-IP" Transport

15.7.1 Motivation to IP Evolution in Mobile Backhaul

Continuous evolution of 3GPP standards going beyond GSM/EDGE and UMTS Rel. 5, and in particular introduction of High Speed Packet Access (HSPA) and soon later Evolved HSPA (I-HSPA), has led to enormous growth of mobile traffic worldwide. However, this unprecedented traffic boom is not driven solely by the emergence of new radio access technologies but also by fierce competition in the mobile market resulting in flat rates as well as the commercial availability of a new generation of handsets with abilities to run advanced applications and services being often internet-based. The coincidence of these abovementioned elements (higher peak throughputs, new pricing policy for data, online applications) also results in a change of characteristics of mobile traffic from voice dominant to data dominant.

Nowadays it is clearly seen that traditional hierarchical (DS0, E1, STM1, . . .) transport networks, which have obviously been designed for mobile voice and low-speed data, are not efficient enough for bursty traffic as they use rigid allocation of basic transport units (e.g. DS0 in case legacy PDH/SDH-based transport) resulting in bandwidth separation for different planes and limited overbooking capabilities. Moreover, the E1/T1 transmission networks do not offer flexible scalability as capacity extensions are made in large steps (e.g. at least 2 Mbit/s in case of E1 lines), which means that the provisioning of extra bandwidth for numerous Base Stations will account for considerable contribution to total network costs (although just a fraction of the installed extra capacity may actually be needed). To summarize, the principles which are fully justified for voice-dominant traffic become too wasteful when it comes to transmission of huge bursts of data. Improvements in transport efficiency leading to lower transport costs are therefore the first and main drivers of mobile backhaul modernization necessary to cope with steadily growing bandwidth demand.

The improvements in transport efficiency are absolutely crucial for network operators to reduce the network costs but, quite recently, especially with the introduction of sophisticated Internet-based and always-on applications, another driver of backhaul modernization has become more and more important, namely user-perceivable network performance. On the one hand, aggressive advertisements set high customers' expectations in terms of peak (i.e. not

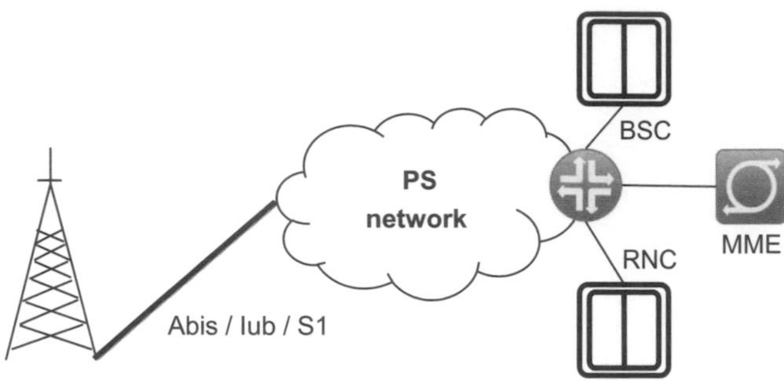

Figure 15.46 "All-IP" transport.

average) throughput rates. On the other hand, provisioning of the advertised speeds to several customers simultaneously would lead to immediate need for E1/T1 capacity extension in most Base Stations as long as legacy transport is in use (even though such extra capacity might not necessarily produce proportional revenues as those are correlated more with transmitted traffic volumes than with peak throughputs). This means that, to be successful, the network operators are supposed to achieve at the same time two (often contradictory) targets related to transport efficiency and customer satisfaction.

Taking these challenges into account the migration to a new transport technology that overcomes existing limitations of legacy PDH/SDH architecture in terms of network capacity and transport costs, and at the same time enables a great experience for mobile users, is inevitable. Meeting these extensive requirements is feasible through deployment of flat, all-IP transport networks (see Figure 15.46).

As a result of the replacement of the legacy transport infrastructure with an IP-based one, several benefits can be observed, which allow mobile operators both to manage "traffic explosion" and to remain profitable:

- Significant bandwidth savings due to optimization of payload structure and introduction of greater flexibility in bandwidth usage (pooling of resources, traffic aggregation and overbooking).
- Far better scalability due to introduction of wideband Ethernet connectivity at Base Station as well as Core side, which permits the adequate evaluation of transport capacity for the actual traffic load and peak-to-average profile.
- Ethernet transport offers a much lower "cost per bit," which contributes to a further reduction in operational expenditure.
- The reuse of common transmission concepts regardless of radio access technology: IP transport can now be applied in GSM/EDGE as well as UMTS/HSPA networks; furthermore, sharing of common transport in a multi-RAT environment leads to synergies that maximize the overall efficiency (statistical multiplexing, common maintenance).
- A simplified transition to flat architecture, which aims to reduce network elements by integrating some functionalities originally executed in dedicated entities that help to reduce end-to-end delay, which is, in turn, a prerequisite for providing real-time services. For instance, the 2G transcoding functionality partly disappears and the rest can be moved to the Media Gateway; another example is part of controller intelligence that is either implemented into Base Station or even entirely integrated, as in LTE eNodeB.

The additional advantage lies in the fact that the features mentioned above do not have a negative impact on user experience. Quite the contrary—it is possible to ensure a similar performance to that achievable in TDM networks by means of proper planning and configuration.

Ultimately, migration to flat all-IP transport network paves the way for global and smooth launch of Evolved Packet Core (EPC). This is not only for the reason that E-UTRAN does not support legacy TDM transport at all, but also because the sooner IP transformation is initiated the more mature packet backhaul is available during LTE introduction, which simplifies its rollout. For instance, from a radio access perspective, it would then be similar to the integration of another technology to the existing 2G/3G installations rather than beginning of something revolutionary.

15.7.2 Transport Aspects in Packet Backhaul

Migration to full packet transport involves replacement of TDM E1/T1 links with Ethernet-based Packet Switched Network (PSN). The PSN introduces some complexity that is not present in TDM-based transport. Hence, specific solutions are leveraged to improve the reliability and control the performance of Ethernet networks. In that sense, IP evolution requires and provides the means to deal effectively with transport impairments, synchronization assurance, higher security threats, and lack of guaranteed bandwidth [7].

The capability of a PSN to serve as a transport medium is expressed in the level of transport impairments understood as packet loss rate, packet delay and delay variation. The presence of such impairments in PSNs, although most undesirable, is unavoidable. However, if packet loss rate, packet delay, and delay variations are kept within a certain range their impact on performance is negligible. Therefore the predefined values of transport impairments must be rigorously maintained within the required service level to preserve performance like that in TDM transport.

Delivery of a synchronization reference signal to every Base Station with the required frequency and time accuracy is necessary to ensure proper operation of both radio interface and transport network. In legacy transport, as long as TDM connectivity is present at each Base Station, the synchronization signal is efficiently distributed over the network by means of E1/T1 lines connected to the controller and, furthermore, to the core. In packet backhaul the synchronization chain does not exist anymore because Ethernet is, by its nature, asynchronous. Thus various synchronization techniques are made available to address this issue; and for now Timing over Packet, which is based on Precision Time Protocol defined in [16], becomes the most preferred one. In this solution there is a master device, which generates timing packets and transmits them towards dedicated Base Station (where a slave device is installed) using a unicast connection. Timing packets traversing the PSN obviously experience delay and delay variation but the slave function implemented in the distant node performs recovery of original clock by means of sophisticated algorithms implemented in it. Each Base Station must have its own connection with the source of time-stamped packets. One of the possible alternatives for the clock recovery algorithms is to apply the basic synchronization concept of traditional TDM-based transport—the transmission of the reference signal over physical layer. In the case of IP transport this is possible by Synchronous Ethernet [17]. In this method, a timing signal is relayed from the primary master to the target Base Station using the synchronization ports of interconnected Ethernet switches. The solution provides an accurate clock signal independent of the PSN load when it is supported at every node between the reference source and the final destination. Other methods relying on external synchronization sources (such as a GPS receiver or dedicated E1/T1 line) are typically seen as too expensive and thus their applicability is rather limited.

Replacement of traditional TDM-based transport with packet backhaul raises higher security concerns, especially when transport is realized through large and public PS networks. In such a scenario, mobile traffic is typically shared with various systems and managed by different operators with different security standards. Global accessibility to packet networks in the context of security is not helpful, either. Hence the ordinary security threats related to eavesdropping or unauthorized data modification are now present not only in the radio interface but also "propagate" over the access part of the networks and the relevant countermeasures must be adapted to IP over Ethernet transport specifics. A set of features that aim at mitigating the risk of attacks and protecting of every type of traffic is available. Firewalls and security gateways are installed in front of important network entities or interconnecting nodes, IPSec provides security measures for user authentication as well as data encryption and integrity, and certification management ensures security associations between network entities and nodes.

Enabling of multiradio access technologies (with very different bandwidth needs) and shared packet transport (where available transport capacity is affected by load fluctuations) brings in some uncertainty on how to guarantee enough bandwidth to handle traffic served by radio interface. This can be ensured by proper Quality of Service (QoS) definition throughout transport network. Such definition should reflect the fact that particular services have different delay tolerance on the one hand and produce different incomes on the other. Therefore packets can be treated with managed unfairness—for instance (depending on operator strategy) voice and real-time data take precedence over non-real-time and best effort data services. In addition to advanced QoS engineering, some congestion control mechanisms are implemented to adjust automatically the data volume to be sent on backhaul depending on PSN overload conditions.

The abovementioned functionalities are fully integrated with regular network devices (both at Base Station side as well as in Core Network) so that transport migration is coordinated with availability of SW and HW features supporting the ongoing network modernization and transport evolution.

15.8 LTE Interworking with Legacy Networks for the Optimal Voice and Data Services

Optimal interoperability of LTE with legacy networks requires efficient intersystem management mechanisms dedicated to both connected and idle modes. Generally, LTE and legacy networks interworking can be considered in two different, although often interwoven, contexts, namely data services and voice services. While interworking in the context of data services can be expected to be seamless cooperation of PS domains of considered systems, interworking in the voice services context is a more complex issue. This is because specific methods of voice services realization in LTE are a consequence of the lack of CS domain in EPS architecture (for further details please refer to Chapter 9).

The purpose of this section is to introduce selected LTE and legacy network interworking techniques that aim to provide optimal performance of both voice and data services. The main focus is on practical deployment aspects. The discussion provided in the following sections is based on the latest versions of the 3GPP specifications available at the time of writing this book. Section 15.8.1 is dedicated to tridirectional intersystem mobility management for optimal data performance. Sections 15.8.2 and 15.8.3 provide discussion on how to assure efficiently continuity of voice services across different technologies. In Section 15.8.4, concise discussion on the Idle Mode Signaling Reduction (ISR) functionality is included.

15.8.1 Intersystem Mobility Management for Data Services

The following functionalities have been defined for enabling PS-related interworking procedures from GERAN to E-UTRAN:

- cell reselection;
- Network Assisted Cell Change (NACC);
- Network Controlled Cell Reselection (NCCR);
- PS handover.

The counterpart features that were standardized for E-UTRAN to perform Radio Access Technology (RAT) change to GERAN are listed below:

- cell reselection;
- RRC connection release with redirection;
- Cell Change Order (optionally with NACC);
- PS handover.

Regarding WCDMA–E-UTRAN interworking, the following procedures were standardized:

- PS handover;
- Cell reselection;
- RRC connection release with redirection.

All these procedures are described in details below.

15.8.1.1 UE-Controlled Cell Reselection

The primary means of changing the serving cell in the GPRS/EDGE network are still based on the reselection procedure—even for the mobiles in Packet Transfer mode where the Temporary Block Flow (TBF) is established and there is ongoing data transmission. This means that during every cell change that occurs during data connection the mobile has to temporarily leave Packet Transfer mode and enter Packet Idle mode. The interruption time during the PS transmission posed by the cell-change procedure may be of the order of a few seconds in an extreme case, if, for example, the source and target cell belong to different Routing Areas. This is one of the key differentiators between handling of the users' mobility inside GSM and inside the LTE network, which has to be considered while analyzing inter-RAT mobility management.

In principle, the UE-controlled reselection algorithm is the same for GSM Idle, GPRS Packet Idle and GPRS Packet Transfer mode.[16] In the pre-Release 8 algorithm for 2G to 3G UE-controlled intersystem cell reselection (the so-called cell ranking algorithm) apart from the minimum CPICH Ec/No and RSCP criteria that the target WCDMA cell has to satisfy,[17] signal-strength difference between GSM serving cell and target UTRAN cell was checked against a predefined threshold. It was concluded, however, from the GSM-WCDMA interworking experience that use of such a direct comparison between the signal strength of both

[16] Support for PBCCH—with the help of which GSM PS/CS differentiation for mobility procedures could be performed—was removed from 3GPP Rel. 9 onwards [23].

[17] The latter criterion is optional for pre-Rel. 5 mobiles.

systems may lead to undesired mobile behavior, like constant changing of the RAT the mobile is camped on. Moreover, the possibility of applying very high values for the signal strength offset parameter (FDD_Qoffset) was often used by the operators in order to effectively prioritize WCDMA over GSM network. This led to the coverage-based inter-RAT reselection [1].

With regard to the WCDMA network, the inter-RAT reselection algorithms are applied for the mobiles that are in Idle, URA_PCH, CELL_PCH and CELL_FACH states. The mobility of the UEs that are in the CELL_DCH state is governed by the handover procedures.

The new concept of so-called absolute priorities was introduced in 3GPP Release 8 to support inter-RAT reselection between GERAN, UTRAN and E-UTRAN including the intra-E-UTRAN/UTRAN interfrequency cell reselection. In this algorithm, direct comparison between signal levels of different RATs is no longer required. The new idea is based on an unambiguous cell reselection priority order between different RATs without predefined offsets. In 3GPP Release 8 for the target GERAN or E-UTRAN cell with higher priority than the serving one only minimum signal strength criteria has to be satisfied for this cell in order to trigger the inter-RAT reselection towards it. In case of a target 3G cell, the value for the minimum quality (CPICH Ec/No) is also checked. In 3GPP Release 9, minimum quality criteria (based on RSRQ) were also introduced for target LTE cells. The mobile can reselect a lower priority cell only if no suitable higher priority cells can be found and moreover no optimal cell can be found in the serving layer/RAT. In the case of LTE, in 3GPP Release 8 reselection to lower priority cells is triggered by the low RSRP level of the serving cell. In 3GPP Release 9, as an option, it could be triggered by the low RSRQ level.

More flexibility was added to the time aspect of the inter-RAT procedures—the time during which all the inter-RAT reselection criteria have to be met was changed from a fixed 5 s (as in case of GSM-WCDMA cell reselection based on cell ranking) to parameterized time.

The typical example of inter-RAT reselection (details in [18]) from a low-priority GSM cell to a high-priority LTE cell is presented in Figure 15.47.[18] The counterpart priority-based inter-RAT reselection in the opposite direction (details in [19]) is presented in Figure 15.48.[19]

Information about the priority of different layers is available in:

- System Information (SI) type 2 quarter in GSM [20];
- SIB19 in WCDMA [21];
- SIB3, SIB5, SIB6, SIB7 and SIB8 in LTE [22].

However, apart from common priorities to be used during the inter-RAT reselection process it is also possible to set the mobile-specific individual priorities that could differ from the common list of priorities distributed in the BCCH. Such dedicated priorities may be sent in:

- Packet Measurement Order or Packet Cell Change Order [23] and Channel Release [20] messages in GSM;
- Utran Mobility Information [21] message in WCDMA;
- RRC Connection Release [22] message in LTE.

[18] It was assumed that THRESH_E-UTRAN_high_Q was not provided to the mobile meaning that LTE RSRP is evaluated instead of LTE RSRQ while in GSM.

[19] It was assumed that threshServingLowQ is not included in SIB3, meaning that reselection to lower priority layers is triggered by a low LTE RSRP value instead of a low LTE RSRQ.

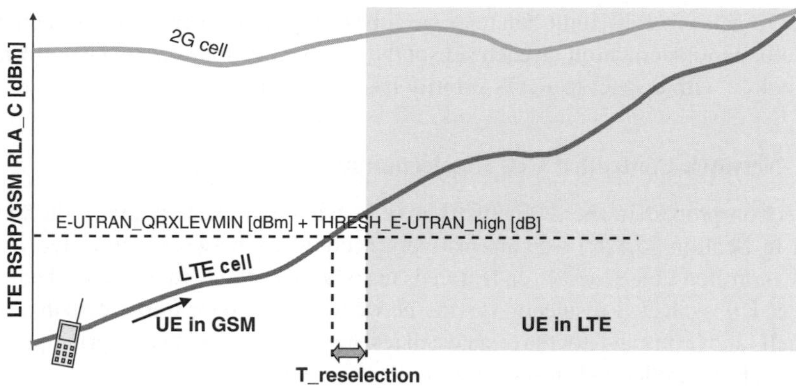

Figure 15.47 Inter-RAT cell reselection from low priority GSM cell to high priority LTE cell.

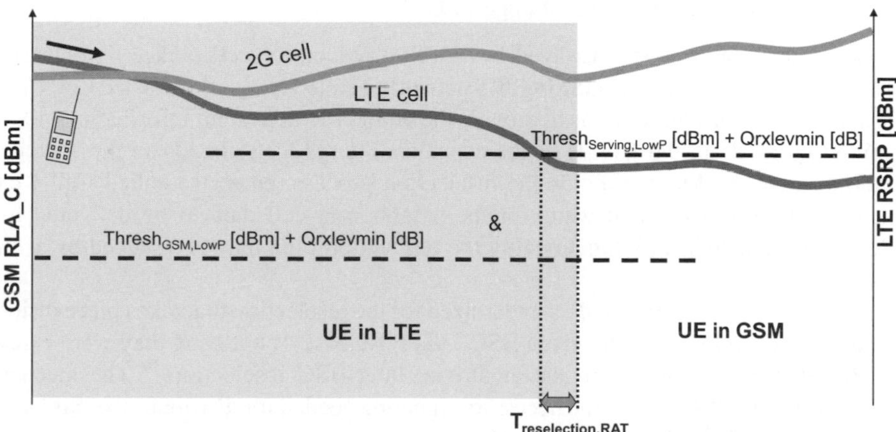

Figure 15.48 Inter-RAT cell reselection from high priority LTE cell to low priority GSM cell.

In this way, certain per-user control over traffic routing may be ensured.

However, the legacy algorithm based on the cell ranking has not been totally ruled out from inter-RAT procedures, even for E-UTRAN-capable mobiles, leading to possibly complicated parameterization that has to be handled by the operator of a multi-RAT network. In the UE that is camped on UTRAN network, legacy cell-ranking algorithms and priority-based algorithms may be used in parallel—the former in the direction towards GSM and the latter towards E-UTRAN. Moreover, cases may arise where in UTRAN the legacy cell ranking algorithm is responsible for handling the reselection towards GERAN whereas, for GERAN to UTRAN, a priority-based reselection algorithm is used in the UE. In case of cell reselection from GERAN to E-UTRAN the situation is less problematic because it has been made mandatory by 3GPP to perform the inter-RAT reselection by the E-UTRAN-capable mobile camped on GSM solely in accordance with the priority-based algorithm, also towards UTRAN.

In order to achieve consistent behavior from priority-based inter-RAT reselection procedures, careful parameterization on each side of the involved systems has to be performed—the common policy with respect to RATs priority has to be implemented first.

15.8.1.2 Network Controlled Cell Reselection in GSM

The reselection process in the 2G network may be triggered exclusively by the mobile—as described in Section 15.8.1.1—or alternatively it could be network controlled if optional Network Controlled Cell Reselection feature is introduced in the network. The clear benefit of NCCR over UE-evaluated reselection is the network control over traffic distribution across different cells and layers and not the related cell reselection delay—that is in principle similar to the one for UE-controlled cell reselection.

The mobile station may be ordered to leave GSM and to enter E-UTRAN by performing the network-controlled cell-reselection procedure. During this procedure, the Packet Cell Change Order message towards an LTE target cell could be included in a similar way to that in which it was defined earlier for the Inter-System NCCR towards WCDMA.

15.8.1.3 Network Assisted Cell Change in GSM

The functionality that was standardized by 3GPP to reduce the cell outage time caused by generic reselection procedure in GSM/GPRS network is called Network Assisted Cell Change. The primary aim of this feature was to provide the set of crucial System Information messages transmitted normally in the BCCH of the considered, target cell already in the source cell. In a normal case, an MS withholds the reselection process, enters so-called Cell Change Notification (CCN) mode, and waits for the neighboring cell data to be delivered to the mobile. With the help of this functionality the reselection time may be reduced by as much as 50% [7].

These procedures were originally standardized for the reselection that takes place inside the area under the supervision of the given BSC (3GPP Release 4) but soon they were extended within 3GPP Release 5 to cover also nontrivial inter-BSC reselection.[20] The mechanism approved by the 3GPP to convey the necessary signaling needed for distribution of SIs between BSCs is called RIM (RAN Information Management) [24].

Further extension of 3GPP specifications to cover the NACC procedure for the intersystem reselection cases from UTRAN to GERAN was included in 3GPP Release 6 [25]. Subsequently, this mechanism was inherited by the LTE architecture. System Information messages are provided to mobiles reselecting from UTRAN/E-UTRAN to GSM, but these messages are not distributed for UTRAN or E-UTRAN cells to the mobiles that are about to perform inter-RAT cell reselection from a GSM network. In the latter case the mobile could enter CCN mode, indicating the target 3G or LTE cell, but it is not supported with any kind of SI data about the target cell.

15.8.1.4 Mobility Management Events in E-UTRAN

Mobility procedures inside the LTE network and for the network changes from LTE to GERAN or UTRAN are handled by evaluating certain events that are related to both measurement

[20] Original GSM architecture did not foresee the direct interface between two BSCs.

actions (i.e. what kind of cells should be monitored) and the actual handover/cell change/redirection procedure. Thresholds for triggering given events could be absolute (typically for coverage handovers) or relative (typically for power budget handovers). For all thresholds, additional hysteresis values may be applied to prevent a ping-pong effect. Apart from thresholds related to either signal strength or signal quality, events' procedures are also governed by certain timers (so called time-to-trigger parameters) that introduce appropriate time hysteresis in the mobility procedures. The shorter the timer is, the more responsive is the mobility algorithm. However, too short a timer leads to a ping-pong effect and unnecessary handover actions. All the events have to be considered jointly in order to ensure proper mobility management. More details about event-based evaluation of network measurements are provided in [22].

15.8.1.5 PS Handover

The handover for PS services, which was initially introduced for WCDMA from its first release, was standardized for GSM late in 3GPP (in Release 6) and it is not widely deployed. Hence, although this procedure was included in 3GPP Release 8 for the inter-RAT procedures between GSM and LTE in both directions, it is not expected to be broadly used in the early LTE deployments for handling GSM-LTE interworking. For further details related to this functionality in the GSM environment, please refer to [7].

In contrast to GSM, the PS handover in WCDMA is the principal functionality that assures seamless mobility for data transactions. This functionality was inherited by LTE architecture and it is the only available option to change the network directly from the UTRAN CELL_DCH state to the LTE RRC_CONNECTED state and vice versa.

Typically, the measurement process of target cells is started in the LTE network once the RSRP or RSRQ of the serving LTE cell falls below a predefined threshold triggering an A2 event (the serving becomes worse than the threshold) by the mobile. Two mobility events are specifically designed to be used to trigger the actual system change to another RAT if the inter-RAT measurement has already started:

- B1 (Inter-RAT neighbor becomes better than threshold);
- B2 (serving becomes worse than threshold1 and the inter-RAT neighbor becomes better than threshold2) [22].

Once the decision about the PS handover to the target RAT is made by the eNodeB, the required network resources on the target cell are reserved in advance in the handover initiation phase. This is the principal feature of PS handover [13]. Only after establishing of the resources in the target network is the Mobility from E-UTRA Command message sent to the mobile and the actual handover takes place.

15.8.1.6 Cell Change Order to GSM

Apart from PS handover there are two other means to change the network from E-UTRAN to GSM in RRC_CONNECTED mode: cell change order and RRC connection release with redirection. The former is an embodiment of the network-controlled cell-reselection procedure. Cell-change order may be triggered without prior measurements of GSM cells [22]. This option is especially applicable for 1 : 1 deployment in the same band where GSM and LTE cells are collocated. If such a measurement procedure is configured, the list of GSM cells (i.e. ARFCNs related to the BCCH of the potential GSM target cells) to be measured prior to the possible RAT change is delivered to the mobile. Like measurements of the WCDMA network, an A2 event is typically used to trigger the measurements of GSM cells. Generally, measurements of inter-RAT layers—especially GSM—should be triggered by the network early enough in order to collect the signal strength samples for GSM cells and decode BSIC of these cells while still being in LTE coverage. However, if measurement gaps are needed for the measurement campaign due to the capability of the UE, they have to be scheduled by the network, potentially at the cost of the user perceived throughput—during the measurement gap the mobile does not receive or transmit any user data. If the signal strength of the serving cell has increased above the given threshold during the measurement campaign (A1 event), measurements of GSM cells are stopped.

Due to the characteristics of the GSM measurement process performed by the mobile (please refer to [26] for further details) it is important to keep the list of 2G cells to be measured by the UE as short as possible. Too long a list may unnecessarily delay the reselection process because measurement gaps are shared in concurrent mode between measurement campaigns related to various RATs if they are triggered by the network. As a consequence of this fact, the time needed, for example, for performing the BSIC decoding of the GSM cell, is increasing, together with an increasing number of UTRAN carriers that should be measured prior to the PS handover to WCDMA [26]. An excessive interference level in the GSM network may reduce the ability of the UE to decode the BSIC of the given GSM cell correctly. On the other hand, too restrictive a list may lead to a suboptimal Cell Change Order (CCO) decision where the mobile will be forced to reselect a cell that is not perfect from the radio conditions point of view while better cells would be not measured. Moreover, if there is no dedicated BCCH and TCH band defined in the spectrum used by the operator—that is the given frequency could belong to BCCH layer in one cluster and TCH in another—it is crucial to configure this list properly—that is, to include only ARFCNs used by BCCH in the given area.

B1/B2-related thresholds (described in Section 15.8.1.5) and corresponding timers could be set separately for different sets of ARFCN frequencies. This means that effective prioritization of various GSM cells could be done. One of the optimization strategies in the inter-RAT procedures could be to route the traffic from E-UTRA to the GSM cells (layers) where more advanced PS features are available, like EGPRS, Downlink Dual Carrier or EGPRS-2, to limit the performance gap between source and target system from the throughput and the delay points of view. For a brief description of these features please refer to Section 15.4.6. Note however that, if the measurements are not scheduled before the actual inter-RAT cell change, other mobility events like e.g. A2 (serving becomes worse than threshold) may be also employed to trigger the actual cell change.

Once the decision to perform the CCO is made by eNodeB, one specific target GERAN cell is included in the Mobility from E-UTRA Command message sent to the mobile (with purpose set to “cellChangeOrder” not “handover” as is the case for PS handover). Note that eNodeB may filter the target cell list according to additional criteria like blacklisting of cells or UE capability—hence it is not necessarily the strongest reported GSM cell that will be a final target for the CCO procedure. This network control over the traffic routing is one of the advantages of employing the Cell Change Order procedure for changing the network from LTE to GERAN—instead of the RRC connection release with redirection. A typical sequence of events for a Cell Change Order in the GSM procedure for a mobile that does not support CS services is presented in Figure 15.49.

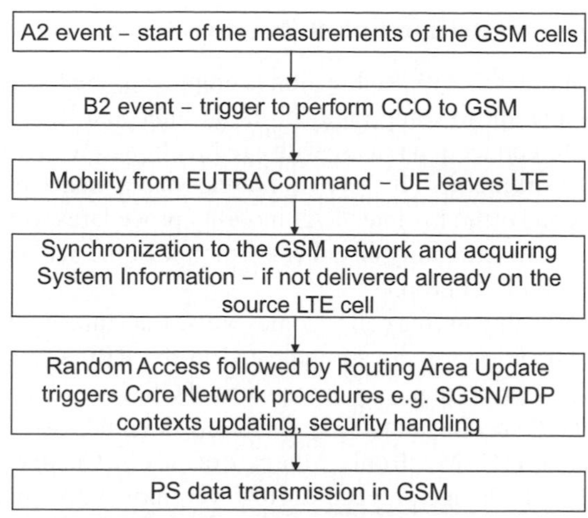

Figure 15.49 Cell Change Order to GSM procedure.

The network could send SI assistance data in the Mobility from E-UTRA Command message—that is, a set of SI related to the target cell. The set of SI for inter-RAT change assistance is not standardized by 3GPP but SI1, SI3 and SI13 could often be transmitted [23], as in the intra-GSM case. The mobile would have to acquire this SI before performing packet access or entering packet transfer mode [23] if they were not provided in the source cell. Actual content of SI of GSM cells could be either retrieved with the help of RIM or made available in the source eNodeB in automatic or manual way (via the O&M process).

Once receiving Mobility from EUTRA Command, the UE is allowed to leave LTE without sending any prior acknowledgement message. Note that if S3/S4 interfaces are not deployed in the network, a legacy Gn interface could be used to connect the 2G/3G SGSN with LTE Core (3GPP 23.401 annex D [13]). In such a scenario, from the point of view of the legacy 2G/3G SGSN, the MME will be perceived as another SGSN while the PGW will be treated like a GGSN during the Cell Change Order procedure. After leaving the LTE network, the mobile has to tune to the GSM frequency, synchronize to the network and listen to the SI (if this is needed). Sending the SI inside the Mobility from EUTRA Command permits a reduction in the cell outage time—even 1 s could be gained if NACC is employed together with CCO because there is no need to read the SI after camping on the GSM cell. If all the necessary messages are known to the UE, the Channel Request on RACH is sent by the mobile in order to have the radio resources granted for making the Location Area Update (LAU) if the mobile is CS capable. Subsequently, random access procedure has to be repeated in order to perform a Routing Area Update. If Gs interface between MSC and SGSN is available and a Combined LA/RA update is possible, this time could be reduced by approximately 2 s—since the LAU message is embedded in the Routing Area Update (RAU) message and there is no need to perform the separate LAU. The expected overall time needed to perform CCO from LTE to GSM (i.e. between receiving Mobility from EUTRA Command and sending RAU Complete in GSM) may be even longer than several seconds. If, for any reason, the UE fails to connect to the GSM network and the T304 timer, the value of which is provided inside Mobility from EUTRA Command, expires, the UE sends RRC a Connection Reestablishment Request message with cause "Handover failure" to try to revert back to LTE.

15.8.1.7 RRC Connection Release with Redirection

If the RRC Connection Release with Redirection is employed, the UE is ordered to enter the RRC_IDLE state and attempt to camp on a cell in an indicated RAT via the cell selection procedure [18]. The RRC connection release with a redirection to WCDMA or GSM should be used as a kind of last resort method, if there are no reasonable means to keep the connection inside the LTE network and orthodox inter-RAT mobility procedures were not triggered due to lack of support for certain functionality on either the mobile or the network side or due to other reasons, for example rapid field drop.

Usually an A2 event (i.e. the serving cell becomes worse than threshold) is used to trigger the actual cell change. Redirection targets are indicated in the RRC Connection Release message [22] where UTRAN carrier frequency or sequence of GSM carrier frequencies is included.

Originally, in the 3GPP specifications [22], the network could assist the UE by providing certain SI regarding the target GSM cell only in the case of the Cell Changer Order enhanced with the Network Assisted Cell Change procedure. However, mostly to enhance the CS fallback procedure that may be also based on redirection procedure (cf. Chapter 15 (*See* CS Fallback Via RRC Connection Release With Redirection)), the possibility of providing SI assistance data was extended in 3GPP Release 9 to RRC connection release with redirection case. Since the exact GSM cell the mobile will redirect to is not known in advance during the redirection procedures, instead of delivering System Information of a single GSM cell, a set of SI related to up to 32 GSM cells may be in such a case provided—together with their ARFCN and BSIC (called Physical Cell ID in E-UTRAN specifications). Once the pair of ARFCN and BSIC matches with the GSM cell the mobile redirects to, the appropriate SI could be used to speed up the access procedure to the new cell.

The RRC connection release with redirection could be also used to move the UE out of the WCDMA network. In the "Redirection info" in the RRC Connection Release message sequence of GSM or E-UTRAN frequencies could be given [21]—together with some supplementary information like the BSIC of the GSM cell or list of blacklisted cells for the LTE network. After receiving this message the mobile tries to camp on any of the applicable cells. As an additional option, inter-RAT redirection information could be also included by the network in RRC Connection Reject message—sent as the network response to RRC Connection Request. Such procedure could be triggered if, due to some reasons, the requested RRC connection could not be accepted in UTRAN.

Counterpart procedure is also available in the GSM network. Once "Cell selection indicator after release of all TCH and SDCCH" is included in Channel Release message, the mobile uses this information in cell selection algorithm—information about desired target UTRAN as well as E-UTRAN cells could be included there [20].

15.8.1.8 Summary

An example of the relationship between various event thresholds mentioned in Section 15.8.1 is presented in Figure 15.50. For simplicity reasons, LTE RSRQ or WCDMA CPICH EcNo triggered events are not presented.

Typically, the measurements and triggers for leaving the LTE network are scheduled in sequential manner and the availability of a more favorable layer from the inter-RAT strategy point of view is checked firstly. In Figure 15.50, higher priority was assigned to WCDMA target cells in comparison with GSM cells. If both the cell-change order and the RRC connection release with redirection are supposed to be used for the inter-RAT mobility towards GSM, cell-

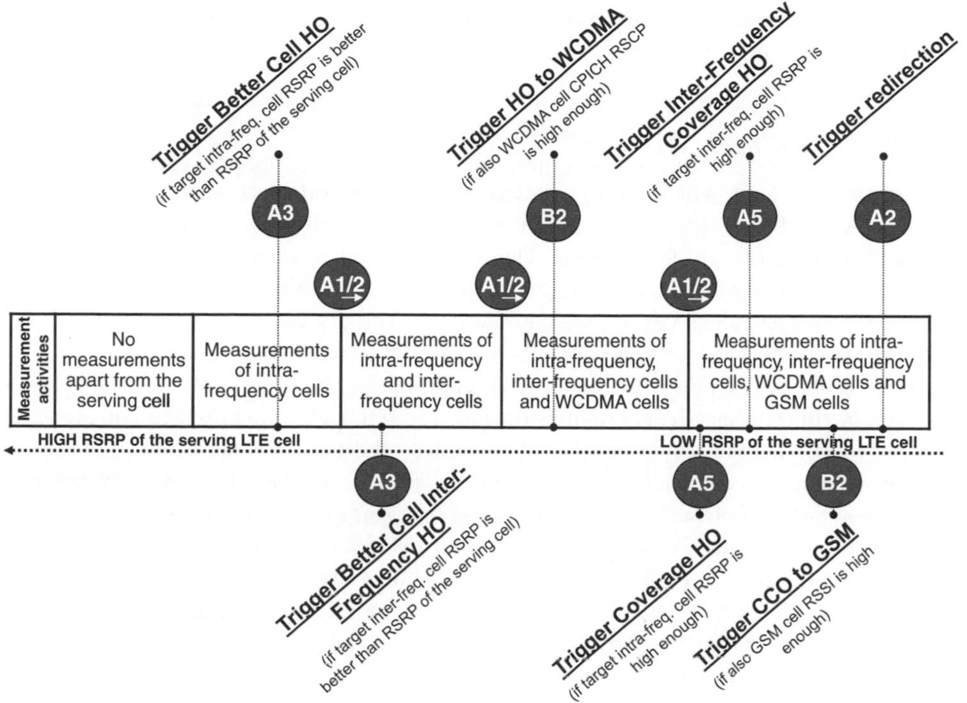

Figure 15.50 Various event thresholds—summary.

change order related measurements should usually be started first. Connection release with redirection may be useful if RSRP is dropped rapidly once, for example, the LTE coverage is ending. Note that the differentiation between various potential target cells for intra-RAT as well as inter-RAT mobility from E-UTRAN could be applied not only via appropriate setting of certain thresholds but also time to trigger values delivered to the mobile.

In order to assure consistent behavior with respect to interworking functionalities the network planning and optimization has to be performed jointly for all involved RATs. For instance, in case of GSM-LTE interworking following alignments may be considered:

- between idle and dedicated mode inside LTE—for example the relationship between RSRP leading to the inter-RAT reselection from LTE to GSM and CCO specific threshold;
- between the minimum 2G signal strength requirements related to access to the GSM cell and CCO parameters in order not to try to force the reselection to a GSM cell that is too weak;
- between LTE and GSM reselection parameters that govern the network change from GSM to LTE (via a common reselection process based on the absolute priorities as described in Section 15.8.1.1) to avoid a ping-pong effect;
- between CCO-related parameters and intersystem reselection parameters from GSM to LTE also to avoid a ping-pong effect.

Another important aspect to be considered while analyzing inter-RAT mobility is handling of QoS during such procedures. One-to-one mapping between an EPS bearer and a PDP Context was standardized [13]. Some rules related to R99 and R8 QoS parameters were provided in

3GPP standards but the exact implementation of this mapping is however operator-specific. For more details regarding QoS handling please refer to [13].

15.8.2 CS Fallback

Voice services in LTE, which incorporates no CS domain in its architecture, attract special attention. A number of different alternatives have been developed by either 3GPP or other organizations like the VoLGA Forum. These are covered in detail in dedicated chapters throughout the book. In this section, in the context of interworking between LTE and legacy systems, CS Fallback deserves special consideration. The discussion is focused on practical deployment aspects and interdependencies with data-centric mechanisms.

The CS Fallback (CSFB) functionality, introduced by 3GPP in Release 8 and further enhanced in subsequent releases, allows operators to reuse CS domains and corresponding RANs of legacy networks for provisioning of voice (and other CS services) to multimode terminals attached to EPS. CS Fallback provides a mechanism for service-based redirecting of CSFB-capable mobiles from E-UTRAN towards other RATs that support CS-domain services (e.g. GERAN or UTRAN). It may be triggered by either mobile-originated or mobile-terminated calls. The CS Fallback functionality requires that E-UTRAN coverage is overlapped by target RAT coverage. A dedicated interface between the MME of EPS and MSC/MSC Server of CS core network is necessary for interworking procedures such as combined EPS/IMSI attach or combined Tracking Area/Location Area update. The so-called SGs interface has been defined for this purpose [27]. The combined EPS/IMSI attach and combined TA/LA update procedures ensure that a UE is registered not only to EPS but also to the appropriate location area of CS domain of the underlying network so that CS services are available. To ensure proper handling of data services during CS Fallback, cooperation between EPS and the target PS core network is necessary. The intersystem interworking architecture can be based either on the 3GPP Release 8 S3/S4 interfaces or on the pre-Release 8 Gn/Gp interfaces. It must be noted, however, that the Gn/Gp-based interoperability architecture entails some modifications of particular interworking procedures, like routing and tracking area updates, data forwarding, and so on, which in turn may result in suboptimal performance from both the network and the user viewpoint. Details on the Gn/Gp interoperation architecture can be found in 23.401 Annex D [13].

When analyzing the CS Fallback functionality in the context of its deployment in a life network, the main focus is automatically placed on the performance of voice services in terms of call setup time. Nevertheless, activation of the CS Fallback, which by definition leads to temporary switch-over from E-UTRAN to another RAT, may influence the performance of PS services in terms of data session interruption time and user-data throughput. Furthermore, discussion of aspects of the deployment of CS Fallback should cover not only user-perceived QoS topics but also those that relate to the impact on overall network performance. All these aspects are discussed in more detail below.

15.8.2.1 Voice Performance

Since the CS Fallback procedure forces a UE to switch from E-UTRAN to a CS-capable RAT, total voice call establishment time includes an additional delay compared to call setup executed directly in a given RAT. This delay depends mainly on how long it takes to redirect a UE from E-UTRAN and how quickly the CS call setup procedure can be executed in the target cell. These delays, in turn, are strongly determined by the target RAT technology as well as the mobility management mechanisms on which CS Fallback itself is based. There are three different mechanisms that can be used to fall back UE to another RAT: inter-RAT PS handover, cell-change order (including Network Assisted Cell Change), and RRC connection release with redirection [27].

CS Fallback Via Inter-RAT PS Handover

CS Fallback via inter-RAT PS handover is the most advanced option and provides the shortest call setup delays. In this CS Fallback approach, whenever a UE needs to be moved to another RAT due to pending CS call establishment, the standard inter-RAT PS handover from E-UTRAN procedure is invoked. The target cell is determined either based on UE measurement reports or based on configuration settings only [27]. With the first option, a UE is requested to perform target carrier measurements before leaving the LTE cell, which introduces an additional delay to the call setup time. To minimize the time spent on target cell measurements, the number of target frequency carriers should be kept to a minimum, especially for GERAN layers. This additional delay could be avoided by triggering EPS/IMSI-attached terminals in ECM-CONNECTED mode to measure and report inter-RAT neighbor cells proactively, so that actual target-RAT cells measurements are available when CS Fallback is attempted. With the target cell selection based on UE measurement, current signal propagation conditions are taken into account so that the most suitable cell can be chosen. The second alternative, so-called blind target cell selection, allows avoiding additional call setup delay but can be applied only in coordinated scenarios with one-to-one overlapping between E-UTRAN and target RAT cells. The applicability of the first method is not limited to coordinated scenario. In case E-UTRAN coverage is overlapped by several target RAT layers, a prioritization mechanism can be used to prefer certain technologies or frequencies to others.

The next steps of handing over of a UE towards a selected target cell are similar as in case of a normal inter-RAT PS handover procedure described in [13]. The main difference is that, having switched to the target cell, the UE attempts to establish a CS connection. The procedure of CS call establishment follows rules specific to the target RAT. In UTRAN, CS call can be set up in parallel to PS connections established upon successful completion of an inter-RAT handover procedure without any PS session interruptions and additional voice call setup delays. In GERAN, such a smooth establishment of a CS connection can take place only if both the network and the UE support Dual Transfer Mode and enhanced CS establishment in DTM (see [28]). In such a scenario (Figure 15.51), the CS call setup procedure can be executed over already allocated packet resources and without leaving packet transfer mode. If the network and the UE support DTM, but not enhanced CS establishment in DTM, CS call setup cannot be done while in packet transfer mode. Thus voice call establishment entails temporary intermission of ongoing PS sessions and switching to packet idle mode. Having left packet transfer mode, the UE attempts to establish a CS connection via the standard random access procedure [27]. This results in longer voice call setup times as compared with UTRAN or GERAN with enhanced CS establishment in DTM scenarios. After successful CS call setup, packet transfer mode can be re-established so that the PS data session is continued in parallel to the voice call. Data session interruption time is then determined by the duration of the CS call setup procedure. Upon re-entering packet transfer mode the UE performs the remaining steps of the inter-RAT handover as in a standard non-CS Fallback case. Having completed voice call, the UE follows standard mobility management procedures. Whether the UE stays in the target RAT or moves back to E-UTRAN, it depends on parameterization of those procedures, for example Cell Selection Indicator information element included in

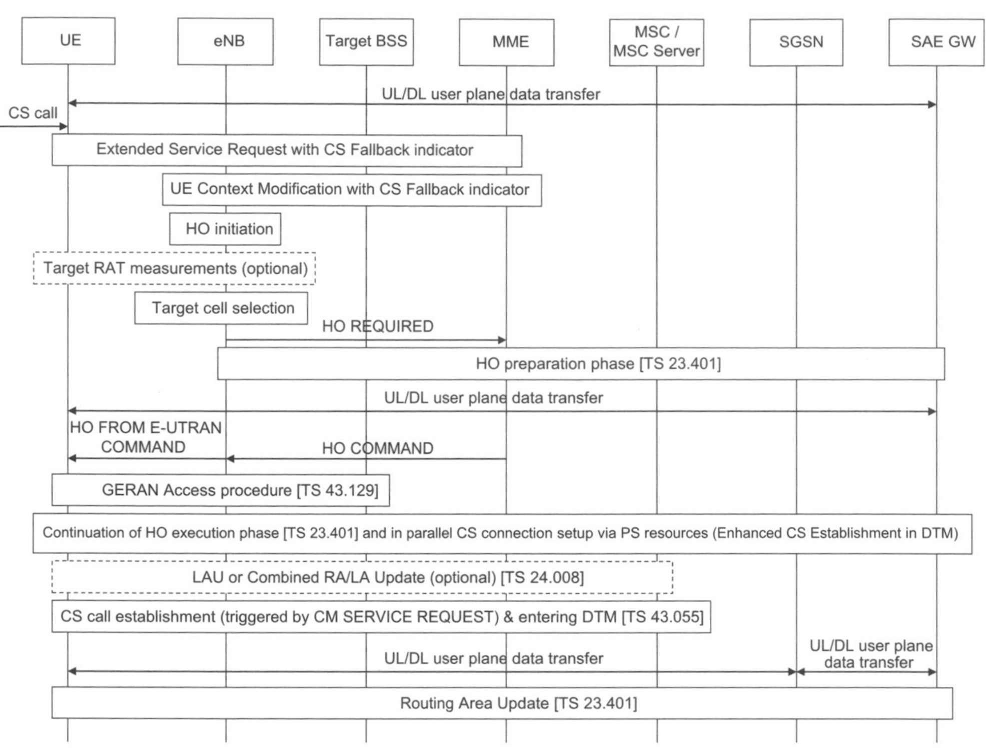

Figure 15.51 CS Fallback to GERAN via inter-RAT PS HO for mobile originated call in ECM-CONNECTED state; DTM and enhanced CS establishment supported.

Channel Release message may force the UE to reselect to E-UTRAN as soon as a suitable cell is available [20].

If either the network or the UE does not support Dual Transfer Mode, packet transfer mode setup during inter-RAT PS handover must be left not only for the time during which a CS connection is established but for the whole CS call duration. Having switched to packet idle mode, a UE triggers a CS call setup procedure by sending a channel request on a random access channel [27]. The need to go through the random access procedure introduces an additional delay to the voice call setup time. Another consequence of the lack of DTM support is that all active PS sessions must be put on hold until voice call is completed. This is realized via a so-called suspend procedure that must be triggered by the UE before requesting a dedicated traffic channel for the CS call (before sending CM Service Request with service type set to originating call establishment or Paging Response message). Consequently, all the remaining standard steps of inter-RAT PS handover are abandoned. The Suspension Request message is sent by the UE to the BSC over a Standalone Dedicated Control Channel (SDCCH) so that no PS resources are involved. The BSC forwards the Suspension Request to the SGSN. If inter-RAT inter-working architecture is based on the legacy Gn/Gp interfaces, upon receiving Suspend message from the UE the SGSN follows the intra-SGSN suspend procedure. It means that PS sessions are suspended internally within the SGSN and without any interaction with the P-GW [29]. If inter-RAT interworking architecture is based on S3/S4 interfaces, being triggered by the UE, the SGSN attempts the suspend procedure towards the S-GW and the P-GW. In consequence the deactivation of GBR bearers and suspension of non-GBR bearers are triggered. In this case, as opposed to the Gn/Gp scenario, the P-GW should discard packets dedicated to the suspended UE [29]. When the CS connection is completed, suspended PS sessions may be resumed via procedures described in [29]. The UE may start the resume procedure either from a legacy RAT or having switched back to E-UTRAN. Whether the UE stays in the target RAT or moves back to LTE, it depends on parameterization of existing inter-RAT mobility management mechanisms. With appropriate configuration of those mechanisms the UE may be forced to reselect back to LTE as soon as a suitable cell can be found.

Regardless of the target RAT type and capability, the CS call setup procedure may need to be preceded by a location area update. The LAU procedure is performed by the UE if Location Area Identity (LAI) of the new cell towards which it has been handed over differs from the one stored in the UE [27]. This procedure introduces an additional delay to the voice call setup time and hence, to minimize the total delay, it is crucial to ensure that LAI stored by the UE is up-to-date and corresponds with the LAI of the target cell. Whether the Location Area that the UE is registered to, as a result of the combined EPS/IMSI attach or the combined TA/LA update procedure, matches to the LA of the target cell depends on the tracking area and location area plans as well as the tracking area list management algorithms implemented in the MME. To minimize the impact of LAI mismatch on the CS Fallback performance, the mechanism based on triggering location area update just after UE is handed over to the UTRAN/GERAN cell and without waiting for current LAI information, can be considered. This solution shortens call setup time for those terminals that are not registered in correct LA but introduces unnecessary delay for those registered in proper LA. Hence it is kind of trade-off and is worth of considering if the risk of LAI mismatch is relatively high due to, for example, lack of alignment of TA and LA plans. Location area update-related aspects are discussed further in Chapter 15 (*See* Interdependencies with Fundamental Procedures).

Discussing CS Fallback via inter-RAT PS handover, it should be noted that this solution requires standard PS handover functionality to be supported by both the target RAT and the terminal. This, in practice, narrows the usability of CS Fallback via inter-RAT PS handover to scenarios in which the target RAT is UTRAN and not GERAN in case of which PS handover is rather rarely deployed. Furthermore, if CS Fallback is triggered when a UE is in ECM-IDLE mode, inter-RAT handover mechanism may not be efficient way for CS Fallback realization as there is no real need to establish PS connections in the target RAT. Hence, other inter-RAT mobility procedures, discussed in a subsequent part of this section, may be regarded as more practical ones.

CS Fallback Via Cell Change Order

In scenarios where an inter-RAT PS handover is not practicable and a target RAT is GERAN, CS Fallback may be based on the inter-RAT cell change order (CCO) mechanism. In such a case, each mobile-originated or mobile-terminated call triggers the CCO procedure. As in the standard CCO case, the CS Fallback target cell must be provided to the UE. Target cell selection can be based either on UE measurement reports or on configuration settings. The advantages and disadvantages of these two alternatives are analogous to those discussed for a CS Fallback via inter-RAT PS handover scenario. In case of measurement-based target cell determination, the number of frequency carriers to be measured should be kept to a minimum to avoid long delays introduced by this step to the total CS call-establishment time. Furthermore, if E-UTRAN coverage is overlapped by more than one GERAN layer, some prioritization mechanisms can be used to prefer certain frequencies to others.

Having switched to the target cell, the UE must acquire defined minimum amount of SI before it is allowed to attempt the initial access procedure. This results in an additional delay in call setup time. By employing the Network Assisted Cell Change (NACC) feature, the UE may be provided with target cell system information while it is still in the LTE cell. The UE may use this information when accessing the target cell and hence avoid scanning the BCCH for system information. If not all necessary system information has been delivered to the UE, it still needs to acquire missing system information before initiating the random access procedure in the new cell. Further discussion on CCO and NACC is provided in Section 15.8.1.6. Having collected all necessary system information, the UE attempts to establish a CS signaling connection so that the voice call setup procedure can be continued. Here, as in the case of CS Fallback via inter-RAT PS handover, the call setup procedure may need to be preceded by a location area update. Thus the discussion on LAU aspects provided hereafter is also relevant to CS Fallback via the CCO scenario. Furthermore, as during a normal CCO procedure, a routing area update must be performed by the UE as soon as possible. Whether the RAU procedure is performed in parallel with CS connection depends on the DTM capability of the UE and the network. If the DTM is supported by both the network and the UE, RAU can be realized over the CS signaling channel using GPRS Transparent Transport Protocol (GTTP) and without the need for PS connection establishment. Subsequently, the UE may enter DTM mode so that CS and PS services are available simultaneously (Figure 15.52). Having successfully completed the RAU procedure, the SGSN sends an inquiry to the MME (via either S3 or Gn interface) about EPS bearers of the UE so that LTE originated data sessions can be continued utilizing GERAN PS resources. The data session interruption time depends, in such a scenario, mainly on two factors: the time a UE needs to tune to indicated frequency and acquire necessary system

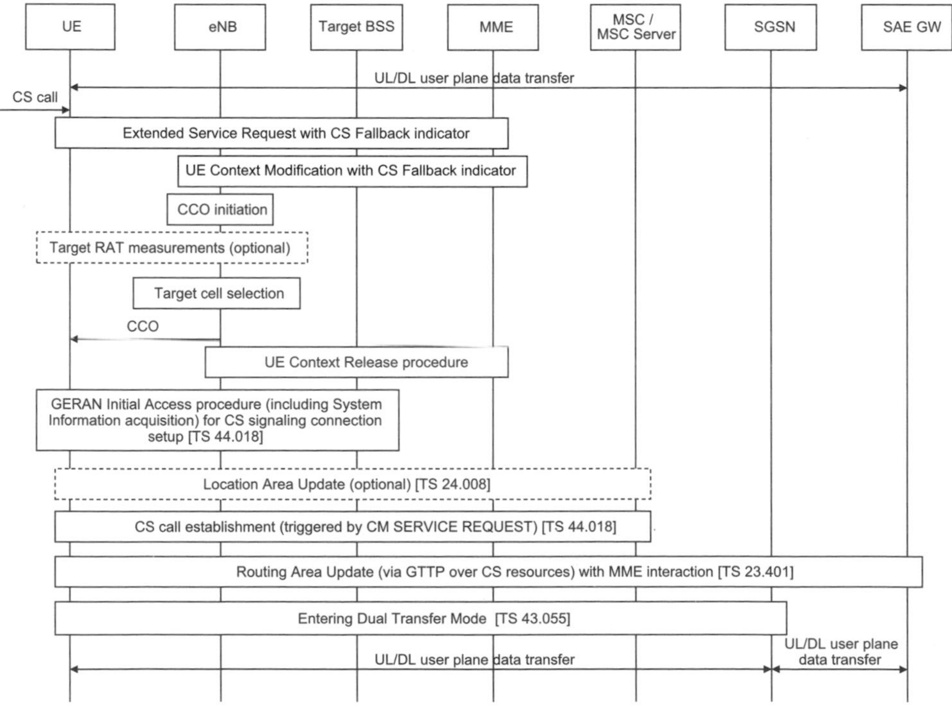

Figure 15.52 CS Fallback to GERAN via CCO without NACC for mobile originated call in ECM-CONNECTED state; DTM supported.

information as well as duration of the RAU procedure. If the UE had no ongoing session in EPS when the CSFB was triggered, no PS resources are assigned to the UE in GERAN cell. Upon completion of CS call, the UE starts to follow the standard mobility management procedures. To force reselection towards LTE cells, appropriate parameterization that favors E-UTRAN cells over other RAT cells should be applied.

If either the UE or the network does not support the DTM, the RAU procedure cannot be realized over CS signaling links and hence requires a dedicated PS connection to be set up. Consequently the routing area update and any PS session establishment must be put on hold until the CS call is completed. Due to the temporary unavailability of PS services, the UE must trigger the suspend procedure [29] before continuing with voice call setup. The Suspension Request message is transferred over CS signaling link (no need for PS connection setup) to the BSC, which in turn forwards this message to the SGSN. The SGSN triggers the suspend procedure towards MME via either S3 or Gn IF. Since the MME may be unable to retrieve full GUTI based on the combination of P-TMSI and RAI received from the SGSN, and hence may not be able to determine which UE context should be suspended, in Release 9, the 3GPP has defined the mechanism for triggering suspension internally within EPS upon ordering the UE to fall back to non-DTM environment [27]. Having triggered the Suspend message the UE may finally continue with the CS call setup by sending either CM Service Request (with service type set to originating call establishment) or Paging Response via the SDCCH channel. As soon as a voice call is terminated, the UE may start the resume procedure by performing RAU, if it remains in GERAN, or TAU if switches back to LTE. In such a scenario, the main contributor to the total data session interruption time is the voice call duration.

CS Fallback Via RRC Connection Release With Redirection

In scenarios where neither inter-RAT PS handover nor inter-RAT cell change order is applicable, the only option for CS Fallback realization is the RRC connection release with redirection. This solution is applicable to any target RAT. In case of RRC connection release with redirection, the UE is provided with the list of frequency carriers of potential target cells and not with the one exact cell towards which fallback should be performed. This realization does not require any target RAT measurements performed before leaving an LTE cell. The list of potential targets can be determined based on configuration parameters and UE capabilities only. However, if any target RAT measurements (e.g. periodic measurements) are available upon triggering CS Fallback, they may be reused by the algorithm of targets list generation. Such an approach, on the one hand, allows the UE to leave LTE cell as soon as CS Fallback is triggered and without any delay caused by target RAT measurements. On the other hand, this involves the UE scanning the indicated spectrum and selecting a suitable cell in target RAT by itself. This time spent on searching for a suitable target cell is the main concern in the context of voice call setup time. To keep this at a minimum, the frequencies that are indicated to the UE should be selected in an optimal way so that they correspond to the most suitable target cells. Having selected the target cell, the UE must, analogously to CS Fallback via CCO scenario, acquire necessary system information before it is allowed to start the random access procedure. The resultant additional delay in the voice call establishment may be avoided with the help of Release 9 enhancement—the so-called Multi Cell System Information [27]. This functionality enhances RRC connection release with redirection mechanism so that a UE may be provided additionally with physical identities and system information corresponding to potential target cells. Having selected one of the cells for which the necessary system information has been acquired before leaving LTE cell, the UE may start the CS call setup procedure immediately. The call establishment procedure itself is executed according to

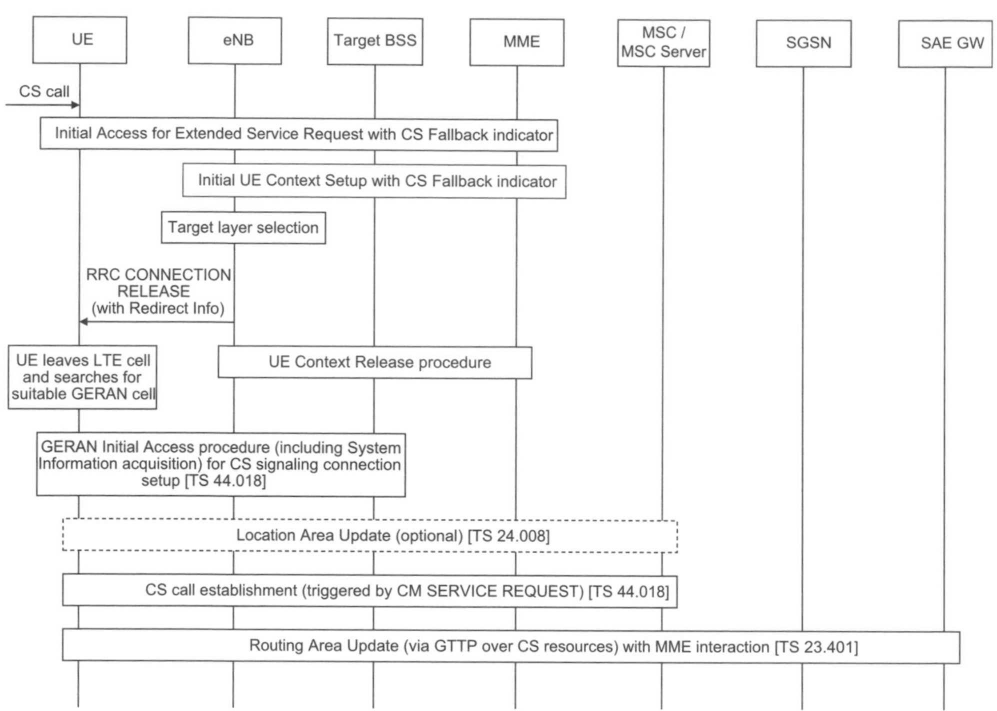

Figure 15.53 CS Fallback to GERAN via RRC connection release with redirection for mobile originated call in ECM-IDLE state; DTM supported; Multi Cell System Information not supported.

standard rules specific to given RAT. Figure 15.53 presents the CS Fallback procedure to GERAN. If the UE-stored LAI does not match the LAI of the new cell, the voice call setup must be preceded by a location area update. Consequently the voice call establishment time may be protracted. Discussion on how to minimize the probability of the need for LAU is provided in the subsequent part of this section. In parallel to the CS domain activities, a UE should also perform a routing area update. In UTRAN and Dual Transfer Mode GERAN scenarios, there are no issues with executing RAU in parallel to CS activities. Consequently, LTE-originated data sessions (if any) can be continued in the target RAT concurrently with the voice call. Data session interruption time is, in such a scenario, determined mainly by the time of searching for the suitable target cell and acquiring necessary system information as well as RAU execution time.

In the non-DTM GERAN case, RAU and any PS data session setup require the CS connection to be released, and hence can be performed only if the voice call is completed. This entails the suspend procedure to be triggered by the UE before CM Service Request (with service type set to originating call establishment) or Paging Response can be sent. The suspend procedure is executed in the same way as in CSFallback via the CCO case. Furthermore, due to P-TMSI and RAI to GUTI mapping issues, the mechanism for triggering suspension internally within the EPS, upon redirecting UE to a non-DTM-capable GERAN cell may be helpful. The PS session resume procedure that can be attempted on completion of a CS call is triggered by either RAU, if the UE remains in GERAN, or TAU, if the UE switches back to LTE. In this scenario potential data session interruption time depends not only on how quickly the UE searches for the target cell and acquires system information but also on the voice call duration.

As in other CS Fallback scenarios, having completed the CS call, a UE follows standard mobility management procedures and by their appropriate configurations it may be forced to switch back to LTE as soon as a suitable cell is found.

Interdependencies with Fundamental Procedures

The target RAT type and the mechanisms by which CS Fallback operates are not the only factors influencing voice call setup time. The performance of the CS Fallback procedure also depends on the EPS Connection Management (ECM) state of the UE. CS Fallback may be triggered (by a UE or a network) in either ECM-IDLE or ECM-CONNECTED mode. If CS Fallback is attempted while a UE is in ECM-IDLE mode, an additional call setup delay must be considered because the UE needs to switch to ECM-CONNECTED mode. This transition is realized via the standard Initial Access procedure defined in [13].

Another aspect that should be discussed in the context of CS Fallback execution time is the paging procedure. As an intrinsic feature of a mobile terminated call, CS paging must be invoked regardless of the ECM state. An EPS/IMSI-attached UE is paged for incoming call in the RAT it is currently camped on. Without ISR, a CS Fallback-capable UE must always perform a location and routing area update when reselecting from E-UTRAN to other RAT so that the MSC/VLR knows where certain UE should be paged. Knowing the RAT a UE is camping on, the MSC/VLR pages this UE either directly in managed RAT or via MME in E-UTRAN. If the ISR is active a UE can reselect between registered areas without executing tracking, routing, and location area update procedures (see Section 15.8.4). Consequently, the MSC/VLR is not able to determine unequivocally whether a UE is currently camped on an E-UTRAN, UTRAN, or GERAN cell. Therefore, the MSC/VLR sends a paging message via SGs interface to the MME, which in turn handles paging accordingly. If the UE is in the ECM-CONNECTED state, the MME triggers paging in serving cell as in a normal case. If the UE is in the ECM-IDLE state, the MME pages the UE in all tracking areas the UE is registered to and forwards a paging message via the S3 interface to the associated (via the ISR mechanism) SGSN, which in turn initiates paging in the routing area to which the UE is registered [27]. This mechanism guarantees that a UE will be paged for a mobile-terminated call even if the location area update is skipped due to ISR. On the other hand, such an extended paging route may introduce additional delay in voice call setup time.

As already mentioned, one of the factors impacting CS Fallback call setup time (regardless of the actual fallback mechanism) is the potential need for a location area update preceding voice call establishment. To optimize CS Fallback performance and minimize the necessity of location area updates that are obligatory if UE-stored Location Area Identity (LAI) does not match the actual LAI of the new cell, planning of tracking and location area boundaries should be reconsidered. As the LAI stored by the UE is the one to which the UE is registered via combined EPS/IMSI attach or combined TA/LA update, it is essential that a UE executes tracking area update every time the underlying location area is changed. If there is no concept of a tracking area list, the straightforward solution would be to align the boundaries of selected tracking areas in LTE with the boundaries of location areas in target RAT. Nevertheless, with the concept of a tracking area list a terminal can be registered to multiple tracking areas and therefore does not perform combined TA/LA update every time it crosses a TA boundary. In this context, apart from alignment of TA and LA boundaries, some enhancements of algorithms for handling tracking area lists in MME could be considered so that the tracking area list indicated to the UE does not contain tracking areas spread over more than one location area. If a tracking area list covers multiple location areas, the UE might not be provided with the actual LAI of the target cell before leaving the LTE cell, which in turn results in longer call setup time in the target RAT. An analogous issue with the matching of location area identities may be encountered in scenarios where the tracking area covers the location area boundary. Such a configuration may result in ambiguity of TAI-to-LAI mapping tables managed by MME and used for the purpose of a combined EPS/IMSI attach and combined TA/LA update. Consequently terminals may be registered, starting from the combined EPS/IMSI attach procedure, in a location area different from the location area of the cell selected as the CS Fallback target. It must be noted, however, that the alignment of TA and LA boundaries and LA-aware tracking area list handling may lead to nonoptimal tracking area management and consequently to an increase in the signaling load in LTE. To counter this effect, the readjustment of cell reselection parameters could be considered. In scenarios where the LAI mismatch is unavoidable, the impact of LAU on CS Fallback performance may be minimized by configuring the MSC to reduce the frequency of Authentication, TMSI reallocation, and Identity check for terminals being EPS/IMSI attached via SGs interface so that LAU is speeded up.

Another important aspect that must be considered in the context of minimizing the necessity of location area updates preceding CS call setup is provisioning of SGs connectivity between proper MMEs and MSCs/MSC Servers. In other words, the MMEs and the MSCs/MSC Servers that serve a given geographical area must be connected to each other. Furthermore, the TAI-to-LAI-to-MSC/VLR mapping tables from which the MME derives the LAI and VLR numbers must be properly configured in the MME. Otherwise, a UE may fall back to the cell that is served by a MSC/MSC Server in which a UE is not registered—hence CS connection establishment would be delayed until successful LAU completion.

15.8.2.2 Data Performance

As already mentioned, impact on the performance of PS services is, besides voice call setup time issue, another important aspect of CS Fallback deployment in a life network. This, however, is valid only in case of CS Fallbacks triggered while a UE is ECM-CONNECTED state (in ECM-IDLE mode there is no ongoing data session). In this context, data session interruption time and temporary degradation of data throughput are of the main interests. How much user-perceived quality of service is affected depends on the CS Fallback mechanism and the target RAT technology. PS service interruption aspects, including suspend and resume procedures, have already been discussed along with voice call setup delay topics. The impact of CS Fallback on user data throughput may be twofold. First, if CS Fallback is realized via inter-RAT PS handover or cell change order, a UE may be requested to measure target RAT carrier frequencies before leaving the LTE cell. This affects the performance of ongoing data transfers because, during inter-RAT measurements, a UE cannot be scheduled with any data and it cannot transmit in uplink—so-called inter-RAT measurement gaps must be applied (unless a UE is a dual-radio terminal). This effect can be avoided by applying, if possible, a blind cell selection discussed in previous sections. The situation is different in CS Fallback via RRC connection release with redirection, since in such a case no target RAT measurements are needed before leaving the LTE cell, regardless of the network deployment scenario. This, on one hand, does not cause user data throughputs to deteriorate when still in the LTE cell but, on the other hand, it increases data session interruption time. The other aspect that should be considered in the context of the influence of CS Fallback on user data throughput is related to differences in data rates offered by E-UTRAN and target RAT. The higher the differences are, the more perceptible degradation of user data throughput may be expected. Furthermore, insufficient performance from the target system may result in dropping of the data sessions originally running over guaranteed bit rate EPS bearers. In this context, target RAT specific enhancements that aim to minimize the performance gap, like those discussed in Section 15.4.6, can be considered.

15.8.2.3 Impact on the Overall Network Performance

Performance of the CS Fallback functionality may be analyzed not only from the user perspective, as in previous sections, but also from the network viewpoint. When deploying CS Fallback in their networks, the operators may wonder if and how the overall performance of LTE and underlying target systems may be affected. As CS Fallback reuses some fundamental procedures like, for example, paging, initial access with NAS signaling, UE context setup/modification/release, RRC connection release with redirection, the increase of signaling traffic on radio, and S1-MME interfaces may be expected. Furthermore, the signaling load on S3/S4 or Gn/Gp interfaces may grow as well, especially in scenarios with high utilization of inter-RAT PS handover mechanism. How much the signaling traffic will increase, it depends on CS Fallback usage ratio. With relatively low usage of this feature, its impact on the overall network performance may be hardly noticeable. From the target RAT point of view, CS Fallback activation may result mainly in an increase of CS call setup-related signaling load and CS traffic itself. If the target RAT is UTRAN or DTM-capable GERAN, the growth of PS traffic and corresponding signaling may be observed as well. Furthermore, the increase in the number of location and routing area update procedures and related signaling may be expected. The impact of CS Fallback on the target system's performance depends on how much additional traffic is offered to the target RAT as a consequence of the CS Fallback activation. The lower this additional traffic is, the less noticeable is the influence of CS Fallback on the target system.

15.8.2.4 Single Radio Voice Call Continuity

In the discussion on voice-related intersystem interworking aspects, the Single Radio Voice Call Continuity (SRVCC) feature should be mentioned. Standardized in 3GPP Release 8 and further enhanced in subsequent releases, the SRVCC functionality complements the IMS-based Voice over LTE solution by introducing a mechanism for transferring IMS-anchored voice calls from E-UTRAN to the CS-capable RAT. Hence, service continuity may be sustained in scenarios with non-continuous E-UTRAN coverage underlain by UTRAN and/or GERAN. The SRVCC is an inter-RAT PS-to-CS handover that involves both radio access and core network interactions. The SRVCC handover is only applicable to voice-related EPS bearers. Non-voice related EPS bearers must be handled with inter-RAT PS handover mechanisms defined in [13]. Whether SRVCC and inter-RAT PS handover can be processed in parallel depends on the UE and the target RAT capabilities. Further details on the SRVCC procedure can be found in Chapter 9, VoLTE. Hereafter, discussion on how SRVCC activation may influence the network performance is provided.

The SRVCC feature aims to improve the quality of voice services by saving voice calls from being dropped due to running out of E-UTRAN coverage. Hence, dropped call statistics and user-perceived quality of voice services may improve after SRVCC activation. Introduction of SRVCC may, however, entail some compromise in terms of non-voice service performance. In scenarios where inter-RAT PS handover of non-voice-related EPS bearers and SRVCC cannot be executed simultaneously, the later one takes precedence and only inter-RAT PS-to-CS handover is executed. Consequently, re-establishment of PS resources in the target cell for continuation of non-voice sessions is deferred, which results in longer service interruption times. If the target RAT is UTRAN, SRVCC and inter-RAT PS handover can be executed in parallel only if both the target network and the UE support the PS handover functionality. If the target RAT is GERAN, SRVCC and inter-RAT PS handover can be performed simultaneously only if both the target network and the UE support the DTM Handover. The DTM Handover, standardized in 3GPP Release 7 [28], provides GERAN with means for simultaneous execution of CS domain and PS domain handovers [28]. It efficiently combines standard CS and PS handover procedures so that CS and PS resources are allocated in the target cell before the terminal leaves the source cell. Originally, the DTM Handover was meant for GERAN-to-GERAN and UTRAN-to-GERAN transitions but it can also be reused in case of handovers from E-UTRAN. It should be noted here that pure DTM capability, even complemented with PS Handover, without DTM Handover is not enough to allow for simultaneous SRVCC and inter-RAT PS handover execution towards GERAN cell. It only allows for the establishment of PS connections in parallel to a CS call and for performing a routing area update over a CS signaling channel (using GTTP protocol) [28].

If inter-RAT PS handover cannot be processed in parallel with SRVCC, only CS resources are pre-allocated to the UE and smoothly accessed after switching to the target cell. PS resources for non-voice related EPS bearers can be established in the new cell after successful execution of the routing area update (with MME interaction), which in turn may be attempted upon completion of the SRVCC handover [30]. Such a procedure results in protracted interruption time for nonvoice services if these were active upon the triggering of the SRVCC handover. Moreover, if the target RAT is GERAN and DTM is not supported by either the network or the UE, PS connection setup must be postponed until the CS call is completed. Consequently, the suspend procedure towards the source MME (as defined in [29]) must be executed. When the CS call is completed, the suspended PS sessions may be resumed via corresponding procedures described in [31].

Having completed the CS call in UTRAN or GERAN, the UE follows standard idle or PS mode (if any PS session remains active) mobility management procedures. With appropriate parameterization of those procedures, the UE may be forced to camp back to E-UTRAN as soon as a suitable cell is available. At the time of writing this book, no dedicated mechanism for transferring IMS anchored voice calls from a CS-capable RAT back to E-UTRAN has been standardized. However, work on the so-called Reverse SRVCC mechanism has already started within 3GPP. It is planned that this feature will incorporate handovers from UTRAN/GERAN CS domain to E-UTRAN for voice calls either previously handed over from E-UTERAN to UTRAN/GERAN CS domain (via SRVCC) or directly setup in the UTRAN/GERAN CS domain but anchored in IMS [3]. Such functionality will complement existing intersystem interworking mechanisms and will allow for more efficient traffic management strategies aiming at enhancing the performance of coexisting systems.

15.8.3 Idle Mode Signaling Reduction

Smooth LTE and legacy systems interworking requires mobility management features to be complemented with an efficient intersystem location management mechanism. For that purpose the ISR functionality has been standardized by 3GPP [13]. This enhancement requires S3/S4-based intersystem interworking architecture so that the S-GW is the anchor for UE mobility. The ISR mechanism is a kind of evolution of the LTE specific tracking area list concept to the inter-RAT scenario. With the tracking area list concept, UE can be registered to multiple TAs and therefore does not need to perform TA update every time it crosses TA boundary (unless it camps on a cell of tracking area it is not registered to). With ISR, a UE can be simultaneously registered to multiple tracking areas of E-UTRAN and routing area of UTRAN or GERAN. As long as a terminal moves between RATs but within the registered areas, neither RAU nor TAU is performed (with the exceptions of periodic updates). Hence, the signaling load due to inter-RAT transitions may be reduced. This, however, is possible only at the cost of increased paging-related signaling, which, in the case of ISR, has to be initiated in both SGSN and MME. Without ISR, when an idle-mode UE switches between RATs, the routing area update (if the target is GERAN/UTRAN) or the tracking area update (if the target is E-UTRAN) must be executed, which in turn entails deregistration of a UE from the source RAT. In frequent inter-RAT transition scenarios (e.g. LTE islands on top of GERAN/UTRAN), the "ping-pong" effect may generate so much overall additional signaling traffic that its reduction, even at the cost of increased paging load, is justifiable.

The ISR functionality does not influence the standard rules of RAU and TAU procedures associated with UE mobility within a given RAT. Routing and tracking area updates must be executed on moving to an area to which the UE is not registered or on expiry of periodic update timers. These are managed independently for 2G/3G and E-UTRAN. It must be noted, however, that as long as the UE switches between routing areas served by the same SGSN, ISR remains active so that, when changing to E-UTRAN, TAU is needed only if the current TA is not on the multiple TA list to which this UE is registered. In a similar way, the tracking area update procedures within E-UTRAN do not deactivate ISR as long as the serving MME is not changed—RAU is not required if the current RA matches with the one with which the UE is already registered. If the serving MME or the serving SGSN is changed, the ISR is deactivated and tracking or routing area updates following inter-RAT transitions cannot be avoided. The ISR is also deactivated as a result of an implicit detach performed by the MME or the SGSN.

Idle Mode Signaling Reduction is also applicable to CS Fallback-capable terminals—that is, entering the GSM or WCDMA network as such does not lead to ISR deactivation [27]. Potential paging messages will be routed (from the MSC via the MME to the SGSN) to both registered tracking areas and the routing area. However, any location area update or combined RAU/LAU procedure triggered by a CSFB-capable mobile leads to ISR deactivation. For further details related to ISR, please refer to [13, 27].

References

[1] 3GPP TS 36.104 (2010) *Base Station (BS) radio transmission and reception*, V. 8.11.0, 3rd Generation Partnership Project, Sophia-Antipolis.

[2] CEPT (2010) Report 40. Report from CEPT to the European Commission in response to Task 2 of the Mandate to CEPT on the 900/1800 MHz bands, ECC.

[3] 3GPP TS 36.104. (2010) *Base Station (BS) radio transmission and reception*, V. 10.1.0, 3rd Generation Partnership Project, Sophia-Antipolis.

[4] 3GPP TS 36.101. (2011) *Base User Equipment (UE) radio transmission and reception*, V. 10.1.1, 3rd Generation Partnership Project, Sophia-Antipolis.

[5] 3GPP TS 45.005 (2010) *Radio transmission and reception*, V. 9.5.0, 3rd Generation Partnership Project, Sophia-Antipolis.

[6] Holma, H. and Toskala, A. (2009) *LTE for UMTS. OFDMA and SC-FDMA Based Radio Access*, John Wiley & Sons, Ltd, Chichester.

[7] Säily, M., Sébire, G., Riddington, E. (2010) *GSM/EDGE: Evolution and performance*, John Wiley & Sons, Ltd, Chichester.

[8] 3GPP TS 45.004. (2010) *Modulation*, V. 9.1.0, 3rd Generation Partnership Project, Sophia-Antipolis.

[9] 3GPP TS 44.006. (2008) *Mobile Station—Base Station System (MS—BSS) interface; Data Link (DL) layer specification*, V. 6.8.0, 3rd Generation Partnership Project, Sophia-Antipolis.

[10] 3GPP TS 24.008. (2010) *Mobile radio interface Layer 3 specification; Core network protocols; Stage 3*, V. 10.1.0, 3rd Generation Partnership Project, Sophia-Antipolis.

[11] 3GPP TS 23.251. (2011) *Network Sharing; architecture and functional description*, V. 10.1.0, 3rd Generation Partnership Project, Sophia-Antipolis.

[12] 3GPP TS 23.122. (2010) *Non-Access-Stratum (NAS) functions related to Mobile Station (MS) in idle mode*, V. 10.2.0, 3rd Generation Partnership Project, Sophia-Antipolis.

[13] 3GPP TS 23.401. (2011) *General Packet Radio Service (GPRS) enhancements for Evolved Universal Terrestrial Radio Access Network (E-UTRAN) access*, 3rd Generation Partnership Project, Sophia-Antipolis.

[14] 3GPP GP-110332. (2011) *Full support of Multi-Operator Core Network by GERAN*, 3rd Generation Partnership Project, Sophia-Antipolis.

[15] 3GPP TS 22.042. (2011) *Network Identity and Time Zone (NITZ); Service Description; Stage 1*, V. 10.0.0, 3rd Generation Partnership Project, Sophia-Antipolis.

[16] IEEE 1588. (2008) *Standard for a Precision Clock Synchronization Protocol for Networked Measurement and Control Systems*, V. 2, 3rd Generation Partnership Project, Sophia-Antipolis.

[17] ITU-T (2006) Recommendation G.8261, Timing and synchronization aspects in packet networks: Ethernet over Transport aspects—quality and availability targets, International Telecommunications Union, Geneva.

[18] 3GPP TS 36.304. (2010) *User Equipment (UE) procedures in idle mode*, V. 10.0.0, 3rd Generation Partnership Project, Sophia-Antipolis.

[19] 3GPP TS 45.008. (2011) *Radio subsystem link control*, V. 10.0.0, 3rd Generation Partnership Project, Sophia-Antipolis.

[20] 3GPP TS 44.018. (2010) *Radio Resource Control (RRC) protocol*, V. 10.0.0, 3rd Generation Partnership Project, Sophia-Antipolis.

[21] 3GPP TS 25.331. (2010) *Radio Resource Control (RRC) protocol*, V. 10.2.0, 3rd Generation Partnership Project, Sophia-Antipolis.

[22] 3GPP TS 36.331. (2010) *Radio Resource Control (RRC) protocol*, V. 10.0.0, 3rd Generation Partnership Project, Sophia-Antipolis.

[23] 3GPP TS 44.060. (2010) *Radio Link Control/Medium Access Control (RLC/MAC) protocol*, 3rd Generation Partnership Project, Sophia-Antipolis.

[24] 3GPP TR 44.901. (2009) *External Network Assisted Cell Change (NACC)*, V. 9.0.0, 3rd Generation Partnership Project, Sophia-Antipolis.

[25] 3GPP TS 25.413. (2011) *Radio Access Network Application Part (RANAP) signaling*, V. 10.0.1, 3rd Generation Partnership Project, Sophia-Antipolis.

[26] 3GPP TS 36.133. (2010) *Requirements for support of radio resource management*, V. 10.1.0, 3rd Generation Partnership Project, Sophia-Antipolis.

[27] 3GPP TS 23.272. (2011) *Circuit Switched (CS) fallback in Evolved Packet System (EPS)*, V. 10.2.1, 3rd Generation Partnership Project, Sophia-Antipolis.

[28] 3GPP TS 43.055. (2007) *Dual Transfer Mode (DTM); Stage 2 (Release 7)*, V. 7.6.0, 3rd Generation Partnership Project, Sophia-Antipolis.

[29] 3GPP TS 23.060. (2011) *General Packet Radio Service (GPRS); Service description; Stage 2*, V. 10.3.0, 3rd Generation Partnership Project, Sophia-Antipolis.

[30] 3GPP TS 23.216. (2011) *Single Radio Voice Call Continuity (SRVCC); Stage 2 (Release 10)*, V. 10.0.0, 3rd Generation Partnership Project, Sophia-Antipolis.

[31] 3GPP TR 23.885. (2011) *Feasibility Study of Single Radio Voice Call Continuity (SRVCC) from UTRAN/GERAN to E-UTRAN/HSPA; Stage 2 (Release 10)*, V. 1.2.0, 3rd Generation Partnership Project, Sophia-Antipolis.

[32] 3GPP TS 48.008 v10.0.0 “Mobile Switching Centre – Base Station System (MSC-BSS) interface; Layer 3 specification”, January 2011.

Index

The LTE/SAE Deployment Handbook, First Edition. Edited by Jyrki T. J. Penttinen.